G. HOLZER – T. HORST – P. SVATEK (Hrsg.)

DIE LEIDENSCHAFT DES SAMMELNS

Streifzüge durch die Sammlung Woldan

ÖSTERREICHISCHE AKADEMIE DER WISSENSCHAFTEN
PHILOSOPHISCH-HISTORISCHE KLASSE

EDITION WOLDAN

Band 3/1

Herausgegeben von Christine Harrauer

ÖSTERREICHISCHE AKADEMIE DER WISSENSCHAFTEN
PHILOSOPHISCH-HISTORISCHE KLASSE

GERHARD HOLZER – THOMAS HORST –
PETRA SVATEK (Hrsg.)

Die Leidenschaft des Sammelns

Streifzüge durch die Sammlung Woldan

Verlag der
Österreichischen Akademie
der Wissenschaften

ÖAW

Wien 2010

Vorgelegt von w. M. Hermann Hunger in der Sitzung am 18. Juni 2010

Logo der Edition Woldan:
Detail eines Holzschnitts vom Titelblatt der Cosmographie
des Gemma Frisius (Antwerpen 1530).

Umschlagbild:
Die Wohnung des Wiener Sammlers Erich Woldan
(Ausschnitt aus einem Foto im Archiv der Sammlung Woldan)

Umschlaggestaltung:
Reisenbauer / Wallisch / Holzer

Die verwendete Papiersorte ist aus chlorfrei gebleichtem Zellstoff hergestellt,
frei von säurebildenden Bestandteilen und alterungsbeständig.

ISBN 978-3-7001-6894-2

Druck und Bindung: Prime Rate kft., Budapest

http://hw.oeaw.ac.at/6894-2
http://verlag.oeaw.ac.at

Vorbemerkung

Im Sommer 1989 konnte die Österreichische Akademie der Wissenschaften (ÖAW) mit der *Sammlung Woldan* eine der bedeutendsten und wertvollsten mitteleuropäischen Privatkollektionen alter Karten, Atlanten, Globen, Reisebeschreibungen, geographischer Werke und topographischer Ansichten übernehmen.

Der Wiener Privatgelehrte Prof. Dr. Erich Woldan (* 7.10.1901, † 8.1.1989) hatte seit früher Jugend mit beharrlicher Konsequenz und außergewöhnlicher Sachkenntnis *Geographica* im weitesten Sinn gesammelt und im Laufe seines Lebens eine Spezialbibliothek höchsten Niveaus zusammengetragen[1]. Sie umfasst rund 11.000 Titel in etwa 20.000 Bänden und Blättern. Woldans Bibliothek enthält nicht nur eine ganze Reihe von Rara und Rarissima, sondern auch einige Druckwerke, die nach dem heutigen Forschungsstand als Unikate zu gelten haben.

Dass sich die *Sammlung Woldan* heute im Besitz der Österreichischen Akademie der Wissenschaften befindet, ist dem verstorbenen wirkl. Mitglied Univ.-Prof. Dr. Günther Hamann, Ordinarius für neuere Geschichte an der Universität Wien und international angesehener Experte auf den Gebieten der Entdeckungs- und Wissenschaftsgeschichte, zu verdanken: Einerseits konnte er Erich Woldan dazu bewegen, seine Bibliothek testamentarisch der ÖAW zu übertragen, andererseits gelang es Hamann, die von ihm im Rahmen der Kommission für Geschichte der Mathematik, Naturwissenschaften und Medizin der ÖAW ins Leben gerufene *Dokumentationsstelle für Geschichte der Naturwissenschaften und Geographie* in räumlicher und personeller Hinsicht so auszustatten, dass die Übernahme und fachgerechte Betreuung der *Sammlung Woldan* gewährleistet war.

Erich Woldans ausdrücklicher Wunsch war es, die zu seinen Lebzeiten in nur sehr eingeschränktem Maße benutzbare Bibliothek nach seinem Ableben allen Interessenten, und zwar sowohl Forschern als

---

1 Vgl. J. Dörflinger, In memoriam Erich Woldan, in: Der Globusfreund 38/39 (1990), 195-202; I. Kretschmer, Professor Dr. Erich Woldan, 1901-1989. Vierzig Jahre ehrenamtlicher Bibliothekar der Österreichschen Geographischen Gesellschaft, in: Mitteilungen der Österreichischen Geographischen Gesellschaft 132 (1990), 342 - 345.

auch Studierenden, zugänglich zu machen. Dass diesem Wunsch nun entsprochen werden kann, verdanken wir besonders der Initiative von Prof. Dr. Johannes Dörflinger, der als geschätzter Mitarbeiter Prof. Hamanns und persönlicher Freund Woldans schon zu dessen Lebzeiten Objekte aus der Sammlung bearbeitet hat. Durch zahlreiche einschlägige Publikationen und viele Praktika, die er im Rahmen seiner Lehrveranstaltungen mit dieser Sammlung gestaltet hatte, machte er viele dieser Zimelien weltweit bekannt. Er war es auch, der in vielen seiner Freunde und Schüler, die an diesem Band mitgearbeitet haben, die Begeisterung für die Geschichte der geographischen und kartographischen Erforschung der Erde geweckt hat.

Inhalt

Band 1

Einleitung 11

Ananda ABEYDEERA, Taprobane: Mapping an Island of Paradoxes
A Realistic Depiction of Taprobane in Antiquity becomes a Mythical Representation in Christian Middle Ages 15

Barbara BAUER, „Das Schiffsleben ist doch ein elendes!"
Johann August Miertschings Reisetagebuch der Arktisexpedition unter dem Kommando von Robert M'Clure (1850 – 1854) 41

Kurt BRUNNER, Die Karten der *Schedelschen Weltchronik* der Sammlung Woldan 63

Antal András DEÁK, Ein kleines, seltsames Buch in der Sammlung Woldan
Sigmund von Birken: *Der Donau-Strand* 81

Helmuth GRÖSSING, Eine Erkundung über eine Erkundung
Einige landeskundlich-geographische, vornehmlich deutsche Flüsse betreffende Beobachtungen im Werk des Konrad Celtis 95

Paul D. A. HARVEY, Medieval local maps from German-speaking lands and central Europe 113

Otto HAT, Die kartographische Darstellung des Weinviertels
Unter besonderer Berücksichtigung der Niederösterreichkarte von Georg Matthäus Vischer und der ersten beiden Landesaufnahmen 133

Bernd HAUSBERGER, P. Joseph Stöcklein, der *Welt-Bott* und der Ruf der Exotik 153

Markus HEINZ, Versuch einer *umständlichen Historie* der Historie *der Land-Charten* 185

Gerhard HOLZER, … *i nga tini mea o to tatou nei motu* …
Ferdinand von Hochstetter als Kartograph in Neuseeland 197

Thomas HORST, Die Globen der Sammlung Woldan: Erdkugeldarstellungen als Quelle für die Geschichtswissenschaft 233

Tafeln / Plates zu Band 1 297

Band 2

Helga HÜHNEL, Erzherzogliche Gipfelstürmer
Handschriftliche Reisenotate Erzherzog Rainers über Expeditionen auf den Schneeberg und in angrenzende Gegenden Anno 1802 und 1805 321

Marianne KLEMUN, Die Österreichisch-Ungarische Nordpolexpedition (1872-1874)
Euphorie der Gemeinsamkeit und die Positionierung der Akademie der Wissenschaften (Wien) im Vorfeld der Expeditionsplanung 343

Walter KOCH, Epigraphisches zur *Schedelschen Weltchronik* 365

Ivan KUPČÍK, Die Karte des March-Flusses mit der Einzeichnung des Oder-Donau-Kanals von ca. 1719 387

Peter H. MEURER, Das Woldan-Exemplar des Kölner *Itinerarium Europae* 397

Jan MOKRE, Grundbesitz für eine Karte
Die Geschichte der Virginia- und Maryland-Karte des Augustin Herrman von 1673 411

Mary SPONBERG PEDLEY, A Map for Benjamin Franklin 425

Franz REITINGER, *Strom der Zeiten*
Zur Kartierung von historischen Entwicklungen 445

Günter SCHILDER, Neue Horizonte
Die erste niederländische Reise nach Ost-Indien (1595-1597) 463

Petra SVATEK, Die *Austriae Chorographia* des Wolfgang Lazius . 485

Zsolt Győző TÖRÖK, Die „ungarische" Woldan-Karte und ihre kartographischen Vorgänger im 16. Jahrhundert
Eine unbekannte Kartentradition in Zentraleuropa zur Zeit der Türkenkriege 505

Franz WAWRIK, Hieronymus Münzer
Arzt, Reisender, Geograph, Diplomat 515

Karl R. WERNHART, Die Bildquellenwerke der Sammlung Woldan
Grundlage für die Entstehung der *Monumenta Ethnographica* – Erinnerungen eines Zeitzeugen 537

Tafeln / Plates zu Band 2 549

Einleitung

Die Sammlung Woldan zählt zu den bedeutendsten österreichischen Sammlungen geographischer Literatur, in der Reiseberichte, landeskundliche Beschreibungen und viele Landkarten und Atlanten zu finden sind.[1] Dieser Sammelband hat sich zur Aufgabe gestellt, einen Teil dieser Objekte aufzuarbeiten und damit einem breiteren Leserkreis zugänglich zu machen.

Die Idee für diesen Doppelband aus der Edition Woldan geht auf die Initiative des Münchener Kartographiehistorikers Thomas Horst zurück, der die Sammlung Woldan erstmals 2002 während seines Studiums bei Univ.-Prof. Dr. Johannes Dörflinger kennengelernt hatte und seit damals ein oftmaliger und engagierter Benützer dieser Sammlung geblieben ist. In einem Gespräch mit dem Sammlungsleiter Mag. Gerhard Holzer und der Wiener Historikerin Dr. Petra Svatek Anfang des Jahres 2008 entwickelte sich die Idee, junge Kartographiehistoriker – meist Dörflingerschüler – für Beiträge zu einem Sammelband über die Sammlung Woldan zu gewinnen. Diesem Projekt schlossen sich namhafte Fachvertreter aus Österreich, Deutschland, Großbritannien, Ungarn, den Niederlanden, den USA und Sri Lanka an. Einige andere, wie die bedeutende österreichische Kartographiehistorikerin Univ. Prof. Dr. Ingrid Kretschmer und der Dörflingerschüler Dr. René Tebel, mussten ihre ursprünglich geplante Mitarbeit aus gesundheitlichen oder persönlichen Gründen wieder zurückziehen.

Dank der Unterstützung durch die Leiterin der Bibliothek der Österreichischen Akademie der Wissenschaften und Herausgeberin der *Edition Woldan*, Univ.-Prof. Dr. Christine Harrauer, konnte dieses Projekt, das längst den Rahmen der Kartographiegeschichte gesprengt hatte, in die Realität umgesetzt werden.

Vierundzwanzig Artikel zu den unterschiedlichsten Sammlungsbeständen vom Mittelalter bis ins 20. Jahrhundert, die durch die Darstellung eines „Bezugsobjektes" aus der Sammlung eingeleitet werden, sollen dem Leser einen Einblick in die Vielfältigkeit dieser ehemaligen Privatsammlung Erich Woldans geben.

---

1 Vgl. G. Holzer, Die Sammlung Woldan, in: Handbuch der historischen Buchbestände in Österreich 2, Hildesheim 1995, 152-154.

Der Großteil der Beiträge befasst sich gemäß der ursprünglichen Zielsetzung mit kartographischen Materialien vom Mittelalter bis in die frühe Neuzeit: Die Bedeutung mittelalterlicher Grundlagen für Renaissancekartographie werden in Beiträgen über die Entwicklung der Darstellung Sri Lankas/Taprobanes auf frühen Weltkarten (Ananda Abeydeera) und über mittelalterliche Regionalkarten in Zentraleuropa (Paul Harvey) analysiert. Die Schedelsche Weltchronik (1493), das umfangreichste illustrierte Buch des 15. Jahrhunderts, wurde im Hinblick auf ihre kartographische Ausstattung (Kurt Brunner, Franz Wawrik), aber auch auf ihre epigraphische Gestaltung hin (Walter Koch) näher untersucht. Andere Beiträge setzten sich mit der Karte *Austriae Chorographia* des Wolfgang Lazius aus dem Jahr 1563, die uns heute nur mehr als Nachdruck aus dem Jahre 1620 erhalten ist (Petra Svatek), der Entwicklung des um 1590 aus der *Kölner Schule* entstandenen Reiseatlas *Itinerarium Europae* (Peter Meurer), der aus der Zeit um 1600 stammenden Ungarnkarte *Descriptio Regni Ungariae* und ihren kartographischen Vorgängern (Zsolt Török), der Karte und der Beschreibung des *Donau-Strand*es von Sigmund Birken aus dem Jahr 1664 (Andras Deák), der Virginia- und Marylandkarte des Augustin Herrman von 1673 (Jan Mokre) sowie mit der kartographischen Darstellung des Weinviertels (Otto Hat) auseinander. Aus dem 18. Jahrhundert wurden die sehr seltene *Mappa des March-Flus* von Norbert Wend von Linck (Ivan Kupcik) und eine der ersten Karten der neu gegründeten Vereinigten Staaten von Amerika von Jean Lattré (Mary Pedley) besprochen. Basierend auf dem Buch *Versuch einer umständlichen Historie der Land-Charten* von Eberhard David Hauber wurde schließlich noch die Geschichte der Disziplin Kartographiegeschichte näher erläutert (Markus Heinz).

Der umfangreichste und übergreifendste Beitrag zu den Cartographica beinhaltet die erstmalige Gesamtaufarbeitung der Globen der Woldan Sammlung (Thomas Horst). Die Liste reicht vom Globus des Jodocus Hondius d. Ä. (1601) über Globen Vincenzo Coronellis, Johann Gabriel Doppelmayrs bis hin zu dem in den späten 1920er Jahren herausgegebenen Schulglobus der Wiener Kartographischen Anstalt Freytag & Berndt.

Weitere Beiträge beschäftigen sich mit den landeskundlich-geographischen Darstellungen des Conrad Celtis (Helmuth Grössing), der ersten niederländischen Reise nach Ost-Indien (1595-1597) sowie den Karten, die im Zuge dieser Reise entstanden waren (Günther

Schilder) und dem wichtigsten Sammelwerk von Missionsberichten der Jesuiten, dem *Neuen Welt-Bott* von Joseph Stöcklein (Bernd Hausberger).

Im Rahmen der Beiträge über Sammlungsobjekte des 19. Jahrhunderts wurden vor allem Reiseberichte analysiert. Beginnend von nahe gelegenen Reisezielen wie in dem Artikel über die handschriftlichen Reisenotate Erzherzog Rainers in die Schneebergregion in den Jahren 1802 und 1805 (Helga Hühnel), der auch einen kleinen Einblick über die wissenschaftliche Erforschung dieses Gebietes im 16. und 17. Jahrhundert gewährt, streckt sich der Bogen mit dem Reisetagebuch von Johann August Miertsching über die Arktisexpedition von Robert M'Clure, die zur Entdeckung der Nordwestpassage führte (Barbara Bauer), und die Hintergründe der Österreichisch-Ungarischen Nordpolexpedition (1872-1874) (Marianne Klemun) über die arktischen Regionen bis ans „andere Ende der Welt", wo über kartographische Tätigkeit des deutsch-österreischen Geologen Ferdinand Hochstetter 1858/59 in Neuseeland (Gerhard Holzer) berichtet wird.

Darüber hinaus enthält der Doppelband noch Beiträge, die sich mit der historisch-graphischen Darstellung des „Stroms der Zeiten" (Franz Reitinger) sowie der Edition völkerkundlich-kulturanthropologischen Bildquellen der Sammlung Woldan in den *Monumenta Ethnographica* (Karl R. Wernhart) beschäftigen.

Der Sammelband wurde mit Karten, Landschaftsansichten und Photographien illustriert. Die jeweiligen Autoren waren bemüht, die Urheberrechte nach bestem Wissen hierfür einzuholen. Sofern keine speziellen Herkunftsnachweise angegeben sind, stammen die Bilder aus der Sammlung Woldan. Falls einzelne urheberrechtliche Ansprüche unberücksichtigt blieben, ersuchen die Autoren um nachträgliche Mitteilung.

Die Herausgeber möchten allen Autoren danken, die sich bereit erklärt haben, in kurzer Zeit ihre Beiträge für diesen Doppelband zu erarbeiten. Unser ganz besonderer Dank gilt Univ.-Prof. Dr. Johannes Dörflinger, der durch seine hervorragende Lehrtätigkeit, seine Persönlichkeit und Kollegialität ein steter Mentor für die Beitragenden dieser Publikation war.

Wien, im März 2010 Die Herausgeber

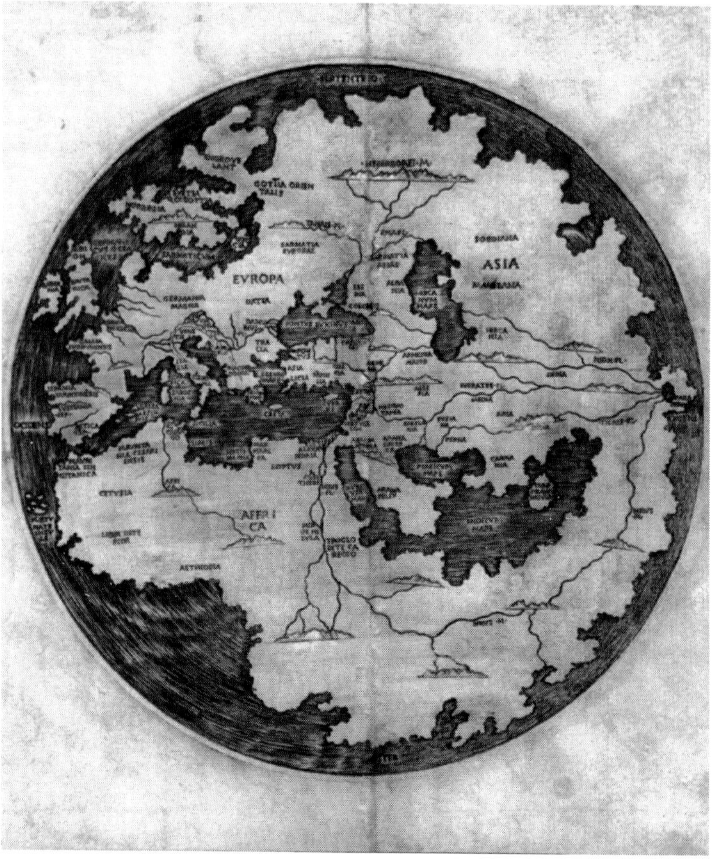

Fig. 1: The so-called Wieder-Woldan-Map, Anonymous. (Venice (?) ca. 1485): Four rivers issue from the Paradise and two of them flow across the landmasses that enclose the Indian Ocean in the bosom of which is the island called Tabrobana Insula. Woldan Collection, Academy of Sciences, Vienna, K-V(Bl): WE 3.

Claudius Ptolemaeus,
Geographia,
Straßburg 1513

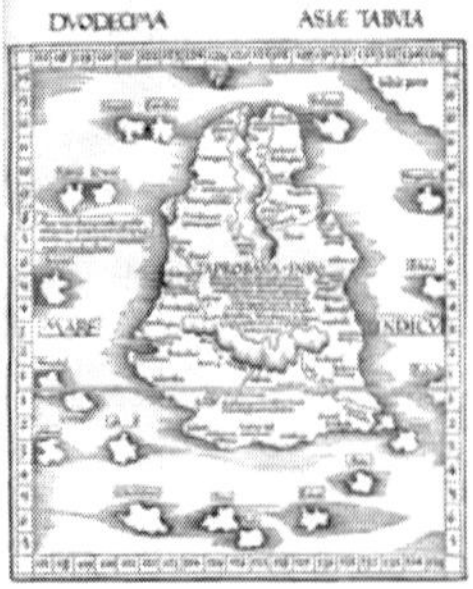

Ananda Abeydeera, Paris

Taprobane: Mapping an Island of Paradoxes

A Realistic Depiction of Taprobane in Antiquity becomes a Mythical Representation in Christian Middle Ages

This essay concerns the depiction of Sri Lanka in medieval *mappaemundi* such as those of Beatus, the Isidore map, the maps of Hereford and Ebstorf, and those in the tradition of Catalan atlases from Majorca and Modena. It will show how these latter straddled a decisive transition, thus helping Fra Mauro to usher in a newer portrayal of the island Saylam (the fore-runner of pre-modern Ceylon, cf. plate p. 316 b).

Known to the ancient Greeks and Romans as Taprobane, Sri Lanka figures as a tiny insular speck on the modern map of the world. Nevertheless, the variety of dimensions, positions and mythical attributes (Fig. 1)[1] bestowed upon it in course of its long history by classical authors and medieval mapmakers had maintained with considerable emphasis, that it was a large land-mass of continental proportions, inhabitated by *Antichthones* (cf. Pliny, *n. h.* 6. 81), counter-balancing the *oikoumene* or the inhabited part of the globe. It appeared as such uninterruptedly in European maps ever since Ptolemy outlined it with the respective coordinates as the largest

---

1 See the Anonymous Medieval Map circa 1500 (Woldan Collection, Vienna), described by E. Woldan, A circular, copper-engraved, medieval world map, in: Imago Mundi 11 (1967), 13-16 and T. Campbell, The Wieder-Woldan Incunable World Map, in: Imago Mundi 42 (1990), 83.

island in the world in the final regional map, the Twelfth Map of Asia, in his Geography. It always bulked large in the popular imagination of both Westerners and Easterners who had access to Ptolemy's data directly or indirectly (Fig. 2).

Fig. 2: Claudius PTOLEMAEUS, Geographi[a]e opus novissima traductione e Gr[a]ecorum archetypis castigatissime pressum: ceteris ante lucubratorum multo prestantius, Strassburg 1513,
Woldan Collection, Academy of Sciences, Vienna, G-V: WE 10.

With the disappearance of the Roman Empire, a long period sets in during which contacts between the East and the West become rare. As it was no longer possible to travel to the East and learn *in situ*, the West was obliged to depend on the bits and pieces of information gathered from the storehouse of Greek and Roman learning. Thus were the Middle Ages gradually led to the conviction that ancient authors always knew all that was necessary and possible to

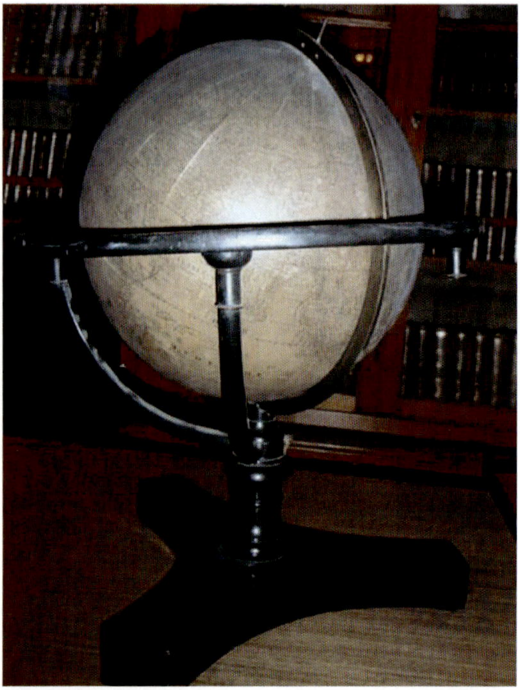

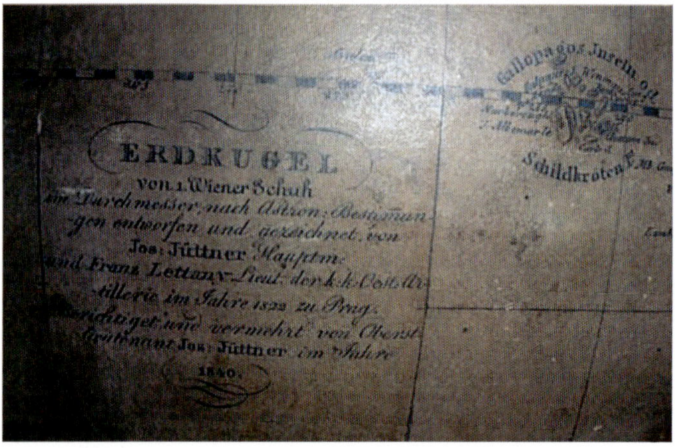

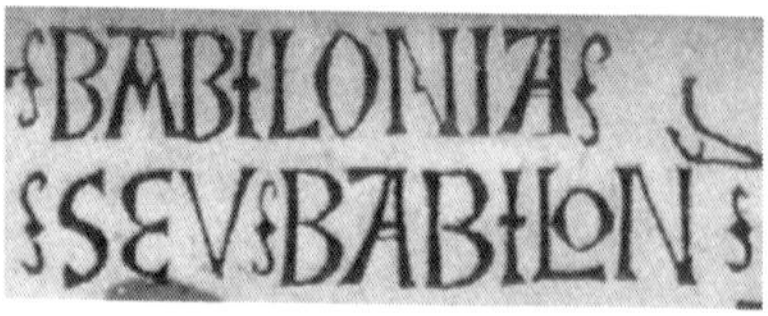

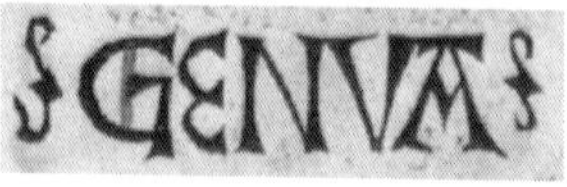

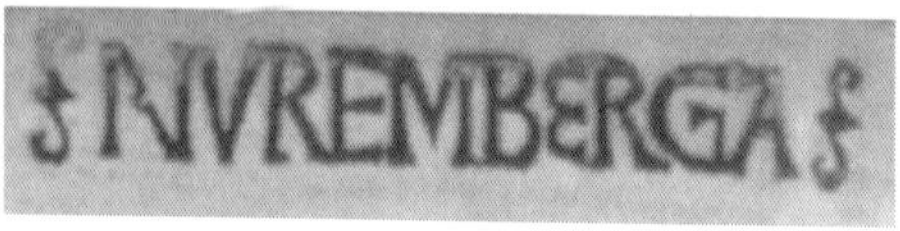

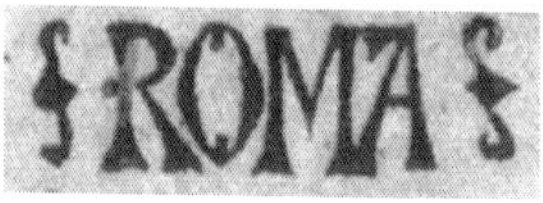

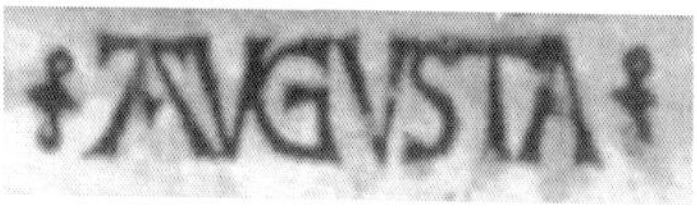

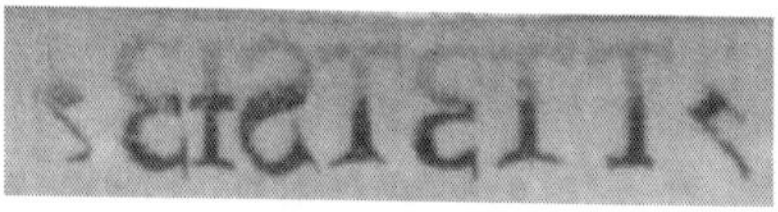

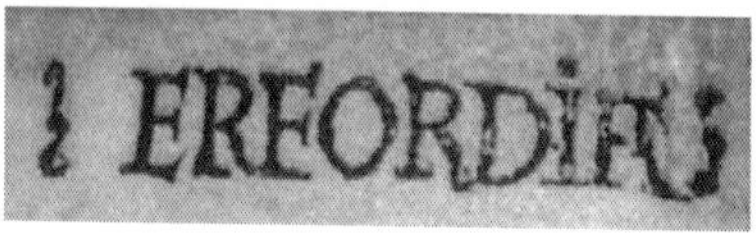

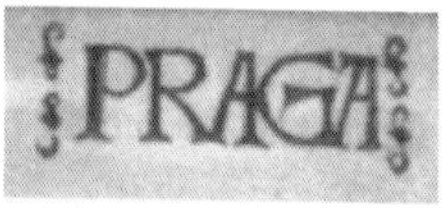

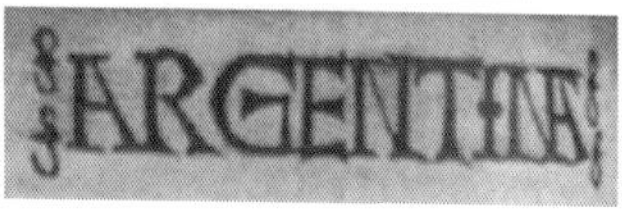

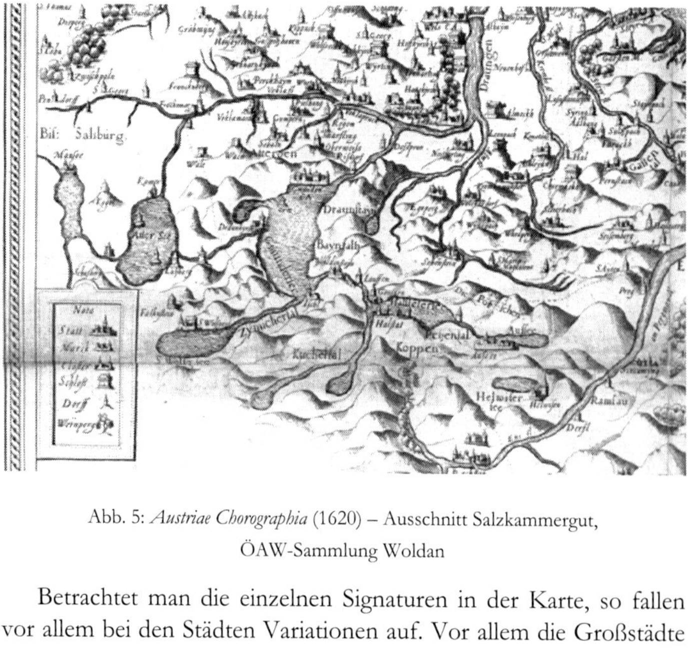

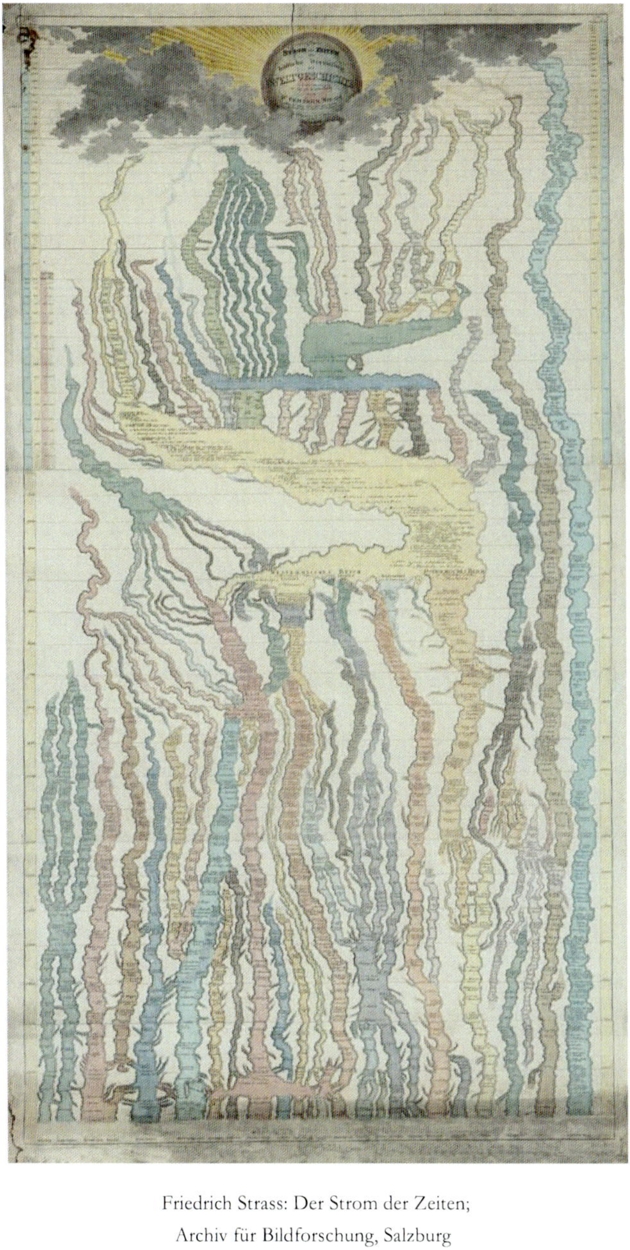

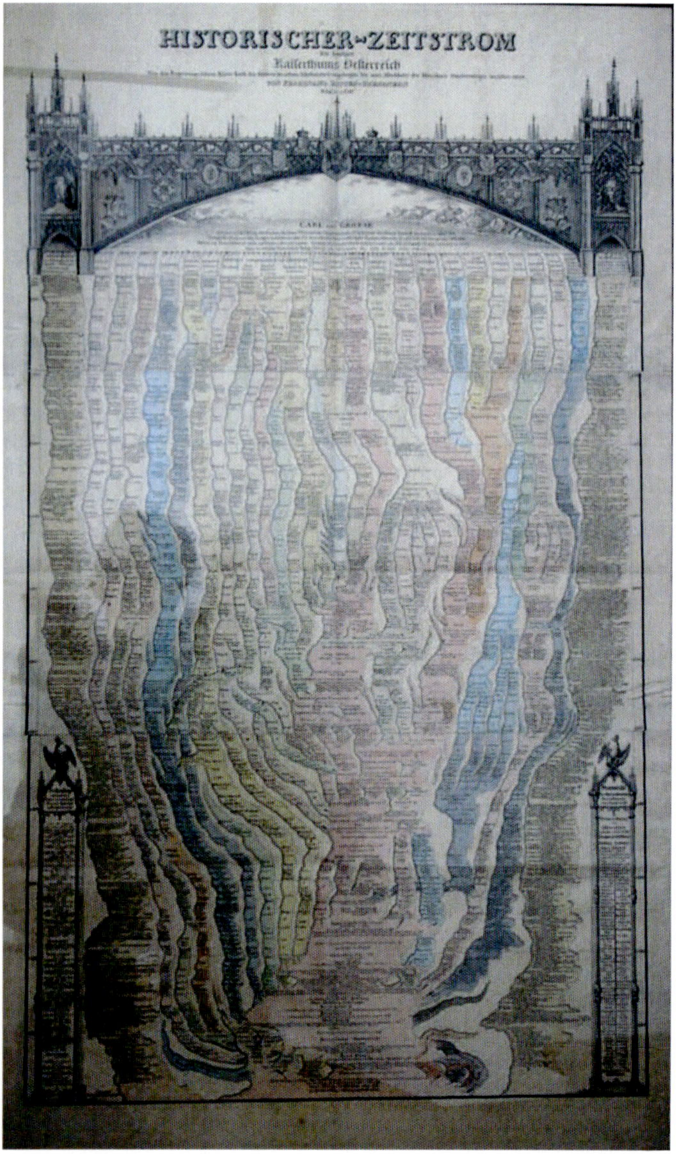

know about Taprobane. Consequent to the rise of Islam, Greek and Latin literature lost its hold on geography of Taporbane and the knowledge of the island became outdated: it was subsequently a legendary island of conspicuously mythical attributes in medieval geographical lore. If Taprobane was out of reach in the reality, it remained also inaccessible in the realm of imaginative geography. Most of the medieval *mappaemundi* such as those of the Hereford and Ebstorf, locate it in the eastern confines of the surrounding ocean in the vicinity of Paradise. What mattered most in these cartographic representations was not the depiction of a precise geographic knowledge of the island, but an imparting of an ideological perception of mythical attributes.

This great admiration for ancient authors, coupled with a fear inspired by the pagan literature, prompted Christian scribes to make compromises and borrow from classical sources. Clerics and scribes found themselves busy adapting, compiling and summarizing from them from the very outset itself of the decline of ancient world. They yearned for these substances gleaned from great works of ancient authors in order to quench their thirst for knowledge. Indeed, from this classical heritage, what they sorted out to pick and choose for such adaptations, compilations and summaries, was what's most entertaining and amusing in other words, things of little depth and rather of fabulous nature. To know about the wide world they turned towards the *Bestiaria*, *Lapidaria* and *Herbaria*: the *Bestiaria* described animals of different countries including of course fabulous animals and monstrous creatures, the *Lapidaria* concerned treatises that described the nature and virtues on precious stones and the *Herbaria*, in the manner of the *Physiologus*, an Alexandrinian compilation of the second century, translated into Latin in the 4th and 5th century.

For all that, the knowledge of Taprobane was not so much sought in the works of Strabo (15, 1 17), Pliny (*n. h.* 6, 81), and even much less in Ptolemy (7, 4) than in the *Collectanea rerum memorabilium* of Julius Solinus, a compiler of the third century. The *Collectanea*, now embellished with new anecdotes and adjusted to cater to the taste of the changing times was going to be the principal source of inspiration for a number of medieval authors. The island of Taprobane thus found its way into the *Etymologiae* of Saint Isidore bishop of Seville, a leading ecclesiastic figure. Isidore of Seville (ca. 560-636) is the one who launched Taprobane from the Iberian Peninsula on its protean career by drawing scholarly attention to its description in two passages: 14, 3, 5 and 14, 6, 122.[2]

In the course of the thirteenth century, other encyclopedic works emerged and flourished, continuing to find inspiration from Solinus's compilations. The Imago mundi of Honorius d'Autun (Honorius Augustodunensis) who also speaks of Taprobane in this work[3] was the most widely-known encyclopedic work because of the influence it exerted over a long period of time and over a vast number of countries of cultural and linguistic diversity. At the heyday of great medieval period when the encyclopedic compilations thrived particularly at the beginning of the thirteenth century people derived a great pleasure from the recitations of Solinus' and Honorius' translations. Pierre de Beauvais is considered to be the first to have translated Solinus into French and his *Mappemonde* dedicated to Robert, the count of Dreux, forms part of such works which largely catered to the popular taste of his day.

Here is Pierre de Beauvais's version of Taprobane: *Plus loin, l'Inde la grande, pays très chaud, baigné par l'Océan, avec l'île de Taprobane, où il y a pas moins de dix cités; les gens sont gigantesques et vivent très longtemps; ils ont une langue à part, que nul n'entend, et ils n'entendent celle de personne; pour faire du commerce avec les marchands étrangers, ils étalent leurs produits sur le rivage, et les trocs se font sans paroles. Il y a deux hivers et deux étés par an, et toujours de la verdure. Solin raconte que les montagnes sont d'or dans ces régions, mais personne n'y va, à cause de la multitude des serpents.*[4]

Another well-known work was the *Image du monde* of Gossuin, a poem of 6 600 verses. Its success was such that 4 000 new verses were soon added to it. Sixty-nine manuscripts of the work preserved to our day bears witness to its popularity. The passage relative to Taprobane is very similar to the description found in the *Mappemonde* of Pierre de Beauvais. However Gossuin steps back less than Pierre de Beauvais, notably with regard to the wealth of the island: *Après vient l'Inde, qui prend son nom d'un fleuve au Nord, entourée par la grande mer,*

---

2 J. André – Jean Filliozat, L'Inde vue de Rome. Textes latins de l'Antiquité relatifs à l'Inde, Paris 1987, 322-324.

3 Honorius Augustodunensis, De imagine mundi, libri tres, lib. I, Migne, Patrologia latina 172, 123: *Hoc India ab Occidente clauditur. In quo etiam est sita Taprobanes insula, decem civitatibus inclyta. Haec duas æstates et duas hiemes uno anno habet, et omni tempore viret.*

4 C.-V. Langlois, La vie en France au Moyen Age 3, Genève 1970, 127-128.

avec l'île de Taprobane. Ce pays a deux étés et deux hivers par an, mais si tempérés qu'il y a toujours de la verdure, des fleurs et des fruits. C'est la contrée de l'or et des pierres précieuses, dont les dragons et griffons, monstres capables d'emporter un homme armé à cheval, défendent les gisements dans les montagnes. Lieux si doux, et si délectables, et si 'espirituels' que c'est un paradis à habiter.[5]

However the encyclopedia par excellence, the work of reference of the age is the *Li Livre dou Tresor* or *La Naissance de toutes choses* (The Book of the Treasure or The Birth of All Things) written in French by the Italian Brunetto Latini. This work is a very good illustration of what occurs to the original source in the course of adaptation. Brunetto Latini locates farther East the silent trade described by Pliny in his description of Taprobane, and what is more he gives the impression as if it is actually taking place in his time:

Yonder live men, the Seres, who from the leaves and barks of retting trees make wool of which they clothe themselves. They are live amiably and peacefully among themselves, but refuse to enter into contact with foreigners. They do not want any of our merchandise and sell theirs to our merchants without uttering a word: on the bank of a river they display them with labels on which the prices are marked: our traders carry them away after having deposited there the price requested.[6]

This passage also sheds light on the type of errors which frequently appear in medieval translations and adaptations based on classical accounts. Solinus, in his work, described Taprobane a few lines apart from the Red Sea. From this Brunetto Latini arrives at the conclusion that the island is situated in the Red Sea: *Au-delà vivent des hommes, les Sères, qui de feuilles et d'écorces d'arbres rouies font une laine dont ils s'habillent. Ils sont aimables et paisibles entre eux, mais refusent d'entrer en relations avec les étrangers. Ils ne veulent point de nos marchandises et vendent les leurs à nos marchands sans dire un mot: ils les déposent sur la rive d'un fleuve avec des étiquettes où le prix est marqué: les nôtres les emportent après avoir mis à la place le prix.*[7]

On reading these passages taken from the leading compilations, one may wonder whether the imagery of Taprobane had perpetuated itself, almost unchanged from antiquity to the Middle Ages.

5 Langlois, La vie en France au Moyen Age 3, 167.
6 Ibd., 360-361.
7 Ibd., 360-361.

The symbolism which permeates the Christian thought from its beginning is quite explicit in the *Etymologies* of Isidore of Seville, needed to appear beneath the thing, - *res* -, the sense - *verbum* -, that it was necessary to find out. The conscience that the medieval man possessed of himself, of his relationship with the universe of God necessitated in order for him to become orderly and clear, to find his unity, a symbolic support: *Et toi, chrétien selon Dieu, partage aussi l'Écriture du Vieux Testament en deux parties, à savoir selon l'histoire et selon la signification spirituelle: sépare la vérité de sa figure, distingue les choses de l'esprit de celles du corps, conserve le sens spirituel qui vivifie, afin de ne pas mourir de faim parce que la lettre aurait pourri au temps d'hiver, c'est-à-dire au jour du jugement dernier, car l'Apôtre déclare: 'La loi de l'esprit n'est pas la loi du corps; la lettre tue et l'esprit vivifie'.*[8]

The vision of the world is thus integrated into this eschatological perspective in which what was significant were the facts and not their veracity which was not necessary. It is in this spirit, it seems to me, that we need to understand the medieval vision of the island.

If Taprobane is inaccessible in the reality, it is also inaccessible in the imaginary world of the West. It is a closed world, *surrounded by the great sea*, a zone of mystery imagined as a garden, of which one is unable to say whether it is Paradise or it is the foretaste of hell.

These three authors agree in depicting and find a sort of consensus in presenting an abundance of wealth in Taprobane: it is the *country of gold, of pearls and of precious stones.* It is an immense source of wealth yet inaccessible because it is defended by diabolical monsters, serpents, dragons, griffons and lion-bodied birds.

The prohibition that falls either on a sacred space of which the most known example of the Middle Ages was the Paradise, barred to men by a curtain of flames, either on a space considered impure, but always avoided by the community of men. It is the case of Taprobane: *no one dares go there, because of the vast number of serpents, the greater part of the island is desert because of the heat.* To attempt entering the space of Taprobane constitutes a grave transgression that will cause the death of the culpable: even «an armed man riding a horse» cannot escape the diabolical creatures. In other words, it is the hell, the eternal death.

---

8 G. Bianciotto (Ed.), Bestiaires du Moyen Age. Mis en français moderne, Paris 1980, 33-34.

In this forbidden world, in other words, in this fearsome world, the Middle Ages projected the antithesis of its world: the wealth, the enormity, the inhumane, the monstrosity. We had already seen the legendary wealth of the island and the presence of monstrous creatures such as the dragon and the griffon. The enormity, apprehended by the west is encountered firstly in the luxuriance of the nature: the island is evergreen and the trees and plants there blossom and bear fruits throughout the year because of its two summers and two winters the enormity is then seen in the gigantic stature of its inhabitants.

The giants, the Genesis informs us, are born of the union between the sons of gods and daughters of men. They evoke for the believers the ancient and heroic times of the history, the patriarchs. The giants inhabited the earth and the Lord bestowed upon men hundred and twenty years, in conformity with the most popular mythical conceptions. It is the superhuman, the opposite of the west which to quote the image of Dante[9], diminishes such as *a mantle which shortens itself swiftly* because *the time turns its scissors.*

For Gossuin, this place *so tender and so delectable* is *a Paradise to live*, it is therefore a sacred place. It is the most beautiful venue, the wealth symbolizes the beauty and therefore the good, it is the most pleasant place to live because of its climate. Pierre de Beauvais's description seems to follow very closely this interpretation. On the other hand Brunetto Latini presents rather a paradisiacal land which can only be entered after one has crossed the sea unobstructed, following the flight of a bird of the country: *La mer est le symbole de notre monde. Les navires représentent les justes qui ont traversé sans danger, en toute confiance, les tourmentes et les tempetes du monde, et qui ont convaincu les ondes mortelles, c'est-à-dire les puissances de ce monde.*[10]

Otherwise, it is a space separated by a large river, where elephants and other savage beasts roam. The elephants, tells us Pierre de Beauvais, represent Adam and Eve: *Quand arrive le temps où le mâle est pris du désir de procréer, il se dirige vers l'Orient avec sa femelle, près du Paradis où naquit Adam. Là se trouve un arbre qui est appelé 'mandragore'. La femelle mange la première le fruit de l'arbre, puis elle en donne au mâle afin qu'il en*

9 Dante, La divine comédie, translated H. Longon, Paris 1951, 151.
10 Beauvais, Bestiaires, 26.

mange aussi. Aussitôt qu'il en a mangé, tous deux s'accouplent, et immédiatement la femelle conçoit.[11].

The elephants are there to recall the first sin of mankind that befell Adam and Eve. It is the foretaste of hell which awaits men who live in sin. One can therefore say that Brunetto Latini synthesizes in Taprobane what the world would be of, on the day of the Last Judgment[12]. This place of the Antipodean beings which disconcerts the Middle Ages, is the reflection of the world, the opposite of the west; it is the reverse of the earth where some men are supposed to live with feet opposed to theirs: *L'émotion la plus ancienne et la plus forte chez l'homme est la peur, et la peur la plus ancienne et la plus forte est la peur de l'inconnu.*[13]

In linking the forbidden with the supernatural in the representation of Taprobane, these authors transpose the terror before the abnormal. The anomaly is always fearsome and suspicious, at the same time it is rejected from the familiar universe and integrated into the new world which the Christians await with dread, to put an end to their sufferings on earth. It may be the Paradise, but Paradise is only accessible to the chosen of God, to the others will be reserved the hell. These two possibilities characterize the other world of the Christians. It is with a similar image that the pioneering travelers of the west depart they will soon be adding to it images of other peoples.

Most of the medieval *mappae mundi* reflect this new vision of Christianity. They are of the T-O-type, in keeping with the Roman tradition but the division of the world into three parts is interpreted as the distribution of the world among the three sons of Noah.[14] They are often east oriented with Paradise at the top and with Jerusalem at the centre. They are flat in design and the antipodes are not allowed, the only exception being the Beatus map from the

---

11 Beauvais, Bestiaires, 59.

12 G. Le Clerc, La nature de l'éléphant, Bestiaire symbolique, 14th century manuscript, Paris, Bibliothèque Nationale de France, Ms. Français, 14969 - reproduced by J. Baltrusaitis Réveils et Prodiges: Le gothique fantastique, Paris 1984, 257.

13 H. P. Lovecraft, Epouvante et surnaturel en littérature, Paris 1969, 35.

14 Honorius Augustodunensis, De imagine mundi, libri tres: lib. 1, Migne, Patrologia latina 172, 166 A: *Hujus tempore divisum est genus humanum in tria : in liberos, milites, servos. Liberi de Sem, milites de Japhet, servi de Cham.*

Burgos de Osma (from Osma Cathedral, Spain)[15]. *Sans mappemondes,* wrote the minor friar Paulin de Venice (towards 1270-1344) *je dirais qu'il est impossible d'imaginer ni de comprendre tout ce que les textes tant sacrés que profanes rapportent au sujet des fils de Noé, des fils de leurs fils, des quatre monarchies et de tous les royaumes et provinces.*[16]

Now we are going to look at a few maps directly influenced by works culled and compiled from ancient authors.

The Reconstructed Ravenna Map and a map of the 9th century

The Ravenna-world map (so-called because of the name of the location where it was found) is dated to the late seventh century.[17] Numerous are the reconstructions of this map by modern historians but the information it provides are so imprecise that it is difficult to determine its form (either rectangular, circular or oval). In the Eastern edges of the reconstructed Ravenna map appear two islands: one of which is Paradise, the other in its vicinity is Taprobane.

Fig. 3: An image inspired by Macrobius' Commentary on Scipion's Dream, reedited in the 15th century. Repr. from Cartes et Figures de la Terre, Paris 1980, plate 138.

---

15 Paris, Bibliothèque Nationale de France, Ms. lat. 1366, fol. 25; reproduced by K. Miller, Die ältesten Weltkarten 2: Die Weltkarte des Beatus (776 n. Chr.), Stuttgart 1895, plate 2. – See also P. Barber, The Hereford Mappae Mundi, Hereford Cathedral 1999, 10.

16 Cited by D. Lecoq, Rome ou Jérusalem: la cartographie médiévale entre l'influence antique et l'influence chrétienne, in: Bulletin du comité français de cartographie 121 (Sept. 1989), 25.

17 Cf. J. Schnetz, Cosmographia: Eine Erdbeschreibung um das Jahr 700, Uppsala 1951.

The sketch of a *mappamundi,* dated to the ninth century, has been found contained in the Commentary on Scipio's Dream. This representation of the world was inspired by Macrobius, who flourished in the fifth century and wrote a commentary on Cicero's famous description of the cosmos. In this commentary (re-edited in the fifteen century) the name of Taprobane appears abbreviated to *Tapro* and is shown in the south of India as rectangular in form. (Fig. 3)

The Munich 'Isidore' World Map. A Copy of Hugh of St. Victor's large World Map? (North France, circa 1130-1135)

A *mappamundi* is found in a copy of the *Etymologies* of Isidore of Seville, Cod. Monac. lat. 10058, fol. 154 v, in the Bayerische Staatsbibliothek in Munich. The legend inscribed on the Tapbana (sic!) shown on the northern edges of the *mappamundi* reads: *Taprobana insula plena feris et auibus. Taprobana insula habens ciuitates decem*[18] (i.e. one part of the island is inhabited whereas the other part is full of animals). Although this *mappamundi* is circular, the island's location is identical with that of the preceding one which is rectangular in shape.[19]

The Anglo-Saxon, or Cottonian *mappamundi* illustrating the Roman Tradition and other manuscript maps of the 10th and 11th century

Of Anglo-Saxon origin (Canterbury) this map (preserved in the British Library, London, Cotton MS Tib. B. V. fol. 56) is dated to the eleventh century (ca. 1025-1050)[20] and was found in a manuscript copy of the Periegesis by Priscianus, but has no relevance to the text. Given the early date, the map indicates geographic contours of some sophistication. At the top of the *mappamundi* appears a large island whither the cartographer's knowledge about eastern Asia ends. Between the Indus and the Ganges a large island can be seen upon which bears the following legend: *Taprobane habet X civitates, bis in anno*

18 P. G. Dalché, La «Descriptio Mapae Mundi» de Hugues de Saint-Victor, Paris 1988, 193.

19 Reproduced by P. Barber, The Hereford Mappae Mundi, 11.

20 Ibid., 4.

messe fruges (trans.: There are ten cities in Taprobane and twice a year fruits are harvested there).

The Beatus Map has been drawn at the Saint Sever abbey and it illustrates an eleventh century manuscript of the *Commentary on the Apocalypse of Beatus* who flourished in the eighth century. This is one of the medieval *mappamundi* which has vibrant colors.[21] Let us note the spelling of the name form *Tapaprone* and the legend reads: *Insula tapaprone indie subiacens ad eurum, patens. In longitudine milia passum dccclxxv. Scinditur amne interfluo; tota margaritas et gemmis repleta est; pars eius bestiis et elefantis repleta est. Partem uero homines tenent. Habet enim civitates decem.*

The *mappamundi* that was found in a manuscript of Lambert of Saint Omer is dated to the tenth century and bears the following legend: *Taprobana est insula indiæ ubi dicunt duas esse hiemes et duas æstates, et bis floribus vernare locum.*

The Hereford Mappa mundi by Richard of Haldingham

Of English origin it is dated to the thirteenth century and forms part of those maps which serve the purpose of decoration behind the altars of cathedrals. The Hereford *Mappamundi* distinguishes itself from those listed above because of its dimensions: 160 cm x 130 cm.

The legend on it reads Island of Tapobana. It is situated in the south of the river banks of India, in the part whither commences the Indian Ocean. Every year it has two summers and two winters and twice a year it is adorned with flowers. However the remotest part of it is full of elephants and dragons. The island has ten cities.[22] Over the representation of Tapbana, two monsters are depicted with the mentioned dracones as if to insist on their presence in that part of the island. At the northern extremity is shown a mountain of gold which explains the presence of the two monsters in the role of guardians of treasures[23] (plate p. 316 a).[24]

---

21 Paris, Bibliothèque Nationale de France, Ms. Latin, 8878, fol. 45 ter.

22 Cf. Copie manuscrite de la carte d'Hereford faite d'après la copie de Thomas Ballard, s. l. 1847 (Bibliothèque Nationale de France, Département des cartes et plans, Ge. A. 649).

23 Cf. V. de Santarem, Essai sur l'histoire de la cosmographie et de la cartographie pendant le Moyen Age et sur le progrès de la géographie 2, Paris 1852, 429-430.

24 P. Barber, The Hereford Mappae Mundi, 34.

The Jerome map (of the East), dated to circa 1150

The Indian Ocean covers the upper part of the map (British Museum, Add. 10049, cf. fig. 4) where nine islands are shown, the biggest one being Taprobane marking the extremity of the ocean. A river drawn in the middle divides it into two parts upon which are the indications to the effect that a part of it is inhabited and the other is uninhabited.

Fig. 4: The so-called Jerome map of the East, circa 1150 AD. London, The British Library, Add. 10049. Reproduced from Konrad Miller, Mappaemundi II, Stuttgart 1897, plate 11.

The Sawley World Map (or Henry of Mainz World Map)

The *mappae mundi* of Henry of Mainz, dated to the late twelfth century is in the Corpus Christi College, Cambridge, Ms. 66, p. 2. The map is not superimposed over the body of Christ unlike the Ebstorf *mappamundi* but at the top features a scene of the Last Judgment.[25]

---

[25] Reproduced P. Barber, The Hereford Mappae Mundi, 11.

The Ebstorf *Mappae mundi* dated to circa thirteenth century

It was discovered in 1830 in the Ebstorf monastery but was destroyed when Hannover was subject to bombardment in 1943. It measures 350 cm x 350 cm and was composed of thirty colored leaves of parchment. There does not exist without doubt a work hailing from the Middles Ages, wrote Walter Rosien, neither in art nor in literatures, that could give us in brief an image as lively and as complete as the Ebstorf representation gives of the world.[26] In the background of *mappamundi* is drawn the body of Christ and the terrestrial Paradise is shown as lying in the vicinity of the island of Taprobane. The inscription on the island reads: *This island is situated at the southern extremity of India and it's length is 470 thousand feet. Its river divides branches off into five streams. It is full of pearls and gems…There are ten major cities and numerous other less important ones. This island has two summers and two winters in a year and twice the year round it remains fully verdant.*[27]

The World Map in Antoine de la Salle's La Salade nouvellement imprimée, 1440

The *mappamundi* contained in the work of Antoine de la Salle called *La Salade nouvellement imprimée.* It was written in 1440. This wood-cut map was found in the manuscript the original of which is lost. At the eastern extremity is shown the Terrestrial Paradise. Not too far from it, to the south of India is shown the large island of Tapbana surrounded by a cluster of smaller islands bearing no names.

The Borgia world map

The Borgia world map is dated to circa 1430, and is preserved in the Apostolic Library in the Vatican City. To the south of the Indian Peninsula is Taprobana. Two town-vignettes symbolize cities.[28]

---

26 W. Rosien, Die Ebstorfer Weltkarte, Hannover 1952, 15; cf. also Harvey's article in this volume p. 130.

27 E. Sommerbrodt, Die Ebstorfer Weltkarte nebst Atlas von 25 Lichtdrucktafeln, Hannover 1891, plate 10.

28 Cf. Facsimile reproduced from the original: Il mappamondo Borgiano, Biblioteca Apostolica Vaticana 1980.

The Rudimentum novitiorum World Map, 1475 and the mappamundi in La Mer des Hystoires, 1506

The legend of the *mappamundi* found in *Rudimentum novitiorum* relevant to our discussion reads: *Trapo patane est insula indiae jacens ad austrum exquo oceanus indicus incipit. patens in longitudine octingentis lxx. milibus et .l. passuum in latitudine vero lxxv milia stadiorum. Scinditur autem amnis in ter fluvia tota repleta margaritis et gemmis bestis et elephantibus est repleta: partem vero homines tenent; in hac dicunt in anno duas estates et duas hyemes et bis floribus vernare locum utilissima insula est indiae virens omni tempore et folia eius sempre sunt virencia et nunquam cadunt ut dicit ipse li[ber] xiiii.*[29]

Over the representation of the island both forms of the name *Tabobana* and *Trabobatha* appear. In line with the indication for some countries shown on the map two high buildings are shown in the upper portion. To the left of it is a silver mountain and to the right is a walled enclosure through the outlets of which flow downwards four rivers. Some scholars think that the two persons figuring upright and engaged in a discussion are a Christian and a Jew. But its location however indicates that it is the Paradise with its four rivers irrigating the earth below. The tree of the Moon is juxtaposed against the tree of the Sun. To the right of the Paradise is Persia, shown through the land feature of a hillock. This is the first ever *mappamundi* to have been printed in Europe and the woodcut was by Lucas Brandis (plate p. 317 a).

The mappamundi found in La Mer des Hystoires, 1506 edition, fol. lxiiii v, chapter iiclxxi offers the following legend: *De trapobatane est une ysle en la mer de inde assise vers la partie dassirie qui contient en longueur iiii.c.xxxv. lieues et xliiii. en longueur. Il y a moites riuieres couras p[ar] le trauvers dicelle ysle. Beaucoup de perles et autres pierres précieuses une partie de ceste ysle est pleine d'elephants et d'auttres bestes lauttre parte est habitee des hommes. On dit qu'il y a deux estez et deux yuers en lan. Et que deux fois l'an produit fleurs et feuilles tousiours verdes que iamais ne cheet. Ceste ysle est la tres utiles de toute inde come dit Ysidore ou xiiii.*[30]

---

29 Rudimentum Novitiorum, Lübeck, Lukas Brandis, 1475, folio 144.

30 La Mer des Hystoires, v. 1-2, Lyon 1506.

The Andreas Walsperger World Map, 1448

The *mappamundi* drawn by Andreas Walsperger is preserved in the Apostolic archives of the Vatican City, signature: Pal. Lat. 1362 B. It was drawn about the year 1448 by the Benedictine monk Andreas hailing from Constance. At the western extremity of the *mappamundi* a towering monument rising high is shown to designate the location of Terrestrial Paradise. Almost in front of it is found the island of Taprobana in the Indian Ocean (plate p. 317 b).

The Johannes Leardus World Map, 1448

The map drawn by Johannes Leardus of Venice is dated to 1448. The Terrestrial Paradise is shown as perching atop of the map. The island of Taprobane appears therein as located not far from it in the Indian Ocean (plate p. 317 c).

These *mappaemundi* furnish us already with a sufficient proportion of material in order for us to make an idea of how the image of Taprobane was perceived in the west during medieval times. What mattered most in these cartographic representations was not the depiction of a precise geographic knowledge of the world, but an imparting of an ideological perception. They illustrate more often than not religious works or serve as background to Biblical episodes or religious history (hence the conservation of some of the *mappaemundi* in cathedrals, monasteries and convents). In a word, we may say that they served to function as a symbolic support of the Faith of medieval men in their relationship with God and the Universe. What is significant is the Christian sense instilled into the representation of Taprobane, and not the matter-of-fact veracity of what is said about it by classical authors and geographers.

At that time, if Taprobane is out of reach in the reality, it remains also inaccessible in the imaginary world. Most of the medieval *mappaemundi* such as those of the Hereford and of Ebstorf, locate it in the eastern confines of the circumfluent ocean in the proximity of the Paradise. It is an island, in other words a place approachable with great difficulty, because it is surrounded by the ocean, the symbol of terrestrial world, a world full of treacherous traps and violent storms. Only the righteous who possessed the wisdom and knew how to wade through dangerous currents and cross the sea unhindered can get to the island.

The island, the very one thus depicted is also a world of mystery imagined as a great garden: a land of dales and vales pregnant with greenery, plants and seeds constantly sprouting, trees blossoming and bearing fruits twice the year by reason of its double winter and double summer. This charming garden is all the more so pleasant because of its lush, luxuriant vegetation: pearls are in plentiful, gems, jewels and precious stones are in profusion all over. This sumptuousness, the wealth symbolizes the beautiful, therefore the good and one has every reason to believe that the island represents the perfection in this world. The number of cities, almost always fixed to ten, comes doubtless also from this willingness to highlight and signify the perfection by a rounded off figure which enters into numerous arithmetical combinations.

However this world of the unknown that catches attention since it is the reverse of the terrestrial life and because it inspires as much fear as ever: it is inhabited by wild beasts, elephants and even by dragons. We learn from the *bestiaries* that the elephant is the symbol of Adam and Eve, and the beast is there to remind us of the original sin. The dragon, the diabolical creature par excellence, is associated with it to recall hell. The unknown, it needs to recall is also the new world which awaits the Christian after the Last Judgment.

In fact, during this eclipse of the west in that part of the world, the East takes over: Chinese monks set out on long journeys in search of their Buddhist origins and above all merchants and Arab geographers. As from the inception of High Middle Ages, Arab traders, who had founded a trading settlement in Colombo about 949, knew the island well enough to call it Serendib, transcription of the Prakrit from of Sihaladipa (Sielediba) or from Sanskrit Sinhaladvîpa. The beauty and the wealth of the island struck their imagination to the extent of prompting them to figure it in the account of Sindbad the Sailors' sixth voyage[31] and believing moreover that Adam lived upon the island after his banishment from the Paradise.

This perception of Taprobane begins to change only with the emergence of the Catalan atlas tradition after the conquest of Spain by Arabs, with the work of a new "Ptolemy" the geographer Al-Idrisi serving as a catalyser.

31 Cf. A. Galland, Les Milles et Une Nuits 1, ed. by G. Picard, Paris 1960, 213-222; first edition: 1706.

Taprobane *à la Catalane*: A Passe-partout Name

Now we are concerned with the representation of Taprobane in the Catalan cartographic tradition as preserved in two surviving maps commonly known as the *Catalan Atlas of Paris* (Fig. 5) and the Catalan Atlas of Modena.

Fig. 5: Detail from the facsimile of the Catalan Atlas (1375) by Freiesleben. Courtesy of the Newberry Library, Chicago.

Catalonia, in the north-east of Spain, was occupied by the Arabs before it was incorporated into the Kingdom of Aragon in 1137. Later, in 1229, the Balearic Islands of Catalonia (including Majorca and Minorca) was annexed to the same Kingdom. During the Arab occupation of Catalonia, both Arabs and Jews were instrumental in promoting growth of a highly significant cartographic school which flourished for several centuries. These cartographers produced maps characterized by elaborate geographic features, rich in toponymy, ethnography, history and in lavish illustrations of cities, fauna, and effigies of foreign potentates.

The Catalan Atlas, dated to 1375, was drawn by a Majorcan mapmaker, sometimes attributed to Abraham Cresques and is of elongated proportions with a breadth of 69 cm and a length of 3,9 meters. The Atlas is composed of twelve sheets mounted on boards like a folding screen of which the first two are devoted to cosmological information while the remaining sheets form the map itself. These folding sheets of the screen can be spread out on a table top and may be viewed by going around the table; the legends upon the map are not oriented in a consistent manner.

For the Asiatic portion of the representation of his illuminated Atlas, the Catalan cosmographer from the Kingdom of Majorca found a great deal of inspiration from Marco Polo's description of the world. The depiction of the pearl-divers and fish-eating men seen in the Indian Ocean region of the Atlas demonstrate how much Marco Polo was a source of information, and the legends which accompany these human figures word for word attest the Catalan cosmographer's indebtedness to Marco Polo. However, even though Marco Polo did describe Seilan, the name by which the island was known to him in his day but did not make the least mention of Taprobane (the Hellenized name of Sri Lanka) in his Description of the World, the Catalan cosmographer could not dispense Taprobane in his Atlas, since it was so well attested by ancient and medieval geographers.

The existence of the name Taprobane was so much a part of the cartographic tradition and the medieval vision of the world that the Majorcan map-maker, contradicting Marco Polo, but complying with the image passed on to him down the ages, included a huge island entitled *Trapobana* in the extreme lower portion of the sixth panel of his world map. In this version of the name, the spelling follows closely the Latin form transmitted by some manuscripts of Isidore of Seville[32], among others, but with a metathesis of *r* posited after the initial *T*. Thus spelt *Trapobana* became a peculiar linguistic feature commonly occurring in the medieval geographic literature of the Iberian Peninsula amounting almost to a Hispanization of the Hellenized name *Taprobane*. Let us point out at the outset that, *Trapobana* was exactly the name-form found in the writings of

32 Isidor de Sevilla, Etymologiarum, bilingual edition (Latin text and Spanish translation), ed. by J. Oroz Reta and M. A. Marcos 2 (Biblioteca de Autores christianos 14/3, 5 and 12), Madrid 1983, 167-168 and 194-195.

Columbus and persistently made all the way to the present day' Spain[33].

The rich configuration of *Trapobana* is even more highlighted by the prominence given to it in the atlas as an imposing immense island marking the boundaries of the eastern world. This island on the distant horizon becomes yet another occasion for the Catalan cosmographer to bring together all sorts of disparate information however loosely related to the Far-East, and present it under the name of *Trapobana.* This interpretation, or rather updating, of *Trapobane* by the Majorcan map-maker brings the island out of its legendary confines and confronts it with a variety of attempts at identification.

In the legend above the island we see how the Catalan cosmographer from Majorca mixes real and fictional indications of a number of diverse countries under the single name of Trapobana. Let us now dismantle the montage of the legend. First of all, the presentation of Trapobane - *This is called by the Tartars Magno Caulij the last in the east*[34]. In this term *Magno Caulij*, Gerini saw the synonym of Menang-Kabau[35], or more correctly *Mânangkabo*, the name of an inland country of Sumatra[36]. At the same time, with neither justification nor proper analysis, Gerini interprets the appellation to the credit of Sumatra[37], just as Cordier, Bouchon, and Tastu did before him: *Illa Trapobana = la Taprobane des anciens, aujourd'hui l'île de Sumatra. Celui-ci* [Abraham Cresque] *aussi bien informé que qui que ce soit, tout en suivant les traces de Ptolémée son maître, le corrige et l'interprète, en plaçant l'Illa Iana* [...]

33 L. M. Meras, Imagens de America, Barcelona 1985, 237: *La mítica isla de Trapobana, según Ptolomeo*

34 H.-Ch. Freiesleben, Der katalanische Weltatlas vom Jahre 1375 nach dem in der Bibliotheque Nationale, Paris, verwahrten Original farbig wiedergegeben, Stuttgart 1977.

35 C. E. Gerini, Researches on Ptolemy's Geography of Eastern Asia, London 1909, 646.

36 J. Crawfurd, A Descriptive Dictionary of the Indian Islands and Adjacent Countries, Benares 1974, 273-276, s.v. *Menangkabau* (reprint of the 1856 edition).

37 ILLA TRAPOBANA (for Taprobana = Sumatra island), cf., Gerini, Researches, 647.

là où se trouve Ceylan et l'Illa trapobana à la place qu'occupe aujourd'hui l'île de Sumatra.[38]

There existed two Taprobanes, one of the Ancients and the other of the Middle Ages

Henry Cordier for his part writes, *Tastu, a raison de dire 'aujourd'hui l'île de Sumatra' mais il a tort de la marquer 'la Taprobane des anciens', et surtout de dire dans ses notes 'On le voit, je n'adopte pas ici le sentiment de savants tels que d'Anville, Gosselin, Barbié du Bocage, MM. le baron Walckenaer et Huot, continuateur de Maltebrun, lesquels veulent que l'antique Taprobana, ne soit autre que l'île de Ceylan. Je me range du côté du géographe catalan'. Tout le monde sait aujourd'hui qu'il y a eu deux Taprobane, celle des anciens, Ceylan, et celle du moyen âge, Sumatra, de même qu'il y a eu deux Java, Java mineure, cette même Sumatra, et Java majeure, Java meme.*[39]

Trapobana, a Hybrid Medley of Japan and Korea

Yule, who saw in this *Trapobana* a hybrid mixture of Japan and Korea[40], derives the term Kaoli or Cauli, as we have it in the Catalan Atlas, from a Chinese name designating Korea.[41] Indeed, Marco Polo uses the name Cauli in his travel account to denote one of the provinces surrendered to the Great Khan in the aftermath of a revolt led by a certain Nayan. Pauthier on the other hand, seems to ignore Yule's interpretation when he says: *Cauly must have been the country of*

---

38 Cf. Notices et extraits des manuscrits de la Bibliothèque nationale et d'autres bibliothèques, 1841, 138-139.

39 H. Cordier, L'Extrême-Orient dans l'Atlas Catalan de Charles V roi de France, in: Bulletin de géographie historique et descriptive 1895, 41-42.

40 H. Yule, Cathay and the Way Thither 1, Nendeln/Liechtenstein 1967, 303 (reprint of the 1913-1916 edition).

41 Kaoli is the Chinese name Koli of Corea (Ibid. 3, 129, footnote 2). – Cf. also Ibid 1, The Book of Ser Marco Polo, London 1903, 345: Kauli is properly Corea, probably here district on the frontier thereof, as it is improbable that Nayan had any rule over Corea. The Corean kingdom proper could not be a part of the prince's appanage. Marco Polo might mean the northern part of Corea, which submitted to the Mongols in A. D. 1269, with sixty towns, and which was subordinated entirely to the central administration in Liao-yang. As to the southern part of Corea, it was left to the king of Korea, who however, was a vassal of the Mongols, (Palladius, 32.). The king of Corea (Korye, Kao-li) was in 1288 Choung wang (1274-1298); the capital was Syong-to, now Kâi syeng (k'ai-ch'eng).

Kao-kiu-li, the primitive designation of the lou or the Circuit of Thoung-ning, a dependency of the same government.[42] Korean oceanographer Sangbok Hahn, in a paper presented at the International Seminar on the Korean Culture and the Silk Roads, took *Cresquez's Caulij* legend as an account of Korea.[43] All these, references and attributions therefore, demonstrate that the ancient Taprobane evidently has nothing to do with the name Caulij of the Majorcan cosmographer.

As early as 1858, one of the pioneer historians of cartography Joachim Lelewel read the contentious terms as *Magna cavillatio* and attempted to interpret them as "the place where you were hood-winked and where, there are big deceivers" (thinking the terms ought to be read as *magni cavilli*?). Entering the much debated issue, he stated that the Catalan Taprobana should rather be the Malay Peninsula.[44] Again, as recently as 1993, Mireille Pastoreau unhesitantly identified *illa Trapobana* of the Catalans with Sumatra: *The island of Sumatra, bottom right, ends the painting where myths and reality get mixed up.*[45]

Let us now move on to the next indication in the legend: In this island there are peoples different from all others. On some mountains in this island, there are men of great stature, twelve ells tall, like giants and very dark and who do not think, but eat white foreign men, if they can get hold of them. In his interpretation of this Trapobana, the Majorcan cosmographer attributes anthropophagy to the Taprobanians certainly through a borrowing from Marco Polo, who refers to cannibalism while describing the kingdoms of Ferlec and Dagroian, located on the large island of Sumatra, and also in his description of the Andaman Islands. Marco Polo says, They are very cruel men; they eat all men whom they can catch as long as they are not their own men. Apparently, it is Polo's description that the Majorcan cosmographer followed, but with a few additions from his

---

42 S. D. Hahn, World Maps Compiled in Aid of the East-West Exchanges, in: Korean Culture and the Silk Roads, Seoul 1991, 172: Taprobane in Ptolemy's world map in the middle of the Indian Ocean, indicates present Sri Lanka (Ceylon). However, Cresques' Taprobane indicates Korea by note.

43 Ibid., 172: Taprobane in Ptolemy's world map in the middle of the Indian Ocean, indicates present Sri Lanka (Ceylon). However, Cresques' Taprobane indicates Korea by note.

44 Cf. J. Lelewel, Géographie du Moyen Age, Brussels 1852, 58 and note 108.

45 Cf. Indes merveilleuses: l´ouverture du monde au XVIe siècle, Paris, Bibliothèque nationale 1993, 27.

own fantasy, mixing together a variety of Marco Polo's notes and arbitrarily squeezing them into his Trapobana.

Finally we arrive at the last indication of the legend, the only statement which corresponds to the legendary fame of the ancient Taprobane as interpreted by Isidore of Seville: In this island there are two summers and two winters, and twice a year the trees and plants flower. This is the utmost isle of the Indies. It is rich in gold and silver and abounds in precious stones. With a few variants, this is borrowed from the *Etymologiarum* of Isidore of Seville, who takes the legendary notions of the climate, flora, and harvesting seasons of India to be found in Pliny, Solinus and Martianus Capella and arbitrarily applies them to Taprobane of hallowed antiquity. The cartographic iconography of this novel interpretation of Taprobane is nuanced by the figure of an elephant coming from a *Bestiary* and also a distinctive feature from a *Lapidary* - a ruby held by a king, perhaps the king about whom Marco Polo speaks in his description of Ceylon: He possesses the grandest ruby that ever was seen, being a span in length, and the thickness of man's arm, brilliant beyond description.

The Majorcan cosmographer has made a hotchpotch, bringing together information clearly, concerning Korea, Sumatra (through the association of little kingdoms found on this island), the Andaman Islands, and Ceylon, all under the banner of *Trapobana.* As for the place-names such as Melaro, and Malao, it is without doubt that they come through borrowings from Marco Polo, possibly with a little twist of Catalan spelling, betraying their Malay origin as for instance the term Menlay (a place in southern Malaya). Much less impressive features of the Majorcan representation *Trapobana,* compared with the very ornate figure of the white elephant and the picturesquely turbaned eastern potentate, are banally serrated the chain of mountains that surround the island and the citadel vignette accompanying the legend *Aquesta ciutat és deserta per serpents"* (trans.: This city has been abandoned because of snakes). With the implied notion of a threat of reptiles to human habitat, we conclude our observations on the Catalan version of Taprobane, introduced by the Majorcan into Medieval cartography, where it acquired multifarious identities befitting an *All Encompassing Eternally Verdant Island in the Sun.* Thanks to the Majorcan cosmographer, the legendary Taprobane is now on its way to absorb a geographic pliability hitherto unknown in the Western cartographic tradition.

Reaching Dominica and Guadeloupe on his second voyage in 1493, Columbus noted in his journal the appearance of snakes. This herpetological indication leads us to believe that he was probably familiar with cartographic documents similar to that of the Catalan Atlas, which records serpents along with gold and silver in Taprobane. The Catalan Atlas references to double seasons and snakes in Taprobane are echoed in the writings of Christopher Columbus, who in Hispaniola and the Northern Coast of Venezuela, believed he had reached that fabled isle.

When, after his third voyage, Columbus claimed to have reached Taprobane, his adversary, the "doctor" Cisneros, maintained that the islands of the Indies discovered by Columbus were not in the East but some islands in the Atlantic Ocean already known to so many classical writers. Hence any novelty. In the face of these criticisms by the "doctor" Cisneros the Catholic Monarchs in their letter of 16th August 1494, queried Columbus as to the climate of the islands just discovered.[46]

In his letter of 26th February 1495 addressed to their Highnesses, Columbus emphasizes the climatic condition of the country as has been characterized by the Catalan Atlas, affirming that *We have proved that this land does bear fruit twice a year.*[47] With this assertion, Columbus not only responded to the query of the Court but also aimed at disproving and silencing the arguments of his enemies, who were increasingly questioning the true nature of the islands of the Indies just explored.

Another celebrated Catalan *mappaemundi*, a circular one on parchment, and made *circa* 1450, is in the Biblioteca Estense at Modena, from which it is known as the *Catalan Atlas of Modena* or simply as the *Este map*. On this map, Taprobane is placed at the far southern corner of the Indian Ocean and it is drawn, unlike any other feature on the map as a perfect triangle, left blank save a mountain range that surrounds its borders in a manner very similar to

---

[46] Cf. M. F. de Navarette, Collecion de los viages y descubrimientos, que hicieron por mar les Españoles 2, Madrid 1825, 154-155.

[47] J. Gil, Mitos y utopías del Descubrimiento: Colon y su tiempo, Madrid 1992, 77.

the *Illa Trapobana* of the Catalan Atlas of 1375. The legend attached to this oddly geometrical island reads: *Yla apellada trapobana* [sic] *en la qual a de agrestas montanyes en lo qual abita gent molt diferent del mon que no han nanguna raho los quals son forts e grans com azagants e san nagres e si poden prendre nangun home de la terra ferma manjanlos se.* (trans.: Island called Trapobana [sic], where there are wild mountains, in which live people very different from others: they are strong and as big as giants, and are black, and if they can capture any people from the mainland, they eat them.). The Este map thus provides a strikingly graphic demonstration of how, for the Catalan mapmakers Taprobane becomes a *passe-partout*, a blank space into which could be foisted assorted exotic geographical and various other information.

Yet its draughtsman or the illuminator, despite the fact that he was able to finish drawing the vignettes of preening mermaids and full bellied vessels in the Indian Ocean, failed for some unknown reasons -either because he lacked inspiration or did not have the time necessary, to complete the illumination of the space the mapmaker reserved for the legendary island of Taprobane. It was obviously meant to encompass a dazzling iconographic feature of conspicuous excellence. What a pity! We missed thus another marvel of the East.

Furthermore, in the Indian Ocean, the Catalan Atlas of Modena features without much ado two islands beneath India, one called Silan and the other named Jaua. The latter is without a shadow of doubt instead the *Illa Jana* of the Catalan Atlas.[48] In fact this latter *Illa Jana* of the Majorcan cosmographer is the one that was taken by Bouchon and Tastu for Ceylon, but the Modena cartographer however gives the correct spelling *Jaua* for the Catalan *Iana* thus distinguishing it clearly from Silan that he locates facing *Jaua*, so as to exclude all eventual risks of assimilations and confusions.

Jean Michel Massing thought that the location of the island of Jana *in India* and its geographic position, suggested that it is instead Ceylon[49]. Not far from Jana, a sober depiction of an island which Freiesleben thought stands for Haynan. A rhomb line emanating

---

48 Carta Catalana, Biblioteca Estense at Modèna, C. G. A. 5d – Cf. G. H. Kimble, The Catalan World Map of the R. Biblioteca Estense at Modena, London 1934, plates 3 and 4.

49 J. M. Massing, Observations and Beliefs: The World of the Catalan Atlas, in: J. A. Levenson (Ed.), Circa 1492: Art in the age of exploration, New Haven and London 1991, 26-33.

from a wind-rose is drawn across the island slightly obliterating a letter which renders the reading difficult leaving room for some to suppose that it stands for Haynan. On the other hand Java is found spelled Jana in the Spanish translations of Marco Polo, for example in the version used by C. Columbus which leads us to suggest that Jana in the Catalan atlas is a mere corruption of Java and it is as Java that we need to consider that term.

These two outstanding Catalan cartographic documents from Spain itself are perhaps the closest that we can get to the kind of pictorial depiction of Taprobane that would have been available to the cosmographers and royal experts who assessed the projects of Columbus's explorations of the Indies and counseled the King and the Queen when they were holding the Court at Barcelona, the port-city of Catalonia.

Epilogue

It is indeed not only a great pleasure but also an honor to me to be invited by the editors (whom I met at the recent ICHC held in Copenhagen) of the present volumes to contribute to this publication dedicated to Professor Johannes Dörflinger, the most helpful and kind scholar I had had the good fortune to come across at the bi-annual gatherings of the International Conference on the History of Cartography. Ever since the time he took the generous initiative to host me during a UNESCO-Austria sponsored fellowship (1988-1989) in the history of cartography at the Institute of History, University of Vienna, until now we remained good friends. The exceptional courtesies extended to me during my unforgettable *Viennese Days* were such that I was a guest of both him and his wife, including their children, bearing witness to the loyalty and trustworthiness he helped me cultivate in that warm and friendly atmosphere. As a token of this long-lasting affection nourished by mutual appreciation, I offer him this humble contribution of mine to the world's knowledge even though consisting of a precious few pages, nevertheless, being the results of research conducted in depth, linked to a quest and a passion to understand my country's pictorial representations viewed and imagined by people in map-producing cultures and their enigmas in the past which prompted me to undertake several years of relentless work, the initial part of which I carried out at the Austrian National Library (ÖNB) and at other repositories of maps and globes in Vienna while enjoying an unusually congenial stay at his institute.

About the Author

Ananda Abeydeera took a Ph.D. from the University of Paris-Sorbonne in 1990 and works as a foreign research associate of the Research Centre on Travel Literature of the same university. After graduating from the University of Sri Lanka he held a teaching position there for several years. He also taught at the Ecole des Hautes Etudes en Sciences Sociales (EHESS) and at the School of Oriental Languages and Civilizations (INALCO) in Paris as a visiting lecturer. He is the author of numerous contributions on Sri Lanka in ancient, medieval and pre-modern travel literature and in the history of cartography. He was a member of the international team of specialists of the five-months long UNESCO Maritime Route Expedition from Venice to Hakata (1990-1991).

Contact: ananda.abeydeera@hotmail.fr

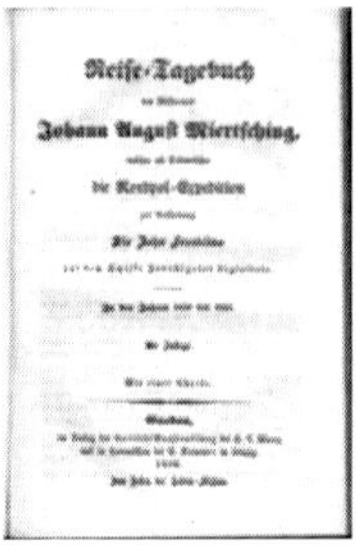

Johann August Miertsching,
Reise-Tagebuch,
Gnadau 1856

Barbara Bauer, Wien

„Das Schiffsleben ist doch ein elendes!“

Johann August Miertschings Reisetagebuch der Arktisexpedition unter dem Kommando von Robert M’Clure (1850-1854)

Unter den über 4.300 geographischen Werken, die die Sammlung Woldan beherbergt, befinden sich einige Reiseberichte, die die Suche nach einer Nordwestpassage zum Thema haben. Einen davon möchte ich auf den folgenden Seiten vorstellen: das Reisetagebuch von Johann August Miertsching, der 1850 bis 1854 an jener britischen Marineexpedition teilnahm, die schließlich die lang gesuchte nordwestliche Durchfahrt entdeckte.[1]

Die Suche nach der Nordwestpassage, dem Seeweg, der über die arktischen Gewässer Nordamerikas vom Atlantik in den Pazifik führt, beschäftigte Großbritannien ab dem 16. Jahrhundert. 1845 segelten zwei Schiffe unter dem Kommando von John Franklin ins arktische Eis, um die Passage zu finden, doch die Reise wurde statt des von der Admiralität und der britischen Öffentlichkeit erwarteten Triumphes eine der größten Tragödien der Seefahrt.

1 J. A. Miertsching, Reisetagebuch des Missionars Johann August Miertsching, welcher als Dolmetscher die Nordpol-Expedition zur Aufsuchung Sir John Franklins auf dem Schiff Investigator begleitete. In den Jahren 1850-1854, Gnadau 1855 [bzw. 2. Auflage, Gnadau 1856, Sammlung Woldan, R-I: PO/Ark 47]. Zitat in der Überschrift: Ebd., 13.

1848 schickte die Admiralität die ersten Suchexpeditionen nach Franklin aus. Im Laufe der folgenden zehn Jahre wurden über zwanzig Expeditionen in die Arktis unternommen, um Spuren der Verschollenen zu finden. Eine davon war die Expedition unter dem Kommando von Richard Collinson, die im Jänner 1850 von England abreiste; sie nahm die Route über Cape Horn, um von der Beringstraße aus in das arktische Insellabyrinth vorzudringen. Collinsons Expedition bestand aus zwei Schiffen, der *Enterprise* und der *Investigator.* Erstere wurde von Collinson selbst befehligt, letztere stand unter dem Kommando von Commander Robert John LeMesurier M'Clure, einem zweiundvierzigjährigen Iren, dessen Karriere bis dahin nur langsame Fortschritte gemacht hatte. Die *Investigator* hatte eine Besatzung von sechsundsechzig Männern an Bord.

Um allen Hinweisen auf den Verbleib von Franklins Schiffen nachgehen zu können, war es notwendig, einen Übersetzer für die Eskimosprache mitzunehmen. Die britische Admiralität wandte sich an die Mission der vereinigten mährischen Brüder in London mit der Bitte, der Expedition einen ihrer in Labrador tätigen Missionare als Dolmetscher zur Verfügung zu stellen. Die Wahl fiel auf den zweiunddreißigjährigen Sachsen Johann August Miertsching, der sich gerade in Deutschland aufhielt. Miertsching war bis auf den Naturwissenschaftler Berthold Seemann, der 1848 die *Herald*-Expedition begleitete, der einzige Deutsche, der an einer Franklin-Suchexpedition der britischen Marine teilnahm. Er wurde in den Offiziersrang aufgenommen und vorläufig auf der *Investigator* einquartiert, weil auf der *Enterprise* bei der Abreise der Expedition keine Kajüte frei war. Collinson wollte, dass der Übersetzer während eines kurzen Aufenthalts in der Fortescue Bay (Patagonien) auf die *Enterprise* umzog, doch Commander M'Clure setzte sich dafür ein, dass Miertsching bis Honolulu an Bord der *Investigator* bleiben konnte. *Mir war dies sehr lieb, denn der Umzug von Schiff zu Schiff bei starkem Regen und hohen Wellen hätte viel Beschwerlichkeit gehabt und meine Sachen wären alle naß geworden*, schreibt der Missionar.[2]

2 Miertsching, Reisetagebuch, 11 sowie vii, 3; L. Neatby (Ed.), Frozen Ships. The Arctic Diary of Johann August Miertsching 1850-1854, London 1967, xf.; L. P. Kirwan, A History of Polar Exploration (Penguin Books 1705), Harmondsworth 1962, 33; T. B. Collinson (Ed.), Journal of H.M.S. Enterprise, on the Expedition in search of Sir John Franklin's Ships by

Nach der Umfahrung von Cape Horn wurden die beiden Schiffe getrennt, und bei seiner Ankunft auf Honolulu musste M'Clure erfahren, dass die *Enterprise* einen Tag zuvor abgesegelt war. Er setzte in der Folge alles daran, sie einzuholen und nahm, anstatt wie vorgesehen über Kamtschatka zu fahren, die schnellere, aber gefährliche und damals kaum bekannte Route über die Aleuten. Dank dieser Abkürzung erreichte die *Investigator* den Kotzebue Sound lange vor der *Enterprise* am 29. Juli, wo sie die bei Point Barrow stationierte *Plover* und die unter dem Kommando von Kapitän Kellett stehende *Herald* traf. Auf keinem der beiden Schiffe wusste man etwas von Collinson. Kellett bat M'Clure, wenigstens vierundzwanzig Stunden zu warten, doch M'Clure weigerte sich. *Da aber unser Kapitain mit Recht glauben konnte, daß die schneller segelnde Enterprize unbemerkt hier vorbei gesegelt sei, so antwortete er* [...] *durch Signale: Kann nicht warten!*, schreibt Miertsching.[3] In Wahrheit ging M'Clure zu Recht davon aus, dass er die *Enterprise* überholt hatte; er wollte jedoch nicht warten, weil er sich von Collinson schäbig behandelt fühlte.[4] Ein weiterer Grund für M'Clures eigenmächtiges Vordringen ins arktische Eis war vermutlich, dass er die späte und vielleicht letzte Chance auf Ruhm ergriff, die sich ihm bot.

Kapitän Collinson erreichte die Beringstraße erst Ende August, redete sich ein, M'Clure ohnehin nicht mehr einholen zu können, und verbrachte den Winter in Hongkong. Als er 1851 ins Eis fuhr und feststellte, dass die *Investigator* überall schon vor ihm gewesen war, wurde Collinson bewusst, dass er sich dadurch, dass Miertsching auf M'Clures Schiff geblieben war, nicht mit den Inuit verständigen konnte und somit seine Suche nach Franklin kaum mehr eine Aussicht auf Erfolg haben würde.

Behring Strait 1850-55, London 1889, 50; S. Osborn (Ed.), The Discovery of the North-West Passage by H.M.S. Investigator, Captain R. M'Clure, 1850, 1851, 1852, 1853, 1854, London 1856, 20.

3 Miertsching, Reisetagebuch, 17; Neatby (Ed.), Frozen Ships, 33; A. Cooke – C. Holland, The Exploration of Northern Canada 500 to 1920. A Chronology, Toronto 1978, 187.

4 Der Schiffsarzt Doktor Armstrong gibt das in seinem Reisebericht offen zu, vgl. A. Armstrong, A personal narrative of the Discovery of the North-West Passage; with numerous incidents of travel and adventure during nearly five years' continuous service in the Arctic regions while in search of the expedition under Sir John Franklin, London 1857, 76-78.

Während die Fahrt der *Enterprise* weitgehend in Vergessenheit geriet, wurde die Expedition der *Investigator* berühmt, vor allem deshalb, weil M'Clure im Oktober 1850 die Nordwestpassage fand. Der weitere Expeditionsverlauf war allerdings weniger erfolgreich und endete beinahe in einer Katastrophe von Franklin'schen Ausmaßen, weil das Eis in der Bucht, in der die *Investigator* ihren zweiten Winter verbrachte, nicht mehr aufbrach. Als bereits alles verloren schien, wurde die Besatzung im Frühjahr 1853 von Kapitän Kellett, der nach der Rückkehr der *Herald* nach England mit einer Expedition unter dem Oberkommando von Sir Edward Belcher erneut in die Arktis aufgebrochen war, gerettet. Die *Investigators* kehrten im Herbst 1854 nach England zurück.[5]

M'Clures Expedition ist durch mehrere Berichte gut dokumentiert. Das ist insofern überraschend, als es allem Anschein nach Bemühungen seitens des Kommandanten gab, die Aufzeichnungen anderer Expeditionsteilnehmer verschwinden zu lassen. Beim Umzug auf Kelletts Schiffe durfte niemand seine Aufzeichnungen mitnehmen. *Daß ich meine Schreibereien, die mir viel Mühe gemacht, und die mir über Alles werth sind, zurücklassen muß, schmerzt mich sehr*, schreibt Miertsching.[6] M'Clure versprach dem Übersetzer, ihm sein Tagebuch bei einer späteren Gelegenheit mitzubringen, hielt sein Versprechen aber nicht: *Mein Tagebuch hatte der Kapitain nicht mitnehmen können; denn wenn dasselbe auch nicht schwer war, hätte er doch dann auch die Schriften und Journale der andern Offiziere mitnehmen müssen, und das wäre zu viel geworden.*[7] Die Begründung, die M'Clure Miertsching gab, überzeugt nicht, weil der Übersetzer, der kein Mitglied der Marine war, nicht mit anderen Offizieren gleichgesetzt werden konnte – M'Clure hätte für ihn ohne weiteres eine Ausnahme machen können. Als im Frühjahr 1854 eine Schlittengruppe unter Kelletts zweitem Navigationsoffizier Frederick Krabbé die *Investigator* aufsuchte, konnte dieser

5 Kirwan, History, 189f.; Cooke – Holland, Northern Canada, 187f.; F. Skead, Private Journal kept on board H.M.S. Enterprise, 14 October 1849 – 8 March 1852, Scott Polar Research Institute, MS 1161 D, 30, 33, 35; P. Berton, The Arctic Grail. The quest for the North West Passage and the North Pole, 1818-1909, New York 2000, 295.

6 Miertsching, Reisetagebuch, 139; Neatby (Ed.), Frozen Ships, 201 (Fußnote); Berton, Arctic Grail, 263.

7 Miertsching, Reisetagebuch, 149.

keine Tagebücher außer dem des ersten Leutnants finden.[8] Die wahrscheinlichste Erklärung für das rätselhafte Verschwinden der Aufzeichnungen ist, dass M'Clure, der bereits an seinem eigenen Mythos als Entdecker der Nordwestpassage bastelte, und Berichte anderer Expeditionsteilnehmer, die von Hunger, Leid, Verzweiflung, Skorbut, Auspeitschungen, Streitigkeiten und Fehlentscheidungen erzählten, unterdrücken wollte. Sein Vorhaben war nur teilweise erfolgreich: Zumindest die beiden Schiffsärzte und der Matrose James Nelson konnten ihre Aufzeichnungen retten.[9] Von M'Clures zweitem Leutnant Samuel Gurney Cresswell blieben zwar keine Bordjournale, aber einige Briefe, die er am Anfang der Reise an Familienmitglieder schrieb, und die Aufzeichnungen, die er auf seiner Heimreise nach England führte, erhalten.[10] Kapitän M'Clures Journale und Briefe befinden sich heute in der Bibliothek der Royal Geographical Society in London.[11]

Drei Berichte über die *Investigator*-Expedition wurden veröffentlicht. Der „offizielle" Reisebericht der Expedition, der ebenso wie Miertschings Tagebuch in den Beständen der Sammlung Woldan zu finden ist, erschien 1856 unter dem Titel *The Discovery of the North-West Passage by H.M.S. Investigator, Captain R. M'Clure, 1850, 1851, 1852, 1853, 1854, ed. by Commander Sherard Osborn, from the logs and journals of Capt. Robert LeM. M'Clure.* Der Eindruck, dass es sich bei dem Buch um den vom Kapitän verfassten Bericht handelt, ist aller-

---

8 F. Krabbé, Journal of the Proceedings of Her Majesty's Sledge Newton, detached from Her Majesty's Ship Resolute, between 3d April and 13th June 1854, under the Command of Mr. F. J. Krabbé, Master, in: British Parliamentary Papers 1855a, Sledge Journies, Part II, No.16, London 1854-55, 715; Miertsching, Reisetagebuch, 166, 177.

9 Das Tagebuch des stellvertretenden Schiffsarztes Henry Piers befindet sich in den Beständen des National Maritime Museum, Greenwich (Journal of Henry Piers, Assistant Surgeon in H.M.S. Investigator, 14 Dec. 1849-31 Mar. 1853, on the Franklin Search expedition, National Maritime Museum, JOD/102). – Das Archiv des Scott Polar Research Institute in Cambridge beherbergt eine Kopie des zweibändigen Journals von James Nelson (J. H. Nelson, Voyage of H.M.S. Investigator, in search of Sir John Franklin, and resulting in the discovery of the North-West Passage. Journal, 1850-54, v. 1-2, Scott Polar Research Institute, MS 748/1-2; BJ Xerox).

10 Publiziert in: D. Harrod (Ed.), War, ice and piracy. The remarkable Career of a Victorian Sailor. The Journals and Letters of Samuel Gurney Cresswell, London 2000.

11 The R. J. LeM. McClure Collection, Royal Geographical Society, ar SSC/106.

dings falsch: Das Buch wurde von Sherard Osborn geschrieben, der offiziell nur als Herausgeber auftritt. Commander Osborn nahm an der Expedition von Edward Belcher teil und war bei den von ihm geschilderten Ereignissen erst ab Juni 1854 dabei. M'Clure übergab ihm nach der Rückkehr der Expeditionen nach England seine Tagebücher zur Bearbeitung, und Osborn, der sein schriftstellerisches Talent schon mit einem Reisebericht über eine frühere Franklin-Suchexpedition bewiesen hatte,[12] verfasste eine spannende Abenteuergeschichte, die M'Clure und seine tapfere Crew verherrlicht. Mit der Realität hat Osborns glatt geschliffene, beschönigende Erzählung wenig zu tun.

Der Schiffsarzt der *Investigator*, Doktor Alexander Armstrong, veröffentlichte 1857 einen Reisebericht, der als Entgegnung auf das Buch Osborns zu verstehen ist. Armstrong stellt Osborns Heldenepos eine realistische Schilderung der Reise entgegen, die die Leiden der *Investigator*s und die an Bord herrschende Verzweiflung und Hoffnungslosigkeit nicht verschweigt; er äußert sich außerdem sehr unfreundlich über M'Clure und dessen Mannschaftsführung. Armstrong und M'Clure kamen nicht miteinander aus, vielleicht, weil sie andere Prioritäten setzten. Für den Schiffsarzt kam das Wohlergehen und die Gesundheit der Crew an erster Stelle, während für den ehrgeizigen M'Clure ab dem zweiten Winter nur noch die Befahrung der von ihm entdeckten Nordwestpassage zählte. Der Arzt wirft dem Kapitän in seinem Reisebericht Größenwahn, Skrupellosigkeit, falsche Entscheidungen und bewusste Täuschung vor.

Miertsching, dessen Bordjournal mit der *Investigator* im Eis verloren ging, rekonstruierte sein Tagebuch auf Kelletts Schiffen aus der Erinnerung mit Hilfe einiger geretteter Notizen und Kapitän M'Clures Aufzeichnungen, die dieser ihm zur Verfügung stellte. Er informiert seine Leser darüber, dass sein Reisebericht erst ab Mai 1853 ein authentisches Tagebuch ist. Obwohl er die Rekonstruktion für mangelhaft hielt, ist sein Bericht der ehrlichste, humorvollste und im Bezug auf das Alltagsleben ergiebigste, der über die Reise der *Investigator* erschien. Dieses Bordtagebuch wurde 1967 in einer englischen Übersetzung von Leslie Neatby herausgegeben. Miertsching

12 S. Osborn, Stray Leaves from an Arctic Journal, or, eighteen months in the Polar regions in search of Sir John Franklin's Expedition in the years 1850-1851, London 1852.

selbst veröffentlichte 1855 eine Überarbeitung. In der 1855 publizierten Version fehlen einige Passagen des Bordjournals, vor allem jene, die Streitigkeiten zwischen den Offizieren der *Investigator* betreffen. Manchmal wurden Formulierungen abgeschwächt oder Namen weggelassen, und mitunter fasste Miertsching mehrere Einträge zu einem zusammen oder datierte ein wenig anders. Außerdem ließ er bei der Überarbeitung seines Bordtagebuchs für die Publikation einige kritische Bemerkungen über die *Plover* und Belchers Expedition weg.[13]

Miertschings Reisetagebuch nimmt unter den Quellen zur Reise der *Investigator* eine besondere Stellung ein, weil es sich um den Bericht eines Außenseiters handelt, der zwar im Unterschied zu den meisten anderen Expeditionsmitgliedern aufgrund seiner Missionarstätigkeit in Labrador mit dem Leben im hohen Norden, aber nicht mit dem Leben auf See vertraut war. Miertsching berichtet daher über Aspekte des Alltagslebens, die in anderen Journalen übergangen werden, weil sie Marinemitgliedern ohnehin selbstverständlich schienen. Aus seinen Tagebüchern bekommt man einen guten Einblick in die tägliche Routine, die Religionsausübung und den Umgangston an Bord; man erfährt aber auch viel über das Fortkommen der Schiffe, die Gefahren der Eisnavigation und über die Schwierigkeiten, die sich durch die große Kälte im Alltag ergaben. Während Osborns Buch alles Negative ausblendet und Armstrong aufgrund seiner Abneigung gegen den Kapitän nicht immer als unvoreingenommen gelten kann, bleibt Miertschings Schilderung ausgewogen; er wollte M'Clures Unternehmungen weder idealisieren (wie Osborn) noch kritisieren (wie Armstrong), sondern berichtet offen und unbekümmert über die Vorkommnisse der Reise. Er freute sich über die Erfolge der Expedition, verschweigt aber den hohen Preis, den die Besatzung zahlen musste, nicht. Wenn es um Daten und Messergebnisse geht, verlässt man sich allerdings besser auf Doktor Armstrongs Bericht, da Miertsching dazu neigt, Phantasie-

13 Vgl. zum Beispiel Neatby (Ed.), Frozen Ships, 37, 218, 229, 231, 235 sowie Miertsching, Reisetagebuch, 149f. Das Vorwort zum publizierten Reisebericht (i-x) wurde von einem namentlich nicht genannten Mitglied der Herrnhuter Bruderschaft verfasst. Anschließend folgt ein Inhaltsverzeichnis (xi). Die Reiseerzählung umfasst fünf Kapitel; den Anhang (195f.) bilden zwei Tabellen, die Jagderfolge Miertschings während der Reise dokumentieren.

zahlen anzugeben, und manchmal mit der Reihenfolge von Ereignissen durcheinander kommt; er stellt auch einige Gerüchte, die er aufgeschnappt hatte, als Tatsachen dar.[14]

Die Landschaften und Tiere, die die Expeditionsmitglieder sahen, vergleicht Miertsching in seinem Reisetagebuch mit den ihm bekannten in Labrador. Er berichtet, dass Nordlichter zwar häufig vorkamen, aber nicht so schön wie in Labrador waren, die Luftspiegelungen jedoch viel stärker waren als jene, die er kannte. Das Aussehen der Küsten von Prince Albert's und Banks Land erinnerte ihn an die Umgebung von Herrnhut, nur dass alles Grün fehlte. Auch das Aussehen und die Kleidung der Inuit, die man traf, vergleicht Miertsching mit dem Erscheinungsbild der Inuit von Labrador. Der Missionar erfüllte seine Funktion als Übersetzer überaus gut: Er konnte sich mit allen Stämmen, denen die Expeditionsmitglieder begegneten, trotz der zum Teil unterschiedlichen Dialekte verständigen und es gelang ihm stets, das Vertrauen der anfangs oft misstrauischen und ängstlichen Bevölkerung zu gewinnen. Besonders die Inuit von Cape Bathurst schlossen ihn ins Herz und wollten ihn gar nicht mehr fort lassen; der Häuptling bot dem Missionar Schlitten, Hunde, Zelt und seine Tochter an, um ihn zum Bleiben zu bewegen. Die Miertsching entgegengebrachte Sympathie wurde von ihm erwidert; bei der Lektüre des Reisetagebuchs fällt auf, dass Miertsching die Inuit mochte und sich freute, wenn er Gelegenheit bekam, sich mit ihnen zu unterhalten. Während sich die Offiziere der verschiedenen Franklin-Suchexpeditionen oft abfällig über die eingeborene Bevölkerung äußern, sah Miertsching nicht auf die Inuit herab und schrieb vor allem über ihre positiven Eigenschaften. *Diese Eskimos beschämen manche Christen, indem sie ihre Waaren lieber lange Tagereisen weit schaffen, um sie für Sachen, die ihnen nützlich sind, einzutauschen, statt in der Nähe verderbliches Wasser (Branntwein) dafür nehmen zu müssen,* berichtet er zum Beispiel über die Bewohner von Point Warren.[15] Als um das Seelenheil seiner

[14] Vgl. zum Beispiel Miertsching, Reisetagebuch, 157, 167. Einmal schreibt Miertsching von *Mr. Omaneys* [sic] *Zelt*, obwohl Kapitän Ommanney weder an M'Clures noch an Belchers Expedition teilnahm (170); Kelletts Winterhafen gibt er fälschlich mit Winter Harbour statt Dealy Island an (136). Bei den Todesfällen kommt Miertsching mit den Namen der Verstorbenen und den Sterbedaten durcheinander (138-140).

[15] Miertsching, Reisetagebuch, 37; vgl. auch 27-29, 35, 40-42, 82-84.

Mitmenschen besorgter Missionar versuchte er stets, den Inuit von Gott zu erzählen. *Von einem höheren, göttlichen Wesen haben sie nur mangelhafte Begriffe*, meint er: *Doch giebt es ihrer Meinung nach zwei verschiedene Wesen, ein gutes und ein böses, in zwei verschiedenen Ländern, wohin sie selbst nach dem Tode kommen; Manche in das eine, Andere in das andere Land, je nachdem sie hier in diesem Leben gewesen sind.*[16] Für längere Gespräche mit den Inuit blieb Miertsching jedoch nie Zeit, da M'Clure und seine Offiziere lediglich an Informationen über Franklin interessiert waren und es stets eilig hatten, weiterzukommen.

Da es Miertsching nicht um eine objektivierte Reisebeschreibung ging, sondern um eine subjektive Schilderung seiner Eindrücke, erfährt man aus seinem Tagebuch nicht nur viel über den Reiseverlauf, sondern auch einiges über den Autor selbst. Miertsching war nicht weltfremd, aber manchmal etwas naiv; seine Moralvorstellungen waren streng. Er war botanisch interessiert,[17] ein sehr eifriger und erfolgreicher Jäger[18] und ein geschickter Handwerker: Er stellte zum Beispiel Blechkessel für die Schlittenreisen her und baute sich auf Belchers Schiff *North Star* eine Kajüte nebst Einrichtung.[19] Der Missionar wollte sich nützlich machen und half, wo immer er konnte. Am Anfang der Reise war es allerdings schwierig für ihn, sich seinen Mitreisenden verständlich zu machen, weil er nicht Englisch konnte. In seinem Tagebuch meint er, dass seine Schiffskameraden alle nett zu ihm waren, doch er fühlte sich anfangs [...] *sehr fremd in meinem schwimmenden Haus und der unbekannten Gesellschaft.*[20] Ab Mitte Februar lernte er täglich zwei Stunden Englisch, und am 10. Juni vermerkte er: *Das Lesen englischer Bücher geht immer besser, doch giebt es noch immer viele Worte, die ich nicht verstehe und bei denen ich das Lexikon zur Hand nehmen muß.*[21]

---

16 Miertsching, Reisetagebuch, 30.

17 Ebd., 37, 92, 97. Im Juli 1852 schrieb er: *Ich habe auf den verschiedenen Ländern bis jetzt 3785 Exemplare von Pflanzen, Gräsern und Moosen gesammelt und will diesen Monat damit fortfahren.* (118).

18 In seinem unveröffentlichten Tagebuch lobt er sich selbst: [...] *with these northern ptarmigan and white hares no skilled sportsman is as renowned as I* (Übers. nach Neatby (Ed.), Frozen Ships, 106).

19 Miertsching, Reisetagebuch, 73, 132, 174.

20 Ebd., 4 sowie 2.

21 Ebd., 15 sowie 6. Ein wenig Deutsch sprach an Bord der *Investigator* nur der Matrose Charles Anderson, ein Afroamerikaner, der Koch auf einem Schiff gewesen war, das deutsche Auswanderer nach Amerika gebracht hatte. Er

Das Leben auf dem Schiff behagte dem Missionar überhaupt nicht. *Dies ist Alles für mich so neu, daß ich mich kaum hineinfinden kann,* schreibt er, *Ich vermuthete etwas ganz Anderes auf unsrer gefahrvollen Reise. Ach Herr und Heiland, habe Geduld! Hilf Herr, auf welche Weise es auch sein mag. Ich bin sehr niedergeschlagen.*[22] Obwohl Miertsching nicht an Seekrankheit litt, empfand er die Bewegung des Schiffs als unangenehm, besonders bei starkem Wellengang: *Man kann weder stehen noch sitzen, und selbst liegend muß man sich mit Knie und Ellbogen feststemmen, um nicht aus der Koje geworfen zu werden. Dazu ist Alles im Schiffe naß.*[23] Gegen Feuchtigkeit und Kälte war Miertsching empfindlich; er berichtet, dass *die Betten* [...] *einen häßlichen Modergeruch angenommen hatten, der Kopfweh verursacht* [24] und beschwert sich über die Unbequemlichkeit und Enge des Schiffs. Außerdem litt er stark unter Heimweh und hatte Sehnsucht nach seiner Heimat Deutschland und seiner Missionsstation in Labrador.[25] Besonders am Anfang der Reise hatte Miertsching oft das Bedürfnis, allein zu sein, und zog sich in seine Kabine zurück. Der Wunsch, in Ruhe gelassen zu werden, hing sicherlich zum Teil auch mit seinen mangelnden Englischkenntnissen und mit den Konflikten zusammen, die der Kapitän mit seinen Offizieren austrug und aus denen sich Miertsching heraushalten wollte. M'Clure war ein strenger Kapitän, der seine Crew selbst bei kleinen Vergehen schwer bestrafte und kein Vertrauen in seine unerfahrenen Leutnants hatte; seine Wutanfälle waren gefürchtet. Aus Samuel Gurney Cresswells Briefen erfährt man, dass der Kapitän den

---

war bei der Mannschaft als Spaßmacher beliebt, schreibt Miertsching, aber *Leider ist er einer der ganz Gottvergessenen Menschen.* (8).

22 Miertsching, Reisetagebuch, 6.

23 Ebd., 12f. An einer anderen Stelle seines Tagebuches schreibt er ähnlich: *Man kann weder stehen, noch sitzen, noch liegen. Dabei ist Alles: Kleider, Wäsche, Betten durch und durch naß, und im Schiffe herrscht ein widerlicher Geruch.* (13).

24 Ebd., 66.

25 Ebd., 5, 31 und 76. Die schönsten Momente waren für Miertsching jene, in denen er Post von seinen Mitbrüdern bekam. Als Kelletts und M'Clures Mannschaften im Frühjahr 1854 Briefe erhielten, war Miertsching traurig, dass für ihn keine dabei waren: *Ich war sehr niedergeschlagen, und es kam mir vor, als hätte man mich ganz aufgegeben. Doch mitten in meiner Verzagtheit wurde ich beschämt. Denn man brachte mir ein Packet von London an mich adressirt, welches einen ganzen Jahrgang der Missionsblätter (Periodical accounts) enthielt. Ich freute mich innig, etwas von meinen Brüdern und Schwestern zu hören* [...] *und sagte dem unbekannten Absender in meinem Herzen viel tausend Dank.* (167).

ersten Leutnant William Haswell systematisch schikanierte; später kam es auch zu Streitigkeiten zwischen M'Clure und Cresswell. Als am 24. Jänner 1850 bei einem Sturm die oberen Masten brachen, während der wachehabende Offizier Haswell gerade kurz seinen Posten verlassen hatte, schreibt Miertsching von einem *furchtbaren Aufruhr.*[26] Bei einer anderen Gelegenheit meint er: *Gott sei Dank, daß der gestrige Tag vorbei ist. Es war der unangenehmste, den ich bis jetzt auf dem Schiff erlebte. Ich hielt mich so viel wie möglich in der Kajüte, fern von den aufgereizten Leuten, auf.*[27] In Miertschings unveröffentlichtem Tagebuch kommen Bemerkungen, die sich auf Streitigkeiten zwischen Kapitän und Offizieren und auf Meinungsverschiedenheiten der Offiziere untereinander beziehen, viel häufiger vor als im publizierten Tagebuch; allerdings gibt Miertsching keine Information über die Ursachen der Konflikte.[28]

Sherard Osborn charakterisiert Miertsching in seinem Reisebericht als *cheerful and contented under all circumstances.*[29] Wenn man Miertschings Aufzeichnungen liest, hat man allerdings nicht den Eindruck, dass der Missionar „fröhlich und zufrieden" war; im Gegenteil, er war – auch wenn er das vielleicht nicht immer nach außen zeigte – oft traurig und pessimistisch gestimmt. Schon der Aufbruch der Expedition am 20. Jänner 1850 war für ihn Anlass zu trüben Gedanken: *Werden wir Europa wiedersehen? und wann?* […] *Der Blick in die Zukunft ist dunkel! Eine Nordpolexpedition, so reich an Hindernissen und Gefahren! Wie lange werden uns die Eisfelder hoch oben im Norden festhalten? Vielleicht 2 Jahre, oder länger, oder gar auf immer?* [30] Dass Miertsching trotz seiner negativen Einstellung nicht in Depressionen verfiel, verdankte er seinem Glauben. Wenn es ihm schlecht ging oder das Schiff in Gefahr geriet, suchte und fand er Trost bei Gott: *Doch dem hohen und erhabenen, allmächtigen Herrn und Heiland, unter dessen*

---

26 Miertsching, Reisetagebuch, 5. Im unveröffentlichten Tagebuch schreibt er bei dieser Gelegenheit von einer *wahren Rebellion* (Neatby (Ed.), Frozen Ships, 9) sowie Samuel Gurney Cresswell an seine Eltern, H.M.S. Investigator, Sunday [January 1850], in: Harrod (Ed.), War, ice and piracy, 61f.; Samuel Gurney Cresswell an Joseph Cresswell, H.M.S. Investigator, at Sea, Thursday, in: Ebd., 62; Samuel Gurney Cresswell an seine Eltern, H.M.S. Investigator, off Flaxman Is, August 21st, 1850, in: Ebd., 69.

27 Ebd., 14.

28 Vgl. Neatby (Ed.), Frozen Ships, 9f., 15, 21f.

29 Osborn (Ed.), Discovery, 98.

30 Miertsching, Reisetagebuch, 2f.

Schutz wir Unwürdigen die gefahrvolle Reise beginnen, ist dies Alles bekannt, Er thue, was Ihm wohlgefällt! [31]

Mit seiner ausgeprägten Frömmigkeit stand Miertsching anfangs ganz allein da. Er war von der „Gottlosigkeit" und der mangelnden Disziplin der Seeleute entsetzt. Laut Miertsching verging am Anfang der Reise kaum ein Tag, an dem nicht einer der Besatzung im Arrest war; am 6. April 1850 vermerkte er die erste Auspeitschung, [...] *eine abscheuliche Strafe, den Rücken so zu zerfleischen, aber es scheinen solche Strafen bei so rohem Gesindel nothwendig zu sein.*[32] Am 9. April schreibt Miertsching von einer Schlägerei, und anlässlich des Aufbruchs von Honolulu vermerkte er: *Der Lärm, das Singen, Pfeifen und Fluchen der halb Berauschten erinnerte mich unangenehm an das mir Bevorstehende.*[33] Die lauten Vergnügungen der Matrosen behagten ihm nicht, und auch die Feiern, die mit Extraportionen und Grog gefeiert wurden, waren nicht nach seinem Geschmack. Besonders angewidert äußert er sich über das Neptunsfest, das am 5. März 1850 anlässlich der Überquerung des Äquators gefeiert wurde. Alle Matrosen, die zum ersten Mal den Äquator passierten, wurden mit Teer angestrichen, der mit einem Stück Fassreifen wieder abgeschabt wurde. *Ich will dies Fest nicht genauer beschreiben, denn ich möchte es lieber ganz vergessen,* meint Miertsching; *Man sollte meinen, daß solche in steter Gefahr auf dem Wasser lebende Menschen ernster sein müßten, als die in größerer Sicherheit auf dem Lande lebenden, aber es ist gerade umgekehrt, denn habe ich je freche und gottlose Leute gesehen, so waren es diese. Das strenge Schiffsgesetz ist kaum hinreichend, Ordnung zu erhalten.*[34] Über das Weihnachtsfest schreibt er missbilligend: [...] *kein Hosianna ertönt in dieser Eiswüste. Die lärmende Feier dieses Tages sprach mich nicht an, ich war deshalb den Tag meist außerhalb des Schiffes, und hatte auf meinen einsamen Wanderungen ein Gefühl der wahren Weihnachtsfreude.*[35] Weil auf der *Investigator* nur unregelmäßig Gottes-

---

31 Miertsching, Reisetagebuch, 3; vgl. auch 51, 90 und 92f.

32 Ebd., 9. Laut M'Clures Aufzeichnungen fand die erste Auspeitschung schon am 8. März 1850 statt: Der Koch wurde wegen meuterischen Betragens bestraft (R. J. Le M. M'Clure, Journal of H.M.S. Investigator, Robert McClure commanding, in search of Sir John Franklin, January – September 1850, 22, in: The R. J. LeM. McClure Collection).

33 Miertsching, Reisetagebuch, 21.

34 Ebd., 7f. Als der Äquator ein zweites Mal überquert wurde, meint Miertsching: *Ich war froh, dass jenes Neptunsfest nicht abermals gefeiert wurde.* (15).

35 Miertsching, Reisetagebuch, 69. *How and in what manner this day and above all the Christmas festival were here celebrated I do not think fit to insert in my journal, but it will*

dienste stattfanden, hielt Miertsching einsame Morgensegen in seiner Kajüte und sang allein zu seiner Gitarre religiöse Lieder. *O wie schön ist's am Sonntag, weder Musik, Tanz, noch Lärm ist zu hören*, schrieb er in sein Tagebuch: *Wenn ich nur einen christlichen Freund in meiner Umgebung hätte!* [36]

Nach der ersten Zeit der Abschottung nahm Miertsching auf dem Schiff eine unaufdringliche, aber sehr beharrliche Missionarstätigkeit auf. Ende Mai 1850 verteilte er religiöse Schriften an die Matrosen, worauf ihn der Kapitän auslachte [...] *und sagte, seine Leute seien nicht so einfache Menschen, wie meine Eskimos u.s.w.*[37] Während die Offiziere auf Miertschings Bemühungen belustigt reagierten, fielen seine Worte bei den Matrosen auf fruchtbaren Boden. Die gefährlichen Situationen, in die das Schiff im Eismeer geriet, trugen sicherlich viel zu seinem Erfolg bei. Nach überstandenen Gefahren war die Crew bei Gottesdiensten stets andächtiger, und wenn die *Investigator* aus Stürmen unbeschädigt hervorging oder Eispressungen standhielt, machte das großen Eindruck auf die Besatzung. Miertsching nutzte die an Bord herrschende ernste Stimmung und begann, in seiner Kabine für alle, die kommen wollten, „Privatversammlungen" abzuhalten, in denen er aus der Bibel vorlas, das Gelesene erklärte und mit den Anwesenden gemeinsam betete. Die Zusammenkünfte gaben der Crew auf der gefährlichen Reise ins Ungewisse Halt und Trost und erfreuten sich besonders im Winter, als es nicht viel zu tun gab, wachsender Beliebtheit. Die Offiziere und der Kapitän nahmen nicht teil, schienen aber, meint Miertsching, nichts dagegen zu haben, dass er welche veranstaltete. In seinem unveröffentlichten Tagebuch schreibt der Missionar allerdings bei einer Gelegenheit, dass M'Clure den Geiger zum Tanz aufspielen ließ, weil es ihm nicht richtig schien, dass die Matrosen Traktate lasen – der Kapitän scheint also anfangs

---

abide in my memory as long as I live, meint Miertsching in seinem unveröffentlichten Journal, (Übers. von Neatby (Ed.), Frozen Ships, 97).

36 Miertsching, Reisetagebuch, 7. Auch das Osterfest feierte Miertsching allein (9). Manchmal fielen die Messen auf der *Investigator* wegen schwieriger Eisnavigation oder anderer wichtiger Beschäftigungen aus (33, 43, 112).

37 Ebd., 15. Im unveröffentlichten Tagebuch schreibt Miertsching, dass M'Clure ihm sagte, er solle seine Traktate lieber gefallenen Frauen schenken, vgl. Neatby (Ed.), Frozen Ships, 41.

versucht zu haben, Miertschings Bemühungen entgegenzusteuern.[38] Später kam es auch bei M'Clure zu einem Sinneswandel, denn Miertsching wirkte nicht nur auf die Crewmitglieder ein. In seinem Tagebuch berichtet der Missionar von langen, interessanten Unterhaltungen mit dem Kapitän, in denen es oft um religiöse Themen ging. Die Ansichten der beiden waren zuerst, wie Miertsching zugibt, sehr verschieden, doch mit der Zeit machte sich Miertschings Einfluss bemerkbar und M'Clure begann, regelmäßig in der Bibel zu lesen.[39] Miertsching „bemutterte" den Kommandanten: Er fertigte Stiefel für ihn an, besuchte ihn, wenn er krank war, und las ihm etwas vor. In seinem Reisetagebuch nennt er ihn oft *meinen theuren Kapitain*. Umgekehrt mochte auch M'Clure Miertsching gern: Er kümmerte sich von Anfang an um den Übersetzer, lud ihn jeden Tag auf ein Glas Wein ein, schenkte ihm ein Gewehr und bot ihm an, seine nass gewordenen Bücher und seine Gitarre in der Kapitänskajüte aufzubewahren. Als die *Investigator* in wärmeren Breiten unterwegs war, stiegen die beiden jeden Abend auf das Verdeck und betrachteten auf dem Rücken liegend die Sterne.[40] Miertsching dürfte der einzige gewesen sein, der mit M'Clure wirklich gut auskam. Während der Kommandant sich bei seinen Offizieren und der Crew hart und unnachgiebig zeigte, scheint er bei Miertsching, der kein Marinemitglied war, das Gefühl gehabt zu haben, offen über seine Sorgen und Probleme sprechen zu können, ohne dass seine Autorität darunter litt. Weil Miertsching außerhalb der strengen Hierarchien der Marine stand, war er der einzige Mensch an Bord, bei dem sich M'Clure aussprechen konnte. Vor seinen Untergebenen musste der Kapitän, auch wenn die Situation ausweglos schien, optimistisch und zuversichtlich auftreten – er durfte keine Schwäche zeigen, weil die Disziplin an Bord sonst zusammengebrochen wäre. Nur Miertsching

38 Miertsching, Reisetagebuch, 78, 103, 121, 141 sowie Neatby (Ed.), Frozen Ships, 34.

39 Miertsching, Reisetagebuch, 14f., 22, 51, 105f., 136 sowie Neatby (Ed.), Frozen Ships, 79, 139. Als sich einmal eine günstige Gelegenheit zum Fortkommen bot, bestand M'Clure sogar darauf, eine Messe zu halten, bevor das Schiff weiterfuhr, was selbst der fromme Miertsching nicht für sinnvoll hielt (Ebd., 144).

40 Miertsching, Reisetagebuch, 6, 8, 16, 31, 73, 88.

wusste, wie es dem Kapitän wirklich ging: *Es thut mir Leid um unsern vielgeprüften, werthen Kapitain, er muß sich zwingen, guten Muth zu zeigen.*[41]

Der Missionar kam mit allen Expeditionsmitgliedern gut aus; nur Doktor Armstrong, der ihn in seinem Reisebericht nie beim Namen, sondern immer nur *the interpreter* nennt, konnte ihn nicht leiden, eine Abneigung, die auf Gegenseitigkeit beruhte. Armstrong wurde zwar von der Crew bewundert, scheint aber bei fast allen Offizieren unbeliebt gewesen zu sein. [42] In Miertschings Bordjournal findet man die kryptische Bemerkung, dass Armstrong im Februar 1851 von den anderen Offizieren *gehörig gedemüthigt* wurde. Ein anderes Mal berichtet der Missionar von Streitigkeiten zwischen Armstrong und M'Clure, im Zuge derer die *Falschheit* des Doktors ans Tageslicht gekommen sei.[43]

Auch wenn Armstrong, wie Miertsching es in seinem Bordtagebuch ausdrückt, nicht *the best of comrades* war[44] – seine Aufgabe, sich um die Gesundheit und das Wohl der Besatzung zu kümmern, nahm er sehr ernst. In seinem Reisebericht schreibt er ausführlich über die aufgetretenen Krankheiten und ihre möglichen Ursachen. Auch Miertsching berichtet in seinem Journal regelmäßig über den Gesundheitszustand an Bord, wenn auch nicht so ausführlich wie Armstrong. Der Missionar litt während der Reise einige Male an Erkältungen, Zahnschmerzen und aufgrund der Feuchtigkeit seiner Kabine an Rheuma: *Den ganzen Tag zittere ich vor Frost und muß mich dann am Abend in das feuchte, dumpfige Bett legen, wo ich vor Schmerzen wenig Ruhe finde. Ach, wenn wird doch dies elende Schiffsleben ein Ende haben!*, beklagt er sich zum Beispiel in einem auf 9. Oktober 1852 datierten Eintrag.[45] Rheuma, Erkältungen und Schneeblindheit traten bei der Besatzung der *Investigator* häufig auf, und in einigen Fällen mussten nach Erfrierungen Amputationen vorgenommen werden. Zwei

---

41 Miertsching, Reisetagebuch, 116; vgl. 121: *Der Kapitain zeigt mit großer Seelenstärke vor den Offizieren und der Mannschaft den besten Muth und die größte Hoffnung, aber in seinem Innern sieht es ganz anders aus.*

42 Vielleicht mit Ausnahme Cresswells, der in einem Brief an seinen Vater schrieb: *The more I see of Armstrong the more I think him a good fellow of the first water* […] (Samuel Gurney Cresswell an Francis Cresswell, H.M.S. Investigator, Plymouth, January 17th, 1850, in: Harrod (Ed..), War, ice and piracy, 63), sowie Neatby (Ed.), Frozen Ships, xvi.

43 Neatby (Ed..), Frozen Ships, 103, 158.

44 Ebd., 103.

45 Miertsching, Reisetagebuch, 123 sowie 49, 66 und 146.

Expeditionsmitglieder litten an psychischen Erkrankungen. Im August 1852 schreibt Miertsching zum ersten Mal davon, dass der Schiffssteward Mark Bradbury geisteskrank geworden war, und im Eintrag vom 18. Oktober desselben Jahres berichtet er, dass auch einer der Offiziere, der Maat Robert Wynniatt, vom Wahnsinn ergriffen war. Während Francis Leopold M'Clintock, ein Offizier Kapitän Kelletts, Bradburys Erkrankung mit übermäßigem Alkoholkonsum in Verbindung brachte, sah der Matrose James Nelson die Ursache für den psychischen Zusammenbruch seines Kameraden in der aus dem Alkoholmissbrauch resultierenden Degradierung vom Steward zum einfachen Matrosen. Über die Ursache der Erkrankung Wynniatts meint Nelson: *I* [...] *imagine it probably originated in the great disappointment experienced, consequent on the vessels not breaking out last season* [...][46] Wenn man Miertschings Tagebücher liest, erhält man allerdings eher den Eindruck, dass die Ursache für den Zustand des Maats in den Konflikten und zwischenmenschlichen Spannungen an Bord zu suchen war. Besonders aus dem Bordtagebuch wird deutlich, dass Wynniatt dem Missionar gegenüber keinerlei Aggressionen zeigte, aber gegen Doktor Armstrong und insbesondere gegen Kapitän M'Clure wilde Drohungen ausstieß. Er versuchte mehrmals, M'Clure tätlich anzugreifen und kündigte an, den Kommandanten und die Offiziere umbringen und das Schiff anzünden zu wollen. Der Umstand, dass sich Wynniatts Zustand auf Kelletts Schiffen schnell besserte, deutet ebenfalls darauf hin, dass der Maat die angespannte zwischenmenschliche Situation auf der *Investigator* nicht verkraftete. Wie man aus Nelsons Journal erfährt, wurde Wynniatt in England wieder gesund, während Bradbury in geistiger Umnachtung starb.[47]

Im Frühjahr 1852 brach auf der *Investigator* Skorbut aus, eine Folge der unzureichenden Lebensmittelrationen. Der Proviant, der Collinson und M'Clure zur Verfügung stand, war von äußerst schlechter Qualität. Weil immer wieder verdorbene Lebensmittel weggeworfen werden mussten, waren auf der *Investigator* schon zu Beginn des zweiten Winters Kürzungen der Rationen notwendig, zuerst auf zwei Drittel, dann sogar auf die Hälfte. Im dritten Winter

---

46 Nelson, Voyage 1, 226 und F. L. M'Clintock, Logbook of H.M.S. INTREPID, 1852-54, National Maritime Museum, MCL/12, 273f. sowie Miertsching, Reisetagebuch, 121 und 124.

47 Nelson, Voyage 2, 77-79; Miertsching, Reisetagebuch, 125f.; Neatby (Ed.), Frozen Ships, 174f., 178, 180.

wurden die Portionen noch einmal verkleinert und bestanden laut Miertsching pro Mann und Tag aus zwanzig Lot[48] Brot oder Zwieback, acht Lot Fleisch, drei Lot Gemüse und drei Lot Schokolade, Tee und Zucker.[49] Die Schiffsärzte befürchteten, dass die unzureichenden Rationen schlimme Folgen für die Gesundheit der Männer haben würden, doch Doktor Armstrongs Proteste wurden vom Kapitän ignoriert. *Wir müssen in dieser traurigen Einöde verbleiben, bis es dem Herrn gefällt uns zu erlösen*, schreibt Miertsching; *Schon jetzt läßt sich viel Unzufriedenheit bei der Mannschaft wahrnehmen; man sieht die Hungrigen den Kehrichthaufen umwühlen, der noch vom vorigen Winter her auf dem Eise liegt, um irgend etwas zu finden, den nagenden Hunger zu stillen.*[50] M'Clures Befehl, dass geschossenes Wild abgeliefert werden musste, wurde zuweilen umgangen, indem die Jäger ihre Beute gleich, nachdem sie erlegt worden war, roh verzehrten. Man aß auch Füchse und Lemminge, und wenn ein Rentier geschossen worden war, wurden selbst Eingeweide und Haut verwertet; letztere wurde als Sauerfleisch zubereitet. Nur zu Weihnachten durften sich die Expeditionsmitglieder satt essen.[51]

Nach dem dritten Winter waren alle Expeditionsmitglieder entkräftet, abgemagert und skorbutkrank. M'Clure, der mittlerweile geradezu besessen davon war, die Nordwestpassage zu befahren, versuchte, den schlechten Gesundheitszustand seiner Mannschaft vor Kapitän Kellett zu verheimlichen, doch eine ärztliche Untersuchung, die Kellett im Frühjahr 1853 anordnete, brachte die erschreckende Wahrheit ans Tageslicht. Kellett befahl entgegen der Proteste M'Clures, die *Investigator* im Eis zurückzulassen. Als Mitte Juni die letzten *Investigator*s auf Kelletts Schiffe transferiert wurden, schildert Miertsching ihre Ankunft: *Es war ein trauriger Anblick, den ich nie vergessen werde. Auf jedem der 4 Schlitten lagen 2 Kranke festgebunden, Andere ganz entkräftete wurden von ihren etwas stärkeren Kameraden geführt, wieder Andere hielten sich an die Schlitten, und diese wurden von einer Mannschaft gezogen, die zum Theil so schwach war, daß sie alle 5 Minuten kraftlos*

48 Ein Lot(h) entspricht etwa 15,6 Gramm.

49 Miertsching, Reisetagebuch, 60, 109, 111, 115-117, vgl. M'Clure, Journal Investigator, 12; Osborn (Ed.), Discovery, 131; Collinson (Ed.), Journal Enterprise, 252.

50 Miertsching, Reisetagebuch, 123; Armstrong, Personal Narrative, 487, 531f., 565.

51 Miertsching, Reisetagebuch, 123, 128, 131.

niederfielen und vom Kapitain und ihren Gefährten wieder aufgerichtet werden mußten. Es war ein Bild des unsäglichsten Elends. Wäre unser barmherziger Herr nicht selbst in's Mittel getreten und hätte uns zur rechten Zeit Hülfe gesendet [...] *wir wären Alle elend umgekommen!* [52]

Insgesamt verbrachte M'Clures Mannschaft vier Winter im Eis. Bei den Winterarrangements orientierte man sich, wie alle britischen Expeditionen des 19. Jahrhunderts, an William Edward Parry, der 1819/20 als Erster in der Arktis überwintert hatte. Im Einklang mit den Maßnahmen seines berühmten Vorgängers ließ M'Clure die obersten Maste abnehmen und eine Plane aus Segelleinwand montieren, die das Verdeck wie ein Zelt überdachte. Auf dem Deck wurde eine Schneeschicht angebracht, die mit einer Mischung aus Schnee, Sand und Wasser bedeckt wurde. Miertsching hielt diesen *Polarcement* für eine neue Erfindung, aber schon Parry hatte die Schneeschicht mit einem solchen Gemisch übergießen lassen, damit sich eine harte Oberfläche bildete. Rund um das Schiff wurde das Eis geebnet und eine Schneemauer gebaut, die die Kälte abhalten sollte.[53]

Auch bei der Organisation von Winterunterhaltungen fungierte Kapitän Parry als Vorbild für alle Expeditionen, die nach ihm in die Arktis fuhren. Parry hatte für seine Männer Theateraufführungen und Maskenbälle veranstaltet, eine Schule eingerichtet und Zeitungen herausgegeben. M'Clure versprach seiner Crew zu Beginn des ersten Winters zwar, Belustigungen und Vergnügungen für sie einzurichten, blieb aber im Vergleich zu anderen Expeditionsleitern weitgehend

---

52 Miertsching, Reisetagebuch, 149; vgl. Armstrong, Personal Narrative, 570f. Wie irreführend M'Clures Angaben über den Zustand seiner Mannschaft waren, wird aus seiner Korrespondenz mit Kellett deutlich: Commander M'Clure an Captain Henry Kellett, C.B. H.M.S. Resolute, Dealy Island, 2d May 1853, in: British Parliamentary Papers 1854a, VII, No. 3, London 1854, 91f. sowie Robert John LeMesurier M'Clure an Henry Kellett, H.M.S. Investigator, Mercy Bay, May 23rd 1853, in: H. Kellett, Letterbook of H.M.S. RESOLUTE, 1853-54, National Maritime Museum, MCL/42, 21.

53 Miertsching, Reisetagebuch., 55f. und 58; W. E. Parry, Journal of a voyage for the discovery of a North-West Passage from the Atlantic to the Pacific, performed in the years 1819-20 in His Majesty's Ships Hecla and Griper, London 1821, v, 101f., 117; Ders., Journal of a second voyage for the discovery of a North-West Passage from the Atlantic to the Pacific, performed in the years 1821-22-23 in His Majesty's Ships Fury and Hecla, under the orders of Captain William Edward Parry, R.N., F.R.S., and commander of the expedition, London 1824, 120, 128, 376f.

untätig. Auf der *Investigator* gab es lediglich eine Abendschule, in der die Matrosen Schreiben, Lesen, Rechnen und Navigation lernen konnten, und manchmal deklamatorische Unterhaltungen, Tänze oder Konzerte. Tagsüber vertrieben sich die Matrosen die Zeit mit Ballspielen, Lesen, Kartenspielen oder dem Ausbessern ihrer Kleidung; jeden Tag mussten sie einige Stunden auf dem Eis spazieren gehen. *Das Schiffsleben geht maschinenmäßig fort,* schreibt Miertsching über die tägliche Routine im Winter, *ein Tag ist wie der andere, und jede Arbeit geschieht jeden Tag zu derselben Zeit* [...] *Solche, die sich nicht privatim beschäftigen, die nicht schreiben, lesen, zeichnen oder stricken, sind übel daran.*[54]

Die zweite Überwinterung war noch trostloser als die erste. Die Tage waren von Langeweile und Hunger geprägt, und bei der Beleuchtung musste man sparen: [...] *die meiste* [...] *Zeit mußten wir im Finstern versitzen* [...], schreibt Miertsching.[55] Die Hauptbeschäftigungen waren Spazierengehen und Schlafen. *Wölfe heulen um das Schiff und ihr Geheul macht die finstere, kalte, todte Nacht noch öder und trauriger*, berichtet der Missionar. Die Schule war nicht mehr so gut besucht wie im ersten Winter, [...] *denn wenn man den Tag über mit leerem Magen auf den Bergen durch Schnee und Eis herum läuft, so fühlt man sich am Abend ganz abgemattet und hat keine Lust zum Lernen.*[56] Im dritten Winter gab es keine Schule mehr, und die Matrosen hatten mehr Zeit, über die sie frei verfügen konnten.

Ihren vierten Winter in der Arktis verbrachten die *Investigator*s auf Kelletts Schiffen. Obwohl die Lebensmittelrationen wegen des Zuwachses der Schiffsmannschaft reduziert werden mussten, waren die Wintermonate nicht mehr ganz so deprimierend. Die neue Umgebung und die größere Gesellschaft brachten Abwechslung in den eintönigen Alltag. Kellett und seine Offiziere waren bei der Organisation von Unterhaltungen für die Crews wesentlich engagierter und einfallsreicher als M'Clure. Im August 1853, als man auf den Aufbruch des Eises wartete, wurden zum Beispiel Wettkämpfe im Laufen, Sackhüpfen, Ringen, Springen, Ziehen und

---

54 Miertsching, Reisetagebuch, 71f. und 66f.; Parry, Journal of a voyage, 106, 127; W. E. Parry, Second Voyage, 123f., 378; Ders., Journal of a third voyage for the discovery of a North-West passage from the Atlantic to the Pacific, performed in the years 1824-25 in His Majesty's Ships Hecla and Fury, under the orders of Captain William Edward Parry, London 1826, 49-51.

55 Miertsching, Reisetagebuch, 112.

56 Ebd., 111 und 114.

Lastentragen veranstaltet. Im Winter fanden Theateraufführungen und Zaubervorführungen statt, und die Offiziere hielten für die Mannschaft Vorträge. Weil die Verbindung zwischen den Schiffen oft aufgrund von Stürmen unterbrochen war, wurde ein elektrischer Telegraph installiert, der laut Miertsching viel Unterhaltung bot: *So wurde vermittelst desselben von einem Schiff zum andern Schach gespielt und dergleichen mehr.*[57] Der Platzmangel auf den überfüllten Schiffen führte allerdings bald zu zwischenmenschlichen Irritationen. *So freundlich unsre Leute auch anfangs aufgenommen wurden, so ist unter den mancherlei Unbequemlichkeiten, die durch den Zuwachs der Schiffsmannschaft entstanden sind, das Mitleid und die Theilnahme allmählich kühler geworden, und die unvermeidlichen Beschwerden machen die eigentlichen Bewohner des Schiffes mitunter verdrießlich*, schreibt Miertsching. *Die Mannschaft vom Investigator lebt zurückgezogen, sie halten zusammen und verrichten ihre Arbeit gut und pünktlich, ohne zu schimpfen und zu fluchen. Sie geben keine Gelegenheit zu Unannehmlichkeiten, und wenn in den freien Abendstunden die Uebrigen sich mit geräuschvollen Vergnügungen die Zeit verkürzen, so versammeln sich unsre Pietisten, wie sie spottweise genannt werden, auf unserm Schiff Intrepid, wo wir schon manche angenehme und gesegnete Privat-Erbauung gehabt haben.*[58]

Im Sommer 1854 beendete Edward Belcher die Expedition. Er ließ vier seiner fünf Schiffe im Eis zurück, zog alle seine Männer ab und gab die Suche nach Collinson und Franklin auf. Miertsching hatte nach allem, was er auf M'Clures Expedition durchgemacht hatte, genug von Polarexpeditionen; als die Expeditionsmitglieder die Arktis hinter sich ließen, schrieb er: *Seit dem 27. Juli 1850, wo wir* [den Polarkreis] *in der Behringstraße überschritten, waren wir in dem eisigen Polarkreise wie gefangen. Ich hoffe, nie wieder diesen Kreis betreten zu müssen.*[59] Dennoch war er stolz, an jener Expedition teilgenommen zu haben, die die Nordwestpassage gefunden hatte: *Wie merkwürdig diese meine Lebensführung! Wie hätte ich es mir einst träumen lassen, daß ich die Nordpolexpedition mitmachen sollte, der es beschieden war, die 300 Jahre lang vergeblich gesuchte nordwestliche Durchfahrt zu entdecken! Bis hierher hatte der Herr in Gnaden geholfen, möchte es Ihm doch gefallen, mich einst noch den Tag des*

57 Miertsching, Reisetagebuch, 161; vgl. 152, 157-160.
58 Ebd., 160 und 167.
59 Ebd., 184.

Wiedersehens meiner Lieben in der theuren Heimat erleben zu lassen! O daß er alle seine Absichten mit mir und durch mich ganz erreichen möchte! [60]

M'Clures Expedition sollte für Miertsching nicht nur die einzige Arktisreise, sondern auch der letzte Aufenthalt im Norden sein. Die Herrnhuter Brüdergemeinde schickte ihn nicht zurück zu den Inuit von Labrador, die er so lieb gewonnen hatte, sondern nach Südafrika, um die Hottentotten zu missionieren. Miertsching setzte sich schließlich in seiner Heimat Sachsen zur Ruhe, wo er 1877 im Alter von neunundfünfzig Jahren starb.[61] Sein Reisebericht über die M'Clure-Expedition ist das einzige Buch, das er je veröffentlichte; es lässt die Reise der *Investigator* noch heute lebendig werden, und jeder, der es gelesen hat, wird den freundlichen Missionar nicht so schnell wieder vergessen.

Quellen- und Literaturverzeichnis

Armstrong A., A personal narrative of the Discovery of the North-West Passage; with numerous incidents of travel and adventure during nearly five years' continuous service in the Arctic regions while in search of the expedition under Sir John Franklin, London 1857.

Berton P., The Arctic Grail. The quest for the North West Passage and the North Pole, 1818-1909, New York 2000.

Collinson T. B. (Ed.), Journal of H.M.S. Enterprise, on the Expedition in search of Sir John Franklin's Ships by Behring Strait, London 1889.

Cooke A. – Holland C., The Exploration of Northern Canada 500 to 1920. A Chronology, Toronto 1978.

Harrod D. (Ed.), War, ice and piracy. The remarkable Career of a Victorian Sailor. The Journals and Letters of Samuel Gurney Cresswell, London 2000.

Kellett H., Letterbook of H.M.S. RESOLUTE, 1853-54, National Maritime Museum, MCL/42.

Kirwan L. P., A History of Polar Exploration (Penguin Books 1705), Harmondsworth 1962.

Krabbé F., Journal of the Proceedings of Her Majesty's Sledge Newton, detached from Her Majesty's Ship Resolute, between 3d April and 13th June 1854, under the Command of Mr. F. J. Krabbé, Master, in: British Parliamentary Papers 1855a, Sledge Journies, Part II, No.16, London 1854-55, 707-722.

---

60 Miertsching, Reisetagebuch, 144f.

61 Neatby (Ed.), Frozen Ships, xvi.

M'Clintock F. L., Logbook of H.M.S. INTREPID, 1852-54, National Maritime Museum, MCL/12.

M'Clure R. J., Journal of H.M.S. Investigator, Robert McClure commanding, in search of Sir John Franklin, January – September 1850, in: The R. J. LeM. McClure Collection, Royal Geographical Society, ar SSC/106.

M'Clure R. J.. – Kellett H., Commander M'Clure an Captain Henry Kellett, C.B. H.M.S. Resolute, Dealy Island, 2d May 1853, in: British Parliamentary Papers 1854a, VII, No. 3, London 1854, 91-93.

Miertsching J. A., Reisetagebuch des Missionars Johann August Miertsching, welcher als Dolmetscher die Nordpol-Expedition zur Aufsuchung Sir John Franklins auf dem Schiff Investigator begleitete. In den Jahren 1850-1854, Gnadau 1855.

Neatby L. (Ed.), Frozen Ships. The Arctic Diary of Johann August Miertsching 1850-1854, London 1967.

Nelson J. H., Voyage of H.M.S. Investigator, in search of Sir John Franklin, and resulting in the discovery of the North-West Passage. Journal, 1850-54, v. 1-2, Scott Polar Research Institute, MS 748/1-2; BJ Xerox.

Osborn S. (Ed.), The Discovery of the North-West Passage by H.M.S. Investigator, Captain R. M'Clure, London 1856.

Parry W. E., Journal of a voyage for the discovery of a North-West Passage from the Atlantic to the Pacific, performed in the years 1819-20 in His Majesty's Ships Hecla and Griper, London 1821.

Parry W. E., Journal of a second voyage for the discovery of a North-West Passage from the Atlantic to the Pacific, performed in the years 1821-22-23 in His Majesty's Ships Fury and Hecla, under the orders of Captain William Edward Parry, R.N., F.R.S., London 1824.

Parry W. E., Journal of a third voyage for the discovery of a North-West passage from the Atlantic to the Pacific, performed in the years 1824-25 in His Majesty's Ships Hecla and Fury, under the orders of Captain William Edward Parry, London 1826.

Skead F., Private Journal kept on board H.M.S. Enterprise, 14 October 1849 – 8 March 1852, Scott Polar Research Institute, MS 1161 D.

Zur Autorin

Barbara Bauer, geb. 1973, Studium der Geschichte, Anglistik und Rechtswissenschaften an der Universität Wien, Mag. iur. 2000, Mag. phil. 2002 (Diplomarbeit bei Univ.-Prof. Dr. Johannes Dörflinger), Dr. phil. 2009 über *Die Arktisreisen von Francis Leopold McClintock*, derzeit beschäftigt in der Verwaltung der Wirtschaftsuniversität Wien.

Kontakt: Barbara.Bauer@wu.ac.at

Hartmann Schedel,
Buch der Croniken und geschichten,
Nürnberg 1493

Kurt Brunner, München

Die Karten der *Schedelschen Weltchronik* der Sammlung Woldan

Einleitung

In der Sammlung Woldan befindet sich auch ein kürzlich restauriertes Exemplar vom *Buch der Chroniken und geschichten mit figuren und pildnussen von anbeginn der welt bis auf diese unnsere Zeit* von Hartmann Schedel, das 1493 in Nürnberg bei Anton Koberger gedruckt wurde. Das Buchwerk ist umfänglich, es hat ein Format von 33 x 47 cm und umfasst beinahe sechshundert Seiten.[1] Von der sog. *Schedelschen Weltchronik* wurden siebenhundert Exemplare gedruckt, davon dürften heute noch rund dreihundert Exemplare existieren.[2] Die *Schedelsche Weltchronik* ist die deutsche Übersetzung des einige Monate davor beim selben Verleger in gleicher Ausstattung erschienenen Werks *Liber Chronicarum*.[3] Hartmann Schedels Weltchronik zählt zu den bedeutendsten verlegerischen Leistungen des Inkunabelzeitalters.

Beide Buchwerke sind mit 1.800 Holzschnitt-Illustrationen von Michael Wolgemut und Wilhelm Pleydenwurff ausgestattet, und zwar mit 29 doppelseitigen Stadtansichten und zwei Karten.

---

1 Siehe die kürzlich erschienene prächtige und kommentierte Faksimilierung: H. Schedel, Weltchronik. Kolorierte Gesamtausgabe von 1493. Einleitung und Kommentar von Stephan Füssel, Köln 2005.

2 Gegenwärtig liegen Antiquariatspreise bei bis zu 200.000 Euro.

3 Von der lateinischen Auflage wurden 1.400 Ausgaben gedruckt; sie ist mit über 650 Seiten umfänglicher als die deutsche.

Bei den beiden Karten handelt es sich um

- eine Weltkarte und
- eine *Germania-Karte*[4] des Hieronymus Münzer.

Im Exemplar der Wiener Sammlung Woldan ist in der vorderen Innenseite des Einbands zudem ein etwas beschädigtes Wandkarten-Fragment einer *Germania-Karte* mit der deutschen Ostseeküste eingeklebt. Dieses erweist sich als Abkömmling einer Etzlaub-Karte.[5]

Die Weltkarte (Abb. 1) ist ptolemäischen Ursprungs; die Germania-Karte (Abb. 4) beruht auf der verloren gegangenen Karte des Nikolaus Cusanus, entstanden um 1450. Die zusätzliche Germania-Karte der Sammlung Woldan (Abb. 7) ist vom Etzlaub-Typ. Die beiden Germania-Karten beruhen also auf den ersten beiden modernen Germania-Karten.[6]

Die Weltkarte

Die Karte (Format 24,5 x 33,5 cm) ist auf den Blättern 12v und 13r in den Text integriert. Links davon, auf Blatt 12v sind in einer senkrechten Leiste sieben Fabelwesen abgebildet.[7] Um die Karte (Abb. 1) sind als Ausschmückungen die Söhne Noahs Japhet, Sem und Ham angeordnet. Links unten werden in einem lateinischen Text die Hauptwinde beschrieben. Im Kartenrahmen sind diese zwölf Winde eingetragen. Die Karte ist sehr spärlich beschriftet; im Zentrum der Karte findet sich *Hierusalem* als Verweis zur Heilsgeschichte und zu den mittelalterlichen Weltkarten. Die weitere

---

4 Vgl. zum politisch und historisch neutralen Begriff „Germania-Karte" P. H. Meurer, Mappae Germaniae. Die schönsten und bedeutendsten Deutschlandkarten von 1492 bis 1803, Neustadt an der Saale 1984, 7: [...] *die Gattung der Germania-Karten* . [...] *zeigen das Territorium des alten Deutschen Reiches und damit Mitteleuropa von Flandern bis Ostpreußen und von Schleswig bis zur Adria* [...].

5 Dies ist nicht ganz ungewöhnlich, so ist in einem der Exemplare der *Schedelschen Weltchronik* der Bayerischen Staatsbibliothek, München (vgl. BSB, 2 Inc.c.a. 2918-331, a) – dem Handexemplar Hartmann Schedels – die *Romweg-Karte* von Erhard Etzlaub, die *Ursprungskarte* unseres zusätzlichen Kartenausschnitts der Sammlung Woldan, eingeklebt.

6 Siehe hierzu umfassend P. H. Meurer, Corpus der älteren Germaniakarten. Ein annotierter Katalog der gedruckten Gesamtkarten des deutschen Raumes von den Anfängen bis um 1650, Alphen aan den Rijn 2001.

7 Solche Fabelwesen sind in der Chronistik stets vertreten.

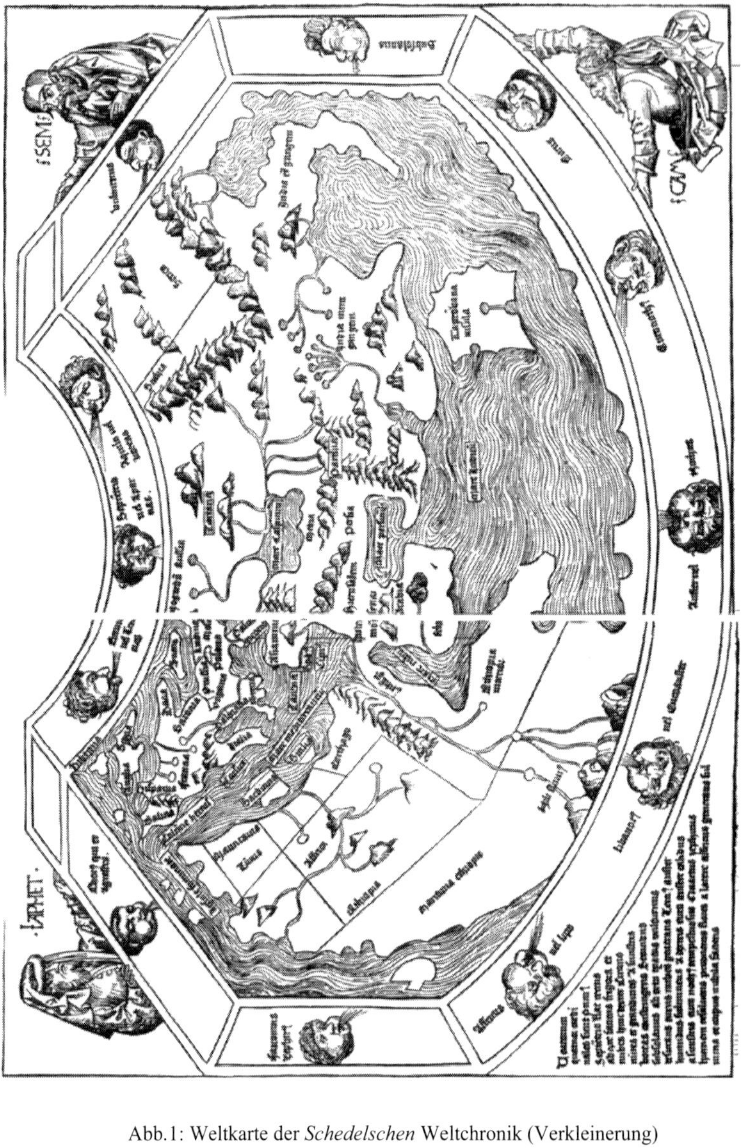

Abb.1: Weltkarte der *Schedelschen* Weltchronik (Verkleinerung)

Abb.2: Weltkarte der gedruckten Ptolemäus-Ausgabe von Bologna 1477 (Verkleinerung)

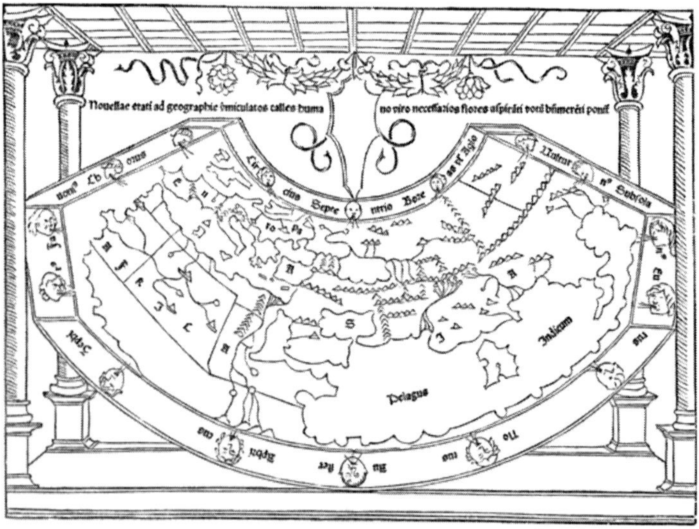

Abb.3: Weltkarte aus *De situ orbis* des Pomponius Mela, 1482 von Erhard Ratdold gedruckt (Verkleinerung)

Kartenbeschriftung ist sehr karg. Meeresflächen sind mit wellenartig angeordneten Linien ausgefüllt. Flüsse sind doppellinig geschnitten; sie haben fast alle ihren Ursprung in großen Quellseen. Gebirgszüge sind in der „Maulwurfshügel-Manier" eingetragen.

Als Vorlage dieser ptolemäischen Weltkarte wird häufig jene der 1482 – also nur wenige Monate vor der Weltchronik – gedruckten Ulmer Ptolemäus-Ausgabe genannt. Aber schon ein erster visueller Vergleich ergibt, dass dies nicht sein kann. Unsere Weltkarte in der *Schedelschen Weltchronik* ähnelt vielmehr jener der ersten acht Handschriften (erste Rezension bzw. Handschriften A) des Nicolaus Germanus[8], die zwischen 1460 und 1466 entstanden.[9] Eine dieser Handschriften, der Codex Neapolitanus latinus V.F. 32 der Bibliotheca Nazionale Neapel war Vorlage für die 1477 gedruckte Ptolemäus-Ausgabe von Bologna (Abb. 2). Allerdings zeigen diese Ptolemäus-Ausgaben noch eine ältere Version der afrikanischen Westküste. Direkte Vorlage der Weltkarte war eine Weltkarte in der 1482 von Erhard Ratdolt in Venedig gedruckten Ausgabe von *De situ orbis* des römischen Geographen Pomponius Mela[10] (Abb. 3), die sich auch in der Bibliothek von Hartmut Schedel befand.

---

8 Der Benediktinermönch Nicolaus Germanus, der um 1420 in Reichenbach bei Regensburg geboren wurde, wirkte zwischen ca. 1460 und 1466 in Florenz, vgl. hierzu L. Böninger, Die Deutsche Einwanderung nach Florenz im Spätmittelalter (The medieval Mediterranean 60), Leiden 2006, 334-348.

9 Die Weltkarte baut hier noch auf die erste ptolemäische Weltkartenprojektion mit geradlinigen Meridianen auf; erst die späteren Handschriften (zweite und dritte Rezension bzw. Handschriften B und C) des Nicolaus Germanus weisen die zweite ptolemäische Weltkartenprojektion mit den gekrümmten Meridianen auf.

10 Siehe C. Reske, Die Produktion der Schedelschen Weltchronik in Nürnberg. The Production of Schedel´s Nuremberg Chronicle (Mainzer Studien zur Buchwissenschaft 10), Wiesbaden 2000, 180 und V. v. Loga, Die Städteansichten in Hartmut Schedels Weltchronik, in Jahrbuch der der Königlichen Preußischen Kunstsammlungen 9 (1888), 93-107, hier: 105 sowie 184-192.

Abb.4: Die Germania-Karte von Hieronymus Münzer der *Schedelschen Weltchronik* (Verkleinerung)

Die Germania-Karte des Hieronymus Münzer

Die Germania-Karte des Hieronymus Münzer[11] hat ein Format von 31 x 47 cm und ist als Doppelblatt ohne Blattnummer auf den Blättern 286v und 287r am Schluss des Bandes zu finden.

Der Ursprung der Karte

Diese Karte der *Schedelschen Weltchronik* hat indirekt die Karte des Gelehrten und Kirchenpolitikers Nikolaus Cusanus, die um 1450 entstanden sein muss, zur Grundlage.[12] Der Inhalt des nicht erhaltenen Kartenmanuskripts dieser ersten modernen Karte Mitteleuropas ist jedoch in handgezeichneten und gedruckten Karten erhalten geblieben.

So fertigte der bis 1496 in Florenz wirkende Henricus Martellus Germanus[13] um 1490 drei Abschriften der Cusanus-Karte als nordorientierte Kartenmanuskripte im ungefähren Maßstab 1:4 Mio. an. Seine Germania-Karten der sog. Redaktion A oder Martellus-Redaktion der Cusanus-Karte finden sich als lateinische Handschriften[14]:

11 Vgl. F. Wawrik, Von den Anfängen der österreichischen Kartographie bis zur Zweiten Türkenbelagerung Wiens (1693), in: I. Kretschmer – J. Dörflinger – F. Wawrik (Hrsg.), Österreichische Kartographie. Von den Anfängen im 15. Jahrhundert bis zum 21. Jahrhundert (Wiener Schriften zur Geographie und Kartographie 15), Wien 2004, 11-73, hier: 26f.; Zur Biographie des Hieronymus Münzer vgl. Ders., Hieronymus Münzer – Arzt, Reisender, Geograph, Diplomat (im zweiten Band des vorliegenden Sammelbandes, S. 515-536) sowie W. Iwanczak, Die Kartenmacher. Nürnberg als Zentrum der Kartographie im Zeitalter der Renaissance, Darmstadt 2009.

12 Nach P. H. Meurer, Zur Systematik der Cusanuskarte: Überlegungen aus der Sicht der rheinischen Landeskunde. in: Kartographische Nachrichten 33 (1983), 219-225 ist [...] *über sie ist wenig mehr bekannt, als die Tatsache, dass es sie gegeben hat.*

13 Martellus Germanus wurde vermutlich als Heinrich Hammer in Schlüsselfeld bei Bamberg geboren und wirkte von 1448 und 1496 in Florenz, vgl. L. Böninger, Die Deutsche Einwanderung, 2006.

14 Siehe hierzu Peter H. Meurer, Corpus der älteren Germaniakarten, 2001 und Kurt Brunner und Thomas Horst, 850 Jahre München – 550 Jahre München in Karten, in: Kartographische Nachrichten 59 (2009), 83-91, hier: 84f.

- im Codex Florentianus Becchiana Latinus, Biblioteca Nazionale Florenz,
- in der Bibliothèque du Musée Condé Chantilly,
- in einer unsignierten Ptolemäus-Ausgabe der Universiteitsbibliotheek Leiden.

Dem Leidener Exemplar – das heute wegen eines Wasserschadens nahezu unlesbar ist[15] – folgt sehr genau eine erste gedruckte Germania-Karte.[16] Diese um 1490 etwas grob in Kupfer gestochene Karte stammt vom Florentiner Kartographen und Verleger Francesco Roselli und hat einen ungefähren Maßstab von 1:4 Mio.[17] Der Kupferstich ist eine Quelle weiterer Karten, so auch für die Germania-Karte von Hieronymus Münzer der Weltchronik von Hartmann Schedel. In einigen Teilen weicht Münzers Germania-Karte von dieser Vorlage allerdings ab, nicht aber in den beiden näher zu betrachtenden Gebieten Voralpenland und Ostseeküste. Möglicherweise hat Hieronymus Münzer nicht nur den Kartendruck von Francesco Roselli, sondern auch die Kartenmanuskripte von Henricus Martellus Germanus oder gar deren Vorlage aus der Hand des Nicolaus Cusanus nutzen können.

Die 1492 in Eichstätt gedruckte sog. *Eichstätter Cusanus-Karte* der Redaktion B der Cusanus-Karte bzw. die Entwurfszeichnung des Nicolaus Germanus kannte Hieronymus Münzer offensichtlich nicht.

Der Holzschnitt von 1493

Die von Michael Wolgemut etwas grob in Holz geschnittene rechteckige Karte ist die erste in ein Buch eingebundene gedruckte Germania-Karte. Sie reicht aber weit über diesen Raum hinaus: im Norden bis zu den britische Inseln und Skandinavien, im Osten zum Schwarzes Meer (Abb. 4). Die Karte ist schlicht gehalten und bleibt ohne Kartenrahmen, am Kartenrand sind die Bezeichnungen der Himmelsrichtungen angebracht: *Mitnacht*, *Orient*, *Mitag* und *Occident.*

---

15 Der Karteninhalt ist durch eine Nachzeichnung von August Wolkenhauer vom Jahre 1907 erhalten, vgl. A. Herrmann, Die ältesten Karten von Deutschland bis Gerhard Mercator, Leipzig 1940, Karte Nr. 2.

16 Siehe P. H. Meurer, Corpus der älteren Germaniakarten, 2001.

17 Das einzige erhaltene Exemplar befindet sich in der Biblioteca Centrale, Florenz.

Abb.5: Ausschnitt Alpenvorland aus der Germania-Karte von Francesco Roselli, um 1490 (Verkleinerung)

Abb.6: Ausschnitt Ostseeküste aus der Germania-Karte von Francesco Roselli, um 1490 (Verkleinerung)

Ähnlich wie bei der Weltkarte der *Schedelschen Weltchronik* sind die Meeresflächen durch wellenförmige Linien gefüllt. Umfangreich sind Fließgewässer eingetragen; dabei kommt es zu Bifurkationen, so entwässern Saale und Main – wie auch in der Roselli-Karte – sowohl über Rhein und Elbe. Ähnliches kommt auch in der Darstellung der Alpen vor. Die Gebirgszeichnung erfolgt in der „Maulwurfshügel-Manier“; die Gebirgszüge bleiben aber namenlos. Die Alpen erhalten bereits eine differenzierte Gliederung in Längs- und Quertäler. Beschriftet sind Länder, Flüsse und Siedlungen. Die wenigen Ortsnamen sind willkürlich angeordnet und bleiben ohne Ortssignaturen.

Das Alpenvorland

Abb. 5 gibt in einem Ausschnitt das Alpenvorland aus der Germania-Karte des Francesco Roselli um 1490 wieder.

Unsere Karte des Hieronymus Münzer (Abb. 4) weist die Ländernamen *Svevia*, *Bavaria* und *Austria* auf. Die Donau (*Danubia fl*) fließt wie in den Ptolemäus-Ausgaben nach einem – nahezu rechtwinkeligen – Knick bei Ulm fast geradlinig nach Osten, also ohne ein Donauknie bei Regensburg und nördlich von Budapest. Als südliche Zubringerflüsse sind eingetragen und beschriftet: Iller (*Iler fl*), Lech (*Licus fl*), Isar (*Isara fl*), Inn (*Enos fl*), Mur (*Mur Fl*), und Drau (*Drava fl*); gegenüber der Roselli-Karte fehlt die in den Inn fließende Salzach und die Leitha.

Als Siedlungen – ohne Ortssignatur – sind für Schwaben und Altbayern Augsburg (*Augusta*), München (*Monichen*)[18], Regensburg (*Ratisbona*), und Passau (*Bassau*) eingetragen. In Österreich sind es Salzburg (*Saltzburg*), Steyr (*Steyer*), Wien (*Wienna*), Wiener Neustadt (*Nova ciui-tas*) und Judenburg (*Iudenburg*); im Alpenraum findet sich noch Villach.

Die deutsche Ostseeküste

Die Deutsche Ostseeküste in Münzers Germania-Karte verläuft wie in der Roselli-Karte (Abb. 6) ziemlich geradlinig von Lübeck

---

18 In der Roselli-Karte und gleichfalls in der ihr als Vorlage dienenden Leydener Handschrift (genauer: in der Nachzeichnung von August Wolkenhauer vom Jahre 1907, vgl. A. Herrmann, Die ältesten Karten) ist München (*monaco*) übrigens doppelt eingetragen; siehe hierzu K. Brunner – T. Horst, 850 Jahre München, 84.

(*Lubick*) in östlicher Richtung; allerdings nicht exakt in östlicher Richtung, wie in der vierten Europa-Karte der Ptolemäus-Ausgaben. Wie in Ptolemäus-Ausgaben ist aber Dänemark stark nach rechts geknickt. Weiterhin fehlt die Insel Rügen.

Eingetragen sind die Küstenstädte Rostock und Danzig (*Dantzg*). Da die Namen keinen Ortssignaturen zugewiesen sind, ist eine genaue Platzierung nicht möglich. So kommt der Kartenname Greifswald (*Grispalt*) direkt an die Küste und Stettin (*Stetin*) ins Landesinnere. Die Oder (*Ader fl*) fließt ins Stettiner Haff, die Weichsel (*Vistula fl*) in die Danziger Bucht. Die Seen der Mecklenburgischen Seenplatte sind zu einem See zusammengefasst, die (unbeschriftete) Havel führt fälschlicherweise zur Oder. Im Landesinneren finden sich die Gebietsbezeichnungen *POMERN*, *Marchia* und *Marchia Nova*, sowie *Prvssia* und *Germania Magna*.

Skandinavien

Münzers Germania-Karte der *Schedelschen Weltchronik* beinhaltet in ihrem oberen rechten Teil die Ostsee; am nordwestlichen Rand sind stark zusammengedrückt Irland, Britannien[19], Island und die skandinavische Halbinsel dargestellt. Die Roselli-Karte (Abb. 6) und die ihr zugrunde liegende Handschrift der Universiteitsbibliotheek Leiden zeigen aber lediglich die Ost- und Nordseeküste und den südlichen Teil Dänemarks. Die Darstellung Skandinaviens stammt wohl von der Nordeuropa-Karte, einer *Tabula moderna*, der 1482 gedruckten Ulmer Ptolemäus-Ausgabe.[20]

Eine erste *Tabula moderna* Nordeuropas befindet sich bereits 1428 im *Codex Nanceanus latinus 441* der Bibliotheque Municipale, Nancy. Anreger und Auftragsgeber dieser Nordeuropa-Karte war der französische Kardinal Guillaume Fillastre; er verwertete das Wissen des dänischen Humanisten Claudius Clavus.[21] Dieser folgten noch bis 1490 zehn weitere Handschriften mit Nordeuropa als *Tabula moderna*

---

19 Schottland ist von England durch eine Meerenge getrennt.

20 Die Ulmer Ptolemäus-Ausgabe hat die Handschriften des *Codex Wolfeggianus latinus* des Nicolaus Germanus, der in der Schlossbibliothek in Wolfegg/ Württemberg verwahrt wird, als Grundlage.

21 Vgl. J. Babicz, Nordeuropa in den Atlanten des Ptolemaeus, in: D. Unverhau – K. Schiezel (Hrsg.), Das Danewerk in der Kartographiegeschichte Nordeuropas, Neumünster 1993, 107-126.

in lateinischen Ptolemäus-Codices. Diese Handschriften haben etwas unterschiedliche Inhalte und können in drei Versionen eingeteilt werden[22].

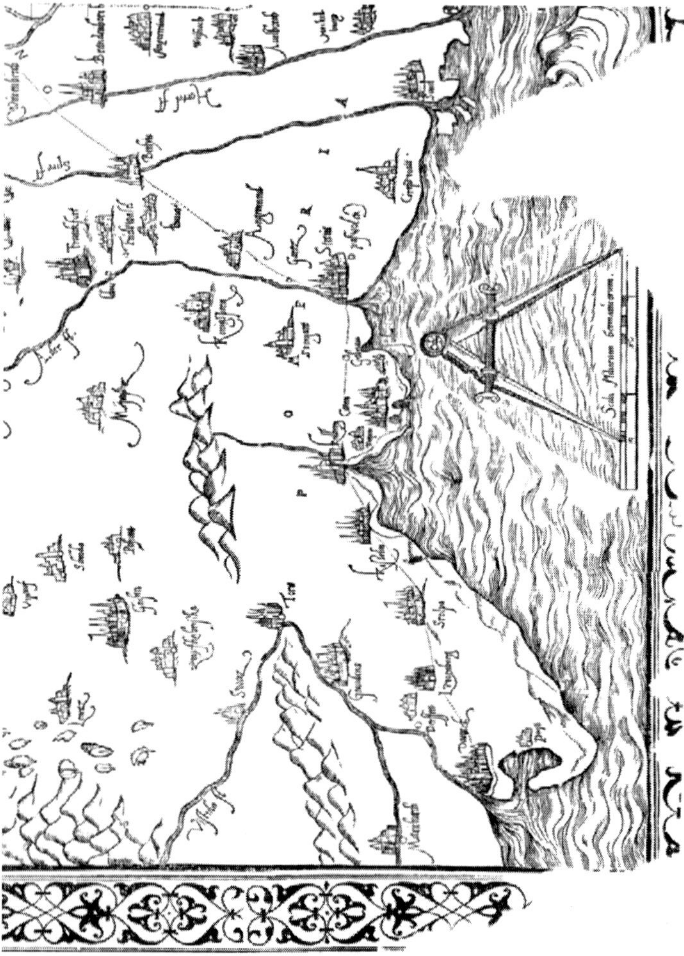

Abb.7: Kartenfragment der Schedelschen Weltchronik der Sammlung Woldan, Wien (Verkleinerung)

22 Vgl. K. Brunner, Nordeuropa-Darstellungen in den Ptolemäus-Ausgaben am Ende des 15. Jahrhunderts, in: W. Scharfe (Hrsg.), 6. Kartographiehistorisches Colloquium Berlin 1992, Berlin 1994, 19-29.

Das eingebundene Kartenfragment

Das auf der vorderen Innenseite des Exemplars der *Schedelschen Weltchronik* der Sammlung Woldan quer eingeklebte Kartenfragment (Format 30 x 40 cm) zeigt südorientiert die deutsche Ostseeküste (Abb. 7) und erweist sich als Abkömmling der Romweg-Karten des Erhard Etzlaub.

Der Ursprung des Kartenfragments

Unabhängig und losgelöst von der vorausgegangenen Germania-Karte des Nikolaus Cusanus und ihren Abkömmlingen entsteht kurz vor 1500 eine zweite wichtige Darstellung Mitteleuropas. Der Nürnberger Arzt, Kompassmacher und Astronom Erhard Etzlaub schuf die Karte *Das ist der Rom-Weg von meylen zu meylen* [...] *durch deutsche lantt.* Diese rechteckige, südorientierte Plattkarte wurde als Holzschnitt gedruckt (Maßstab ca. 1:5 Mio.) und in ihr sind die wichtigsten Pilgerwege (ausgewiesen mit punktierten Linien) für das *Heilige Jahr 1500* nach Rom eingetragen. Die *Romweg-Karte* ist in mehreren Exemplaren erhalten, die von drei Zuständen des Druckstocks gedruckt wurden. Ihr folgt die Karte *Das seyn dy lantstrassen durch das Romisch reych* [...]. gedruckt von *Georg glogkendon zu Nurnbergk 1501.* Diese *Landstraßen-Karte* ist in vielen Teilen identisch mit der *Romweg-Karte*, ist jedoch im Westen bis nach Mittelfrankreich und geringfügig im Süden erweitert[23].

Mit der südorientierten Romweg-Karte von Erhard Etzlaub entstand vor fünfhundert Jahren schlagartig ein moderner Kartentyp mit bemerkenswerter Genauigkeit[24]. Abb. 8 zeigt das Alpenvorland von der Donauquelle bis Wien. Das Donauknie bei Regensburg ist nun erstmals in einer Karte eingetragen. Die Deutsche Ostsseeküste mit

---

23 Vgl. K. Brunner, Erhard Etzlaubs Karte, Die Landstraßen durch das Römische Reich, in: R. Buzin – T. Wintges (Hrsg.), Kartographie 2001 - multidisziplinär und multimedial. Beiträge zum 50. Deutschen Kartographentag, Heidelberg 2001, 43-54 sowie H. Krüger, Erhard Etzlaub´s „Romweg"-Map and its Dating in the Holy Year of 1500, in Imago Mundi 8 (1951), 17-26 und A. Wolkenhauer, Der Nürnberger Kartograph Erhard Etzlaub, in: Verhandlungen des 16. Deutschen Geographentages zu Nürnberg, Wiesbaden 1907, 124-146.

24 Siehe K. Brunner, Erhard Etzlaubs Karte, 2001.

Abb.8: Ausschnitt Alpenvorland aus *Romweg-Karte* von Erhard Etzlaub, vor 1500 (Verkleinerung)

Abb.9: Ausschnitt Ostseeküste aus *Romweg-Karte* von Erhard Etzlaub, vor 1500 (Verkleinerung)

Mecklenburg und angrenzenden Gebieten (Abb. 9) ist ziemlich fehlerhaft, vor allem bei der Darstellung der Wasserläufe. So münden Spree und Havel in die Ostssee statt über die Elbe zur Nordsee[25].

Derivate der Etzlaub-Karten

Der Inhalt der Romweg-Karte von Erhard Etzlaub wirkte stark nach. Als erste Abkömmlinge sind Karten von Martin Waldseemüller zu benennen: die *Tabula moderna Germanie* seiner Ptolemäus-Ausgabe, die 1513 und 1520 in Straßburg erschien sowie die großformatige Wandkarte *Carta Itineraria Europae*, gedruckt 1511 sowie 1520.

Abkömmlinge der Etzlaub-Karten waren mehrfach auch Karten des französischen Verlegers Jérôme de Gourmont; hierzu sind auch Wandkarten zu zählen[26].

Der Holzschnitt um 1545

Das auf dem vorderen Buchdeckel eingeklebte Kartenfragment der *Weltchronik* von Hartmut Schedel der Sammlung Woldan (Abb. 7) gehört zu einer der als Holzschnitt gedruckten Wandkarten von vier Stöcken auf vier Blättern von Jérôme de Gourmont. Das einzige bekannte vollständige Exemplar dieser *VNIVERSAE GERMANIAE DESCRIPTIO* befindet sich in der British Library, London[27].

Unser Fragment (Blatt 3) beinhaltet die Deutsche Ostseeküste von Marienburg (*Marienburch*) bis Rostock. Die Küstenlinie weicht stark von jener der *Romweg-Karte* ab und ist offensichtlich richtiger als

---

25 Siehe hierzu W. Scharfe – H. Scheerschmidt (Hrsg.), Berlin-Brandenburg im Kartenbild. Wie haben uns die anderen gesehen? Wie haben wir uns selbst gesehen? (Staatsbibliothek zu Berlin – Ausstellungskataloge N. F. 42), Berlin 2000.

26 Vgl. Meurer, Corpus der älteren Germaniakarten, 2001, 204f.

27 Eine inhaltlich weitgehend identische Wandkarte (*Beschreibung des weith Berümpten Deutschlandt*) von 1569 desselben Verlegers befindet sich in der Bayerischen Staatsbibliothek in München; eine Reproduktion ist Beilage zu H. Zimmerer, Hans Sachs und sein Gedicht von den 110 Flüssen des deutschen Landes mit einer zeitgenössischen Karte, Jahresbericht der Geographischen Gesellschaft München 16, München 1896, 1-50 – Vgl. auch F. Grenacher, The „Universae Germaniae Descriptio“ Map of Jérôme de Gourmont, in: Imago Mundi 14 (1959), 55-63.

jene. So ist die Danziger Bucht mit dem Haff besser herausgearbeitet; als in der Vorlage. Die Insel Rügen fehlt wiederum.

Die Anzahl der Siedlungen ist stark reduziert. Die Anzahl der Städte auf dem Romweg von Danzig über Stettin und Berlin entspricht jedoch der *Romweg-Karte*. Gleichfalls stark reduziert ist die Anzahl der Wasserläufe. Wie in der *Romweg-Karte* münden Havel und Spree fälschlich in die Ostsee. Die Landschaft, die zur Küste reicht, ist mit *Pomeria* beschriftet.

Im Bereich des Ostseeraums am unteren Kartenrand findet sich eine Maßstabsleiste für deutsche Meilen (*Scala Miliarium Germanicorum*) mit einem geöffneten Zirkel verziert. Hieraus kann ein Kartenmaßstab von ca.1:1,5 Mio. hergeleitet werden. Der Kartenrand ist mit einer Ornamentbordüre ausgestattet.

Schluss

Die Existenz der *Schedelschen Weltchronik* im Bestand der Sammlung Woldan zeigt einmal mehr ihre Bedeutung. Das eingeklebte Kartenfragment macht Hartmut Schedels *Buch der Chroniken und geschichten mit figuren und pildnussen von anbeginn der welt bis auf diese unnsere Zeit* dabei zu einer Rarität.

Um die Erschließung der Sammlung Woldan und die wiederholten Hinweise auf ihre Bedeutung hat sich besonders auch Johannes Dörflinger im hohen Maße verdient gemacht.

Literaturverzeichnis

Babicz J., Nordeuropa in den Atlanten des Ptolemaeus. in: Unverhau D. – Schiezel K. (Hrsg.), Das Danewerk in der Kartographiegeschichte Nordeuropas, Neumünster 1993, 107-126.

Böninger L., Die Deutsche Einwanderung nach Florenz im Spätmittelalter, Leiden 2006.

Brunner K., Nordeuropa-Darstellungen in den Ptolemäus-Ausgaben am Ende des 15. Jahrhunderts, in: Scharfe W. (Hrsg.), 6. Kartographiehistorisches Colloquium Berlin 1992, Berlin 1994, 19-29.

Brunner K., Erhard Etzlaubs Karte, Die Landstraßen durch das Römische Reich, in: Buzin R. – Wintges T. (Hrsg.), Kartographie 2001 – multidisziplinär und multimedial. Beiträge zum 50. Deutschen Kartographentag, Heidelberg 2001, 43-54.

Brunner K. – Horst T., 850 Jahre München – 550 Jahre München in Karten, in: Kartographische Nachrichten 59 (2009), 83-91.

Fischer J., Die Karte des Nicolaus von Cues (vor 1490). Die älteste Karte von Mitteleuropa. (Kartographische Denkmäler der Sudetenländer 1), Prag 1930.

Grenacher F., The „Universae Germaniae Descriptio“ Map of Jérôme de Gourmont, in: Imago Mundi 14 (1959), 55-63.

Herrmann A., Die ältesten Karten von Deutschland bis Gerhard Mercator, Leipzig 1940.

Iwanczak W., Die Kartenmacher. Nürnberg als Zentrum der Kartographie im Zeitalter der Renaissance, Darmstadt 2009.

Kreuer J. W., Die kartographische Wende von Schedel zu Martellus (Descriptio und Karte), in: Unverhau D. (Hrsg.), Geschichtsdeutung auf alten Karten. Archäologie und Geschichte (Wolfenbütteler Forschungen 101), Wiesbaden 2003, 129-146.

Kreuer W., Tabulae Mundi. Frühe historische Kartentafeln bis zum Beginn der Neuzeit, Essen 2004.

Krüger H., Erhard Etzlaub´s „Romweg“ – Map and its Dating in the Holy Year of 1500, in Imago Mundi 8 (1951), 17-26.

Loga V. v., Die Städteansichten in Hartmut Schedels Weltchronik, in Jahrbuch der der Königlichen Preußischen Kunstsammlungen 9 (1888), 93-107 und 184-192.

Meurer P. H., Zur Systematik der Cusanuskarte: Überlegungen aus der Sicht der rheinischen Landeskunde. in: Kartographische Nachrichten 33 (1983), 219-225.

Meurer P. H., Mappae Germaniae. Die schönsten und bedeutendsten Deutschlandkarten von 1492 bis 1803, Neustadt an der Saale 1984.

Meurer P. H., Corpus der älteren Germaniakarten. Ein annotierter Katalog der gedruckten Gesamtkarten des deutschen Raumes von den Anfängen bis um 1650, Alphen aan den Rijn 2001.

Reske C., Die Produktion der Schedelschen Weltchronik in Nürnberg. The Production of Schedel´s Nuremberg Chronicle (Mainzer Studien zur Buchwissenschaft 10), Wiesbaden 2000.

Rücker E., Die Schedelsche Weltchronik. Das größte Buchunternehmen der Dürerzeit. Mit einem Katalog der Stadtansichten (Bibliothek des Germanischen Nationalmuseums Nürnberg zur deutschen Kunst – und Kulturgeschichte 33), München 1973.

Scharfe W. – Scheerschmidt H. (Hrsg.), Berlin-Brandenburg im Kartenbild. Wie haben uns die anderen gesehen? Wie haben wir uns selbst gesehen? (Staatsbibliothek zu Berlin – Ausstellungskataloge N. F. 42), Berlin 2000.

Schedel H., Weltchronik. Kolorierte Gesamtausgabe von 1493. Einleitung und Kommentar von Stephan Füssel, Köln 2005.

Wolkenhauer A., Der Nürnberger Kartograph Erhard Etzlaub, in: Verhandlungen des 16. Deutschen Geographentages zu Nürnberg, Wiesbaden 1907, 124-146.

Zimmerer H., Hans Sachs und sein Gedicht von den 110 Flüssen des deutschen Landes mit einer zeitgenössischen Karte (Jahresbericht der Geographischen Gesellschaft in München, 16, 1894-1895) München 1896, zgl. Programm des königlichen Maximilian – Gymnasiums für das Schuljahr 1895/96, München 1896.

Sämtliche Abbildungen sind der Faksimilesammlung des Lehrstuhls für Kartographie und Topographie, Universität der Bundeswehr München entnommen.

Zum Autor

Kurt Brunner, geb. 1945, Univ.-Prof. Dr.-Ing., Lehrstuhl für Kartographie und Topographie der Universität der Bundeswehr München.

1960-1964 Lehre beim Bayerischen Landesvermessungsamt, danach Ingenieurschulstudium der Kartographie, 1968-1972 Studium des Vermessungswesens an der TU München, 1972-1979 Wissenschaftlicher Mitarbeiter am Lehrstuhl für Kartographie und Reproduktionstechnik der TU München, 1977 Promotion, 1979-1988 Professor für Kartographie an der Fachhochschule Karlsruhe, seit 1988 Universitätsprofessor für Kartographie und Topographie an der Universität der Bundeswehr München, 1999/2000 Gastprofessor am Institut für Geographie und Raumforschung der Universität Wien, seit 2004 Mitglied der Kommission für Glaziologie der Bayerischen Akademie der Wissenschaften. Arbeitsschwerpunkte: Topographische Kartographie, Hochgebirgskartographie, Geschichte der Kartographie.

Kontakt: Institut für Geodäsie der Universität der Bundeswehr München, Werner-Heisenberg-Weg 39, D – 85577 Neubiberg;
kurt.brunner@unibw-muenchen.de

Sigmund von Birken,
Der Donau-Strand,
Nürnberg 1664

Antal András Deák, Esztergom

Ein kleines, seltsames Buch in der Sammlung Woldan

Sigmund von Birken, *Der Donau-Strand* [1]

Es hat mir große Freude bereitet, für Herrn Professor Johannes Dörflinger diesen Beitrag zu verfassen, da er einer meiner Förderer war und mich bei meinen kartographischen Forschungen besonders am Beginn moralisch unterstützt und ermutigt hat. Dafür möchte ich ihm sehr herzlich danken. Zuerst hatte ich die Absicht, alle Donaukarten der Sammlung Woldan zu bearbeiten. Wegen der großen Fülle an Karten war mir das aber leider nicht möglich. So wählte ich ein aus kulturhistorischer Sicht sehr interessantes und seltenes Stück der Sammlung, Siegismund von Birkens *Donau-Strand* von Birken und dessen Kartenbeilage, aus.

Ungarn, 1663-1664

Das Erscheinungsjahr des *Donau-Strandes* (1664) steht im Zeichen des großen türkisch-christlichen Krieges von 1663/1664. In diesem Jahr zogen die Türken mit 150.000 Soldaten gegen Wien, konnten die

---

1 Auf der Universität Paussau, Forschungsstelle Frühe Neuzeit läuft ein DFG-Projekt „Birken-Edition". Man darf also erwarten, dass in absehbarer Zeit der gesamte Text dieses Werkes einer breiten Öffentlichkeit zugänglich gemacht wird. Siehe: http://www.phil.uni-passau.de/projekt-birken/FFN_DFG.html (12.4.2010)

Stadt allerdings nicht einnehmen. Ein Jahr später versuchten sie wieder bis Wien vorzudringen. Diese bedrohliche Lage verlieh dem Buch von Birkens Aktualität, denn der Verfasser forderte wie viele andere bedeutende Europäer den vereinten Auftritt der europäischen Könige gegen die Osmanen. Zu dieser Zeit entstand die Heilige Liga (5. März 1664), welche die Wiedereroberung der von den Türken besetzten europäischen Länder und Gebiete zum Ziel hatte. Ihre Mitglieder waren die Länder des Habsburgischen Reiches – darunter auch Ungarn –, Venedig und Polen. Die deutschen Fürstentümer stellten Soldaten zur Verfügung.

Ende Juli 1664 standen sich die vereinigten christlichen Heere und die Türken an den beiden Ufern des Flusses Raab bei Szentgotthárd feindlich gegenüber. Am 1. August griffen die Türken unerwartet an, indem sie den Fluss überschritten. Der kämpferische Einsatz der christlichen Truppen war von Erfolg gekrönt, ein Siebentel des türkischen Heeres wurde vernichtet. Neben dem militärischen war vor allem der moralische Erfolg für die christlichen Truppen von großer Bedeutung, denn.mit dieser Schlacht wurde bewiesen, dass die Türken besiegbar waren.

Der *Donau-Strand*

Die türkische Gefahr, die türkischen Feldzüge gegen Ungarn und Wien und die europäischen Machtstreitigkeiten verliehen dem Buch von Birkens noch mehrere Jahrzehnte lang Aktualität.[2] Es wurde immer wieder mit erweitertem Inhalt neu aufgelegt[3], denn man ergänzte den Grundtext mit den Erzählungen der neuesten Kriegsereignisse. Dies hatte zur Folge, dass immer weniger Illustrationen integriert wurden. Konnte man in der ersten Ausgabe (1664) noch dreißig Kupferstiche von verschiedenen ungarischen Festungen[4] und

2 Man denke in erster Linie an die Belagerung von Wien im Jahre 1683 und an den 17 Jahre lang dauernden Befreiungskrieg.

3 Die Erstauflage von 2000 Stück war rasch vergriffen. Es folgen über 20 Auflagen, so schon 1665, 1674, 1683, 1684.

4 Pressburg, Győr, Pápa, Komorn, Neuhäusel, Tata, Palota, Gran, Fülek, Visegrád, Novigrad, Vác, Ofen und Pest, Alt-Ofen, Veszprém, Stuhlweissenburg, Új-Zrínyivár, Kanizsa, Sziget, Tokaj, Kassa, Eger, Hatvan, Szolnok, Grosswardein, Gyula, Temesvár, Sziszek, Sabác, Belgrád.

eine aus drei Blättern zusammengesetzte, große Landkarte (53 x 121 cm) der Donau finden, so sind die späteren Ausgaben an Illustrationen viel ärmer und auch die große Karte wurde nicht mehr gedruckt. In der Ausgabe von 1665 fehlen zum Beispiel die Burgdarstellungen zur Gänze. In der von Homann im Jahre 1688 herausgegebenen Variante, die unter dem Titel *Der Vermehrte Donau-Strand* erschienen war, sind zwei, auf ähnliche Weise die ganze Donau darstellende, aber kleinformatige Landkarten (19 x 22 cm) und nur 13 Abbildungen von Festungen enthalten.

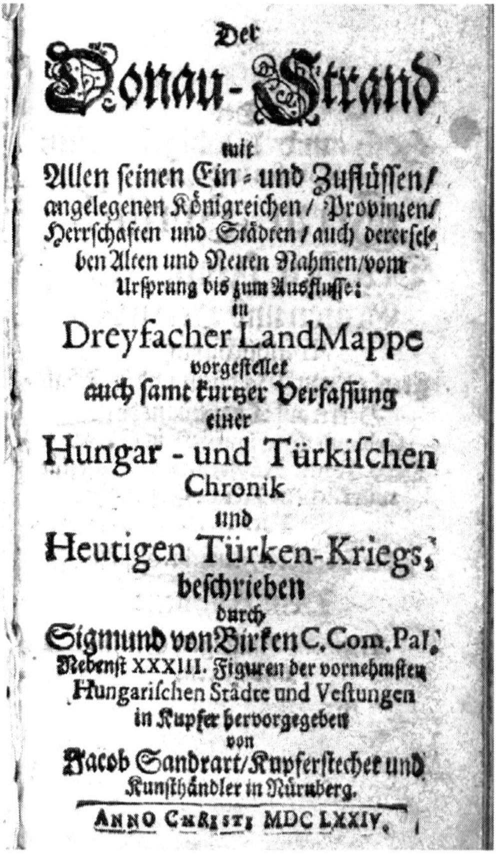

Der
Donau-Strand
mit
Allen seinen Ein- und Zuflüssen/
angelegenen Königreichen/ Provinzen/
Herrschaften und Städten/ auch derersel-
ben Alten und Neuen Nahmen/ vom
Ursprung bis zum Ausflusse:
in
Dreyfacher LandMappe
vorgestellet
auch samt kurzer Verfassung
einer
Hungar- und Türkischen
Chronik
und
Heutigen Türken-Kriegs,
beschrieben
durch
Sigmund von Birken C. Com. Pal.
Nebenst XXXIII. Figuren der vornehmsten
Hungarischen Städte und Vestungen
in Kupfer hervorgegeben
von
Jacob Sandrart/ Kupferstecher und
Kunsthändler in Nürnberg.
ANNO CHRISTI MDCLXXIV.

Abb.1: Titelblatt des *Donau-Strand* Exemplars aus der Sammlung Woldan

Das Exemplar des *Donau-Strandes* aus der Woldan-Sammlung

Der Titel des Werkes lautet mit seinem vollen Namen: *Der Donau-Strand mit Allen seinen Ein- und Zuflüssen angelegenen Königreichen Provinzen Herrschaften und Städten auch derselben Alten und Neuen Nahmen vom Ursprung bis zum Ausflusse: in Dreyfacher Land Mappe vorgestellet auch samt kurtzer Verfassung einer Hungar- und Türkischen Chronik und heutigen Türken-Kriegs, beschrieben durch Siegmund von Birken C. Com. Pal. Nürnberg: Jacob Sandrart, 1674*. (Abb. 1)[5].

Am Ende der Einleitung (*Vor-Erinnerung*) befindet sich eine – gewissermaßen die Rolle der *captantia benevolentiae* erfüllende – kleine Landkarte (12 x 13 cm), welche die obere Donaustrecke mit ihren Nebenflüssen – mit Donaueschingen im Zentrum – bis Dütlingen darstellt. Da schreibt Birken mit nicht geringem Stolz: *Von seinem Ursprung / ist über andertshalb Tausend Jahre her / viel Dings / aber wenig gewisses / geschrieben worden: und ist zu bewundern / dass unter so vielen alten und neuen Geographen kein einiger gewesen / der sich bemühet hätte / diesen Brunnen in rechten Augenschein zu nehmen / und eine Abbildung desselben / nebenst einer wahren ausführlichen Beschreibung / vor den Tag zu legen. Diesen Mangel zu ersetzen / hat / auf freundliches Ersuchen / H. Martinus Menradt / Burger und Kunstmahler zu Hüfingen / einen Abriss / beydes der Landschafft / und des Ursprungs Orts / zu Pappier gebracht und deren dieser in der Mappen / jener auf dem Titelblat dieses Werkleins / dem wehrten Leser in Kupfer vor Augen gestellet wird.* [6]

In der weiteren Folge nennt er sowohl antike (Ptolemäus, Plinius, Tacitus) als auch zeitgenössische Autoren (Gerhard Mercator, Andreas Althamerus, Paulus Henznerus, Philippus Cluverus), welche sich bereits Vorstellungen über die wahre Quelle der Donau gemacht hatten. Als Beispiel nennt Birken in diesem Zusammenhang die bayerische Stadt Donaueschingen. Er erwähnt, dass es Schriftsteller gibt, die eher den *Brige* und den *Brege* für die Quelle der Donau halten, weil sie weit entfernt vor Donaueschingen entspringen; mit eigenen Worten: *Weil diese beyde Wasser / sonderlich die Brege / weit über der Donau entspringen; auch / da sie mit ihr sich vereinigen/ starke Glüsse sind*

---

5 ÖAW-Sammlung Woldan: K-I: EU 142 (2)

6 S. v. Birken, Der Donau-Strand, Nürnberg 1674, 2.

/ und sie annoch als einen Bach antreffen: so sind etliche in den Gedanken / diese zwey Flüsse seyen die rechten und wahren Quellen des Donau-Strandes. Wiewohl die Anwohnere von keinem andern Ursprung / als dem izt-beschriebenen / wissen: so sind doch starke Mutmassungen / die solches widersprechen.[7]

Nachdem von Birken zur Quelle der Donau Stellung nahm, beschreibt er die wichtigsten Orte, welche die Donau in ihrem Lauf berührt. Als erste ungarische Stadt nennt er Pressburg. Danach beschreibt er die zwei Schüttinseln. Bei Altenburg (Magyaróvár) geht er auf das Türkenproblem ein: *Als der Grosse Türk Solimannus Anno 1529. auf Wien zuzoge / hat er diesen Ort mit Sturm / oder / wie andere wollen / durch Übergab einbekommen / und die Besatzung / so 300 Böhmen waren / aufgehoben.*[8]

Dieser Ort Altenburg bildete in der Erzählung von Birkens sozusagen eine Zäsur. Denn ab jetzt nehmen die Ereignisse der Geschichte der Türkenkriege einen breiten Raum ein. Er versäumt aber nicht, die Trajanus-Brücke auf der unteren Donau zu beschreiben: *Diese Trajanus-Brück stunde auf 20. aus Quakerstucken aufgemauerten Pfeilern / derer Höhe / nur von der Oberfläche des Wassers an / 150. Schuhe / die Breite aber 60. Schuhe gewesen; und waren sie oben mit Schwibbogen geschlossen / stunden auch jeder bey 180 Schritte von dem andern. Diese Zahlen zusammen gerechnet/ bringen auf 4.000 Schritte / und also eine Teutsche Meil; ist demnach dieser Bau / wegender Breite und Tieffe des Stroms / wohl unter die Welt-Wunderwerke zu zehlen. Sonsten ware an der Pfeiler einem / diese Uberschrift zu lesen: PROVIDENTIA AUGUSTI / VERE PONTIFICIS / VIRTUS ROMANA QUID / NON DOMAT / SUB JUGUM ECCE RAPITUR / DANUBIUS.*[9]

---

7 Birken, Der Donau-Strand, 15; 1702 ließ Luigi Ferdinando Marsigli mit Hilfe von Entfernungs- und Höhenvermessungen klären, welche der Quellen als die wahre Quelle der Donau anzusehen ist.

8 Birken, Der Donau-Strand, 50.

9 (Ebd., 101). Übers.: *Dank der Fürsorge des Kaisers, des wahren Brückenbauers – was bezwingt römische Tatkraft nicht? – wird, sieh her, der Danubius in Joch gezwungen.*

Die große Landkarte der Donau

Titel der Karte

Die Inschrift der beigegebenen Landkarte verrät uns Folgendes: *Danubius Fluviorum Europae Princeps: cum omnibus accessoriis Fluminibus, et quae alluit Regnis, Provinciis, Dynastiis, Urbibus eorumque Nominibus priscis ac recentioribus / A Fonte ad Ostia Ex variis Historicor[um] et Geographor[um] Monumentis opera ac studio Sigismundi Betulii Com. Pal. C. collectum et delineatum, sculpsit et Excudit Iacobus Sandrart Chalcographus Norimbergae A. C. MDCLXIV.* (Abb. 2)[10] Inventarnummer: ÖAW Sammlung Woldan K-V(Bl): EU 142 (1) *Autor*: Sigismundus Betulius.

Näheres über den Inhalt der Karte

Diese Landkarte ist sehr reich an Details und erzielte vor allem durch die nicht geographische Aussagen weite Verbreitung, denn sie diente vor allem der Propaganda! Wir kennen aus dieser Zeite keine ähnliche Landkarte, die mit Hilfe der Visualisierung die westlichen Machthaber zu überzeugen versucht, dass die Türken ganz leicht nach Westen vordringen und so ganz Europa gefährden könnten. Es sei also die Zeit gekommen, die Kräfte gegen die türkische Expansion zu bündeln. In der oberen Ecke des dritten Kartenblattes, unter dem Flügel eines Adlers sagt er im Mund der symbolischen Gestalt der Donau Folgendes: *Huc animos!* | *Gloriosissime Imperator,* | *Potentissimi Reges,* | *Serenissimi Principes ac Proceres* | *Christiani!* | *Fallor AN DUBIUS nunc DANUBIUS quo se vertat,* | *ad vos respicit.* | *Hactenus, qua Ister, Turcicus audiit,* | *misere Binominis:* | *Vos obtestatur, ope Vestra ut posthac* | *audiat uno nomine Christianus.* | *Fidem, Fortunam, Fortitudinem Vestram implorat:* | *immitis sub Barbaro aerumnae pertaesus,* |

---

10 Übers.: *Der Danubius, der Fürst der europäischen Flüsse, samt allen seinen Nebenflüssen, und alles was er benetzt in Königreichen, Provinzen, Fürstentümern, Städten, mit deren alten und heutigen Benennungen, von seiner Quelle bis zu seiner Mündung, auf Grund der Aufzeichnungen von Historikern und Geographen; in sorgfältiger Arbeit von Sigismundus Betulius, dem Hofpfalzgrafen, gesammelt und als Landkarte skizziert; geschnitten und gedruckt von Jacob Sandrart, dem Nürnberger Kupferstecher, 1664.* Betulius ist die latinisierte Namensform für „von Birken" (*betula* lat. = Birke)

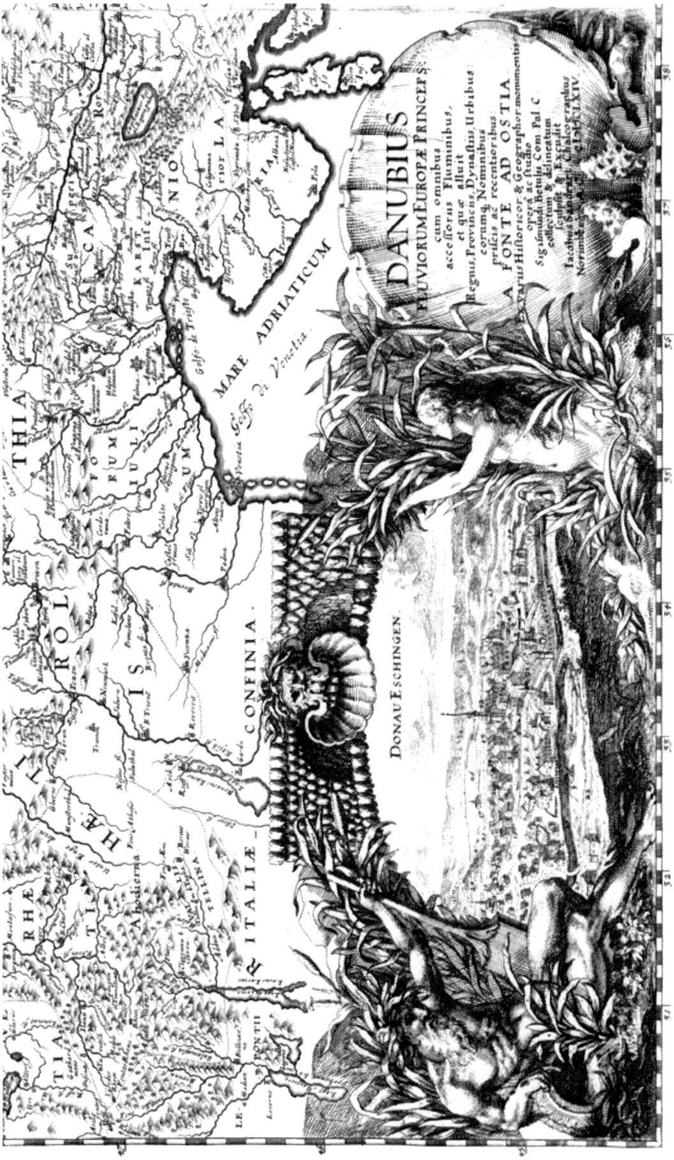

Abb. 2: Titelvignette der großen Landkarte der Donau

sub mite Christi jugum postliminio, redire gestit, | Tantum terrarum Turca Christo Vestro eripuit, | Christiani! | Quod Vestrum est repetite, Posteros Vestros in integrum restituite. | Sit Vobis labor omnibus idem: | Regna Provincias, Urbes recuperare; | Religionem sub Tyrannide, Artes liberales sub Barbaris | Christianos in Carcere, ad aratrum, ad remos | misere gementes redimere; | Utramque Danubii ripam Christo adserere. | Sic vincat, ope Leonum, Invictissimus Leopoldus | Faxit Deus Homo Christus! (Abb. 3)[11]

Abb. 3: Die „Propaganda"-Vignette im rechten oberen Kartenteil

[11] Übers.: *Auf Leute! Glorreicher Kaiser, mächtige Könige, hochgeborene Fürsten und christliche Adelige! Täusche ich mich, oder hegt Danubius Zweifel, wohin er sich wenden soll – er sorgt sich um euch. Er, der im Türkengebiet auf den Namen Ister hört, hat auf trauriger Weise doppelten Namen. Er beschwört Euch, durch Eure Hilfe in Zukunft als christlicher Strom nur auf einen einzigen Namen hören zu müssen. Er fleht zu Euch um Treue, Glück und Tapferkeit: Er ist der herben Drangsal unter dem Barbaren überdrüssig, sehnt sich danach, unter das milde Joch Christi zurückzukehren. So riesige Gebiete hat der Türke Eurem Christus geraubt, Ihr Christen! Alles, was Euch gehört, sollt Ihr für Eure Nachkommen wiedererobern! Mit Leib und Seele sollt Ihr daran gehen: die Reiche, Provinzen, Städte wieder zu gewinnen, die von dem Tyrannen unterdrückte Religion, die unter dem Barbarenjoch entkräfteten freien Künste, die Christen, die im Gefängnis, vor dem Pflug, an den Rudern kläglich stöhnen, zu befreien, beide Donauufer der Christenherrschaft anzugliedern. So möge der unbesiegte Kaiser Leopold mit Hilfe der Löwen siegen. Christus, Gott und Mensch, möge dies bewirken!*

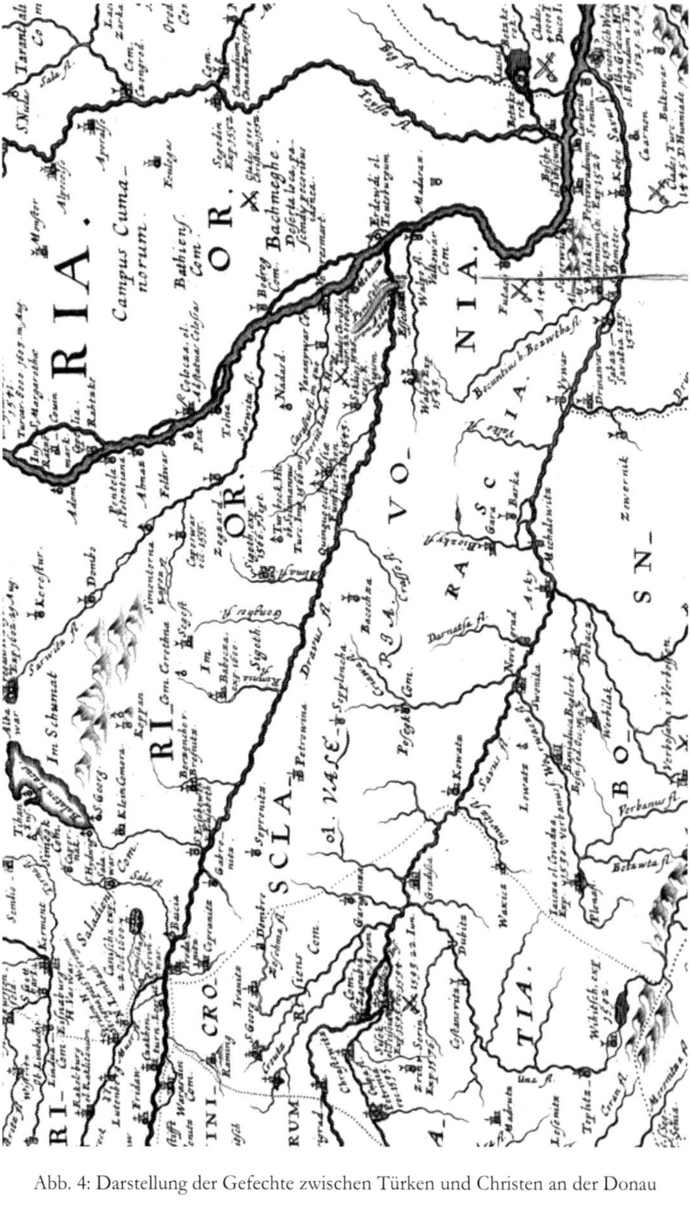

Abb. 4: Darstellung der Gefechte zwischen Türken und Christen an der Donau

Die Landkarte zeigt uns auch einige historische Ereignisse: Von der Mündung der Donau bis zum Ungarischen Altenburg zeichnete der Kartograph die Position der Schlachten mit den Türken ein. Dabei versäumte er nicht, auch die Anzahl der Gefallenen auf beiden Seiten anzugeben. So zum Beispiel bei *Várna: Clades pr. 3000 Tuc. p. 1000 Christian*;[12] bei Drinápoly: *Clades Turc. 1443. 24. Dec. D. Hunniade*[13], usw.

Der *Donau-Strand*, das heißt das Donauufer, ist voll von Kriegsereignissen (vgl. Abb. 4), die vor 1526 meistens mit dem Sieg der Christen endeten, danach aber meistens mit dem Sieg der Türken. Durch die Eroberung von Neuhäusel und Ungarisch Altenburg kommen die Osmanen an die Schwelle von Wien. Die europäischen Politiker wussten sehr wohl, dass Wien die Pforte des christlichen Westens war. Mit dieser konzentrierten, bildhaften Darstellung des Vorrückens der Osmanen wollte von Birken sie zum Handeln aufrütteln.

Die kartographische Auswirkung des *Donau-Strandes*

Kaum zwei Jahrzehnte waren vergangen, da erschienen in Nürnberg zwei weitere kartographische Produkte: Der Buchhändler Peter Paulus Bleul gab 1687 ein kleines Buch (19 x 13 cm) heraus, dessen Art und Weise der Beschreibung der Städte und Festungen an von Birken erinnert. Als Beispiel sollen hier die Zeilen über Gran (Esztergom) erwähnt werden: *Selbiger Orth nun war vor Zeiten die Residenz der Erz-Bischöffe dieses Bistums, so aller jährlichen in die 150 000 Cronen einzutragen pflegte,* [...] *hat starke Maueren und Bollwerck, und kann an vier Örtern den Feind einen Widerstand leisten. Das Schloss, worin die Tom-Kirche S. Adalberts liegt auf einem puren Stein-Felsen, wohin auch man vor diesem das Wasser mit grosser Mühe geleitet. Ausser der Stadt, ist der sogenannte Thomas Berg mit einem Blockhaus versehen, unter dem Berg aber, worauf das Schloss stehet, und zwar an dem Fluss ermeldten Berges, der Wasser-*

12 Die Türken wurden von König László geschlagen, 3.000 türkische, 1.000 christliche Gefallene, 10. November 1444.

13 Die Türken wurden von Hunyadi geschlagen, 24. Dezember 1443.

Thurm, neben einem ziemlich festen Städtlein, und dann über der Donau noch ein anders ziemlich festes Städtlein, Gokarn oder Barkan genannt.[14]

Dem einleitenden Text von 32 Seiten folgen 29 Donau-Sektionskarten. Sie stellen den Fluss vom Ursprung bis zur Mündung dar.[15] Auf dem Titelblatt steht: *Der wegen des höchst-tapferen Kayser-Adlers Helden-Thaten Sieg-berühmte Donau-Fluss oder eine genaue Darstellung aller derer Königreiche, Provintzen, Gespanschafften und Städten, so an und um die Donau herummer liegen da dann in sonderheit die bis hierher durch die Christliche Waffen herrlichst betrittene Siege und eroberungen in dem Königreich Hungarn kürtzlich bemercket werden. Allen Liebhabern der Novelln Insonderheit aber denen so anietzo zu Felde ziehen und grosse Landcarten nicht bequem bey sich führen können, zu lieb in so kleinem Format herausgegeben von einem Liebhaber der Erdbeschreibung. Nürnberg, in Verlegung Peter Paulus Bleul Kunst- und Buch-Händlers 1687.*

Auf dem Titelblatt wird der Name des Zeichners der Landkarte verschwiegen. Von ihm erfährt man nur so viel, dass er *ein Liebhaber der Erdbeschreibung* ist. Auf dem Verso findet man aber vom Kartographen eine *Voran-Sprach an den geneigten Leser*, aus welchem auch sein Name zu entnehmen ist: *Hiermit überreich ich dir den so berühmten Donau-Fluss, in Hoffnung du werdest selben mit so einem Gemüth auf und annehmen, mit was vor einem ich selbien ausgefertiget, er ist zwar lange vor schon an dass Tage-Licht getreten, alleine nie so, dann der meiste Theil solchen berühmten Fluss in gar grossen Carthen vorgestellett, welche deswegen gar umbequem auf der Reise zu führen gewesen, ich aber habe ihn in 28. kleine Carthen zerteilet, damit mann ihn also gar bequem bey sich in der Taschen, zu Felde führen könnte* [...] *Joh. Clemens Molitor.*[16]

Die Landkarte war für Soldaten und Reisende gedacht, was auch in ihrem Format zum Ausdruck kam, sodass man sie. auch bei Feldzügen ohne Schwierigkeit mitnehmen und gebrauchen konnte.

---

14 In Esztergom wurde zur Zeit der Renaissance ein „wunderbares" Wasserwerk gebaut, das bis zur Belagerung der Stadt 1683 das Wasser einer Quelle – mit der Wasserkraft der Quelle – in die Burg 60 m hochpumpte. In diesem zitierten Text finden wir die einzige Nachricht über die Existenz des Werkes von einem westlichen Autor. Die meisten Dokumente, die es erwähnen, sind türkisch, denn die Burg und auch das Wasserwerk waren zwischen 1543-1683 in der Hand der Türken.

15 Die Sektionen verzeichnen auch die Länder von der Schweiz und Oberitalien bis zum Schwarzen Meere samt deren Landeswappen.

16 *Molitor* ist offensichtlich auch eine latinisierte Nennform (molitor lat. = Müller.

Die einzelnen Sektionen wurden mit Buchstaben versehen: es gibt 14 „A" und 15 „B" Sektionen.[17] Diese Distinktion ist jedoch wertlos, weil – in der großen Eile – die Generalkarte nicht mitgebunden wurde. Diese Eile ist auch bei unterschiedlichen Angaben in Bezug auf die Sektionen erkennbar. Dahinter könnte als Ursache durchaus die Rivalität zwischen den beiden Nürnberger Buchhändlern gestanden sein. Bald erschien nämlich ein ganz ähnliches Werk mit dem Titel *Die Donau / der Fürst aller Europaeischen Flüsse*, dessen Verleger der Nürnberger Buchhändler Johann Hoffmann war. Das Werk ist gleichfalls als Taschenatlas konzipiert, der die Donau mit ihrem Einzugsgebiet darstellt.[18] Die Generalkarte wurde auf drei Streifen geteilt, und dem entsprechend die Sektionen mit „A", „B", „C" Buchstaben bezeichnet. Zu den einzelnen Buchstaben gehören je 10 Sektionen. Es befinden sich also in dem Buch 30 Donau-Sektionen. Die Konzeption, der Ort und die Zeit (1688) der Ausgabe sind fast dieselbe wie die des oben erwähnten Werkes und doch handelt es sich andere Karten! Wer sie gezeichnet hat, konnte nicht festgestellt werden. Auf dem Titelblatt ist, wie bei dem genannten Konkurrenzwerk der Kartograph gleichfalls nicht genannt, nur *von Einem Liebhaber,* nicht mehr!

Der ganze Verlauf der Donau wurde ferner auf der Karte von Gerardo Valk *Fluminum princeps Danubius ab ipsa scaturigine in Suevis* […] *donec* […] *Pontum in Euxinum se exonerat* abgebildet. Zudem entstanden zu dieser Zeit auch die Donau-Karten von L. F. Marsigli und J. C. Müller.[19]

Nachdem die Türken allmählich zurückgedrängt wurden, hörte auch das Interesse an dem *Donau-Strand* auf. Damit bestätigte sich der enge Zusammenhang zwischen der Donaukarte und den Türkenkriegen. Von Birken erkannte, dass die Donau eine nicht unwesentliche Rolle in dem Jahrhunderte dauernden Türkenkrieg spielte. Auch die entscheidende Schlacht, welche die Expansion der

---

17 Die Angabe *28 kleine Carthen* stimmt nicht, tatsächlich sind es 29.

18 Titelblatt: *Die Donau / der Fürst aller Europaeischen Flüsse. Das ist: Eine genaue Darstellung / aller Der um und an der Donau gelegenen Königreiche / Fürstenthümer / Länder und Städte / beneben einer kurzen Beschreibung selber / Deme / Zu des Lesern Behuff / annoch ein Register aller der jeniger Oerter / so in denen Charten mit begriffen seynd / beygefüget worden. An das Tage:Licht gegeben von Einem Liebhaber. Nürnberg / In Verlegung Johann Hoffmann/ Buch und Kunsthändler 1688.*

19 Vgl. L. F. Marsigli, Danubius Pannonico-Mysicus, Amsterdam/Hague 1726.

Osmanen nach Westen beendete, wurde an der Donau (bei Zenta) ausgetragen. Die Donau diente nämlich für die von Konstantinopel Richtung Wien ziehenden türkischen Heere als ein wichtiges Versorgungselement; der einzige sichere Weg für den Nachschub an Gütern aller Art war nämlich der Fluss. Mit seinem Buch samt Donaukarte beweist von Birken, dass er dies so immer erkannt hatte. Die zahlreichen Auflagen des Buches *Donau-Strand* und die darauf basierenden anderen Beschreibungen und Donaukarten beweisen die Brisanz dieser Erkenntnis. Es handelt sich dabei nicht nur um kulturhistorische Kuriositäten, sondern um wichtige Propaganda-Mittel im Dienste der Vereinigung der europäischen Mächte gegen die Türken.

Zum Autor

Antal András *Deák*, geb. 1941, Studium an der Eötvös Lóránd Universität zu Budapest; 1970-1983 tätig als Lehrer, seit 1983 Mitarbeiter des Donaumuseums; Forschungsgebiete: Technik- und Kartographiegeschichte des 17. und 18. Jahrhunderts, insbesonders die Tätigkeit von Luigi Ferdinando Marsigli, Johann Christoph Müller, Samuel Timon, Matthias Bél, Samuel Mikoviny und Stephan Széchenyi (19.Jh.).

Kontakt: deak@mail.dunamuzeum.org.hu

Franciscus Irenicus – Konrad Celtis,
Germaniae exegeseos,
Nürnberg 1518

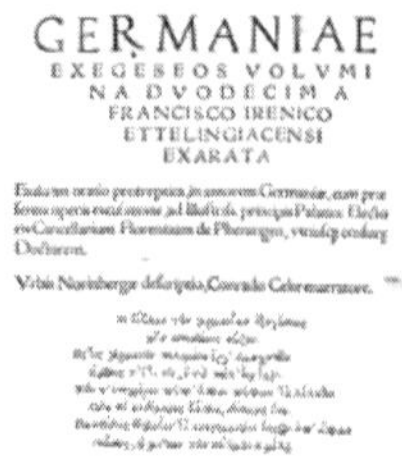
GERMANIAE
EXEGESEOS VOLVMINA DVODECIM A
FRANCISCO IRENICO
ETTELINGIACENSI
EXARATA

Helmuth Grössing, Wien

Eine Erkundung über eine Erkundung

Einige landeskundlich-geographische, vornehmlich deutsche Flüsse betreffende Beobachtungen im Werk des Konrad Celtis

CANDIDE LECTOR!

Ich möchte meinem Freund, dem Wissenschaftshistoriker, Geographen und Kartographen Johannes Dörflinger, vornehmlich aus des großen deutschen Humanisten Konrad Celtis (1459-1508) Elegien *Quatuor libri Amorum,*[1] die ich 2003 nach vielen Jahren der Hingabe an das Werk in eine deutsche Nachdichtung gebracht habe,[2] aber auch aus anderen Dichtkunstwerken des deutschen Erzhumanisten von Celtis' landeskundlich-geographischer Erkundung[3] des deut-

1 K. Celtis, Quattuor libri Amorum secundum quattuor latera Germaniae, hrsg. von Felicitas Pindter, Leipzig 1934.

2 Bislang unveröffentlicht.

3 Erforschung wäre ein zu modernes Wort für des Celtis Denkleistung. – Celtis hatte wohl ein ausgeprägtes Interesse an der Geographie, worüber L. Spitz, Conrad Celtis. The German arch-humanist, Cambridge 1957, 58/59 feststellt: *It was characteristic* […] *of his* [Celtis'] *interest in geography that he namend his proposed societies after the river, forest or city where they would gather.* Dieses geographische Interesse verband der deutsche Erzhumanist mit einer intensiven Sammeltätigkeit, die manches wertvolle Stück (etwa die TABULA PEUTINGERIANA) einbrachte. In diesem Zusammenhang fällt dem Autor Prof. Dr. Erich Woldan (1901-1989) ein, dessen Name mit der großartigen,

schen Landes, darunter auch von einer (wohl fiktiven) Schifffahrt, die zum sagenhaften *Thule* führen sollte, erzählen.

Vor Jahren[4] habe ich darauf hingewiesen, dass die Geographie und Länderbeschreibungen des deutschen Erzhumanisten schon einmal von Theodor Geiger[5] im 19. Jahrhundert abgehandelt worden sind. Zuletzt hat Peter Luh[6] unter Verwendung der sehr aufschlussreichen und ernsthaft um Lösung geographischer Details im Werk des Konrad Celtis bemühten Ausführungen Geigers die ganze Thematik erneut aufgegriffen und manche Erkenntnis hierzu beigesteuert[7].

Hier sollen vor allem einige ausgewählte und auch nicht Vollständigkeit anstrebende, aber markante Einzelheiten zu landeskundlich-geographischen Darstellungen Celtis' vorgeführt werden, die sich vor allem auf die damaligen Grenzflüsse und nördliche Meeresbegrenzung Deutschlands – die Weichsel, die Donau, der Rhein und die Nord- und Ostsee und einige von deren Zubringern – beziehen. Drei Flusslandschaften und das Nordmeer also sind die Begrenzer deutschen Landes (*quatuor latera Germaniae*).

Theodor Geiger hat bereits die auf den Rändern die jeweiligen Bücher der *Amores* illustrierenden Holzschnitte richtig gedeutet.[8] Auf dem Holzschnitt von *Elsula Alpina* z.B. werden 70 Längeneinheiten des *Tractus Germaniae a Danubio per alpes* der *Extensio Danubii a fonte usque in Hamoburgum ultimam urbem Germaniae* zu 100 Einheiten gegenübergestellt. Wie Geiger feststellt,[9] ist die Entfernung vom Ursprung der Save zum Ursprung der Donau ungefähr 420 km (Luftlinie, wie man hinzufügen müsste); die von der Donauquelle im

---

in der Österreichischen Akademie der Wissenschaften verwahrten Sammlung sowie der Buchreihe, in welcher dieser Band erscheint, verbunden ist.

4 H. Grössing, Humanistische Naturwissenschaft. Zur Geschichte der Wiener mathematischen Schulen des 15. und 16. Jahrhunderts (Saecvla Spiritalia 8, Baden–Baden 1983).

5 T. Geiger, Conrad Celtis in seinen Beziehungen zur Geographie. In: Münchener Geographische Studien 2 (1896), 5-40.

6 P. Luh, Die unvollendete Werkausgabe des Conrad Celtis und ihre Holzschnitte (Europäische Hochschulschriften, Reihe XXVIII, Kunstgeschichte 377) Frankfurt/M. 2001.

7 Bemerkenswert sind vor allem Luhs feinsinnige Interpretationen der Bildinhalte.

8 Geiger, Conrad Celtis, 5-40 und Luh, Werkausgabe, 161f.

9 Geiger, Conrad Celtis, 34; auch Luh, Werkausgabe, 161.

Schwarzwald bis Hamburg etwa 600 km. Das Verhältnis von 70:100 entspricht einigermaßen genau jenem von 420 km : 600 km.

Die hauptsächlich linksseitigen Nebenflüsse der Donau werden von Celtis zumeist nur kursorisch behandelt. Die Regen (*Regus fluvius*), aus dem Norden kommend,[10] gesellt sich in Regensburg zur eiligen Donau.[11] Lech (*Licus fluvius*) und Inn (*Aenos*) werden im literarischen Werk des Celtis auch erwähnt. Der Inn und die Ilz[12] bereichern bei Passau mit ihrem Gewässer die Donau.[13] Die zur Donau hin entwässernde Naab (*Navus fluvius*) wird in den *Amores*[14] wie in der *Germania generalis*[15] besungen.

Main, Saale, Eger und Naab entspringen alle auf der Höhe (*iugum*) des sog. Herzynischen Waldes (*silva Hercynia),* womit in der Antike das östlich des Rheins und nördlich der Donau gelegene Mittelgebirge gemeint war, das ursprünglich eine größere Ausdehnung (etwa bis zu den Karpaten) hatte. Im speziellen spricht Celtis hier das deutsch-böhmische Fichtelgebirge an, das Quellgebiet der vier genannten Flüsse. In zehn Versen vereinigt der Dichter die Beschreibung seiner Heimat Franken[16] sowie eine kurze Auflistung der vier Flüsse:

Francia Germano mihi stirps et origo poetae
Hercyniae medio cincta beata sinu.
Quam Moenus mediam perlambit vilifer oram,
Cuius quadrifluo nascitur unda iugo:
Pinifer est, cuius de vertice flumina quattuor
Quattuor in partes orbis amoena cadunt.
Moenus ad occasum Francorum fertur in oras,
Sala sed Arctois perditur Albis aquis,

---

10 Am. II/4, Verse 47-48.

11 K. Celtis, Libri Odarum quattuor. Hrsg. von F. Pindter. Leipzig 1937; zitiert als Od. II/5, Vers 14.

12 Die Ilz entspringt im Nationalpark Bayerischer Wald und mündet in Passau in die Donau.

13 Am. II/13, Verse 11-12: *Inde Ratisponam Patavinaque moenia radis | Oenus et ilsus ubi te locupletat aquis.*

14 Am. II/10, Vers 30.

15 K. Celtis, Germania generalis. Hrsg. von F. Pindter, in: Quattuor libri amorum, Leipzig 1943, Nr. 5, Vers 31.

16 Geboren am 1. Februar 1459 im Weinhauerort Wippfeld am Main in der Nähe von Schweinfurt.

Egra sed Eoi petit ardua regna Boemi,
Sed Nabus Histrinis insinuatur aquis.[17]

Im 1. Buch der *Amores*, dessen Titelholzschnitt überschrieben ist mit *Hasilina Sarmatae*[18] sind die Längeneinheiten der *Distantia fontis Danubii et Vistulae*, also die Entfernung der Quellen von Donau und Weichsel, zur *Distantia fontis Vistulae et hostiorum eius*, die Entfernung von der Weichselquelle in den Schlesischen Beskiden bei Wisła bis zu deren Deltamündung, mit 150 : 100 angegeben, was im absoluten Maß wiederum sehr realistisch ist.

In der Ode *Ad Hasilinam erotice et epithymice* [19] spendet Celtis dem damaligen östlichen Grenzfluß Deutschlands, der Weichsel, ein schönes, poetisches Lob. Ein ebenso dichterisch verbrämtes Elogium von deutschen Gewässern trägt er in einer anderen Ode vor.[20] Hasilina, des Dichters Geliebte und Muse in Krakau, ist schöner als alle Frauen

quas Rhenus habet flumine nobili[21]
Aut quas Danubius pontibus invidens
Vel quae Codoneo barbara colligit
Littore glaesum.[22]

---

17 Am. II/10, Verse 21-30. Deutsche Nachdichtung von Helmuth E. Grössing:
Franken ist mir, dem deutschen Poeten, Wurzel und Stamm
Von dem Mittelgebirg' prächtig im Bogen umrahmt.
Rebentragend beleckt der Main den mittleren Abschnitt
Dessen Quelle entspringt auf dem vierflüssigen Joch:
Fichtenbesät, und vier sprudelnde Flüsse fließen ins Tal
Von diesem Scheitel herab in vier Teile des Lands.
Main, zum Westen gewendet, fließt in fränkische Lande,
Saale aber versiegt nordwärts in Wassern der Elb'.
Eger strebt hin zu dem steinigen Lande des östlichen Böhmen
Und die Naab speist sich ein Wassern des Donaustroms.

18 Hasilina von Rzytonic und Kepstein, die einzig historisch nachweisbare Geliebte des Dichters. Siehe H. Grössing, Celtis in Krakau, in: Aspekte der Bildungs- und Universitätsgeschichte (Schriftenreihe des Universitätsarchivs 7), Wien 1993, 401-408.

19 Od. II/1,Verse 5-8.

20 Od. I/3, Verse 7-10.

21 Die Rheintöchter?

22 Bernstein am Ostseeufer.

Mainz als Stadtdarstellung steht im Zentrum des Titelblattes des 3. Buches der *Amores*; Rhein und Main stellen die Hauptflüsse dar. Das Buch ist der (fiktiven?) Geliebten des Celtis, *Ursula Galla,* gewidmet und beschreibt den westlichen, an Frankreich grenzenden Teil Deutschlands. Im Falle der Proportion von 100 : 30 ist wiederum eine annähernd genaue Entfernungsangabe (Luftlinie) von Rheinquelle zu Rheinmündung (ca. 600 km) gegeben; Mainz und Metz sind etwa 180 km voneinander entfernt. Zu Celtis Zeit war Metz die westlichste deutsche Reichsstadt, gehörte aber seit der Mitte des 16. Jahrhunderts zu Frankreich.[23]

Den Mainzer Stadtbach, *Cia fluvius*, deutsch Gonsbach oder Käsbach genannt, erwähnt Celtis in der ersten Elegie des dritten Buches der *Amores*:[24]

Mens mihi Rhenanam fuerat descendere in urbem
Cui Cia cum Mogano nomina clara dabant.

Es geht um das römische Mogontiacum, für das die alte keltische Gottheit Mogano und das Flüsschen Cia, das auch im betreffenden Holzschnitt eingezeichnet ist, namengebend waren.[25]

Im 3. Buch der *Amores* handelt Celtis auf die übliche didaktische Frageart, vom Rhein ausgehend, eine Reihe von niederdeutschen und niederländischen Flüssen ab, die er mit den Landschaften, die sie durchfließen, in Verbindung bringt:

Qua Lona,[26] *Rura,*[27] *suis Lipia*[28] *quaque vadis,*
Qua Sara[29] *Trevericos laetus despumat in agros*
Sellaque[30] *cum Mosa*[31] *nomina iunctus habet* […].

---

23 Siehe dazu im Speziellen: Luh, Werkausgabe, 164.
24 Am. III/1,Verse 9-10.
25 Luh, Werkausgabe, 165, fragt verwundert, ob Celtis die Kultstätte der keltischen Gottheit Mogano beim römischen Mogontiacum bekannt gewesen war.
26 Vielleicht die Lenne, ein Nebenfluß der Ruhr in Nordrhein-Westfalen. Ist bei Graesse-Benedikt-Plechl, Orbis Latinus, nicht verzeichnet.
27 Die Ruhr.
28 Lipia, 200 km langer Nebenfluß des Rhein; bei Graesse-Benedikt-Plechl, Bd. II: *Lippia fluvius*
29 Die Saar.
30 Vielleicht die Schelde? Fehlt bei Graesse-Benedikt-Plechl.

An den göttlich personifizierten Rhein, gleichsam den Rhein-Gott, spricht Celtis die Bitte aus, seiner Dame Ursula (*Ursula Galla*), die schwer erkrankt war,[32] ein sicheres Geleit stromabwärts zu dem Thermen von Aachen zu geben, wo sie Heilung oder zumindest Linderung ihres Leidens suchen wollte.

In dasselbe Schema passt auch die Elegie *Ad Danubium, ut puellam descendentem in Pannonias numine suo tueatur,* [33], in der Celtis die (hier wieder personifizierte) Donau bittet, fürsorglich seine Dame von Regensburg nach Wien zu führen, damit sich Elsula in den heißen Quellen von Baden bei Wien kurieren kann.

Auch Drau und Save[34] wurden von Celtis im Rahmen der *Quattuor latera Germaniae* erwähnt und zwar als Nebenflüsse der Donau:

Accedunt amnes Dravusque Savusque sonorus
Qui duo Danubium flumina magna petunt,
In quorum ripis Alpestribus oppida clara:
Frisiacum, Grecium, Labacum et Iulichum[35]
Bellogradum fuerat, Turcorum ubi saeva caterva
Experta a nostris proelia saepe viris.

Schließlich kommt die Etsch aus dem lieblichen Amerina-Tal[36] (*Amerina vallis*), wo Tirol sein Haupt zu den „erhabenen

31 Die Maas (franz. Meuse), fließt durch Frankreich, Belgien und die Niederlande. Graesse-Benedikt-Plechl 2.

32 Siehe H. Grössing, Humanistische Naturwissenschaft,. 287 und zuletzt H. Grössing, „Astra inclinant?" Astrologie in den „Amores" des Konrad Celtis, in: C. Friedrich – J. Telle (Hrsg.), Pharmazie in Geschichte und Gegenwart. Festgabe für Wolf-Dieter Müller-Jahncke zum 65. Geburtstag, Stuttgart 2009, 174-175. Möglicherweise hat Ursula an Syphilis gelitten und Celtis infiziert? – Luh, Werkausgabe, 27 spricht expressis verbis von *Syphilis*, obwohl die moderne Forschung diesbezüglich zurückhaltender ist. Siehe Od. III/17.

33 Am., II/13.

34 Am. II/13, Verse 27-32.

35 Friesach und Graz sind etwas großzügig in die Nähe der Drau verlegt. *Labacum* ist Laibach/Ljubljana. Der Ort *Iulichum* ist bislang nicht eindeutig zuordenbar; für Luh, Die unvollendete Werkausgabe, 162 […] *muß* […] *offen bleiben, ob Celtis die Burg von Schönfeld in Friaul, Cividale, Görz oder aber Krainburg oder Cilli im heutigen Slowenien meint.*

36 Hier ist wohl das Meraner Tal gemeint.

Bergesgipfeln" erhebt und sich dann zu den Tridentinischen Bergen hin wendet und schließlich in die Adria mündet.[37]

In Od. III/17[38] beschreibt Celtis eine schwere, ja katastrophale Überschwemmung des Rheins, die er offensichtlich in Mainz miterlebt und mitgelitten hat. *Neulich* habe der Rhein durch eine außergewöhnliche Überschwemmung *seine Flußarme eingebüßt,*[39] Felder und Äcker überschwemmt und alle Widerstände hinweggerissen.[40] Nachdem die Wassermassen wieder in ihr natürliches Bett zurückgekehrt waren, trat eine pestartige Seuche auf, *welche ihm das Mädchen* (wohl Ursula) *geraubt hat.*[41] Die Ode ist eine poetische Verklärung,[42] welche die Naturkatastrophe verniedlichend ästhetisiert und mit der griechischen Mythologie in Verbindung bringt.

Ob in dieser Ode nicht bereits die Lorelei-Sage vorkommt? - betrachten wir die folgenden Verse:

Wie wenn der wilde Neptun die Schlingen der
Flüsse auflöste, jugendlichem Quell Auslaß
gewährt aus dem Grase,
die dann, sich vorwärts stürzend, zum segelbevölkerten Meer ziehn,
einig sind mit den Nymphen, das Glasschloß des Vaters zu sehen.
So denn sputen die Quellflüsse sich,
zum Hause des Vaters zu
streben, sich stürzend von Welle zu
Welle, zum fürstlichen Hause
und sie schwatzen in ihren
moosumrandeten Quellort,
als wollten die Flüsse mit wehendem Haar
sich bei Südwind bereden, daß sie

---

37 Am. II/13, Verse 33-36
At parte ex alia est Amerina vallis amoena
Qua Tyrolis celsum tollit in astra caput
Quaque Tridentinas Athesis se flectit ad Alpes
Cuius in Adriacum decidit unda sinum.

38 Od. III/13, 16, 17.

39 Od. III/17, Vers 1: *cornua perdidit.*

40 Ebd., Verse 1-3.

41 Titelzeile: *pestis secuta est, quae sibi puellam ademerat.*

42 Der Titel beginnt mit *De convivio Neptuni.* Nachdichtung von H. E. Grössing.

schneller ins väterliche Gemach treten können.
Die in den Klüften sich barg, sie hat das Haupt schon gehoben
und sie beschaut im hohlen Spiegel, sich schmückend, den Körper,
schaut die weiße Frau, vom Lichte umflossen, den Phöbus.
Sitzt unter Felsen, aber verläßt ihre wässrige Höhle,
vorher jedoch tilgt sie die schaumigen Flecken vom Körper,
damit dem Gotte des Meeres sie in Schönheit erstrahle.
Auf dem steinigen Felsen singend schlägt sie mit Armen,
wie wenn der Flügel in flutenden
Wogen weiter könnt' schwimmen und
das rauhe Meer sie bei rückfließendem
Wasser auch trägt. Daher streben die
Ströme, die das Vaterhaus sehen
wollen, frohegestimmt mit ihrem
Gewässer durch Lande und Täler.

Die Lorelei-Sage, die wir von Clemens Brentanos Ballade *Zu Bacharach am Rheine* (1801) her kennen, vor allem aber von Heinrich Heines Gedicht (1824), wurde von Werner Bellmann[43] als Variation des antiken Mythos von der Nymphe Echo gedeutet.[44] Echo, deren Liebe Narzissus verschmäht, wird in einen Felsen verwandelt, von dem ihre Stimme widerhallend klingt. Die Möglichkeit, daß Celtis die Lorelei-Sage in einer anderen Fassung kannte, muß zumindest in Erwägung gezogen werden.

Die Rheinüberschwemmung, die der deutsche Archipoeta wohl in Mainz um die Mitte der neunziger Jahre des 15. Jahrhunderts als Augenzeuge erlebt hatte, wird in der genannten Ode dramatisch und

---

43 W. Bellmann, Brentanos Lore Lay-Ballade und der antike Echo-Mythos, in: D. Lüders (Hrsg.), Clemens Brentano. Beiträge des Kolloquiums im Freien Deutschen Hochstift 1978, Tübingen 1980, 1-9.

44 Vgl. Ovid, Metamorphosen 3, 393-401, insbesondere die Verse 395-399: Die Nymphe Echo magert mehr und mehr ab, ihre Haut wird schlaff, die Körpersäfte verschwinden, nur ihre Knochen und Stimme bleiben übrig; die Knochen werden zu Stein und Fels, man sieht sie nirgends mehr und hört allein ihre Stimme – Clemens Brentano soll eine Versfassung der Metamorphosen Ovids aus dem Jahre 1631 für seine Ballade herangezogen haben.

anschaulich geschildert: Main, Neckar, Mosel und die Maas,[45] und alle die Flüsse, die vom deutschen Mittelgebirge[46] zum Rhein hin entwässern, schwellen den Rhein übermäßig an, dem gedankt wird, weil er seine Pflicht erfüllt,[47] d. h. die Nebenflüsse in sich aufnimmt. Dadurch dass er so stark angeschwollen war und einen höheren Stand als alle seine Nebenflüsse erreichte, mündet er nicht mehr dreiarmig, sondern als breite Wasserfläche ins Meer. Er reißt Brücken weg, herrscht durch seine die Länder bedeckenden Fluten,[48] Groß- und Kleinvieh wird ohne Unterlass[49] fortgerissen, mächtige modernde Eichenbäume werden in den Wäldern entwurzelt.

Hättest gesehen schwimmende Körper in die Wässer gerissen,
solche die mit klagenden Schreien den Himmel verfluchten.
Der da hat einen Baumstamm ergriffen, während er abtreibt
Und er höhnt, sich rettend, den Ansturm des wütenden Flusses;
Der aber, der auf dem Heuhaufen sitzt, treibt in das Meer;
Dieser, mitgerissen, schifft mit dem ärmlichen Kittel,
und er beweinet sein Kind, das ihm die Flut jetzt entführt.
Sehen kannst Du Ziegen, verängstigte Schafe, wie sie
Mit ihren heulenden Hunden sich in den Wassern abmüh'n[50]

Schließlich ist der Rhein dann wieder in sein altes Bett zurückgekehrt, und waren auch die übrigen Flüsse (poetisch gesagt) wieder in ihre „väterliche Höhle"[51] eingetreten. Auf den modernden Feldern, wo unfruchtbares wässriges Gras liegen geblieben war, verwesten allenthalben hässliche Kadaver, der Schmutz der Kloaken breitete sich aus, Schlamm war aus den Alpenregionen herbeigeführt worden, und mit pechschwarzen Ausdünstungen verursachte die

---

45 *Naus Bavarae culta secans pinguia Galliae*, Od. III/17, Vers 27. So auch in der Editio princeps. Es handelt sich um einen Fluß, der die *Gallia Bavara* durchfließt. Es ist anzunehmen, dass es sich sowohl bei *Naus* wie bei *Bavara* um einen Lese- und Setzfehler des Druckers handelt; statt *Naus* wäre *Maas*, statt *Bavarae Batavae* zu lesen.

46 *Silva Hercynia*. In der *Norimberga* befasst sich ein längeres Kapital mit der Beschreibung des Hercynischen Waldes. S. dazu L. Spitz, Conrad Celtis, 40.

47 Od. III/17, Vers 31.

48 Ebd., Vers 35.

49 Ebd.,Vers 36.

50 Ebd.,Verse 41-49.

51 *sub patrium specum*, ebd., Vers 83.

Erde die „Infektion“,[52] wobei die Sonne die Luft erhitzte und schädlichen Nebel erzeugte. Alles in allem: Ein Bild des Jammers!

Als dies alles Phöbus vom purpurnen Himmel aus sah,
nahm er den Bogen mit Köcher und das tötende Schwert
und er schickte die Seuche, erboßt, den Völkern am Rhein.
Hierbei hat er mit Pest und der Seuche mein Mädchen getroffen,

dem Celtis, die Ode abschließend, mit roten Lilien das Grab ziert.

Für Celtis war der größte Fluß der *quatuor latera Germaniae* die Elbe. In dem Epigramm *De fonte Albis*[53] artikuliert er offenbar die Vorstellung von Länge und Materialität der Elbe, vor allem gegenüber Rhein und Neckar. *Unter Dir wird Rhein sein und Neckar, der großart'ge Fluß.*[54] Gegenüber dieser Verherrlichung der Elbe fällt die Donau ab, deren Gewässer Celtis als *rudelinguis* „rauhtönend“(?) bezeichnet, das die Musen scheuen.[55]

Konrad Celtis war zweifellos ein Meister der Fabulierkunst – was ein Privileg jedes Dichters ist; aber er hatte sich selbst eine poetische Gegenwelt aufgebaut, die, wiewohl fiktiv, ob ihrer Kunstfertigkeit in gewisser Hinsicht Realitätsanspruch erheben kann. Dichtung und Wahrheit liegen bei ihm sehr nahe beieinander, es waren wohl vielfach „Abenteuer im Kopf“[56] oder war „Erfundene Wahrheit“,[57]

---

52 Im Sinne der medizinischen Miasmen-Theorie.

53 K. Celtis, Fünf Bücher Epigramme. Hrsg. von K. Hartfelder, Berlin 1881, Nachdruck Hildesheim 1963(zitiert als Ep) I/58. – Siehe auch D. Wuttke, Textkritisches Supplement zu Hartfelders Edition der Celtis-Epigramme, in: Renatae Litterae. August Buck zum 60. Geburtstag, Frankfurt/Main 1973, 105-130, wo über 400 Korrekturen der Lesarten Hartfelders aufgelistet werden.

54 Ep. I/58 , 14.
O fons Arctoas inter clarissimus undas,
Qui tua in Oceanum nomina clara tenes.
Inferior tibi Rhenus erit Necarusque superbus
Phoebus ubi rosis destituetur aquis.

55 Ebd., II/12, 24.
Ostia Danubii celebrarunt carmine Musae
Sed nullas Musas fons rudelinguis habet.

56 W. Wenk, „Abenteuer im Kopf“ (Zu Konrad Celtis, Amores III, 5), in: Wiener humanistische Blätter 32 (1990), 41-57.

57 U. Hess, Erfundene Wahrheit. Autobiographische und literarische Rolle bei Conrad Celtis, in: A. Schöne (Hrsg.), Kontroversen, alte und neue. (Akten

die er uns vorführt, ohne dass wir uns „vorgeführt“ vorkommen müssen; in die er uns aber auch ver-führt kraft seiner Dichtkunst, für die er, der erste deutsche *Poeta laureatus*, nicht umsonst aus kaiserlichen Händen den Lorbeerkranz erhalten hat.

Für diese Befindlichkeit des Dichters, der so intensiv seine Kunst erlebt, dass er sich mit dem, was er schreibt, realitätsstiftend identifiziert, ist jene Elegie des IV. Buches des *Amores*[58] bezeichnend.

Er beschreibt eine Schiffahrt von der Mündung der Elbe zur Insel Thule, bei welcher er in ein Unwetter geraten ist.

Richte her das wenig Gepäck und lege zusammen,
Barbara, was man da braucht, um das Meer zu befahr'n.
Dreimal rief die Mannschaft, es riefen die Trommeln des Meeres,
Auf dem Ozean Glück, auf der See gute Fahrt!
Gieß in die Krüge den Wein und nachdem dann diese gefüllt sind,
Hebe ins Schiff auch den Korb Brotes, das man nun bäckt.
Nimm desgleichen vom Ochsengeräucherten, Rohfleisch zudem
Richte aber auch zu g'salzene Butter und Schmalz
Nicht auch mögen uns fehlen magenstärkende Mittel
Ingwer und was auch das Herz immer zu härten vermag.
Wenn man einmal mit gehisstem Segel die Meere durchschiffet
Und die bewegte See alles von Grund auf bewegt,
Dann wird übler Gestank das ganze Fahrzeug erfüllen
Hebt aus dem Magen hervor, Barbara, was Du drin hast.
Keine Medizin kann Dir die Krankheit erleichtern
Als die See allein, die Dir den Magen entleert.

[…..]

Schon ertönt die Schiffestrompete, Barbara, eile,
Steig auf das Schiff und sei, Glückliche, an meiner Seit'!
Brücke zum Schiffe wird eingezogen, der Anker gelichtet
Auch der Wind kommt schon auf, treibt in die See das Gefährt;
Flut aus sieben der Rumben, die Ufers Anblick stets ändert[59]
Deren Ursach bislang allen verborgen noch ist:

---

des VII. Internationalen Germanistenkongresses (7,1986)), Göttingen 1987, 136-147.

58 Am. IV/14.. Dazu siehe auch Od. IV/3, *De enavigatione Codoni sinus. Nachdichtung von H. E. Grössing.*

59 Am. IV/14, Vers 27: *Aestus septenis vicibus qui litora mutat.* Der Wind peitscht die See aus sieben Richtungen der Windrose (*Rumben*) auf.

Ob's ein gewaltiges Tier ist, das wir das Weltall benennen,[60]
Das jenen Atem erzeugt, speiend aus seinem Mund;
Oder ob unter der Erde sich riesige Höhlen befinden
Allwo mit feurigem Wind kämpfen Wogen und Flut.
Äolus jagt das Meer dann rasch in die luftige Höhe
Und ein anderer Wind hebt es von hier in das All;
Sei es auch, dass die Nähe der Sonne, der Abstand des Mondes
Ändert den Anblick des Meers, gibt ihm verschied'ne Gestalt;
Oder die Wirkkraft der Sterne, verschieden in ihrer Bewegung
Treibt die Meeresflut an, sich zu erneuern sogleich.
Sieben der Sterne wenden sich hin zum östlichen Morgen
Aber der achte, sieh da, wählt den westlichen Weg.
Was es auch sei, es kann diese Wunder der Geist nicht mehr fassen,
Von diesem Abweg daher, kehr'n wir zur Straße zurück!
Jene fahren das Schiff mit Rudern hinweg von der Küste
Dieser da bringt es zum steh'n, heftig sind Ruder bewegt.
Einer entzündet die feuerfspeiend' Bombarde, die dröhnt
Weit über's Meer, und der Rauch deckt den Himmel fast zu.
Wieder ein and'rer erklettert den Mast und bindet die Taue
Gibt die Rahen dann frei, lässt die Segel herab.
Üppige Tücher blähen sich auf, der heitere Südwind
Braust und befiehlt seinem Freund Zephyr, zu fliegen mit ihm.
Schon nach dem Norden richtet sich aus der Schnabel des Schiffes
Da wo um Thule herum starrende Untiefen stehn.
Fröhlich fangen wir an, ein Lied gefällig zu singen,
Das dem allmächtigen Gott widmet die lustige Schar.

[.....]

Wer am Steuer sitzt, der wendet das Schiff so wie er
Will, und er gibt ihm vor Richtung, befohlenen Sinn.
Und mit dem Kompaß[61] *bestimmt er die Lage der Lüfte und Winde*
Teilt jedem einzelnen zu das ihm gebührende Wort
Einmal Zephyrus ihn heißend, dann spricht er vom Wehen des Eurus
Nun ruft er „Boreas" aus, Notus, den feuchten, er nennt.
Auch die Zwischenwinde, die vier, und insgesamt kennt er

60 In der *Germania Generalis*, Nr.1 – F. Pindter (Hrsg.), Conradus Celtis Protucius, Germania Generalis, legt Celtis eine Kosmogonie vor, die er *Origo mundi ex ventre Demogorgonis* bezeichnet. Er prüft nachfolgend in Am. IV/14 mehrere Argumente, die ihm aus dem Wissensstand seiner Zeit geläufig sind.

61 d.h., er bestimmt die Himmelsrichtungen, aus denen die Winde kommen.

Dreißig der Stürme an Zahl [...]
[.....]
Schon sind wir glücklich sechzehn der Stunden dahingefahren
Und schon wandte zum Meer Sonne das Pferdegespann.
Siehe, da heben sich Wolken hoch vom nassfeuchten Süden
Und bedecken mit Schwarz gänzlich die himmlische Höh'.
Alsdann beginnen die Wellen des Meeres höher zu werden
Und die Kronen der Gischt weißen das wütende Meer.
Auf der Oberfläche wälzt der Delphin sich im Kreise
Seehund hebet empor über das Wasser sein Haupt
Auch der Mönchfisch,[62] *bedeckt mit kapuzenförmiger Haut,*
Zeigt sich: und ist Beweis für den Hochwasserstand,
Von seinem Anblick nehmen die Schiffer das folgende Sprichwort:
Wenn der Mönchesfisch schwimmt, bricht bald das Unwetter los.
Losgelassen vom Norden sauset der Boreas her[63]
Und gen den Nordeswind kämpft Eurus hitzig und wild,
Dann speit das wütende Meer den Sand hinein auf die Küsten
s rast die Tartaros-Flut, schwärzer noch als der Styx.
Schwarze Gewitterwolken nähern sich unserem Schiffsmast
Hoch in die Lüfte getürmt, blitzend vom Feuer umrahmt
Und schon schaudert vom heftigen Donner gerühret der Äther

---

62 Am. IV/14, Verse 84-87:
Et moniscus piscis, qui tectus pelle cuculli
Emicat: undosi signa futura maris,
A cuius visu sumunt proverbia nautae:
Enatat ut moniscus, mox freta turbat hiems.
Der durchschnittlich 30-50 cm große und max. 2 kg schwere Mönchsfisch (*Kathetostoma giganteum*) kommt im südlichen Pazifik – bei Neuseeland etwa in einer Tiefe bis zu 600 Metern – vor, ist aber auch in der Nähe aller Kanaren reichlich zu finden. – Der Nürnberger Drucker, Briefmaler und Formschneider Stefan Hamer (er wirkte von 1531 bis 1562) erwähnt, dass 1546 bei Kopenhagen *in der Dänemarksee*, ein Mönchfisch gefunden wurde und gibt von diesem eine bildliche Darstellung (halb Mensch, halb Fisch). Der Mönchfisch Hamers gehört somit in die Kategorie „Monströse Meeresungeheuer". http://www.zeno.org/Kunstwerke/A/Hamer,+Stefan.
Zu diesem Thema im allgemeinen C. Riedl-Dorn, Wissenschaft und Fabelwesen (Perspektiven der Wissenschaftsgeschichte 6, hrsg. von H. Grössing) Wien 1989, 9-19.

63 Die nachfolgenden Verse klingen so dramatisch, als hätte Celtis die Sturmflut auf dem Schiff selbst erlebt. Tatsächlich scheint er, zumindest teilweise, Anleihen von Vergil genommen zu haben: vgl. Aeneis 1, 81-123.

So wenn Zeus, der Gott, rütteln würde den Pol.
Massenhaft Regen fließt aus aufgerissenen Wolken
Segeltücher sind naß durch die stürzende Flut.
Schon beginnt auf dem Meer der Ostwind heft'ger zu blasen
Nimmt zur Gänze die See in Bestand und Besitz.
Hoch in die Luft wird gehoben das Schiff vom kräftigen Sturme
Wird in die Tiefe darauf, abwärts ins Meer eingetaucht.
Mit den Wogen in Regen und Dunkel ist es getrieben
Wie in Acherons Furt wird das Fahrzeug gesenkt.
Alle Sterne verhüllen sich in Schwaden von Regen
Beide Bären versteckt, Ursa seh ich nicht mehr,
Nicht den Kepheus, bedeckt mit der höchsten, päpstlichen Mitra
Hab' ich erblickt, und es war keiner der Sterne zu seh'n.
Da schrie des Schiffes Kapitän: Herab schnell das Segel!
Und befahl auch zugleich: Löst von der Rahe das Tuch,
Alle die anderen aber, das Übel zu mindern, führten die
Ruder, mitten im Meer trieben wir also dahin.
Ausbrach gewaltiges Schreien der Menschen hinauf bis zum Himmel
Ungewiß des Geschicks war ein jeder von uns.
So wie man brüllt in Österreichs Hauptstadt, etwa in Wien,[64]
Wenn sich in Kellern die Schar zutrinkt den heurigen Wein,
Oder wie der Schiffer Geschrei, wenn auf dem Schiff
Fässer voll Rebensafts hin nach Bayern verbracht.

[.....]

Ich gelobte sodann Neptun[65] *mit zitterndem Herzen,*
Wenn er sich zeige, er soll beistehen uns in der Not:
„Diesem Meere entreiß mich, Neptun! will Jahrtagsgelübde
Einlösen Dir in dem Lied, das ich Dir widme und weih',
Füge aber hinzu, dass ich nie mehr Dein Meer will befahren
Reise auch fürderhin mehr nie ins fremde Geland.
Meine Aufgabe war es, Deutschland ganz zu beschreiben
Hab es getan und ich hab es in vier Grenzen geseh'n.

---

64 Wo Celtis seit 1497 an der Universität und später auch in seinem *Collegium poetarum (et mathematicorum)* tätig war und wo wohl auch diese Elegie entstanden ist. Siehe dazu H. Grössing, Celtis in Wien, in: Jahrbuch des Vereins für Geschichte der Stadt Wien 47 und 49 (Festschrift für Max Kratochwill), Wien 1992, 7-16.

65 Auch in der Aeneis (1. Gesang) greift Neptun gegen die von Juno zur Vernichtung der Troianer aufgestachelten Winde rettend ein.

Was mir vom Leben noch bleibt, ihr Götter, gewährt es dem Celtis,
Daß er im eigenen Haus sicher und tätig verbleib'!"
Haben's gehört: in Schlaf gesunken erschien mir von oben
Merkur und sprach zu mir folgende Worte sodann:
„Wenn Du wieder in Deutschlands Gauen, Celtis, verweilest
Und Dich das Schiff dahier wieder an Land hat gesetzt,
Wohlauf, eil' dann sofort in die regenbringenden Alpen
Dort, wo der Flüsse drei einig werden im Strom:
Eisack von Osten, Etsch von daher, wo die Sonne verschwindet
Und der dritte entspringt im Serentina-Tal.[66]
Hier, wo Bozen sich im weitläufigen Tale erstrecket
Und Du das heilige Land, unser Tirol nicht verlässt.
Strebe vielmehr nach Triest, nach Görz auch und ins Friaul
Da wo Bonomus[67] *zuhaus, Bischof von göttlicher Gnad'.*
Oder geh' nach Brixen oder Trient, das ist besser,
Máxmiliánus wirkt hier, Vater der Musen und Herr,
Der seit langem in deutschen Landen die Absicht bekundet
Pegasus' Nymphen zu führ'n und den musischen Chor.
Er, in der Tat, wird Dir geben für immer Besitz und Vermögen
Er auch lässt es nicht zu, dass Dir die Armut zur Plag'
Heil'ge Kamenen will er beschenken mit einem Kolleg
Mathematiker zwei bindet er auch an das Haus,[68]

---

66 Gemeint ist wohl die Rienz (ital. Rienza), der drittgrößte Fluß Südtirols, der im Gemeindegebiet von Toblach entspringt und bei Brixen in den Eisack mündet.

67 Petrus Bonomus (Pietro Bonomo), geb. 1458 in Triest, gest. am 4. Juli 1546 in Triest, war (nach der Terminologie von A. Lhotsky, Quellenkunde zur mittelalterlichen Geschichte Österreichs, Graz-Köln 1963, 441) „Kanzleihumanist" Kaiser Friedrichs III.; 1501-1546 Bischof von Triest, 1522-1523 auch Bischof von Wien.

68 Gemeint ist das Wiener *Collegium poetarum et mathematicorum*, das Maximilian 1501 stiftete. In der Gründungsurkunde des am 1. Februar 1502 eröffneten Collegium ist nur vom *Collegium poetarum* die Rede. Nach K. Mühlberger, Poetenkolleg und Dichterkrönung in Wien, in: Bilder-Daten-Promotionen. Studien zum Promotionswesen an deutschen Universitäten der frühen Neuzeit (R. v. Bruch – L. F. Beck (Hrsg.), Pallas Athene. Beiträge zur Universitäts- und Bildungsgeschichte 24) Stuttgart 2007, 85, Anm. 1, ist die Langform *Collegium poetarum et mathematicorum* in zeitgenössischen Quellen belegt und brächte auch die Intention des Konrad Celtis adäquat zum Ausdruck. – Unter *scientiae mathematicae* verstanden die Humanisten den quadrivialen Fächerkanon und die Naturphilosophie. – Siehe dazu auch F.

Dort denn, wo Wiens berühmte Mauern zur Höh' sich erheben
Und wo der Wienerwald wachsen lässt Bacchus-Geschenk.
Dich wird der Herrscher in diesem Kolleg zum Rektor bestellen
An seiner statt wirst Du selbst alle Geschäfte ausführ'n,
Nämlich, dass Du Poeten bekrönst mit dem göttlichen Lorbeer[69]
Und dies jene auch tun, die Dir folgen im Amt".

[.....]

Da ist ein Ort in nördlicher Gegend, wo Deutschland zu Ende
Wo aus eisigem Meer, Thule,[70] *die Insel, sich hebt*
Neben gefürchteten Klippen und überhangenden Felsen
Macht die Gegend dort rauh steinige, wellige Furt.
Nach einem griechischen Wort nennt man die Inseln „Orkaden"[71]
Weiß auch zu sagen, es sei meineidigen Göttern Exil,
Welche der Anwohner nennt mit veränderten Namen „die Trollen"[72]
Weil sie verdichtet in Luft körperlich anwesend sind.
Aberglaube ist's nicht; nicht eingebildeter Irrtum,
Geister sahen wir dort schrecklichtönend am Fels.
Diese freuen sich wohl, den Menschen zu Diensten zu sein
Streben mit ihrem Geschäft menschliche Wohnstätten an;
Aber da in Worten und Taten sie Unrecht verletzet
Flieh'n sie unsere Art, schaffen uns oftmals Gefahr:

Graf-Stuhlhofer, Das Weiterbestehen des Wiener Poetenkollegs nach dem Tod Konrad Celtis' 1508, in: Zeitschrift für Historische Forschung 26 (1999), 401-405.

69 Celtis erhält vom Herrscher das Recht, Poetae laureati zu küren, wie in einem Epigramm in der *Rhapsodia* (in Ep. Nr. 77, 118) nachzulesen ist:
Hanc laurum dedimus Conrado, insignia vatum,
Caesar, ut heroum fortia facta canat,
Quaque pios vates merita cum laude coronet.

70 Celtis scheint mit *Thule* Island gemeint zu haben, eine im Spätmittelalter und 16. Jahrhundert gängige Ansicht.

71 Orkney Islands (veraltete deutsche Bezeichnung *Orkaden*), nordwestlich vor der Küste Schottlands liegende, zu Schottland gehörige Inselgruppe.

72 Die Trolle, (auch Trolde, Trölle), nord. für Unholde, Riesen, Zauberwesen, stammen aus der germanischen Mythologie, die, als männliches Pendant zu Feen, Elfen oder Hexen, Geisterwesen in Riesen- oder Zwergen-Gestalt sind, die Schaden anrichten können. Auch die isländische Folklore und Literatur kennt die Trolle, die hier allerdings nur positiv gesehen und mit guten Elfen oder Feen verglichen werden. Indem Celtis auch von guten Trollen spricht, die von den Menschen allerdings immer wieder enttäuscht werden, gewinnt die Annahme, dass er mit *Thule* Island meint, an Gewicht.

Einmal rauben sie Ruder, versenken Schiffe im Meere
Anker, die schon fixiert, heben sie oftmals vom Grund
Unwetter wird durch sie oft heftiger noch, ungestümer,
Wütet im Felsengestein und an den Klippen des Meers.

[…..]

Nicht war mehr ferne und weit Thule, von allen gekannt.
Aufgerichtet der Mastbaum, in die Höhe gezogen das Segel
Strebte das Schiff aus der Bucht, fertig in voller Fahrt
Fröhlichen Sinnes verlässt man den Ort, wir stechen in See,
Fahren ein in den Port Thules, den wir erstrebt.

[…..]

Fuhr über bergiges Land, abwärts zum Reiche der Etsch,[73]
Bald schon war mir gegeben des Königs Erlaubnis und Urkund[74]
Somit lege ich dar, wie mein Weg da begann.
Dies auch sagte der Herr: „Leicht ist's, den Poeten zu nähren
Wenn Du es wünschest, so sei, Celtis, Dir auch ein Weib.[75]
Oder willst Du festlich geloben, Priester zu werden,
Und die Armut erstrebst, sei es so, wie Du's wünscht".
„Nicht wünsche ich, o König und Fürst und Zierde des Reiches,
Dieses Lebens Verdruß, Ekel vor meinem Selbst.
Vielmehr wünsch' ich durch's Lied, der Ewigkeit König zu sein
Dessen Taten man kennt allüberall in der Welt.
Dies soll, wenn möglich, gescheh'n, o unbesiegbarer König,
Daß Du ein Obdach mir gibst, und ein Jahresgehalt!"
Nickte bejahend der Herr und bestätigt' mit eigener Hand
Alles, was ich begehrt, hieß mich dann geh'n, reich beschenkt.

NUNC FINIVI FELICITER HOC OPUS IN FAVOREM JOANNIS

---

73 Von der Erwähnung der Trolle abgesehen war *Thule* (Island?) Celtis keiner näheren Beschreibung wert und er begab sich poetisch sofort auf die Reise nach Südtirol.

74 Maximilian beurkundet 1501 in Bozen die Gründung des *Collegium poetarum (et mathematicorum)*. Original im Archiv der Universität Wien.

75 Celtis hat sich Ende des 15. Jahrhunderts in Wien mit Heiratsplänen getragen. Siehe dazu H. Grössing, Celtis in Wien, 12f.

Zum Autor

Helmuth Grössing, geb. 1940 in Leoben, Studium der Fächer Geologie, Geophysik und Paläontologie, ab 1960 Österreichische Geschichte, Alte Geschichte und Philosophie, 1965 Dr.phil., Mitglied des Instituts für Österreichische Geschichtsforschung, Habilitation für Geschichte der Neuzeit (1982), Arbeitsschwerpunkte: Wissenschafts- und Kulturgeschichte, im besonderen Österreichs vom 15. bis 19. Jahrhundert, 2002-2008 geschäftsführender Direktor des Instituts Österreichisches Biographisches Lexikon und Biographische Dokumentation der Österreichischen Akademie der Wissenschaften, Mitglied der Kommission für Geschichte der Naturwissenschaften, Mathematik, und Medizin der Österreichischen Akademie der Wissenschaften. Seit 1992 Präsident der Österreichischen Gesellschaft für Wissenschaftsgeschichte.

Kontakt: heg@inode.at

Hans Weigel,
[Karte der Umgebung von
Nürnberg, ohne Titel],
[Nürnberg] 1559

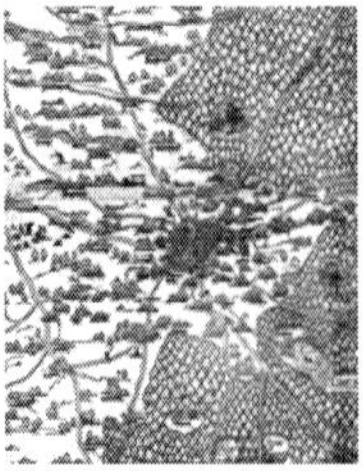

Paul D. A. Harvey, Durham

Medieval local maps from German-speaking lands and central Europe

Maps were not a normal part of everyday life in medieval Europe, and nothing demonstrates this better than the rarity of local maps from before 1500, of maps, that is, of areas known personally to the mapmaker and drawn from local knowledge. In circumstances where we would draw a simple map – to give directions for a journey, to show the boundaries of a property, and so on – people in the Middle Ages would compose a written description, often long and cumbersome, to convey information that might far more easily be presented graphically. From about 1500, a clear dividing line, there was a steady increase in the number of local maps that were drawn, indeed of maps of every kind, so that by the end of the sixteenth century they were understood and used much as we understand and use maps today.[1]

---

1 This can be illustrated from three catalogues of maps published as Bayerische Archivinventare 37, 48, 49: E. Krausen, Die handgezeichneten Karten im Bayerischen Hauptstaatsarchiv sowie in den Staatsarchiven Amberg und Neuburg a. d. Donau bis 1650, Neustadt a. d. Aisch 1973; R. Winkler, Die handgezeichneten Karten des Staatsarchivs Bamberg bis 1780, München 2005; P. Fleischmann, Die handgezeichneten Karten des Staatsarchivs Nürnberg bis 1806, München 1998. No map there is older than the sixteenth century, and the earliest in each catalogue is dated, respectively, 1510, 1521,

Where a local map was drawn before 1500 it must often – perhaps mostly – have been the entirely new idea of some particularly imaginative individual who saw how much more efficient it was to set out topographical information graphically instead of in writing. However, this seldom happened. In two areas of Europe medieval local maps have been systematically recorded, so that we have a reasonably comprehensive picture of what exists. From the Low Countries we have some seventeen maps or closely related groups of maps, the earliest dating from 1307[2], and from England at least thirty-seven have been recorded, four from between 1153 and 1249 and the rest from the mid-fourteenth century onwards.[3]

From two other parts of Europe reports of medieval local maps cannot lay claim to definitive, comprehensive listing; certainly there are more, significantly more, to be discovered. To the 'first harvest' of some ten maps from France that François de Dainville described

---

and 1516; the number of entries from 1500 to 1550 is 29, 8, and 20, and from 1550 to 1600 over 242, 38 and 125.

2 Fifteen are listed in P. D. A. Harvey, Local and regional cartography in medieval Europe, in: J. B. Harley – D. Woodward (Ed.), The History of Cartography 1: Cartography in Prehistoric, Ancient and Medieval Europe and the Mediterranean, Chicago 1987, 464-501, here 499-500; P. van der Krogt, Lokale kaarten van Nederland uit de late Middeleeuwen, in: Caert-Thresoor 27 (2008), 29-42, discusses each with superb colour illustrations (except one that proved untraceable), adding two further groups of maps and pointing out that the dates of several included as before 1500 or excluded as later are by no means certain.

3 Thirty, all those discovered by 1976, are illustrated and discussed in R. A. Skelton – P. D. A. Harvey (Ed.), Local Maps and Plans from Medieval England, Oxford 1986, a further five are included in the list in Harvey, Local and regional cartography, 498-499, and to these should be added a diagram of Cistercian abbeys near Chester, c. 1195 (M.V. Taylor [Ed.], Liber Luciani de laude Cestrie [Lancashire and Cheshire Record Society 64], 1912, frontispiece, p. 59), a map of Pinchbeck Fen, Lincolnshire, mid-15th century (R. Mitchell – D. Crook, The Pinchbeck Fen Map – A Fifteenth-Century Map of the Lincolnshire Fenland, in: Imago Mundi 51 [1999], 40-50, plates 3, 4), and two further maps that may date from before 1500, of part of Romney Marsh, Kent (Canterbury Cathedral Archives DCc/DE 176), and of lands in Stanlow, Cheshire (Kew, The National Archives, E 315/427, ff.29v-30r). I am grateful to Dr A.S. Bendall, Miss Elizabeth Danbury, Dr Maureen Jurkowski, and Dr M.M.N. Stansfield for information on these further items. In addition a date before 1500 seems likely for two building plans that have come to light; see note 16 below.

in 1970 we can now add several others that have been published, most recently an extraordinarily precocious plan of some 200 parcels of land at Fleury-sur-Orne in Normandy in 1477.[4] Systematic search for these maps in French repositories would probably add greatly to our understanding of medieval local maps in general. The same might be said of Italy, where there are almost certainly more medieval local maps than in any other area of Europe; those so far published can do no more than scratch the surface.[5] In contrast, not a single one has been recorded from Wales or Scotland or from the whole of Scandinavia nor, until the very end of the fifteenth century, from Spain or Portugal.[6] In 1991-1993 extensive enquiries of archivists and curators in Spain and Portugal and of historians who worked on medieval records there produced an interesting document showing that King Alfonso X of Castile had a map drawn in settling a dispute between the towns of Toledo and Talavera in 1262.[7] They produced, however, no reference at all to any surviving map; many of those who replied to these enquiries were sure such local maps must exist, but had never actually seen one.

Given our own familiarity with maps it is wholly reasonable for anyone today to suppose that many local maps were in fact drawn in the Middle Ages; it is just that they have failed to survive. This is not the case, as we see clearly when we look at the evidence from England. The private records that were most likely to be kept in the Middle Ages, and to have survived to the present, are those that define or confirm individual rights, in particular rights to landed

---

4 F. de Dainville, Cartes et contestations au XVe siècle, in: Imago Mundi 24 (1970), 99-121; T. Jarry, Autour d'un plan mediéval normand – le plan parcellaire d'Allemagne (Fleury-sur-Orne) de 1477, in: Histoire et Société Rurales 23 (2005), 169-204.

5 As in Harvey: Local and regional cartography, 476-482, 488, 498.

6 M. T. López Fernández – M. J. Sánchez Carrasco, Archivo de la Real Chancilleria de Valladolid – Catalogo de planos y dibujos del Pais Vasco, Madrid 1990, pp. 81-82, describes a map of 1487-97 covering an area of about 100 km. in each direction between Haro (Rioja) and the sea (Planos y dibujos, Carpeta: 30, no. 448). At the 23rd International Conference on the History of Cartography, Copenhagen, 2009, Professor Pilar Chias displayed copies of local maps of 1487 and 1493 from the same archive (Planos y dibujos desglosados, 448, 264).

7 Memorial Histórico Español, 1 (1851), 196. – I am most grateful to Professor Francisco J. Hernández for telling me of this document.

property – and what can do this better than a map? More English medieval maps than we yet know will undoubtedly come to light, but the total number that exists is unlikely to be more than fifty. In contrast, we have well over 1200 cartularies, registers containing copies of documents that mostly concern title to lands;[8] in only twelve of these is there a map. We have many thousands of the original documents – charters and deeds – by which property was conveyed, documents authenticated by wax seals. These seals are fragile and vulnerable, and far more have been lost from existing documents than survive attached to them. Yet there are in England something like 400,000 private seals from before 1500. The figure is a guess, but it is an informed guess and the order of magnitude is certainly correct. Household accounts, the records of what was received, what was consumed or paid out in a noble's household, had no value as evidence of title and can have been preserved only by chance; even so, some 500 survive from medieval England.[9] When we look at these figures and others that could be brought to bear, it is clear that the reason we have only fifty local maps at most is not their failure to survive: they were never drawn. And if this was the case in England, it was true also of other parts of medieval Europe: local maps and plans were drawn only very occasionally and then as an innovative expedient.

Nevertheless, although there was hardly any contact between the individual mapmakers and although most of their productions can have been seen by very few, all these maps conform to the same pattern. Many consist of no more than names with lines that set out boundaries, roads, or rivers. But if a map includes any features above ground level it appears not in ground-plan but as a picture, showing it from the side or obliquely from above. When any more detailed picture of town or landscape is drawn as though from a height unattainable on the ground, we may see this as a map, for the beginnings of cartographic conventions are being brought into play – the artist is not simply copying exactly what he sees before him. These styles correspond to a well-defined stage in the development of topog-

8 G. R. C. Davis, Medieval Cartularies of Great Britain, London 1958.

9 C. M. Woolgar (Ed.), Household Accounts from Medieval England (Records of Social and Economic History, new series, 17, 18), 2 vols, Oxford 1992-1993, 1. 5-6; 2. 691-726.

raphical mapping. They allow room for great variety; local maps from medieval Europe differ enormously one from another. They may or may not be coloured, they may be the merest sketches or drawn with great care and elaboration, they may comprise no more than a few lines and names or they may be immensely detailed, the pictorial element may be no more than a roughly sketched church steeple, as at Clenchwarton (Norfolk) in about 1400, or it may be a picture of an entire tract of country, as on the 1468 map of the Lower Scheldt.[10] All, however, belong to this broad class of topographical map. Above all, none makes any attempt at consistent scale; the concept of exact scale, of representing distances on the ground by strictly proportionate lengths on the map, had yet to be introduced to post-Roman topographical mapping in Europe.

It is against this background that we should see the local maps of this kind that were drawn in German-speaking lands and neighbouring parts of central Europe in the Middle Ages. Very few are known. Setting aside for a moment the plan of Vienna with Bratislava that was copied in the mid-fifteenth century from an original of 1421-1422 and the 1492 printed map of the area around Nuremberg, the following are a complete list:

1. 1429-1430. Geneva, a plan in black ink of streets, walls, and some buildings, by Guillaume Bolomier. Turin, Archivio di Stato, Corte, Paesi, Geneva, categoria 1, mazzo 7, titulo 21. Reproduced and discussed by Matthieu de la Corbière: Le 'Rideau de fer' de Genève ou du bon usage du plan Bolomier, in: Patrimoine et Architecture 14-15 (December 2005), 8-17.[11]
2. ca 1430. Rottweil (Baden-Württemberg) with surrounds, a fully detailed coloured bird's-eye view showing the town besieged by the Emperor Lothar III, possibly copied from a late-fourteenth-century exemplar. Stuttgart, Württembergische Landesbibliothek, Hs. HB VI 110, front flyleaf verso and f.1 r. Manuscript fully described by Johanne Autenrieth: Die Handschriften der ehemaligen Hofbibliothek Stuttgart, 3: Codices iuridici et politici, Wiesbaden 1963, 108; reproduced in colour and discussed by Wolfgang Irtenkauf: Die Rottweiler Hofgerichtsordnung (um

---

10 Harvey, Local and regional cartography, 485, 488-499.

11 I am much indebted to Herr Ralph Ruch for knowledge of this map and for a copy of this article.

1430) (Göppinger Beiträge zur Textgeschichte 74), Göppingen 1981, [vi], [vii], 8-9.[12]

3. 1441. Honau (Baden-Württemberg) and Wantzenau (Bas-Rhin), a diagrammatic plan in black and red ink of estates of Honau Abbey. Strasbourg, Archives départementales du Bas-Rhin, G 4227(8). Reproduced, uncoloured, and briefly described by Franz Grenacher: Current knowledge of Alsatian cartography, in: Imago Mundi 18 (1964), 60-61.
4. 1464. Pomerania, two sketch maps in black ink of coastline, rivers, and settlements showing estates of the Teutonic Order. Krakow, Biblioteka Czartoryskich, MS. 1310, pp. 636, 637. Reproduced and discussed by Bolesław Olszewicz: Dwie szkicowe mapy Pomorza z Połowy XV wieku (Biblioteka 'Strażnicy Zachodniej' 1), Warsaw 1937.
5. 1473. Rattenberg (Tirol), two sketch maps in black ink of parts of neighbouring tributaries of the river Inn with a mountain range, bridges, and a gallows. Munich, Bayerisches Hauptstaatsarchiv, Plansammlung 20865 r and v. Reproduced and discussed by Gerhard Leidel: Die Anfänge der archivischen Kartographie im deutschsprachigen Raum, in: Archivalische Zeitschrift 85 (2003), 90-93, 103-109, 138, 139.
6. 1481. South-west Bavaria, five sketch maps in black ink of rivers, roads, and settlements in the area between Ulm, Günzburg, Mindelheim, Ottobeuren, and Memmingen. Munich, Bayerisches Hauptstaatsarchiv, Plansammlung 3907, 20783, 20784r,v, 20785. Reproduced and discussed by Leidel: Die Anfänge der archivischen Kartographie, 93-100, 103-109, 140-144.
7. 1488(?). 'Mihaloviz' (former Križevci county, northern Croatia), sketch map in black ink naming points along the boundary of this now lost village. Budapest, Magyar Orszáos Levéltár (National Archives of Hungary), DL 101050 (formerly Batthyány archives, Acta Ant. 4.8.290). Reproduced and discussed by Árpád Papp-Váry and Pál Hrenkó: Magyarország régi térképeken, Budapest 1989, 52 f.; discussed by Andrea Kiss: A contribution to research on the historical geography of the first extant 'reambulation' sketch from the Carpathian Basin, in:

12 I am grateful to Frau Popp-Grilli of the Württembergische Landesbibliothek Stuttgart who has kindly provided me with references to relevant work.

Zbornik Odsjeka za povijesne znanosti Zavoda za povijesne i društvene znanosti Hravatske akademije znanosti i umjetnosti 19 (2001), 127-141.[13]

8. 1492. South-west Bavaria, two sketch maps in black ink of rivers, roads, settlements, and some structures in the area between Ulm, Lauingen, Mindelheim, and Memmingen. Munich, Bayerisches Hauptstaatsarchiv, Plansammlung 3764, 10713. Reproduced and discussed by Leidel: Die Anfänge der archivischen Kartographie, 100-109, 145, 146.
9. 1496. Pfuhl (Bavaria), a fully detailed bird's-eye view of river with wooded island, fields, village, and other buildings. Stuttgart, Hauptstaatsarchiv, C 3 Bü 1336, ohne Q. Reproduced in colour and discussed by Thomas Horst: Die älteren Manuskriptkarten Altbayerns (Schriftenreihe zur Bayerischen Landesgeschichte 161), vol. 1, Munich 2009, 33 and 215.[14]
10. 1497-1502. Zurich, a fully detailed coloured bird's-eye view of the city on five panels, attributed to Hans Leu the elder. Zurich, Schweizerisches Landesmuseum, AG 7.1-3, AG 8.1-2. Reproduced, uncoloured, and discussed by Lucas Wüthrich and Mylène Ruoss: Katalog der Gemälde – Schweizerisches Landesmuseum Zürich, Zurich 1996, 40-43; reproduced in colour at http://commons.wikimedia.org/wiki/File:Hans_Leu_Altartafeln_links.jpg and http://commons.wikimedia.org/wiki/File:Hans_Leu_Altartafeln_rechts.jpg.[15]

Others will almost certainly come to light, as well as references to maps that have failed to survive. The map of the northern provinces of Poland that was given to Pope Martin V in 1421 can hardly have been a local map in our definition, but the plan of a watercourse in Tolna county in southern Hungary, mentioned in a charter of 1415,

---

13 Unfortunately the accompanying illustration, said to be this map (p. 139), is in fact a 15th-century map of Staines (Middlesex) in England.

14 I am grateful to Dr Thomas Horst for telling me of this map and its publication.

15 I am grateful to Dr Mylène Ruoss of the Schweizerisches Landesmuseum at Zurich who has kindly provided me with catalogue entries and references.

certainly was.[16] The surviving eighteenth-century copy of a 1316 report on the bounds of Reszege (Szabolcs-Szatmár) in north-east Hungary includes semi-cartographic marginal sketches that we may suppose were copied from the original document.[17] Some maps that might once have been added to our list must now be omitted. The report of a plan of Ulm from about 1480 can only refer to a later map and two maps of Böhringsweiler (Baden-Württemberg) said to date from about 1500 have now been redated to the early seventeenth century on both palaeographical and historical grounds.[18]

In a valuable article in 2005, Gerhard Leidel published for the first time the nine maps in the Bayerisches Hauptstaatsarchiv at Munich that form a substantial part of our entire corpus (nos 5, 6, and 8). He sets them firmly in their administrative and political context, relating them to other records of the boundary problems that brought them into being. Beyond this, he discusses at length – but in the abstract, without reference to other comparable maps – what these maps show of the way landscape was perceived, how it was represented graphically and how this was affected by the thought of the Renaissance. Important as they are, these maps might seem a slender peg from which to hang what amounts to a general discursus on the concepts underlying the early development of topographical mapping, but Leidel makes some helpful and thought-provoking points: the significance of the distinction between perspective and

16 K. Buczec, The History of Polish Cartography from the 15th to the 18th Century, trans. Andrzej Potocki, Amsterdam ^{2}1982, 22-24; Á. Papp-Váry – P. Hrenkó, Magyarország régi térképeken, Budapest 1989, 53; A. Kiss, A contribution to research on the historical geography of the first extant 'reambulation' sketch from the Carpathian Basin, in: Zbornik Odsjeka za povijesne znanosti Zavoda za povijesne i društvene znanosti Hravatske akademije znanosti i umjetnosti 19 (2001), 127-141, here 128.

17 Z. G. Török, Renaissance cartography in East-Central Europe, ca. 1450-1650, in: D. Woodward (Ed.), The History of Cartography 3: Cartography in the European Renaissance, Chicago 2007, 1806-1851, here 1814-1815.

18 R. Oehme, Die Geschichte der Kartographie des deutschen Südwestens, Konstanz 1961, 97; K. Schumm, Inventar der handschriftlichen Karten im Hohenlohe-Zentralarchiv Neuenstein, Karlsruhe 1961, 5. For the revision of these dates I am grateful for the kind assistance and advice of Dr Moegle-Hofacker of Hohenlohe-Zentralarchiv Neuenstein, Dr Schiffer of Hauptstaatsarchiv Stuttgart, and Dr Weig of Stadt Ulm Stadtarchiv; the maps are mentioned, with dates unrevised, in Harvey: Local and regional cartography, 488.

isometric drawing, how towers and mountains enabled landscape to be viewed in near-cartographic form, how, while demonstrably not drawn to scale, the maps show a sense of proportion that one might call an intuitive scale. Consisting, as they do, simply of the lines of rivers and roads with little wording beyond place-names and with only a few features shown pictorially, the maps anticipated neither the artistic nor the scientific approach to representing landscape (*vorkünstlerisch*, *vorwissenschaftlich*), and might be described as protocartography.[19]

Leidel is right to emphasise that what he is putting before us is the early development of 'archival cartography', the functional use of maps in the practical work of government, jurisdiction, and private administration. This was a crucially important aspect of the steadily growing acceptance of maps in the course of the sixteenth century, of their use for every possible purpose. We should, however, make a careful distinction. Because they played such an important part in this process it is helpful to distinguish these archival maps from the rest when we look at the way awareness of maps spread in this period. However, when we look at the way the maps themselves developed there is no reason to separate the archival maps from any others. We best understand the development of local maps and plans, their perception of landscape, their techniques of presentation, when we look at them all together, archival and non-archival; there is no obvious difference in their form or style, or in the concepts that underlay their creation.

One might reasonably question this in looking at our list of maps from German-speaking and neighbouring lands. Taking our first two items, there is the world of difference between the 1429-1430 map of Geneva, with its scratchy ink lines and rough sketches of the town wall and a few buildings, and the view of Rottweil of about 1430, carefully drawn and elaborately painted with full details of buildings inside the wall and of the castle, villages, and landscape outside. We might well suppose that from Geneva we have the sort of plan that was produced for functional administrative use, and from Rottweil a style of presentation used for other purposes. However, this distinc-

---

[19] G. Leidel, Die Anfänge der archivischen Kartographie im deutschsprachigen Raum, in: Archivalische Zeitschrift 85 (2003), 85-146, especially 105-106, 110, 129-131.

tion breaks down when we look at the 1496 map of Pfuhl, made to be shown to a court of justice and it breaks down too when we look at early local maps from other parts of Europe. For instance, a 1460 map of villages on the boundary between France and Burgundy gives us a finely drawn and coloured view of a stretch of landscape, and so too does a fifteenth-century map of Inclesmoor in England.[20] Yet both were drawn for practical administrative purposes; in both origin and context they are unquestionably archival maps. It is worth remembering that while the view of Rottweil may well have been copied from an earlier work of unknown purpose, it is in a functional context that it has come down to us, in a copy of the ordinances of Rottweil's imperial court of justice (*Hofgericht*).[21]

With only these ten maps and plans (or closely related groups) our list may seem meagre compared with those from the Low Countries and England, or with those that surely survive in Italy and France. This, however, is because we have so far been looking only at the manuscript diagrams or picture-maps typical of medieval local maps in other parts of Europe. If we look at three other types of map or plan, rare or non-existent elsewhere, we find that our area makes a particularly important contribution to our European corpus of medieval local maps. These three other types of map partly, and significantly, overlap; they are not three isolated groups, though we shall look at each in turn. They are building plans, maps drawn to consistent scale, and printed maps.

Building plans are almost unknown elsewhere in medieval Europe. From France we have the sketches in the thirteenth-century notebook of Villard de Honnecourt; some may be of vaultings, but it makes little difference conceptually whether what we have is an isometric roof plan viewed from below or a ground plan viewed from above.[22] From England only one building plan can be firmly dated before the sixteenth century, part of Winchester College (Hampshire) from about 1390, but to this we can probably add two others that

20 The French map is reproduced in Harvey, Local and regional cartography, plate 36, the map of Inclesmoor in Skelton – Harvey (Ed.), Local Maps and Plans, plate 12 (ii).

21 Edited, with facsimile, by W. Irtenkauf, Die Rottweiler Hofgerichtsordnung (um 1430) (Göppinger Beiträge zur Textgeschichte 74), Göppingen 1981.

22 H. R. Hahnloser (Ed.), Villard de Honnecourt - kritische Gesamtausgabe des Bauhüttenbuches, Graz, ²1972.

consist solely of lines without wording.[23] We know, however, that in late-medieval England it was not unusual to draw building plans which the architect would show his client, for they are occasionally mentioned in building contracts from 1380 onwards.[24] The Winchester building plan is drawn not to scale but roughly to shape, and measurements are written in. Surviving English building plans of the early sixteenth century are drawn in the same way and this was probably the normal form of the plans drawn in England until the 1540s.[25]

The contrast with central Europe is striking. From Vienna we have an extraordinary collection of nearly 500 elevations, ground plans, and plans of vaulting, for entire churches, for parts of buildings, and for architectural details. They are beautifully drawn, with great precision, and nearly all are to exact (though undefined) scale. Most come from the archives of St Stephen's Cathedral, but the buildings they show are not only in Vienna, for the masons centred on St Stephen's worked in many other places as well and the collection anyway includes plans from other centres, presumably brought by masons who moved from one site to another. Places represented among the drawings thus include Steyr, Bratislava, Prague, Regensburg, Munich, Nuremberg, Constance, Strasbourg, Mainz, and Cologne. Besides this outstandingly important collection, other smaller collections and individual building plans have survived. They are unparalleled in medieval Europe and are of great significance for the history of cartography.[26]

Three points about them are of particular interest. One is that they were drawn by or for the masons themselves and were kept and presumably used by them as a practical tool of their craft. There is no reason to suppose that they were shown to the masons' clients or

---

23 The Winchester plan is identified and discussed by J. H. Harvey in Skelton – Harvey (Ed.), Local Maps and Plans, 141-146. Of the other two, one (London, Society of Antiquaries, MS. 252*) has been identified by J. A. A. Goodall, The earliest English architectural drawing? in: Country Life (15. Nov. 2001), 70-71, as part of Eton College (Buckinghamshire) in about 1449, the other (Kew, The National Archives, E 101/509/19) remains unidentified. I am grateful to Miss Margaret M. Condon for showing me the latter.

24 L. F. Salzman, Building in England down to 1540, Oxford 1952, 17-20, 459-460.

25 P. D. A. Harvey, Maps in Tudor England, London 1993, 95-101.

26 H. Koepf, Die gotische Planrisse der Wiener Sammlungen, Wien 1969.

other members of the public – indeed, they may well have been closely guarded as a secret technique. They are thus quite unlike the plans that English masons made to show their clients. The second point is that the central European building plans, like the English, appear quite late on the scene: the earliest seem to date from the early fifteenth century. We might of course suppose that what we have are merely the first survivors of a much older tradition, that the architects of the Romanesque and early Gothic periods had plans, now lost, of the great churches they built. This is possible, but we should be wary of any historical argument that says 'there must have been'. We know very little of early-medieval building techniques and in default of firm evidence we may reasonably be sceptical of the supposed universal necessity for the drawn building plan; even the masterpieces of early-medieval architecture may well have been built without. The third point is that the earliest elevations in the Vienna collections, dating from the second half of the fourteenth century, are perhaps half a century older than the earliest ground plans. This is interesting but may or may not be significant. The elevations, like the plans, are isometric, drawn to a consistent scale; this, the concept of reproducing in miniature in a constant proportion, is what is really important and it scarcely matters whether the technique is applied to one plane or to another, to the horizontal or to the vertical.

But the building plans were not the only local maps or plans to be drawn to scale in our area. A mid-fifteenth-century plan of Vienna, with a plan of Bratislava in one corner, has a scale-bar of 100 paces or 200 feet, though neither plan actually accords with it. It has been argued on one hand that what we have is a copy of an early-fifteenth-century map to which the scale-bar was added by way of false refinement; and on the other, most ingeniously, that the scale-bar applies only to the distances between suburban churches on what was primarily a route-map for pilgrims coming from Bratislava or beyond.[27] Either way, the map is a quite extraordinary production; it is the concept of consistent scale rather than its achievement that is

---

27 S. Wellisch, Der älteste Plan von Wien, in: Zeitschrift des Oesterreichischen Ingenieur- und Architekten-Vereines 50 (1898), 757-761; M. Kratochwill, Zur Frage der Echtheit des „Albertinischen Planes" von Wien, in: Jahrbuch des Vereins für Geschichte der Stadt Wien 29 (1973), 7-36; R. Härtel, Inhalt und Bedeutung des „Albertinischen Planes" von Wien, in: Mitteilungen des Instituts für Österreichische Geschichtsforschung 87 (1979), 337-362.

so remarkable. It is the first known land map to have a scale-bar, and local maps were first drawn to a consistent scale in the Low Countries only in the 1530s and in England not until about 1541.[28] From Italy the earliest city plan clearly drawn to scale is Leonardo da Vinci's plan of Imola in 1502-1503, but we have some hints of an older Italian tradition of scale-plans of cities: a fourteenth-century plan of Venice that arguably derives from a twelfth-century original drawn to scale, a letter of 1377-1381 that mentions measurements shown on a plan of Florence, fifteenth-century regional maps from northern Italy that are not themselves drawn to scale but include what are, again arguably, tiny scale-plans of cities (one of Verona is particularly striking). In view of this we may well look to Italy for the origins of the map of Vienna, just as much later, in the 1490s, there was probably Italian influence on Konrad Türst's map of Switzerland.[29]

But for the plan of Vienna, just as for Türst's map, we can also look for influences closer at hand, within the German-speaking area. It has been questioned whether the geographical calculations by the early-fifteenth-century monks of Klosterneuburg were made in order to produce maps, but there can be no doubt that by the middle of the century German scholars were studying the geographical positions of places with a view to making small-scale maps of large areas. We have tangible evidence in the Koblenz and Trier fragments of manuscript maps of Germany and, probably, in the maps of Germany, one manuscript and one printed, that are attributed to Nicholas de Cusa; the plan of Vienna may well owe its origin to scholars' interest in calculation and measurement.[30] Certainly we find

---

28 P. S. Teeling, Oud-Nederlandse landmeters, IV, in: Orgaan der Vereniging van Technische Ambtenaren van het Kadaster 7 (1949), 158-170, here 160-161; Harvey, Maps in Tudor England, 31-32.

29 Harvey, Local and regional cartography, 478-479, 498.

30 A. Wolkenhauer, Über die ältesten Reisekarten von Deutschland aus dem Ende des 15. und dem Anfange des 16. Jahrhunderts, in: Deutsche Geographische Blätter 26 (1903), 120-138, here 124-128; D. B. Durand, The Vienna-Klosterneuburg map corpus of the fifteenth century, Leiden 1952; E. Bernleithner, Die Klosterneuburger Fridericuskarte von etwa 1421, in: K.-H. Meine (Ed.), Kartengeschichte und Kartenbearbeitung, Bad Godesberg 1968, 41-44; F. Bönisch, Bemerkungen zu den Wien-Klosterneuburg-Karten des 15. Jahrhunderts, in: K.-H. Meine (Ed.), Kartengeschichte und Kartenbearbeitung, 45-48; P. H. Meurer, Cartography in the German lands, 1450-1650,

the connection between local maps and maps of large areas in the work of Erhard Etzlaub at the end of the century.[31]

Etzlaub produced two printed versions of a map – measured and drawn to scale – of the whole of central Europe as far south as Rome; one version is entitled 'Das ist der Rom Weg'. But in 1492 he also produced a local map, printed by Georg Glockenden, a circle of 16 German miles around Nuremberg, where he lived and worked. This was drawn to scale: it has a scale-bar and below it are instructions explaining how to use this to measure distances on the map with a pair of dividers. The principle of scale and how to use it needed to be explained – but in England the same sort of explanation was needed nearly a hundred years later, in the 1580s. Here we have a clear connection between the general map and the local map; they were constructed by the same man using the same principles, the same concepts and even the same methods, a point made by the verse above the local map:

Ob yemant sprech oder Im selbs trachtet,
Aus wem oder wie man das machet,
Die kunst Geometrey ist genant,
Thut vns des ertreichs leng vnd brait bekant.

Etzlaub was a man of learning, an instrument-maker and a physician. But he was also a sworn surveyor, who undertook various surveys of property for the city authorities of Nuremberg; a map of the imperial forests in the area in 1516 may or may not have been his, but he certainly surveyed a newly bought estate in 1507 and drew official maps of the Nuremberg area in 1516 and 1519. In Etzlaub Germany achieved the union of surveyor and mapmaker that lies behind the whole of the later history of topographical mapping, a union achieved in the Low Countries only in the 1530s and in England not until the 1570s, when surveyors began to produce maps

in: D. Woodward (Ed.), History of Cartography 3, 1172-1245, here 1177-1180, 1183-1188.

31 Etzlaub's debt to Klosterneuburg is clearly brought out by A. Höhn, Franken in der Nürnberg-Karte Etzlaubs und die Daten des Codex Latinus Monacensis 14583, in: Speculum Orbis 3 (1987), 2-8.

instead of just written descriptions of the estates they had surveyed and measured[32].

Etzlaub's map inspired one other local map drawn to scale probably before 1500: a manuscript map, again of the Nuremberg area but without roads, that is in the Landesbibliothek at Coburg.[33] However, it is not only in being drawn to scale that Etzlaub's 1492 map is so interesting: it is no less distinctive in being printed and indeed, apart from the little group of bird's-eye city views that Francesco Rosselli produced at Florence about 1480, it is the earliest printed local map from anywhere in Europe.[34] The earliest printed local map from the Low Countries was published in 1533, the earliest from England in 1559, and these occur in treatises on geography, geometry, and map-making; they are not separately printed maps.[35] Nor was Etzlaub's map an isolated production, a flash in the pan; it soon had successors. One was a printed picture-map, made up from six wood blocks, of the area around Lake Constance; it bears the monogram of the unidentified PW or PPW and has been dated 1505. It may be earlier, but not much as it includes scenes from the Swabian War in 1499.[36] By the time Martin Waldseemüller's map of the upper Rhine was published in 1513, let alone Sebastian Münster's map of the area around Basel in 1538, there was ample precedent in Germany for printing local maps – but a precedent that could be found nowhere else in Europe.

In other parts of Europe that produced local maps in the Middle Ages we find that their geographical distribution was not always even. All the maps from the Low Countries come from a relatively narrow strip along the coast, from Hilversum to Bruges. In England about a quarter of the maps – a quite disproportionate number –

---

32 F. Schnelbögl, Life and work of the Nuremberg cartographer Erhard Etzlaub († 1532), in: Imago Mundi 20 (1966), 11-26; F. Schnelbögl, Dokumente zur Nürnberger Kartographie mit Katalog der Ausstellung anlässlich des 15. Kartographentages der Deutschen Gesellschaft für Kartographie (Beiträge zur Geschichte und Kultur der Stadt Nürnberg 10), Nürnberg 1966, 48, 56, 58; Meurer, Cartography in the German lands, 1194-1195.

33 Schnelbögl, Life and work, 23.

34 Harvey, Local and regional cartography, 477.

35 G. Frisius, Libellus de locorum describendorum ratione, Antwerpen 1533; W. Cuningham, The cosmographical glasse, London 1559.

36 W. Bonacker, Die sogenannte Bodenseekarte des Meisters PW bzw. PPW vom Jahre 1505, in: Die Erde 6 (1954), 1-29.

come from a small area in eastern England. There is no clear reason for this, and in neither case can we speak of a local tradition or style of cartography, for the individual maps vary greatly in form, size, and elaboration. All we can say is that in these limited regions the advantages of maps seem to have been better appreciated than elsewhere. Can we see any similar concentration in German-speaking lands and central Europe? It seems that we can. From central Germany and the Baltic we have nothing but the Teutonic Order's sketch maps of 1464 and apart from a few of the building plans from places as far north as Cologne all the other maps and plans come from the southern part of our area.

However, we should be cautious in concluding that it was only here that people began to see landscape in a cartographic or near-cartographic way and that the Teutonic Order's maps of Pomerania must have been the work of a member of the Order from much further south. At two points the world map that was drawn at Ebstorf (Niedersachsen) in about 1300 clearly draws on immediate local knowledge; in effect it incorporates two local maps within a world picture drawn from many different sources. One is of Reichenau on Lake Constance with its three monastic cells, but the other is of the area around Ebstorf itself, with its representations of Lüneburg, Braunschweig, and Ebstorf and its three martyrs' graves.[37] Here at least, the mapmaker saw the map as showing what was actually on the ground; it was not a remote exercise in learned scholarship. And there is a further source of medieval local maps that historians of cartography have scarcely touched on: the designs of medieval seals.[38] A town's seal might bear a picture of its lord, or some symbol that represented the town, or a simple picture of town walls with or without a distinctive feature of the actual place. But sometimes it presents an oblique view, looking across the walls to a recognisable building within the town; it is in our terms a local map, drawing on near-cartographic conventions. In Toni Diederich's magisterial account of town seals from the Rhineland those of Boppard and Deutz, both from about 1230, and of Siegburg from about 1285

---

37 H. Kugler (Ed.), Die Ebstorfer Welkarte, 2 vols, Berlin 2007, 1. 118-119, 128-131; 2. 61-63, 69, 251, 253, 279-280, 284.

38 I know of only the view of Rome on the golden bull of Ludwig IV that has been considered in this context (P. D. A. Harvey, The History of Topographical Maps, London 1980, 74).

could all be called local maps.[39] Nor is this the only way we see the development of cartography on medieval seals. The thirteenth-century seal of Innsbruck shows a plank bridge on three piers across, presumably, the River Inn, and shows it in plan from above; conceptually it is pure cartography.[40] There is probably much to be learned of early cartography from medieval seal designs, not only in our area but in other parts of Europe too.[41]

Indeed, all that can be offered here is a provisional picture to show the main outlines of development. All work on local maps is best done by scholars living and working in the particular country, even the particular region. Access to local and private archives, to the research of local historians, to journals and other works produced primarily for local circulation, is all most efficiently achieved by those who are on the spot. This article will have fully succeeded if it leads others to carry its work forward to new discoveries, new interpretations.

Bibliography

Autenrieth J., Die Handschriften der ehemaligen Hofbibliothek Stuttgart, 3: Codices iuridici et politici, Wiesbaden 1963.

Bernleithner E., Die Klosterneuburger Fridericuskarte von etwa 1421, in: K.-H. Meine (Ed.), Kartengeschichte und Kartenbearbeitung, Bad Godesberg 1968, 41-44.

Bonacker W., Die sogenannte Bodenseekarte des Meisters PW bzw. PPW vom Jahre 1505, in: Die Erde 6 (1954), 1-29.

Bönisch F., Bemerkungen zu den Wien-Klosterneuburg-Karten des 15. Jahrhunderts, in: K.-H. Meine (Ed.), Kartengeschichte und Kartenbearbeitung, Bad Godesberg 1968, 45-48.

Buczec K., The History of Polish Cartography from the 15th to the 18th Century, trans. Andrzej Potocki, Amsterdam ^{2}1982.

39 T. Diederich, Rheinische Städtesiegel (Rheinischer Verein für Denkmalpflege und Landschaftsschutz Jahrbuch 1984/5), Neuss 1984, 100-107, 198-203, 208-210, 325-328, colour plates 2, 7, black-and-white plates 20, 31, 86.

40 K. Rickelt – C. F. Rickelt, Wappenbuch der Städte und Märkte von Tirol, Innsbruck 1894, 53 and 90.

41 It is interesting, for instance, to compare the ways towns are shown on seals and on maps of large areas. The seal engraver and the mapmaker had the same problem: how to represent a town in a very small space.

Corbière M. de la, Le 'Rideau de fer' de Genève ou du bon usage du plan Bolomier, in: Patrimoine et Architecture 14-15 (December 2005), 8-17.

Cuningham W., The Cosmographical Glasse, London 1559.

Dainville F. de, Cartes et contestations au XVe siècle, in: Imago Mundi 24 (1970), 99-121.

Davis G. R. C., Medieval Cartularies of Great Britain, London 1958.

Diederich T., Rheinische Städtesiegel (Rheinischer Verein für Denkmalpflege und Landschaftsschutz Jahrbuch 1984/1985), Neuss 1984.

Durand D. B., The Vienna-Klosterneuburg map corpus of the fifteenth century, Leiden 1952.

Fleischmann P., Die handgezeichneten Karten des Staatsarchivs Nürnberg bis 1806 (Bayerische Archivinventare 49), München 1998.

Frisius G., Libellus de locorum describendorum ratione, Antwerpen 1533.

Goodall J. A. A., The earliest English architectural drawing? in: Country Life 15 (Nov. 2001), 70-71.

Grenacher F., Current knowledge of Alsatian cartography, in: Imago Mundi 18 (1964), 60-77.

Hahnloser H. R. (Ed.), Villard de Honnecourt – kritische Gesamtausgabe des Bauhüttenbuches, Graz ^{2}1972.

Härtel R., Inhalt und Bedeutung des „Albertinischen Planes“ von Wien, in: Mitteilungen des Instituts für Österreichische Geschichtsforschung 87 (1979), 337-362.

Harvey P. D. A., The History of Topographical Maps, London 1980.

Harvey P. D. A., Local and regional cartography in medieval Europe, in: Harley J. B. – Woodward D. (Ed.), The History of Cartography 1: Cartography in Prehistoric, Ancient and Medieval Europe and the Mediterranean, Chicago 1987, 464-501.

Harvey P. D. A., Maps in Tudor England, London 1993.

Höhn A., Franken in der Nürnberg-Karte Etzlaubs und die Daten des Codex Latinus Monacensis 14583, in: Speculum Orbis 3 (1987), 2-8.

Horst T., Die älteren Manuskriptkarten Altbayerns. Eine kartographiehistorische Studie zum Augenscheinplan unter besonderer Berücksichtigung der Kultur- und Klimageschichte (Schriftenreihe zur bayerischen Landesgeschichte 161), Bd. 1-2, München 2009.

Irtenkauf W. (Ed.), Die Rottweiler Hofgerichtsordnung (um 1430) (Göppinger Beiträge zur Textgeschichte 74), Göppingen 1981.

Jarry T., Autour d'un plan mediéval normand – le plan parcellaire d'Allemagne (Fleury-sur-Orne) de 1477, in: Histoire et Société Rurales 23 (2005), 169-204.

Kiss A., A contribution to research on the historical geography of the first extant 'reambulation' sketch from the Carpathian Basin, in: Zbornik

Odsjeka za povijesne znanosti Zavoda za povijesne i društvene znanosti Hravatske akademije znanosti i umjetnosti 19 (2001), 127-141.
Koepf H., Die gotische Planrisse der Wiener Sammlungen, Wien 1969.
Kratochwill M., Zur Frage der Echtheit des „Albertinischen Planes“ von Wien, in: Jahrbuch des Vereins für Geschichte der Stadt Wien 29 (1973), 7-36.
Krausen E., Die handgezeichneten Karten im Bayerischen Hauptstaatsarchiv sowie in den Staatsarchiven Amberg und Neuburg a. d. Donau bis 1650 (Bayerische Archivinventare 37), Neustadt a. d. Aisch 1973.
Krogt P. van der, Lokale kaarten van Nederland uit de late Middeleeuwen, in: Caert-Thresoor 27 (2008), 29-42.
Kugler H. (Ed.), Die Ebstorfer Weltkarte, 2 vols, Berlin 2007.
Leidel G., Die Anfänge der archivischen Kartographie im deutschsprachigen Raum, in: Archivalische Zeitschrift 85 (2003), 85-146.
López Fernández M. T. – Sánchez Carrasco M. J., Archivo de la Real Chancilleria de Valladolid – Catalogo de planos y dibujos del Pais Vasco, Madrid 1990.
Memorial Histórico Español, 1 (1851).
Meurer P. H., Cartography in the German lands, 1450-1650, in: Woodward D. (Ed.), The History of Cartography, 3: Cartography in the European Renaissance, Chicago 2007, 1172-1245.
Mitchell R. – Crook D., The Pinchbeck Fen Map – A Fifteenth-Century Map of the Lincolnshire Fenland, in: Imago Mundi 51 (1999), 40-50, plates 3, 4.
Oehme R., Die Geschichte der Kartographie des deutschen Südwestens, Konstanz 1961.
Olszewicz B., Dwie szkicowe mapy Pomorza z Poɫowy XV wieku (Biblioteka 'Strażnicy Zachodniej' 1), Warszawa 1937.
Papp-Váry A. – Hrenkó P., Magyarország régi térképeken, Budapest 1989.
Rickelt K. – Rickelt C. F., Wappenbuch der Städte und Märkte von Tirol, Innsbruck 1894.
Salzman L. F., Building in England down to 1540, Oxford 1952.
Schnelbögl F., Life and work of the Nuremberg cartographer Erhard Etzlaub († 1532), in: Imago Mundi 20 (1966), 11-26.
Schnelbögl F., Dokumente zur Nürnberger Kartographie mit Katalog der Ausstellung anlässlich des 15. Kartographentages der Deutschen Gesellschaft für Kartographie (Beiträge zur Geschichte und Kultur der Stadt Nürnberg 10), Nürnberg 1966.
Schumm K., Inventar der handschriftlichen Karten im Hohenlohe-Zentralarchiv Neuenstein, Karlsruhe 1961.
Skelton R. A. – Harvey P. D. A. (Ed.), Local Maps and Plans from Medieval England, Oxford 1986.

Taylor M. V. (Ed.), Liber Luciani de laude Cestrie (Lancashire and Cheshire Record Society 64), 1912.

Teeling P. S., Oud-Nederlandse landmeters, IV, in: Orgaan der Vereniging van Technische Ambtenaren van het Kadaster 7 (1949), 158-170.

Török Z. G., Renaissance cartography in East-Central Europe, ca. 1450-1650, in: Woodward D. (Ed.), The History of Cartography, 3: Cartography in the European Renaissance, Chicago 2007, 1806-1851.

Wellisch S., Der älteste Plan von Wien, in: Zeitschrift des Oesterreichischen Ingenieur- und Architekten-Vereines 50 (1898), 757-761.

Winkler R., Die handgezeichneten Karten des Staatsarchivs Bamberg bis 1780 (Bayerische Archivinventare 48), München 2005.

Wolkenhauer A.,. Über die ältesten Reisekarten von Deutschland aus dem Ende des 15. und dem Anfange des 16. Jahrhunderts, in: Deutsche Geographische Blätter 26 (1903), 120-138.

Woolgar C. M. (Ed.), Household Accounts from Medieval England (Records of Social and Economic History, new series, 17, 18), 2 vols, Oxford 1992-1993.

Wüthrich L. – Ruoss M., Katalog der Gemälde – Schweizerisches Landesmuseum Zürich, Zürich 1996.

About the author

Paul D. A. Harvey, born 1930, is Professor Emeritus of Medieval History in the University of Durham, England. His career has been divided between work as a curator of archives and manuscripts and university appointments in medieval social and economic history, but alongside each he has maintained a continuing interest in the history of maps. His first book was a catalogue of the printed county maps of Warwickshire, but later publications have been on the early history of topographical maps, in England and in general, and on the cartography of the Middle Ages. Most recently he has been working on the medieval maps of Palestine, few in number but of great interest.

Contact: pda.y.harvey@btopenworld.com

Georg Matthäus Vischer,
Archiducatus Austriae inferioris
accuratissima geographica descriptio,
Augsburg 1670

Otto Hat, Groß-Enzersdorf bei Wien

Die kartographische Darstellung des Weinviertels[1]

Unter besonderer Berücksichtigung der Niederösterreichkarte von Georg Matthäus Vischer und der ersten beiden Landesaufnahmen

Eine brauchbare kartographische Erfassung Niederösterreichs erfolgte erst im 17. Jahrhundert durch einen Pfarrer aus Oberösterreich. Die Landstände Niederösterreichs zeigten allerdings erst ihr Interesse, als sie von der gut gelungenen Karte von Oberösterreich erfuhren und Graf Starhemberg Georg Matthäus Vischer protegierte. Dieses Kartenwerk, als Vischer-Karte bekannt geworden, wurde von privaten Verlegern noch lange Zeit hindurch kopiert und in verkleinerter Form veröffentlicht.

Neben den Ständen verlangte vor allem das Militär nach kartographischem Material, denn die Geländeerkundung, ein wichtiger Bestandteil der Kriegsführung, war mühsam, zeitraubend und oft unzulänglich. So kam es unter Kaiserin Maria Theresia zur ersten

---

1 Eine ausführliche Behandlung des Themas findet sich meiner Diplomarbeit, die von Prof. Dr. Johannes Dörflinger betreut wurde (Wien 2004)

Landesaufnahme. Diese, nach Joseph II. benannte Josephinische Landesaufnahme, war jedoch aus militärisch-strategischen Gründen geheim, sodass es von Niederösterreich nur zwei Exemplare in Österreich gibt (im Österreichischen Staatsarchiv und in der Österreichischen Nationalbibliothek).

Die zweite, vermessungstechnisch verbesserte, Landesaufnahme erfolgte unter Kaiser Franz II. (I.), sie wurde daher auch Franziszeische Landesaufnahme genannt. Allerdings ist von dieser nur ein Exemplar in der Kartensammlung des Kriegsarchivs im Österreichischen Staatsarchiv vorhanden (wie die Josephinische Landesaufnahme handgezeichnet und koloriert).

Im Folgenden sollen die Inhalte der oben genannten Karten mit der heutigen kartographischen Situation sowie die beiden Ausgaben der Vischerkarte (1670 und 1697) mit der ersten und zweiten Landesaufnahme verglichen werden.

Lebenslauf des Georg Matthäus Vischer (1628-1696)[2]

Georg Matthäus Vischer wurde am 22. April 1628 in Wenns (Tirol) als Bauernsohn geboren. Er besuchte die Stiftsschule in Stams, nahm aktiv am Dreißigjährigen Krieg teil, bei dem er sich wahrscheinlich seine kartographischen Kenntnisse erwarb, und studierte Theologie.

Als Kaplan war er ab 1654 in verschiedenen oberösterreichischen Orten tätig, ab 1666 als Pfarrer in Leonstein (Oberösterreich). Schon während seiner Zeit als Kaplan beschäftigte er sich mit der Kartographie und fertigte seine erste Karte für den Grafen von Salburg zwischen 1662 und 1666 an.

Vischer bot den oberösterreichischen Ständen an, für sie eine Karte des damaligen *Erzherzogtums Österreich ob der Enns* anzufertigen.

---

2 G. König, Georg Matthäus Vischer, in: I. Kretschmer - J. Dörflinger - F. Wawrik (Hrsg.), Lexikon zur Geschichte der Kartographie 2, Wien 1986, 861.Siehe auch: J. Dörflinger – R. Wagner – F. Wawrik, Descriptio Austriae. Österreich und seine Nachbarn im Kartenbild von der Spätantike bis ins 19. Jahrhundert, Wien 1977, 128; M. Vancsa, Georg Matthäus Vischer. Sein Leben, seine Werke und im besonderen seine „Topographia Austriae inferioris“ – Einleitung zu: Georg Matthäus Vischer, Topographia Austriae inferioris 1672. Original = getreue Neuausgabe veranstaltet vom Verein für Landeskunde von Niederösterreich, Wien 1920, 4f.

Nach längeren Verhandlungen erhielt er 1667 dazu den Auftrag. 1669 erschien die Karte im Stich von Melchior Küsell aus Augsburg auf zwölf Blättern im Maßstab ca. 1:144.000 als *Archiducatus Austriae superioris geographica descriptio.*

Da Vischer die kartographische Tätigkeit zu sehr in Anspruch nahm, musste er 1668 seine Pfarre aufgeben. Mit Graf Starhemberg hatte er einen Gönner gewonnen, der ihn sowohl dem Kaiser (Leopold I.) als auch den niederösterreichischen Ständen für eine niederösterreichische Landaufnahme empfahl. Tatsächlich erhielt er 1669 zu weit besseren finanziellen Bedingungen als in Oberösterreich den Auftrag für eine Niederösterreichkarte. Bereits 1670 erschien sie als *Archiducatus Austriae inferioris accuratissima geographica descriptio*[3] auf 16 Blättern ebenfalls im Maßstab ca. 1:144.000 (118,5 x 174 cm) und von Melchior Küsell in Kupfer gestochen.[4]

Abb. 1: Vergrößerung des Gebietes um Orth an der Donau aus der Niederösterreichkarte von Vischer 1670.

Nachdem er das Landes bereist und Zeichnungen von Orten und Gebäuden (Klöstern, Schlössern, etc.) angefertigt hatten,[5]. veröffent-

3 Österreichische Nationalbibliothek, Kartensammlung, KD 96.081.; ÖAW, Sammlung Woldan, K-III: OE/Inf 36 (1-16)

4 Es wurden insgesamt 1000 Exemplare angefertigt, 200 erhielten die Stände, der Rest (800) stand Vischer zum freien Verkauf zur Verfügung. Siehe Vancsa, Vischer, 5.

5 G. König, Vischer, 861.

lichte Vischer 1672 die *Topographia archiducatus Austriae inferioris modernae* mit 507 Kupferstichen. Auch auf der Landkarte zeichnete er Ortsansichten zur Veranschaulichung ein (siehe Abb. 1 und 2).

Abb. 2: Die Ansicht von Orth an der Donau aus Vischers *Topographia archiducatus Austriae inferioris modernae* 1672

Eine zweite Auflage der Niederösterreichkarte erschien kurz nach Vischers Tod im Jahre 1697, auf der sein Name und sein Porträt entfernt wurden. Allerdings war die Karte mit einem Suchgitter versehen worden. Zudem kamen fehlende Siedlungen hinzu. Ein alphabetisches Ortsregister als Anhang ergänzte die Neuauflage.[6]

Weiters erschienen Karten von Niederösterreich, der Steiermark, von Mähren, Ungarn. Außerdem erhielt er eine Reihe von Gelegenheitsaufträgen privater Kunden für topographische Ansichten (z.B. des Stiftes Admont, von Wien, Graz usw.).

Trotz vieler Aufträge hatte er immer wieder finanzielle Probleme, da er die anfallenden Kosten (Hilfskräfte, Transportkosten, Arbeitsmaterialien) vorfinanzieren musste und die Auftraggeber nur einen Teil der Auflagen bezahlten. Den Rest musste Vischer selbst abset-

6 Österreichische Nationalbibliothek, Kartensammlung, Alb 180-10; ÖAW, Sammlung Woldan, K-V(Bl): OE/Inf 67 (1-20). – Beide Auflagen werden weiter unten genauer beschrieben.

zen. Zweimal bewarb er sich deshalb erfolglos um eine Pfarre in der Steiermark. Von 1684 bis 1688 war er Mathematiklehrer am Edelknaben-Institut in Wien. Die letzten Monate seines Lebens verbrachte er im Stift Kremsmünster (Oberösterreich) als Lehrer für Mathematik, Geographie und Kartographie. Vischer starb vollkommen mittellos und verarmt im Jahre 1696.

Der Generalquartiermeisterstab und die österreichischen Landesaufnahmen

Der Mangel an topographischen Hilfsmitteln ermöglichte dem Militär keine weit reichenden Geländekenntnisse, was einen großen Nachteil für die Feldverrichtungen des Generalstabs bedeutete. Man war auf die Rekognoszierung durch Spähtrupps angewiesen. Die Märsche und Truppenbewegungen, alle Lager und Stellungen im Terrain, wurden durch eigene Zeichner aufgenommen, die nicht der Armee angehörten: *Bis zum Jahre 1763 lag das Kartenwesen in den Händen von Privatpersonen, die mit Unterstützung der Landstände die Aufnahmen besorgten.*[7] Die Militärs erkannten die Notwendigkeit, geeignete Landkarten zu besitzen, die den militärischen Anforderungen entsprachen. Vorerst waren es *Ingenieur-Offiziere, die so genannten Genieoffiziere, die in erster Linie für das Befestigungswesen zuständig waren.*[8] Diese mussten sich aber auch für kartographische Arbeiten verpflichten. Prinz Eugen von Savoyen (ab 1703 „Präsident des Hofkriegsrates", d. h. Kriegsminister) bemühte sich um die Errichtung einer Militär-Ingenieursschule.

Kaiser Karl VI. erließ schließlich 1717 *eine zweckdienliche Anordnung zur Bildung einer Ingenieurakademie.*[9] Erster Direktor wurde der aus Italien stammende Obstlt. Conte Leander Anguissola (1652-1720). Sein Stellvertreter und Nachfolger wurde Johann Jakob (Gian Giacomo) Marinoni (1676-1755). Beide stellten 1706 einen neuen Wiener Stadtplan her. Marinoni entwickelte einen verbesserten Messtisch, der für topographische Aufnahmen bis ins 19. Jahrhun-

7 Siehe J. Paldus, Die militärischen Aufnahmen im Bereich der Habsburgischen Länder aus der Zeit Kaiser Josephs II. Ausgeführt durch den k.k. Generalquartiermeisterstab in den Jahren 1763-1785, Wien 1919, 5f.

8 E. Hofstätter, Beiträge zur Geschichte der Landesaufnahmen, Wien 1989, 22.

9 E. Hofstätter, Geschichte der Landesaufnahmen, 29.

dert verwendet wurde. Er verfasste grundlegende wissenschaftliche Werke zur Landvermessung und trug damit Wesentliches zum Fortschritt der Kartographie in Österreich bei. Die Zöglinge der Ingenieurakademie bekamen *theoretische und praktische Kenntnisse in der Vermessungskunde und im Zeichnen von Landkarten* vermittelt.[10] Marinoni erfuhr zahlreiche Ehrungen und wurde 1727 in den Adelsstand erhoben.[11]

Joseph Liesganig (1719-1799), Jesuitenpater und Professor an der Wiener Universität, führte im Auftrag Kaiserin Maria Theresias verschiedene geometrische und astronomische Berechnungen durch. 1761 unternahm er mit dem Franzosen Cassini de Thury[12] die Vermessung eines Dreiecksnetzes in der Umgebung von Wien. 1762 vermaß Liesganig *die Basis Wiener Neustadt – Neunkirchen* (ca. 12160 m) und 1763 eine *Basis im Marchfeld zwischen Seyring und Mitte der Ortschaften Raasdorf – Glinzendorf* (ca. 12114 m), wobei der Wiener Stephansdom als Nullpunkt diente.[13]

Eine organisatorische Maßnahme im Hinblick auf eine Vereinheitlichung aller militärischen Belange führte 1747 zur Gründung des Militär-Ingenieur-Corps, in dem alle Ingenieuroffiziere zusammengefasst waren. 1758 kam es auf Anregung des Feldmarschalls Daun[14]

---

10 J. Dörflinger, Die österreichische Kartographie im 18. und zu Beginn des 19. Jahrhunderts unter besonderer Berücksichtigung der Privatkartographie zwischen 1780 und 1820 1: Österreichische Karten des 18. Jahrhunderts, Wien 1984, 39.

11 Siehe E. Hofstätter, Geschichte der Landesaufnahmen, 25-29.

12 I. Kretschmer – R. Messner, Landesaufnahme, in: I. Kretschmer – J. Dörflinger – F. Wawrik (Hrsg.), Lexikon zur Geschichte der Kartographie (1), 435: Cesar Francois Cassini de Thury (1714-1784) stellte eine Karte Frankreichs her, *gestützt auf astronomische Beobachtungen und geodätische Triangulation.* Von *1739 bis 1744 wurde das ganze Staatsgebiet Frankreichs mit einem Dreiecksnetz überzogen, an das sich die topographische Aufnahme anschloss. Damit entstand 1750-93 unter Leitung von Cassini de Thury das bedeutendste Kartenwerk des 18. Jhdts.* (Maßstab 1 : 86.400, 184 Blätter in Kupferstich). Nach Cassinis Tod 1784 vollendete sein Sohn das Kartenwerk. Die *Carte de France* war die erste detaillierte Landesaufnahme eines größeren Staates, basierend auf Triangulation und durchgeführt nach modernen Aufnahmemethoden.

13 E. Bernleithner, Die Entwicklung der Kartographie in Österreich, Bonn 1959, 209.

14 Leopold Josef Graf Daun, Fürst von Thiano (1705-1766), Feldmarschall, Hofkriegsrats-Präsident, leitete die Theresianische Militärakademie und war ein hervorragender österreichischer Feldherr. Daun gewann einige Schlachten gegen den Preußenkönig Friedrich II. und erhielt als erster 1757 den neu

durch Kaiserin Maria Theresia zur Schaffung des *Generalquartiermeisterstabes*. Der erste Generalquartiermeister war Feldmarschall Graf Lacy.

Graf Franz Moritz Lacy wurde am 21. 10. 1725 in Petersburg geboren. Mit 18 Jahren trat er in die Armee ein. Seine Tüchtigkeit brachte ihm bald einen grandiosen Aufstieg. 1765 wurde er Generalinspektor der Armee und 1766 als Nachfolger Dauns Präsident des Hofkriegsrates, 1770 Ritter des Goldenen Vlieses und 1774 Staats- und Konferenzminister. Auf Grund seiner Verdienste ernannte ihn Kaiser Franz II. als Nachfolger des Fürsten Kaunitz zum Kanzler des Maria-Theresien-Ordens. Er starb in Wien am 24. 11. 1801.[15]

Das Bedürfnis nach militärischen Karten war nach dem verlorenen Siebenjährigen Krieg besonders bewusst geworden und so *ordnete Kaiserin Maria Theresia auf Antrag der Feldmarschälle Daun und Lacy 1764 die kartographische Aufnahme ihres Reiches*[16] durch den Generalquartiermeisterstab an. Am 2. Mai 1764 hatte Lacy in einem Schreiben an Daun *die Notwendigkeit, die Aufnahme der Länder so rasch wie möglich in Angriff zu nehmen*[17], sehr eindringlich vor Augen geführt.

Mit der Beauftragung des Generalquartiermeisterstabes durch Maria Theresia, die habsburgischen Länder detailliert aufzunehmen und zu mappieren, erhielt die Armee wichtiges Kartenmaterial, das allerdings der Geheimhaltung unterlag und an strenge Vorschriften gebunden war.[18]

Da Kaiser Joseph II. ab 1765 die Leitung in militärischen Angelegenheiten innehatte, wurde die Landesaufnahme des Reiches nach ihm benannt. Er nahm auch Einfluss auf den Fortgang der Mappierung. Vorbild war zweifellos Cassini de Thurys Karte von Frankreich. Es wurde aber auf eine einheitliche Triangulation verzichtet, wodurch die Messtischaufnahmen im einfachen Militärmaß (1 Wr. Zoll: 400 Wr. Klafter = 1:28.800) nicht zusammengefügt werden konnten.[19] Trotzdem zählt die *Josephinische Landesaufnahme* zu

---

gestifteten Maria-Theresien-Orden. Siehe R. Bamberger – M. Bamberger – E. Bruckmüller – K. Gutkas (Hrsg.), Österreich Lexikon 1, Wien 1995, 207.

15 Siehe J. Paldus, Militärische Aufnahmen aus der Zeit Kaiser Josephs II., 4.

16 J. Dörflinger, Österreichische Kartographie, in: I. Kretschmer – J. Dörflinger – F. Wawrik (Hrsg.), Lexikon zur Geschichte der Kartographie 2, 566.

17 J. Paldus, Militärische Aufnahmen aus der Zeit Kaiser Josephs II., 4f.

18 Ebd., 6.

19 I. Kretschmer – Robert Messner, Landesaufnahme, 436.

den herausragendsten kartographischen Leistungen des späten 18. Jahrhunderts. Besonders die Geländedarstellungen sind hervorzuheben. Da die Karten als streng gehütetes militärisches Geheimnis nicht gedruckt werden durften, verwehrte man ihnen die zeitgenössische Anerkennung.[20]

Die Erstellung der militärischen Karten aller Erbländer *erforderte einen Stab an Mitarbeitern und eine planmäßige Organisation.*[21] Von 1764 bis 1785 führten unter Leitung des Generalquartiermeisterstabes Offiziere der Armee und geeignete Fachkräfte der Regimenter mit Fleiß und Ausdauer die Vermessung und Mappierung des Reiches in *5400 prächtig gezeichneten Sektionen*[22] durch. Dazu kamen die militärisch wichtigen Beschreibungen der Gegend, die auf der Karte nicht zum Ausdruck kommen konnten.

Die Kriegskarte des Erzherzogtums Österreich unter der Enns wurde in den Jahren 1773 bis 1781 aufgenommen: Maßstab 1:28.800, 122 Sektionen (42 x 62 cm je Sektion) und drei Bände Beschreibungen. Unter der Direktion von Major von Neu des Generalquartiermeisterstabes erfolgte die Mappierung. 24 Regimentsoffiziere besorgten die Kartenzeichnungen unter der Leitung von Oblt. Dormayer, später Hptm. Hayn, vom Generalquartiermeisterstab. Die Sektionsbeschreibungen führten Hptm. Hayn, Unterlt. Dedovics (Deutschmeister) und Fähnrich Nigel (Khevenhüller) durch.[23]

Im Österreichischen Staatsarchiv werden eine *Kriegs-Charte des Erzherzogthum Österreichs unter der Enns* und die dazugehörigen Beschreibungen – *Description zur Kriegs-Charte des Erzherzogthum Österreich unter der Enns* (3 Bände) aufbewahrt.[24] In der Kartensammlung der Österreichischen Nationalbibliothek befindet sich zudem eine besonders schön ausgearbeitete und kolorierte Zweitfassung, *die wahrscheinlich eigens für den Kaiser angefertigt worden war.*[25]

Der Mangel eines einheitlichen Triangulierungsnetzes bei der Josephinischen Landesaufnahme bewog den Leiter des Generalquartiermeisterstabes Feldzeugmeister Anton Mayer von Heldenfeld

---

20 J. Dörflinger, Kartographie, 566.
21 J. Paldus, Militärische Aufnahmen aus der Zeit Kaiser Josephs II., 8.
22 Ebd., 8.
23 Ebd., 24f.
24 Österreichisches Staatsarchiv Kartensammlung des Kriegsarchivs: B IXa 242.
25 J. Dörflinger – R. Wagner – F. Wawrik, Descriptio Austriae, 183; Österreichische Nationalbibliothek, Kartensammlung, K II 87.

(1765-1842) den Antrag zu stellen, neuerlich eine Landesaufnahme durchzuführen. Erzherzog Karl unterstützte diesen Vorschlag beim Kaiser (Franz I.), der am 2. April 1806 die zweite nach ihm benannte *Franziszeische Landesaufnahme* anordnete. Sie wurde wieder dem k. k. Generalquartiermeisterstab übertragen.[26]

Diese zweite Landesaufnahme basierte auf einheitlichen astronomischen und trigonometrischen Messungen. Für die Veröffentlichung der Karten schuf man das *Topographische Büro des k. k. Generalquartiermeisterstabes*, das *ab 1811 nicht nur den Stich, sondern auch den Druck und den Vertrieb der eigenen Kartenblätter*[27] besorgte. Ab 1819 führte das Büro den Namen *Topographisch-lithographische Anstalt*,[28] ab 1839 hieß es *k. k. Militärgeographisches Institut.* Heute besteht es noch unter der Bezeichnung *Bundesamt für Eich- und Vermessungswesen.*[29]

Die Franziszeische Landesaufnahme stützte sich auf die von Liesganig bereits 1762 zwischen Wiener Neustadt und Neunkirchen vermessene Grundlinie, die als Ausgangspunkt der Triangulierung genommen wurde und 1806 nachgemessen worden war (6410 Wiener Klafter = 12 156,5 m). Die Leitung hatte Hauptmann Ludwig August Freiherr von Fallon inne. Zur Darstellung der topographischen Aufnahme wählte man die Cassinische Projektion. Der Wiener Stephansdom war „Koordinatenursprung" für das Koordinatensystem. Da diese Militärtriangulierung Genauigkeitsmängel aufwies, ging man nach 1817 auf die inzwischen begonnene zivilstaatliche Katastertriangulierung über. Der Maßstab war wie bei der Josephinischen Landesaufnahme 1 Zoll: 400 Wiener Klafter, was 1:28.800 bedeutete. Die Reproduktion erfolgte zunächst in Kupferstich, nach 1818 auch in Steindruck.[30] 1869 wurde die Franziszeische Landesaufnahme unvollendet abgebrochen.

---

26 E. Hofstätter, Geschichte der Landesaufnahmen, 61.

27 J. Dörflinger, Die österreichische Kartographie im 18. und zu Beginn des 19. Jahrhunderts unter besonderer Berücksichtigung der Privatkartographie zwischen 1780 und 1820 2: Österreichische Karten des frühen 19. Jahrhunderts, Wien 1988, 752.

28 E. Hofstätter, Geschichte der Landesaufnahmen, 61.

29 Siehe E. Bernleithner, Entwicklung der Kartographie in Österreich, 216.

30 Ebd., 216.

Die Aufnahme von Ober- und Niederösterreich erfolgte auf 160 Blättern in den Jahren 1807 bis 1819.[31] Ein Exemplar der *Karte des Erz-Herzogthums Österreich ob und unter der Enns astronomisch trigonometrisch vermessen, topographisch aufgenommen, reduzirt und gezeichnet im Jahre 1813 von dem kaiserlich-königlich=oesterreichischen General Quartiermeister Stabe* (31 Sektionen im Maßstab 1:144.000 – Blattformat: 26,5 x 39 cm) ist in der Kartensammlung der Österreichischen Nationalbibliothek aufbewahrt (392.842-D-Kar). Die Herausgabe der Blätter erfolgte von 1813 bis 1823 (Spezialkarte).

1823 erschien eine zweiblättrige *Generalkarte des Erzherzogthums Oesterreich ob und unter der Enns von dem k. k. Generalquartiermeisterstabe. Nach der Specialkarte reduzirt und gezeichnet im Jahre 1823. Gestochen von C. Stein (die Schrift) und A. Withalm (den Terrain)* – (Maßstab 1:288.000, Blattgröße: 64 x 114,5 cm), welche sich ebenfalls in der Kartensammlung der Nationalbibliothek befindet (AB 2 C 4. Kar).

Die Vischer – Karten

Die Karte von 1670 (Abb. 3) – *Archiducatus Austriae Inferioris Accuratissima Geographica Descriptio Authore Georgio Matthaei Visscher*[32] *Tyrolensi* – ist reich verziert, wobei Vischer rechts unten im Porträt und Wappen dargestellt wird. Das Privileg ist wie das Porträt und das Wappen als lorbeerumranktes Medaillon gezeichnet. Der Name des Stechers, Melchior Küsell, ist mit der Jahreszahl 1670 vermerkt. In einer Kartusche in der rechten oberen Ecke befindet sich eine Landschaftsdarstellung der Donau um Klosterneuburg, Korneuburg und Kreuzenstein mit nummerierten Ortsangaben.

Die Dekoration in der linken oberen Ecke zeigt eine Ansicht der Wiener Hofburg mit Porträts Kaiser Leopolds I. und seiner ersten Gattin Margarita Theresia von Spanien (1673 gestorben). Die Dekoration in der linken unteren Ecke weist zwei Männer im römischen Rüstungen auf, die ein Medaillon halten, auf dem der österreichische Bindenschild, das österreichische Wappen mit den

---

31 Ebd., 217; Österreichisches Staatsarchiv, Kartensammlung des Kriegsarchivs: B IXa 196-6.

32 Der Name Vischer ist, wie G. König, Vischer, 862 betont, falsch geschrieben (er könnte mit dem Amsterdamer Kupferstecher und Verleger Claes Jansz. Visscher – auch Nicolaus Joannes Piscator genannt – verwechselt werden).

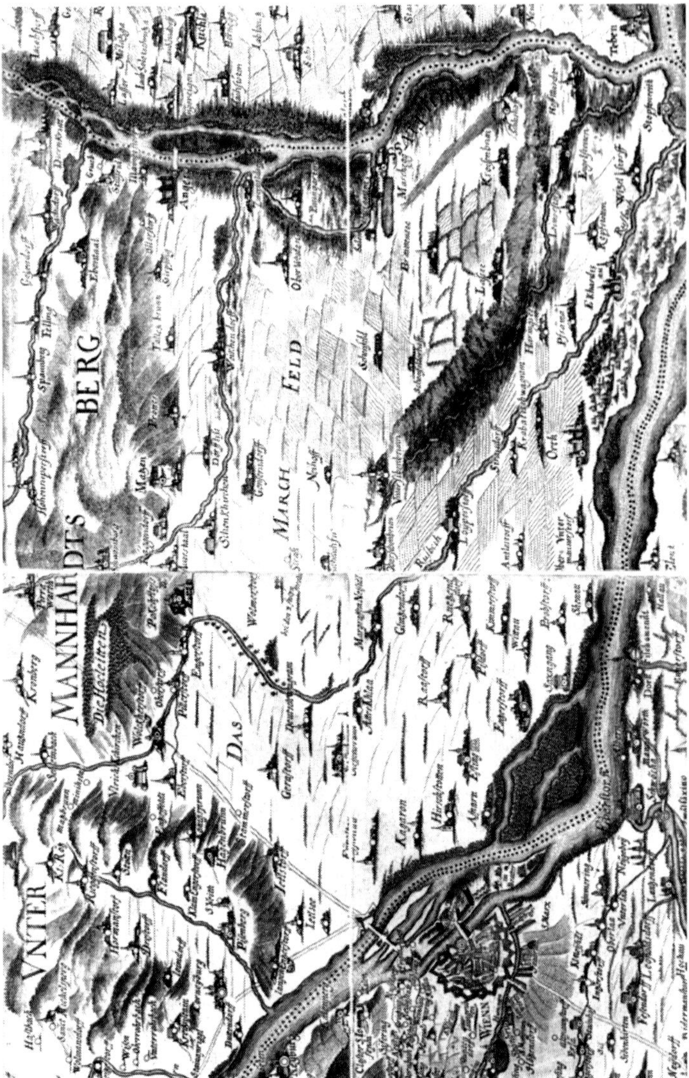

Abb. 3: Das Marchfeld in der Niederösterreichkarte von Vischer 1670

fünf Adlern und darüber der österreichische Herzoghut, zu sehen sind. Umrandet wird die Gesamtkarte von einem Gradstreifen.

Vischer selbst hat sich am Messtisch arbeitend dargestellt. Weiters sind ein Begleiter (Diener oder Mitarbeiter), der eine Tafel mit der Zeichenerklärung hält, und zwei gesattelte Pferde zu sehen. Damit gab Vischer Auskunft über seine Arbeitsweise. Am Boden neben dem Messtisch liegen Zirkel, Gradbogen (Winkelmesser), Messkette und verschiedene Hefte für Skizzen und Aufzeichnungen. Man kann daraus ersehen, dass Vischer die Methoden der Geländeaufnahme gekannt und wohl auch angewandt hatte.[33]

Dass er das Land, das er aufnahm, intensiv bereiste, beweisen seine vielen topographischen Arbeiten. Er zeichnete zahlreiche Ansichten von größeren Orten, interessanten Gebäuden, Burgen und Schlössern. Diese Kupferstiche sind in der *Topographia Austriae inferioris 1672*[34] zu finden. Die meisten Siedlungen des Weinviertels sind als Ortsansichten in Miniaturform auf der Vischer-Karte dargestellt.

Die Zweitausgabe von 1697, *Archiducatus Austriae Inferioris Geographica, et Nouiter Emendata Accuratissima Descriptio,* stammt von denselben Platten. Es wurden aber, wie erwähnt, zahlreiche Ergänzungen gemacht und Veränderungen durchgeführt: Das Suchgitter und ein Ortsregister (*Register der Namben so hierin nach beygesezten Ziffern in dieser Landtaffl des Viertls Vnter Mannharts-Berg in Österreich unter der Enns ordentlich zu finden*) ermöglichen ein leichteres Auffinden der Orte. Obwohl Vischer selbst wahrscheinlich Ergänzungen durch Eintragen neuer Orte vorgenommen hatte, entfernte man sein Porträt und Wappen in der rechten unteren Hälfte sowie den Namen des Stechers Melchior Küsell. An dieser Stelle stehen die Namen Jacobus Hoffmann und Jacobus Hermundt. Widmungen und Wappen (Vertreter der Stände) wurden aktualisiert. Die Dekoration in der linken oberen Kartenecke wurde gleichfalls neu gestaltet und damit der Zeit angepasst. Neben Kaiser Leopold I. ist statt der ersten dessen dritte Gattin Eleonore Magdalena Theresia von Pfalz-Neuburg (und neues Wappenzeichen), an Stelle der drei Engel unter der Krone neben dem Kaiser das Bildnis des späteren Kaisers Joseph

33 E. Hofstätter, Geschichte der Landesaufnahmen, 15.

34 Österreichische Nationalbibliothek, Kartensammlung, KA 537.717-C, ÖAW, Sammlung Woldan, AW-IV: OE/Inf 39.

I. (1705-1711) und neben der Kaiserin das Bildnis Karls VI. (regierte von 1711 bis 1740) neu gestochen worden.

Der erste optische Eindruck der Vischer-Karte ist begeisternd. Das breite Donauband, das Niederösterreich durchzieht, gibt mit den Nebenflüssen eine gute Gesamtdarstellung des Landes, wobei die Geländebereiche (Wälder, Hügel, Berge, Felder), die perspektivisch dargestellt werden, das positive Bild vervollständigen. Man kann sich vorstellen, dass diese Landkarte im 17. Jahrhundert und noch darüber hinaus dem interessierten Betrachter einen unfassenden Überblick verschafft hat. Natürlich stimmen viele Details nicht. Oft sind es große Mängel und Fehler, die wir erst heute erkennen.

Die Donau (*Tonau Fluss*) ist sehr breit (einige Kilometer) mit Überschwemmungsgebiet und Auwäldern gezeichnet; das gleiche gilt auch für die anderen Flüsse. Möglicherweise wollte Vischer die Anschaulichkeit erhöhen? Dies ist keinesfalls störend, im Gegenteil.

Die Landesgrenzen Niederösterreichs verlaufen im Norden und Westen in Wirklichkeit nicht so gerade wie bei Vischer.[35] Die Längenausdehnung West-Ost entspricht mit ca. 160 km der Natur, während die Nord-Süd-Ausdehnung, von der Nordgrenze bis zur Donau (Wald- und Weinviertel), mit 30 bis 60 km nicht stimmt; die wirkliche Länge liegt zwischen 45 km (kürzeste Entfernung von Nordgrenze bis zur Donau) und 70 km (Weinviertel), bzw. 90 km (Waldviertel).

Die Verläufe und Längen der Flüsse/Bäche sind mit vielen Fehlern behaftet, großräumig gesehen aber recht wirklichkeitsnah. Beim Donaufluss fehlt die Marchfeldkrümmung gegen Süden. Die Thaya verläuft im Norden zwischen Drosendorf und Pulkaumündung zu gerade von West nach Ost, das gleiche gilt für den Grenzverlauf mit Mähren. Die Mündung der Thaya in die March liegt zu weit im Norden, tatsächlich knapp nördlich von Hohenau. Die Krümmung vor Hohenau gibt es in Realität nicht. Bei Wien ist eine Brücke über die Donau eingezeichnet, bei Angern eine über die March. Über die Thaya sind im Weinviertel nur zwei Übergänge eingetragen, bei Laa und bei Unter Themenau (nach Lundenburg) – heute zu Tschechien gehörig. Leider sind nicht alle Bäche beschriftet.

---

35 Vgl. Freytag & Berndt, Österreich 2000, Wien, Niederösterreich – Die Straßenkarte – Maßstab 1: 200 000, Wien 2000.

Zwei Poststraßen mit Poststationen gehen von Wien aus: Die erste über *Wolckerstorff* – *Gaunerstorff* (Gaweinstal) – *Kozlstorff* (Ketzelsdorf) nach *Nicolspurg* (und weiter), die zweite über *Langen-Enzerstorff* – *Stokerau* – *Ober Holabrunn* nach *Znäm* (Znaim). Auf der Karte von 1697 wird die Poststraße von *Ober Holabrunn* über *Bulckha* (Pulkau) – *Langau* nach *Fräting* (Vratenin) in Mähren geführt, während die Straße Richtung Znaim entfernt worden war und beim Druck nur mehr andeutungsweise sichtbar ist.

Darstellung der Landschaftsformen: Schraffierungen lassen die Berge und Hügel plastisch erscheinen. Wald wird durch eine Baumsignatur in Ebene und Hügelland dargestellt, wobei die Baumzeichen auch zur Betonung der Hügelform verwendet werden. Die eher runde Baumform soll wahrscheinlich den Laubwald wiedergeben. Im hügeligen Gelände sind viele Weingartensignaturen zu finden, d.h. die Hänge sind mit kurzen Strichen bedeckt. Abhänge werden durch parallele Striche, linear oder gekrümmt, dargestellt. In der Ebene sind rhomben- und trapezförmige Felder gezeichnet, und die einzelnen Flächen mit mehr oder weniger zittrigen Parallelen, manchmal auch gekrümmt, versehen, wobei die Ränder teilweise besonders hervorgehoben werden. Diese Parzellengliederungen sollen wohl eher ein Darstellungsmerkmal sein und entsprechen sicher nicht den tatsächlichen Größen und Grenzen von Ackerflächen.

Die Darstellung der Siedlungen erfolgt im Weinviertel meist durch Ortsansichten (siehe Abb. 1 und 2), wobei Kirchen, wichtige Gebäude, Türme, Schlösser, Burgen, Stadtmauern und Kloster erkennbar sind. Bei wichtigen Orten und Städten ist das Ortsbild sehr wirklichkeitsnah als Vedute dargestellt. Zu allen Ansichten sind noch Signaturen hinzugefügt. Orte ohne Ansicht haben nur eine Signaturkennzeichnung. Bei einer Stadt sind es zwei konzentrische Kreise, bei einem Kloster ein Bischofsstab mit Fahne, bei einem Schloss ein Kreis mit einem Punkt in der Mitte, bei einem Markt ein Kreis mit Querbalken und bei einem gewöhnlichen Dorf ist es nur ein einfacher Kreis. Straßen sind – mit Ausnahme der Poststraßen – keine eingetragen.

Im Viertel unter dem Mannhartsberg sind auf der Vischer-Karte von 1670 insgesamt 447 Orte (85 mit Schloss oder Burg), davon 24 Marktgemeinden (6 mit Schloss/Burg) und 10 Städte (2 mit Schloss) zu finden. Die Städte sind folgende: *Anger* (Angern an der March)[36], *Corneyburg* (Korneuburg), *Enzerstorff* (Groß-Enzersdorf), *Feldtspurg* (Feldsberg – heute Valtice, Tschechien), *Laha* (Laa an der Thaya), *Marchegg* (mit Schloss), *Maissau* (mit Schloss), *Rötz* (Retz), *Schrattentaal* (Schrattenthal) und *Zistertzdorff* (Zistersdorf). Alle Städte sind in der *Topographia Archiducatus Austriae Inferioris Modernae* von 1672 als Kupferstiche enthalten.

Die Josephinische Landesaufnahme von Niederösterreich – *Kriegs-Charte des Erzherzogthum Österreichs unter der Enns*, 1773-1781

Die Darstellung des Geländes erfolgte in Grundrissmanier, es gibt aber keine Höhenangaben. Das Gerippe der Karte bilden vor allem „freistehende Objekte", wie Gehöfte, Mühlen, Kapellen und Wegkreuze, aber auch Brücken, Furten, Wald- und Garten- (Weingärten-)Parzellen.[37] Kirchen, Burgen und Schlösser waren dominierende Orientierungsobjekte.

Hügel, Berge sind durch kurze Striche, teilweise mit Schummerung, die je nach Hanglänge und Neigung kranz- oder girlandenartig aneinandergefügt wurden, dargestellt und durch vermehrte Strichführung erscheinen steilere Hänge dunkler (von hell- bis dunkelgrau, ja fast schwarz). Der Baumbestand in den Waldgebieten ist durch dunkelgraue Baumsignaturen in runder bis länglicher Form und verschiedener Größe mit Schattenwirkung (Lichteinfall von links) gekennzeichnet, wobei je nach Dichte des Waldes diese eng oder weiträumiger gezeichnet sind. Leider sind nur wenige Berge und Wälder beschriftet.

Weingärten sind parzellenförmig mit kleinen Signaturen, die Rebstöcke darstellen sollen, versehen. Diese Weinstöcke (Weingartensignaturen) sehen folgendermaßen aus: ein großes L mit drei kurzen Querstrichen. Sie sind alle gleich groß und mit großer Sorgfalt ausgeführt. Dies gilt übrigens auch für die Baumsignaturen.

---

36 Angern ist nur auf der Vischer-Karte als Stadt ausgewiesen, auf den anderen Karten als Markt!

37 E. Hofstätter, Geschichte der Landesaufnahmen, 45.

Die Karte ist koloriert: Gewässer (Bäche, Flüsse, Seen und Moraste) blau, Gebäude rot, Wiesen grün (hell- bis dunkelgrün), Wälder grau, Weingärten rosa, Sandablagerungen (z. B. Sandbänke in der Donau) hellbraun, Straßen (doppellinig) und Wege (einlinig) dunkelbraun. Grünflächen (Wiesen, Gärten) und Sandflächen sind durch einen Punktraster hervorgehoben, Vertiefungen (natürliche Gräben) sind durch parallel laufende, strichlierte Linien dargestellt, die zum Teil als wasserführend blau angemalt sind. Grünflächen mit blauen Einfärbungen dürften je nach Wetterlage trocken oder wasserführend bzw. morastig gewesen sein. Entlang von Flüssen und Bächen befinden sich meist Grünstreifen mit Baum- oder Strauchsignaturen, nicht alle Bäche sind benannt.

Die Kartenbeschriftung ist auf der Karte, die im Staatsarchiv (Kriegsarchiv) aufbewahrt wird, weniger sorgfältig ausgeführt, während das Exemplar in der Kartensammlung der Nationalbibliothek neben einer besonders sorgfältigen Ausführung auch schöne lateinische Kursivschrift aufweist. Bei Städten und Märkten steht die Bezeichnung Stadt oder Markt, manchmal auch „D" für Dorf.

Zu jeder Aufnahmesektion wurden Landesbeschreibungen angelegt, die alle wichtigen militärischen Hinweise enthalten, *die der Karteninhalt selbst nicht auszudrücken vermag.*[38]

Die Franziszeische Landesaufnahme des Weinviertels

Österreich ob und unter der Enns wurde vom Generalquartiermeisterstab von 1807 bis 1819 in 160 Sektionen aufgenommen und mappiert. Die Einteilung der Sektionen erfolgte in Zonen (von Nord nach Süd) mit arabischen Ziffern bezeichnet und waagrecht als Kolonnen mit römischen Zahlzeichen nummeriert. Das Weinviertel umfasst 32 Sektionen: 39/XIX bis XXIII, 40/XIX bis XXIII, 41/XVIII bis XXIII, 42/XVIII bis XXIII, 43/XIX bis XXIII, 44/XXI bis XXIII, 45/XXII und XXIII.

Die Größe und Lage der einzelnen Sektionen der Franziszeischen Landesaufnahme ist fast identisch mit jenen der Josephinischen Landesaufnahme (gleicher Maßstab). Während die Designation bei der Josephinischen Landesaufnahme aber nicht beschriftet ist, enthalten

[38] E. Hofstätter, Geschichte der Landesaufnahmen, 52.

einige wenige Sektionen der Franziszeischen Landesaufnahme (39/XIX, XX, 40/XX, XXI und 41/XX) Angaben über Häuser, Stallungen, Unterkunftsmöglichkeiten (bequem) für Mann und Pferd. Fast alle Sektionen haben Angaben über den Leiter (*Sous Direction*), wer das Gelände aufgenommen und die Karte gezeichnet hat. Die Berg- und Geländedarstellung erfolgte bei der Josephinischen Landesaufnahme durch sog. Bergstriche, die sich zu „Schraffenstufen" entwickelten.[39]

Der sächsische Kartograph Johann Georg Lehmann (1756-1811) entwickelte Böschungsschraffen, die bei flachem Gelände länger und zarter, bei steilerem Gebiet kürzer und stärker gezeichnet wurden. Durch entsprechendes Variieren der Strichstärken erschienen steilere Hänge dunkel, flachere Hänge heller. Bei der Franziszeischen Landesaufnahme wurde eine modifizierte Form der Böschungsschraffen von Lehmann bei senkrechter Beleuchtung angewandt.[40]

Die Sektionen sind sehr exakt und sorgfältig gezeichnet und mit großen Buchstaben deutlich beschriftet, wobei Stadt, Markt und Dorf im Schriftbild unterschieden werden. Da aber nicht weniger als 18 Militärangehörige des Generalquartiermeisterstabes diese Arbeit (32 Sektionen) durchführten, sind gewisse Unterschiede in der Ausführung festzustellen.

Die Farbgebung ist ähnlich wie bei der Josephinischen Landesaufnahme. Baumbestände werden allerdings nur teilweise durch Signaturen gekennzeichnet, und zwar im Grün- und Ackerland, in den Auen (Donau, March), aber kaum im Berg- und Hügelland. Die Gewässer sind in hellem Blau (dunkelblau umrandet) koloriert, Wiesen und Wälder in abgestuftem Grün, aber in kräftiger Farbgebung (Wälder, Auen auch hellgrau). Die Weingärten sind parzellenmäßig umrandet und rosa koloriert, aber nur spärlich (oder gar nicht) mit Signaturen versehen. Die Häuser (Siedlungen, Einzelgebäude) wurden in kräftigem Rot gehalten und mit Grundstück parzellenartig aneinandergereiht. Die Poststraßen sind besonders hervorgehoben – mit Doppelstrich gezeichnet und rot bemalt.

Neben der Beschriftung von Ortschaften, Meierhöfen, Flüssen, Bächen und Seen wurden auch Augebiete, Wälder, Berge (Hügel), Berghänge (Leiten) mit Namen versehen sowie Remisen, Ziegelöfen,

39 E. Hofstätter, Geschichte der Landesaufnahmen, 84.
40 Ebd., 84.

Abdeckhäuser u. a. eingetragen. In den Jahren 1869 bis 1874 trug man die Eisenbahnlinien und Bahnstationen ein.

Quellen- und Literaturverzeichnis

Bernleithner E., Die Entwicklung der Kartographie in Österreich, in: Berichte zur Deutschen Landeskunde (22/2), Flensburg 1959.

Description zur Kriegs-Charte des Erzherzogthum Österreich unter der Enns, Bd. 1-3, – Österreichisches Staatsarchiv, Kartensammlung des Kriegsarchivs, B IXa 242.

Dörflinger J. – Kretschmer I. – Wawrik F. (Hrsg.), Lexikon zur Geschichte der Kartographie. Von den Anfängen bis zum Ersten Weltkrieg, Bd. 1-2, Wien 1986.

Dörflinger J. – Wagner R. – Wawrik F., Descriptio Austriae. Österreich und seine Nachbarn im Kartenbild von der Spätantike bis ins 19. Jahrhundert, Wien 1977.

Dörflinger J., Die österreichische Kartographie im 18. und zu Beginn des 19. Jahrhunderts unter besonderer Berücksichtigung der Privatkartographie zwischen 1780 und 1820, Bd. 1-2, Wien 1984-1988.

Franziszeische Landesaufnahme von Österreich ob und unter der Enns (1807-1819). – Österreichisches Staatsarchiv, Kartensammlung des Kriegsarchivs, B IXa 196-6.

Generalkarte des Erzherzogthums Oesterreich ob und unter der Enns von dem k.k. Generalquartiermeisterstabe. Nach der Spezialkarte reduzirt und gezeichnet im Jahre 1823. – Österreichische Nationalbibliothek, Kartensammlung, AB 2 C 4.Kar.

Hofstätter E., Beiträge zur Geschichte der österreichischen Landesaufnahmen, . 1. T., Wien 1989.

Kriegs-Charte des Erzherzogthums Österreich unter der Enns. – Österreichische Nationalbibliothek, Kartensammlung, K II 87, und Österreichisches Staatsarchiv, Kartensammlung des Kriegsarchivs, B IXa 242.

Österreichischer Generalquartiermeisterstab, Karte des Erz-Herzogthums Österreich ob und unter der Enns astronomisch trigonometrisch vermessen, topographisch aufgenommen, reduzirt und gezeichnet im Jahre 1813 von dem kaiserlich-königlich=oesterreichischen General Quartiermeister Stabe. – Österreichische Nationalbibliothek, Kartensammlung, 392.842-D-Kar.

Paldus J., Die militärischen Aufnahmen im Bereich der Habsburgischen Länder aus der Zeit Josephs II. Ausgeführt durch den k. k. Generalquartiermeisterstab in den Jahren 1763-1785, Wien 1919.

R. Bamberger – M. Bamberger – E. Bruckmüller – K. Gutkas (Hrsg.), Österreich Lexikon, Bd. 1-2, Wien 1995.

Vancsa M., Georg Matthäus Vischer. Sein Leben, seine Werke und im besonderen seine „Topographia Austriae inferioris". – Einleitung zu Vischer Georg Matthäus, Topographia Austriae inferioris 1672. Originalgetreue Neuausgabe veranstaltet vom Verein für Landeskunde in Niederösterreich, Wien 1920.

Vischer G. M., Archiducatus Austriae Inferioris Accuratissima Geographica Descriptio Authore Georgio Matthaei Visscher Tyrolensi, Augusta Vindelicorum 1670. – Österreichische Nationalbibliothek, Kartensammlung, KD 96.081; ÖAW, Sammlung Woldan, K-III: OE/Inf 36 (1-16).

Vischer G. M., Archiducatus Austriae Inferioris Geographica, et Nouiter Emendata Accuratissima Descriptio (1697). – Österreichische Nationalbibliothek, Kartensammlung, Alb 180-10; ÖAW, Sammlung Woldan, K-V(Bl): OE/Inf 67 (1-20).

Zum Autor

Otto Hat, geb. 1934, bis 1995 Lehrer an Volks-, Haupt- und Sonderschulen sowie Direktor des Polytechnischen Lehrganges in Groß-Enzersdorf, ab 1995/96 Studium der Geschichte und Politikwissenschaft an der Universität Wien, 2004 Abschluss der Diplomarbeit.

Kontakt: Lobaustr. 17, A-2301 Groß-Enzersdorf.

Joseph Stöcklein,
Der Neue Welt-Bott,
Augsburg/Graz 1726-1761

Bernd Hausberger, Mexiko-City

P. Joseph Stöcklein, der *Welt-Bott* und der Ruf der Exotik[1]

1. Stöckleins Sammlung von Missionsberichten

In der Zeit der spanischen Herrschaft gelangten nur wenige Menschen aus dem deutschsprachigen Raum nach Amerika, um aus eigener Anschauung über den Kontinent zu unterrichten. Die meisten von ihnen waren Jesuiten. 1726 erschien der erste Band des wahrscheinlich wichtigsten deutschsprachigen Werks über die Mission der Gesellschaft Jesu. Es besaß einen für seine Zeit

---

1 Eine erste Version dieses Textes erschien als *El padre Joseph Stöcklein o el arte de inscribir el mundo a la fe*, in: K. Kohut – M. C. Torales (Hrsg.), Desde los confines de los imperios ibéricos. Los jesuitas de habla alemana en las misiones americanas, Madrid/Frankfurt 2007, 631-662. – Das neu erwachte Interesse am Welt-Bott belegen zwei weitere Arbeiten aus demselben Jahr: R. Dürr, Der *Neue Welt-Bott* als Markt der Informationen. Wissenstransfer als Moment jesuitischer Identitätsbildung, in: Zeitschrift für Historische Forschung 34 (2007), 441-466; U. Strasser, A Case of Empire Envy? German Jesuits Meet an Asian Mystic in Spanish America, in: Journal of Global History 2 (2007), 23-40. – Mein Zimmernachbar am Colegio de México veröffentlichte Ende 2006 eine Auswahl der spanischen Ausgabe der Lettres Edifiantes: G. Zermeño (Hrsg.), Cartas edificantes y curiosas de algunos misioneros jesuitas del siglo XVIII. Travesías, itinerarios, testimonios, México 2006.

typischen langen Titel,[2] ist aber für gewöhnlich in der Kurzform *Der Neue Welt-Bott* (‚Der neue Weltbote') bekannt. Bis 1761 erschienen fünf monumentale Bände mit Briefen und Berichten aus den Missionsgebieten. Jeder Band ist in 8 Teile gegliedert. Die drei letzten sind ausgesprochene Raritäten. Teil 39 ist gänzlich verloren gegangen, anscheinend in einem Brand direkt nach Drucklegung. Teil 40 hat in einem einzigen Exemplar überlebt.[3]

Der Begründer des *Welt-Bott* war P. Joseph Stöcklein. Stöcklein wurde am 30. oder 31. Juli 1676 in Öttingen in Bayern geboren. 1700 trat er in die Österreichische Provinz des Jesuitenordens ein. Nach Abschluss seiner Studien diente er unter anderem fast zehn Jahre als Geistlicher im Heer Prinz Eugens im Spanischen Erbfolgekrieg und in den Kriegen gegen die Türken in Ungarn und auf dem Balkan. Die letzten Jahre seines Lebens verbrachte er in Graz, wo er am 28. Dezember 1733 starb.[4] Bei seinem Tod war erst die Hälfte der vierzig Teile, die der *Welt-Bott* am Ende umfassen sollte, erschienen. Die Herausgabe übernahmen nacheinander P. Karl Meyer aus Erfurt (Teile 21-24), P. Peter Probst aus Koblenz (Teile 25-28), Franziskus Keller aus Linz (Teile 29-38), der auch eine biographische Skizze Stöckleins verfasste,[5] und schließlich P. Franz Xaver Sacher aus St. Pölten (Teile 39-40).

---

2 *Der Neue Welt-Bott. Allerhand so Lehr- als Geist=reiche Brief, Schriften und Reise=Beschreibungen, welche von denen Missionariis der Gesellschaft Jesu aus Beyden Indien und anderen Uber Meer gelegenen Ländern seit An. 1642. biß auf das Jahr 1726. in Europa angelangt seynd. Jetzt zum erstenmal Theils aus denen Französischen Lettres Edifiantes verteutscht und zusammen getragen von Joseph Stöcklein, gedachter Societät Jesu Priester*, t. 1-5, Augsburg/Graz bzw. Wien 1726-1761.

3 Mitteilung von Michael Müller, Mainz.

4 C. Sommervogel, Bibliothèque de la Compagnie de Jésus 7, Brüssel/Paris 1896, 1699-1701; C. v. Collani, Joseph Stöcklein, in: Traugott Bautz, Biographisch-Bibliographisches Kirchenlexikon, 10, Herzberg 1995, 1516-1517, Internet-Ressource.

5 *Leben, Thaten, Reisen und Mission R. P. Josephi Stöcklein, der Gesellschaft Jesu, Oesterreichischer Provinz, in dem Kayserli. Kriegs-Heer im Reiche und Ungarn Missionarii und Ober=Feld=Caplans, Urheber der Teutschen Verfassung des neuen Welt=Bottens*, in: Welt-Bott, 29, 1755, 141-154.

2. Der *Welt-Bott* in der literarischen Tradition des 18. Jahrhunderts

Die Herausgabe von Reiseberichten, wie sie Stöcklein begann, war keineswegs einzigartig. Nach einigen Vorläufern[6] hatte der Venezianer Giovanni Battista Ramusio mit den drei zwischen 1550 und 1559 erschienen und mehrmals neu aufgelegten Bänden *Delle navigationi et viaggi* den Prototyp der Reiseberichtsammlung vorgelegt. Bald folgten andere seinem Beispiel.[7] In England veröffentlichte 1582 Richard Hakluyt *Divers voyages touching the discouerie of America*[8] und wenig später sein Hauptwerk, *The principall navigations, voiages and discoveries of the English nation*, das 1589 in einem Band und von 1598 bis 1600 in einer erweiterten dreibändigen Ausgabe erschien.[9] Hakluyt übte seinerseits einen großen Einfluss auf die Nachwelt aus und inspirierte Männer wie Samuel Purchas, sein Werk im 17. Jahrhundert fortzusetzen.[10] In Deutschland publizierten Theodor de Bry aus Lüttich, seine Söhne Johann Theodor und Johann Israel de Bry und sein Schwiegersohn Matthäus Merian zwischen 1590 und 1630 vierzehn Bände mit Reiseberichten aus Amerika sowie dreizehn aus Asien.[11] 1587 und 1588 war de Bry in London gewesen und hatte

---

6 D. B. Quinn, Richard Hakluyt, Editor. A Study Introductory to the Facsimile edition of Richard Hakluyt's DIVERS VOYAGES (1582), to which is added a facsimile of A SHORT AND BRIEF NARRATION OF THE TWO NAVIGATIONS TO NEWE FRANCE, Transl. by John Florio (1580), Amsterdam 1967, v. 1-2.

7 G. B. Ramusio, Delle navigationi et viaggi, nel qual si contiene la descrittione dell' Africa et del paese del Prete Janni, con varii viaggi dal Mar Rosso a Calicut, et in sin all' isole Molucche [...], t. 1-3, Venezia 1550-1559.

8 R. Hakluyt, Divers voyages touching the discouerie of America, and the Ilands adiacent vnto the same [...], London 1582.

9 Der längere Titel der 2. Auflage lautet: R. Hakluyt, The principal navigations, voiages, traffiqves and discoueries of the English nation, made by sea or ouer-land, to the remote and farthest distant quarters of the earth, at any time within the compasse of these 1500 yeeres [...], v. 1-3, London 1598-1600.

10 S. Purchas, Hakluytus Posthumus, or Purchas His Pilgrimes, containing a History of the world, in Sea-voyages, and lande-Travells, by Englishmen and others [...], v. 1-5, London 1625-1626.

11 Siehe S. Burghartz (Hrsg.), Inszenierte Welten. Die west- und ostindischen Reisen der Verleger de Bry, 1590-1630/Staging New Worlds. De Bry's Illustrated Travel Reports, 1590-1630, Basel 2004.

auch Hakluyt kennengelernt.[12] In Holland wurde 1619 die erste Reiseberichtsammlung von Michiel Colijn herausgegeben.[13] In Spanien schließlich fehlte zwar eine große Sammlung von originalen Reisebeichten, doch fasste unter anderem Antonio de Herrera die Erfahrungen der so genannten Entdeckung und der Eroberung Amerikas in einer offiziellen Chronik zusammen.[14]

Von den Jesuiten hatte schon der Ordensgründer Ignatius von Loyola regelmäßige Berichte aus ihren Arbeitsgebieten verlangt, und die Briefe des Hl. Franz Xaver aus Asien bildeten das Vorbild für alle spätern Missionare.[15] Gestützt auf die so schnell zusammengetragenen Informationen wurden bald auch erste historische Werke über die Fortschritte der Mission verfasst, so 1601 die *Historia de las misiones* des P. Luis de Guzmán.[16] 1620 erschien mit dem Titel *Auß America, das ist, auß der Newen Welt* eine erste Sammlung jesuitischer Missionsberichte auf Deutsch.[17] Es handelte sich dabei um die Übersetzung eines französischen Buchs aus dem gleichen Jahr.[18] Wahrscheinlich stand die Publikation im Zusammenhang mit den Anstrengungen des P. Nicolas Trigault, der von 1614 bis 1617 aus China kommend mehrere europäische Länder besuchte, um frisches

---

12 S. Burghartz, Transformation und Polysemie. Zur Dynamik zwischen Bild, Text und Kontext in den „Americae" der de Bry, Internet Ressource.

13 M. Colijn, Oost-Indische ende West-Indische voyagien, namelijck, de waerachtighe beschrijvinghe vande drie seylagien, drie jaren achtermalkanderen deur de Hollandtsche ende Zeelandtsche Schepen by noorden Noorweghen, Moscovien ende Tartarien nae de Coninckrijcken van Catthay ende China ghedaen [...], Amsterdam 1619.

14 A. de Herrera, Historia general de los hechos de los castellanos en las islas y tierra firme del Mar Océano, v. 1-4, Madrid 1601-1615.

15 G. Schurhammer – J. Wicki (Hrsg.), Epistolae S. Francisci Xaverii aliaque eius scripta, v. 1-2, Roma 1944-1945, erw. Neudruck 1996.

16 L. de Guzmán, Historia de las misiones que han hecho los religiosos de la Compañia de Jesus para predicar el sancto Evangelio en la India Oriental y en los Reinos de la China y Japon, v. 1-2, Alcalá 1601.

17 Auß America, das ist, auß der Newen Welt. Unterschidlicher Schreiben Extract, von den Jahren 1616. 1617. 1618. Was gestalt Acht Patres Societatis, und zwo andere Ordenspersonen, von deß Christlichen Glaubens wegen ihr Blut vergossen, was auch sonst die Patres Societatis Gott zu Ehren unnd zu Außbreitung seines heiligsten Namens auff sich genommen, Augsburg 1620.

18 Histoire du massacre de plvsievrs religievx de S. Dominqve, de S. Francois, et de la Compagnie de Iesvs, et d'autres Chrestiens, aduenu en la rebellion de quelques Indois de l'Occident contre les Espagnols, Valencienne 1620.

Personal für die Mission anzuwerben.[19] Von den späteren deutschsprachigen Veröffentlichungen der Jesuiten sollen zwei historische Darstellungen genannt werden: die Biographien der jesuitischen Märtyrer aus aller Welt des Provinzials der Böhmischen Ordensprovinz, Matthias Tanner, von 1683,[20] sowie die dreibändige Geschichte der Ausbreitung der katholischen Religion über die Welt des flämischen Jesuiten Cornelius Hazart, in drei Bänden (1678-1701), ins Deutsche übersetzt und erweitert von P. Mathias Soutermans, aus Antwerpen, Mitglied der Österreichischen Ordensprovinz.[21]

Das unmittelbare Vorbild Stöckleins waren die *Lettres édifiantes et curieuses*[22], die von 1703 bis 1743 in Paris veröffentlicht wurden. Ursprünglich von P. Charles Le Gobien initiiert, wechselten ihre Herausgeber im Laufe der Jahre mehrmals. Am Ende umfasste die Reihe 36 kleine Bände, insgesamt wahrscheinlich nicht viel umfangreicher als die fünf voluminösen Folianten des *Welt-Bott.* Stöcklein dachte ursprünglich an eine Übersetzung der in Frankreich veröffentlichten Briefe, doch bald änderte er seine Pläne, wie er selbst beschrieb:

Kaum hatte ich auf guter Freunden Antrib die Hand angelegt / als mir von unterschiedlichen Orten ein sehr zahlreicher Vorrath anderer dergleichen / meistens von Teutschen Jesuitern aus Indien / haubtsächlich in Lateinischer Sprach geschribener Brieffen theils in Originalibus theils in Abschrifften zu dem

---

19 S. Sauer, Gottes streitbare Diener für Amerika. Missionsreisen im Spiegel der ersten Briefe niederländischer Jesuiten (1616-1618), Pfaffenweiler 1992.

20 M. Tanner, Die Gesellschaft JESU Bis zur Vergiessung ihres Blutes wider den Gotzendienst, Unglauben, und Laster, für Gott, den wahren Glauben, und Tugendten in allen vier Theilen der Welt streitend, Prag 1683 (lateinisch: Prag 1675). – Über Hazart, Soutermans und Tanner vgl. Sommervogel, Bibliothèque 4, 181-197, 7, 1406-1408, 1858-1861.

21 C. Hazart, Kirchen-Geschichte, Das ist: Catholisches Christenthum, durch die ganze Welt außgebreitet, Insonderheit bey nächst verflossenen, und anjetzo fliessenden Jahrhundert [...], T. 1-3, Wien 1678-1701 (niederländisch: Antwerpen 1667-1671). Im 2. Bd. wird als Übersetzer P. Mathias Soutermans angegeben.

22 C. Le Gobien (u. a., Hrsg.), Lettres édifiantes et curieuses écrites des missions étrangères par quelques missionnaires de la Compagnie de Jésus, v. 1-36, Paris 1703-1743.

Ende seynd mitgetheilt worden / damit ich aus diesen und jenen ein gantzes Werck verfertigte.[23]

Da Stöcklein seine Ordensbrüder aus den verschiedenen Provinzen der *Assistentia Germanica* der Gesellschaft Jesu aufforderte, ihm Schriften zuzuschicken, die sie für publikationswürdig hielten, erhielt er laufend neue Texte und besaß bald genug Material, um gleich einen zweiten Band seines Werkes anzukündigen.

Die Entstehung des *Welt-Bott* stand nicht nur in der Tradition der Reiseberichtsammlungen. Inspiriert in der antiken Tradition eines Seneca oder der Apostelbriefe des neuen Testaments waren Briefe in der frühen Neuzeit eine beliebte Form philosophischer wie auch literarisch-fiktionaler Texte. Man kann auf Blaise Pascal und seine antijesuitischen *Lettres provinciales* (1656/57) verweisen oder auf Montesquieu und seine *Lettres persanes* (1721). Erwähnenswert sind die fünf Bände der *Cartas eruditas y curiosas* (1742-1760) des Benediktiners Benito Jerónimo Feijoo. Der Titel bezieht sich offensichtlich auf die *Lettres édifiantes et curieuses* der französischen Jesuiten, doch ersetzte der aufgeklärte Feijoo den religiösen, zumindest nicht wissenschaftlichen Begriff der *edificación* (Erbauung) durch die *erudición* (Gelehrsamkeit, Bildung). Im selben Zusammenhang erlebte das literarische Genre des Briefromans einen großen Aufschwung, beginnend mit *Processo de cartas de amores* de Juan de Segura (1548) und besonders seit *Pamela: or, Virtue Rewarded* (1740) von Samuel Richardson bis zu den *Die Leiden des jungen Werthers* (1774) von Johann Wolfgang von Goethe und *Les liaisons dangereuses* (1782) des Pierre Choderlos de Laclos. Die Briefform suggerierte eine direkte, ja intime Kommunikation zwischen Werk und Leser, der sich der Vorstellung hingeben konnte, die beschriebenen Ereignisse dank der authentischen Berichterstattung hautnah miterleben zu können. Diesen Kunstgriff der Literaten haben die Herausgeber von Reiseberichten seit Ramusio vorweggenommen. Sie wollten nicht mehr wiederholen, was die alten Kosmographen geschrieben hatten, sondern neue und „wahre", von Augenzeugen autorisierte Information vermitteln.[24]

---

23 J. Stöcklein, Vorbericht an den Leser, in: Welt-Bott 1/1, o. S.

24 *Sir, having out of a Chaos of confused intelligences framed this Historicall World, by a New way of Eye-evidence* [...] (Purchas, Epistle [...] to prince Charles, prince of Wales, in: Hakluytus Posthumus 1, o. S.); W. Neuber, Die frühen deutschen Reiseberichte aus der Neuen Welt. Fiktionalitätsverdacht und Beglaubigungsstrategien, in: H.-J. König – W. Reinhard – R. Wendt (Hrsg.), Der europäi-

Renate Dürr hat dazu angemerkt, dass Stöcklein die von ihm veröffentlichten Berichte zwar oft stark kürzte und bearbeitete, aber gerade die Anredefloskeln stehen ließ, *weil sie Authentizität verleihen können.*[25]

Gleichzeitig entdeckte Europa seinen Geschmack am Fernen und Exotischen. Ganz neu war auch das nicht. Einmal abgesehen von Vorläufern in der mittelalterlichen Literatur, kann hier zum Beispiel auf den großen Erfolg der Reiseberichte des Amerigo Vespucci oder Hans Stadens gleich am Anfang der europäischen Expansion verwiesen werden. Schon im 18. Jahrhundert, um in einen ganz anderen Bereich der Literatur zu wechseln, erschien in Paris von 1704 bis 1717 die erste europäische Übersetzung von *Tausend und einer Nacht*, ausgeführt von Antoine Galland in zwölf Bänden, die nahezu zeitgleich ins Englische und Deutsche übersetzt wurden.[26] In diesem Kontext wurden Reisebeschreibungen für die verschiedensten Anliegen verwendet, angefangen mit der *Utopia* des Thomas Morus (1516) oder *The Tempest* von William Shakespeare (1611). Im 18. Jahrhundert wurden Reiseromane große Mode. Die berühmtesten Beispiele sind wohl *The Life and Strange Surprising Adventures of Robinson Crusoe, of York* (1719) von Daniel Defoe, die *Travels into Several Remote Nations of the World* [...] *By Lemuel Gulliver* (1726) von Jonathan Swift oder *L'histoire de Scarmentado, écrite par lui-même* (1756) und *Candide ou l'optimisme* (1759) von Voltaire. Keine der vier Erzählungen ist ein Briefroman, aber außer *Candide* haben alle einen Ich-Erzähler und präsentieren sich daher analog zum Brief wie persönliche Berichte eigenen Erle-

---

sche Beobachter außereuropäischer Kulturen. Zur Problematik der Wirklichkeitswahrnehmung, Berlin 1989, 43-64; A. Menninger, Die Macht der Augenzeugen. Neue Welt und Kannibalen-Mythos, 1492-1600, Stuttgart 1995; C. Reichler, Préface. Le secrétaire des saintes, in: C. Reichler (Hrsg.), Lettres édifiantes et curieuses des missions de l'Amérique méridionale par quelques missionaires de la compagnie de Jésus, Paris 1991, 14.

25 Dürr, Der „Neue Welt-Bott", 447.

26 M. Sironval, The Image of Sheherezade in French and English Editions of the Thousand and One Nights (Eighteenth–Nibeteenth Centuries), in: Y. Yamanaka – T. Nishio (Hrsg.), The Arabian Nights and Orientalism. Perspectives from East & West, London/New York 2006, 221f.; E.-P. Wieckenberg, Johann Heinrich Voss und "Tausend und eine Nacht", Würzburg 2002, 19.

bens.[27] All diese Bücher – es ist nicht mehr als eine selektive Auflistung – bilden einen Referenzrahmen für die Marktbedingungen, in denen Stöckleins Werk entstand.

3. Die Herausgeber erklären sich

Auf den Inhalt der im *Welt-Bott* oder anderen Sammlungen zusammengestellten Berichte kann hier nicht näher eingegangen werden.[28] Stattdessen soll versucht werden, die Motivation und die Ziele Stöckleins ausgehend von einem Vergleich seines Einleitungstexts mit denen anderer bereits genannter Werke zu analysieren.

Ramusio stellte 1550 sein Werk in den Zusammenhang mit dem Fortschritt der Wissenschaften. Er hielt fest, dass in verschiedensten Disziplinen, wie der Astronomie, der Philosophie oder der Medizin, im Vergleich zu den Kenntnissen der Autoren der Antike große Fortschritte erzielt worden seien. Nun lieferten die Entdeckungsfahrten der Portugiesen, Spanier und anderer Europäer laufend neue Erkenntnisse, so dass alles, was sich bei den alten Kosmographen lesen ließ, als überholt gelten musste. Ramusio wollte mit den Berichten dieser Reisen den Wissenszuwachs dokumentieren und damit die Basis für eine neue Geographie legen.[29] Zwar wurzelten seine Beweggründe zweifellos in den politischen und wirtschaftlichen Interessen seiner Heimatstadt Venedig, doch sah er sein Werk als Beitrag zu einem allgemeinen, um nicht zu sagen universalen Fort-

---

27 Zum Einfluss der Reiseberichte auf den Roman vgl. L. E. Pennington, Samuel Purchase. His Reputation and the Uses of his Works, in: L. E. Pennington (Hrsg.), The Purchas-Handbook. Studies of the Life, Times and Writings of Samuel Purchas 1577-1626, London 1977, 5-27, hier: 14.

28 Siehe z. B. P. Downes, Die Wahrnehmung des Anderen. Jesuitenmissionare und Indios im ‚Neuen Welt-Bott', in: J. Meier (Hrsg.), Sendung – Eroberung – Begegnung. Franz Xaver, die Gesellschaft Jesu und die katholische Weltkirche im Zeitalter des Barock, Wiesbaden 2005, 355-382.

29 *Ma la cagione che mi fece affaticar volentieri in questo opera fu che, vedendo e considerando le tavole della Geografia di Tolomeo, dove si descrive l'Africa e la India, esser molto imperfette rispetto alla gran cognizione che si ha oggi de quelle regioni, ho stimato dover esser caro e forse non poco utile al mondo il mettere insieme le narrazioni degli scrittori de' nostri tempi che sono stati nelle sopradette parti del mondo e di quelle han parlato minutamente* [...] (Ramusio, All'eccellentiss. M. Ieronimo Fracastoro, in: Ramusio, Delle navigazioni 1, o. S.).

schritt. Er wollte der Welt nützlich sein (*utile al mondo*), indem er die neuen Entdeckungen der Allgemeinheit zugänglich machte.

Einige Jahrzehnte später huldigte Hakluyt in England nicht der Wissenschaft, sondern seiner Nation, wie schon im Titel seines Hauptwerkes deutlich wird. England, so lautete die Botschaft, musste den Platz in der Welt einnehmen, der ihm zustand. Dafür appellierte Hakluyt an den Patriotismus seiner Landsleute, *that desire to aduaunce the honour of our Countrie which ought to bee in euery good man*[30]. Diese Forderung bedurfte keiner Rechtfertigung, ihre Legitimität war eine Selbstevidenz. Für Hakluyt standen sowohl die Ehre wie die Macht und der Wohlstand der Nation auf dem Spiel. Doch noch musste er anerkennen, dass die großen Entdeckungen bisher andere gemacht hatten. Die Engländer suchten lediglich erfolglos nach einer Nordwest- und einer Nordostpassage nach Asien. Um an die von Portugiesen und Spaniern monopolisierten Reichtümer des Orients heranzukommen, blieb ihnen vorerst nur die Piraterie. Die Scham, immer die zweiten zu sein, konnte von wagemutigen Korsaren wie Francis Drake, dem zweiten Weltumsegler, oder all den Seeleuten, die in den Polarmeeren ihr Leben verloren, ohne an ihr Ziel zu gelangen, nur teilweise getilgt werden.[31] England musste auf James Cook warten, damit es endlich eine Persönlichkeit besaß, die man zum größten Seefahrer der Geschichte erklären konnte. Die Fahrten seiner Zeitgenossen dienten Hakluyt allerdings als Ausdruck der Vorsehung. So beschwor er die Ankunft eines neuen Zeitalters, in dem auf die Ära der Spanier und Portugiesen die der Engländer folgen würde,[32] und am Ende wurde er so zum Prediger des kolonialen Expansionismus:

---

[30] Hakluyt, The Epistle Dedicatorie, in: Hakluyt, Divers voyages, 1. Zum historischen Kontext siehe Quinn, Richard Hakluyt , 3-9. Vgl. auch E. Mitsi, "Nowhere is a Place": Travel Writing in Sixteenth-Century England, in: Literature Compass 2 (2005), 1-13.

[31] W. H. Sherman, Bringing the World to England. The Politics of Translation in the Age of Hakluyt, in: Transactions of the Royal Historical Society 14 (2004), 199-207.

[32] *But nowe it is high time for vs to weigh our ancre, to hoise vp our sailes, to get cleare of these boistrous, frosty, and misty seas, and with all speese to direct our course for the milde, lightsome, temperate, and warme Atlantick Ocean, ouer which the Spaniards and Portugales haue made so many pleasant prosperous and golden voyages* (Hakluyt, Preface, in: Ders.: Principal Navigations 1, o. S.), vgl. auch Hakluyt, Epistle, in: Hakluyt, Divers voyages, 1.

Wee reade that the Bees, when they grow to be too many in their own hiues at home, are wont to bee led out by their Captaines to swarme abroad, and seeke themselues a new dwelling place.[33]

Die Wissenschaft war für Hakluyt kein Ziel, sondern ein Mittel, um Konkurrenten zu überholen. Die Spanier und ihre Errungenschaften wie der Lehrstuhl für Nautik an der Casa de la Contratación in Sevilla waren das Vorbild, das es zu übertreffen galt. Mit seinen Büchern wollte Hakluyt potentielle Sponsoren dazu anregen, dieses Vorhaben zu fördern.[34] Er unterstrich den kommerziellen Nutzen eines solchen Unternehmens und den Gefallen, den Gott aus der Ausbreitung der protestantischen Religion in die letzten Winkel der Erde finden würde:

[...] *if we first seeke the kingdome of God, all other things will be given vnto vs, and* [...] *lasting riches do waite vpon them that are zealous for the aduauncement of the kingdom of Christ, and the enlargement of his glorious Gospell.* [35]

Und zur Aussicht, nach Ostasien zu gelangen, merkte er an:

[...] *I take it as a pledge of Gods further fauour both vnto vs and them: to them especially, vnto whose doors I doubt not in time shall be by vs caried the icomparable treasure of the truth of Christianiy, and of the Gospell, while we vse and exercise common trade with their marchants.* [36]

Seine Argumentation, mit all ihren merkantilen Ambitionen, passte perfekt zur diskursiven Konstruktion Englands als protestantische Nation.[37] Hakluyt behauptete, dass für die Spanier die Verbreitung des Glaubens nur ein Vorwand war, während er für die Engländer zum Gefallen Gottes von entscheidender Bedeutung sein musste.

Nicht unähnlich unterstrich der Jesuit Cornelius Hazart, dass die Erfolge der katholischen Mission die Liebe des Himmels und den

---

33 Hakluyt, Epistle, in: Hakluyt, Divers voyages, 1v. Möglicherweise liegt hier ein Einfluss der *Utopia* des Thomas Morus vor; R. Saage, Utopia und die drei Identitäten des Thomas Morus, in: Zeitschrift für Geschichtswissenschaft 53/12 (2005), 1082.

34 Hakluyt, Epistle, in: Hakluyt, Divers voyages, 2v-4; A preface to the Reader, in: Hakluyt, The Principal Navigations 1, o. S.

35 Ebd., 2v.

36 Hakluyt, Dedication to the first edition [1589], in: Hakluyt, Principal Navigations 1, o. S.

37 A. Maccoll, The Construction of England as a Protestant 'British' Nation in the Sixteenth Century, in: Renaissance Studies 18/4 (2004), 582-608.

Vorrang des Katholizismus unter den Religionen offenbarten, denn keine andere hätte Ähnliches vollbracht. Der flämische Pater hoffte, dass die Geschichte dieses Erfolgs als Beweis der Allmacht Gottes und allen Gläubigen zur Erbauung diente.[38]

Die Kluft zwischen einem nationalistischen Autor wie Hakluyt und den Jesuiten wird bei P. Soutermans, dem Übersetzer von Hazart, deutlicher. Er betonte in radikalem Gegensatz zu Hakluyt den transnationalen und globalen Universalismus der Kirche und hielt fest:

Meines Stands Obligenheit ist / sich nicht von den Gräntzen einiges Landes einschrancken zu lassen / sondern mit Apostolischen Eüffer die gantze Welt / wo es möglich / zu umlauffen / und das Paner Christi allenthalben / forderist aber in jene Länder einzuführen / woselbst es zuvor niemahls gesehen oder erkant gewest.[39]

Soutermans unterstrich die Nützlichkeit der Berichte aus fernen Ländern für die Bildung und die Wissenschaft. Der Vorrang der Religion stand für ihn jedoch außer Zweifel. Seit den Zeiten der Entdeckung von Ost- und Westindien war auf Deutsch eine Serie von Werken über die Völker und Länder der Erde erschienen, aber fast keine über die Fortschritte des Christentums unter ihnen. Während für Ramusio das „Nützliche" gerade im „Neuen" bestanden hatte, konstruierte Soutermans einen asymmetrischen Gegensatz zwischen dem „Neuen" (neuem Wissen) und dem „Nützlichen" (dem Religiösen):

Ob nun woll die allgemeine Begierd deß Menschen viel mehr genaigt ist was neües / als was nutzliches zu vernehmen / jenes auch nicht in allweg zu tadeln / so will es doch einen Christglaubigen mehr anständiger und lobsamer seyn / daß er wisse / was Gestalt das Gesetz Gottes / so er vor allen hochschätzen soll / bey denen zuvor unbekandten Völckern eingeführt und außgebreitet worden / als dass ihme bekand sey / was für seltsame Frücht die Baum / was für Kräuter die Erd herfürbringe; wie die wilden Thier geartet / wie die Menschen gekleidet / oder gebildet / die Stätt gebauet / und andere dergleichen ungesehene Sachen beschaffen seynd. Weil nun dieses Wercks Urheber beydes / daß Neüe und das Nutzliche / daß ist / die Beschaffenheit der zuvor unbekandten Länder und Völcker nebst dem alda eingeführten Christen-Thun sehr weißlich vereinbahret /

38 Hazart, Vorrede deß Urhebers, in: Hazart – Soutermans, Kirchen-Geschichte 1, o. S.

39 Soutermans, Dedicatio, in: Hazart – Soutermans, Kirchen-Geschichte 1, o. S.

als hab ich mich desto lieber dieser Mühe-Waltung unterzogen / in Hoffnung bey dem neu begierigen Leser zugleich einen Lust / und Hochschätzung deß wahren Glaubens zu erwecken.[40]

In Prag benützte P. Mathias Tanner in seiner Einleitung fast dieselben Worte wie Soutermans über die universale Berufung der Kirche im Allgemeinen und der Gesellschaft Jesu im Speziellen. In seinem Prolog lässt er jedoch einen anderen Geist erkennen, den man als aggressive Verteidigungsbereitschaft bezeichnen könnte: niemand sonst, so Tanner, hätte auf der Welt soviel Feinde wie die Jesuiten, sowohl unter den Heiden und Ketzern wie in der Katholischen Kirche selbst. Die Opfer, die die Ordensmitglieder auf sich nahmen, dienten dafür als Beweis und bezeugten gleichzeitig den Rang der Gesellschaft Jesu. Zahlreiche Jesuiten verließen ihre Familien und Freunde, um ihr Blut im Martyrium Christus und dem Glauben zu opfern. Doch ihr Bekehrungswerk war dadurch nicht gebremst worden.[41] Die Glorifizierung der Gesellschaft Jesu fehlte auch in den Einleitungen von Hazart oder Soutermans nicht, aber sie ist dort deutlich zurückhaltender.[42] Tanner präsentiert seine Propaganda als Verteidigung gegen die allgegenwärtigen Feinde. Deren Umtriebe rechtfertigten den pathetischen Ton. Gleichzeitig offenbarte Tanner seinen Glauben an die himmlische Vorsehung, denn sowohl die Erfolge wie der Tod der Jesuiten waren Teil der Pläne Gottes.[43]

P. Charles Le Gobien, der Herausgeber des ersten Bandes der *Lettres édifiantes*, richtete sich 1703 direkt an die Jesuiten, genauer an die französischen. Er beschwor den Missionsgeist, *cette ardeur pour les Missions Etrangeres, qui est comme l'ame & l'esprit de nostre Institut*, die Liebe zum Kreuz und die Sehnsucht nach dem Martyrium, *le premier attrait, par lequel Dieu appelle un grand nombre de Missionnaires*[44]. Auffällig ist, mit welchem Nachdruck die Projekte der Jesuiten mit der Politik

40 Soutermans, Erinnerung deß Ubersetzers, in: Hazart – Soutermans, Kirchen-Geschichte 1, o. S.

41 Tanner, Vorrede, in: Gesellschaft JESU, o. S.

42 Vgl. Hazart, Vorred und Erinnerung des Urhebers an den günstigen Leser, in: Hazart – Soutermans, Kirchen-Geschichte 2, o. S.

43 Tanner, Vorrede, in: Gesellschaft JESU, o. S. – Man vgl. die Unterdrückung der Israeliten und die auf die Ägypter losgelassenen Plagen bei 3 Moses, 14:4 und 7:1-5.

44 Le Gobien, Aux jesuites de Frances, in: Letteres édifiantes 1, o. S.

des Königs identifiziert werden. So atmet das Werk, obwohl es die Erfolge der Mission feiert, etwas vom patriotischen Geist Hakluyts. Doch dieser erhob seine Nation, während der französische Jesuit die Monarchie und ihre Diener pries, mit anderen Worten den Staat. Die Gesellschaft Jesu war in der zweiten Hälfte des 17. Jahrhunderts in Frankreich durch schwierige Zeiten gegangen. Jetzt galt es, sich in der französischen Politik neu zu verankern. Die Veröffentlichung der *Lettres édifiantes* waren Teil dieser Strategie.[45]

4. P. Stöcklein bewirbt den *Welt-Bott*

In der Einleitung zum *Welt-Bott* griff P. Stöckleins zahlreiche Elemente seiner Vorgänger auf und gab ihnen an einigen Stellen überraschende Wendungen. Wie in anderen jesuitischen Schriften mischen sich in seinem Text barocke Tradition und Modernität, was manche Beobachter verstört und zu den widersprüchlichsten Interpretationen des Werks der Gesellschaft Jesu geführt hat.

Stöcklein verfasste für den ersten Band des *Welt-Botts* eine Einleitung, die in vier Teile gegliedert war: die *Zuschrifft des Verfassers* (Widmung); die *Allgemeine Vorrede des Verfassers über dieses gantze Werck*; den *Vorbericht an den Leser*; und *Von etlichen zur Schiffahrt und zur See=Kunst gehörigen Worten*, ein umfangreiches Vokabular der Schifffahrt. Im Süden Deutschlands, wo Stöcklein fern der Ozeane lebte und arbeitete, kannte man den für maritime Themen notwendigen Wortschatz nicht, und so bot der Pater seinem Publikum die nötigen Erklärungen.

4.1 *Zuschrifft des Verfassers* (die Widmung)

Die Widmung eines Buches an eine Persönlichkeit von Rang war eine der Konventionen der Zeit. Ramusio hatte seine Sammlung entsprechend seinem wissenschaftlichen Anspruch dem angesehen Gelehrten und Arzt Girolamo Fracastoro gewidmet, der unter anderem der Syphilis ihren Namen gab. Hakluyt verehrte seine Bücher hervorragenden Persönlichkeiten des politischen Lebens Englands und Purchas den *Hakluytus posthumus* sogar dem Prinzen von Wales.

45 Reichler, Préface, 12-13.

De Bry wandte sich im ersten Band seines herausgeberischen Werks an den Herzog von Sachsen, P. Tanner an zwei Schwestern aus böhmischem Adel, und P. Soutermans widmete die deutsche Übersetzung von Hazarts Geschichte Ferdinand Bonaventura von Harrach, der sich unter anderem als kaiserlicher Botschafter in Madrid hervortat. Im 18. Jahrhundert verfiel dieser Brauch, zuerst in Frankreich und mit einiger Verspätung auch in Deutschland.[46] P. Le Gobien setzte den *Lettres édifiantes* keine Widmung mehr voran, sondern nur einen einleitenden Brief an die französischen Jesuiten. Wenig später verzichtete Montesquieu explizit auf diese überkommene Tradition: *Je ne fais point ici l'épître dédicatoire, et je ne demande point de protection pour ce livre: on le lira, s'il est bon; et s'il est mauvais, je ne me soucie pas qu'on le lise.*[47]

Im *Welt-Bott* erzählte Stöcklein, dass er ursprünglich überlegt hatte, sein Werk einer angesehenen Persönlichkeit zu widmen. Doch dann entschied er sich anders. Die Ehre gebührte den alten und neuen Missionaren, und so überreichte er ihnen von seinem Arbeitszimmer in Graz aus symbolisch sein Buch:

An die so wohl im Himmel bereits frohlockende / als auf Erden wider die Hölle in allen Vier Welt-Theilen annoch streitende Missionarios der Gesellschaft Jesu. Wie nicht weniger an andere Heilig-mäßige Personen / derer in gegenwärtigem Buch gedacht wird. Er verband diese Erklärung mit der eindeutigen Metapher der Mission als Schiff, auf deren Hauptmast eine *weiße mit einem Purpur=Creutz verschränckte und über alle Segel gleichsam in den Himmel fliegende Flagge" der Missionare wehte, die „Vatter und Mutter / Brüder und Schwester / Hab und Gut um deren Missionen willen verlassen" hatten. Sie waren dem Kreuzweg Christi gefolgt und saßen jetzt an seiner Seite, „jeglicher auf seinen absonderlichen Ehren=Stul / damit ihr die Völcker richtet / denen ihr geprediget / und die grausame Wütterich verdammet / die euch gemartert oder verfolget haben.*[48]

Stöcklein benützt hier Bilder eines – in diesem Fall religiösen – Heroismus, wie er für die Barockliteratur typisch war.[49] Er wollte

---

46 G. Schramm, Widmung, Leser und Drama. Untersuchungen zu Form- und Funktionswandel der Buchwidmung im 17. und 18. Jahrhundert, Hamburg 2003, 536.

47 A. Lefèvre (Hrsg.), Montesquieu: Lettres persanes [1721], Paris 1873, 1.

48 Stöcklein, Zuschrifft des Verfassers, in: Welt-Bott 1/1, o. S.

49 M. Disselkamp, Barockheroismus. Konzeption ›politischer‹ Größe in Literatur und Traktatistik des 17. Jahrhunderts, Tübingen 2002.

sowohl die Gesellschaft Jesu wie ihr Missionswerk durch den Schweiß und das Blut hervorheben, das ihre Mitglieder vergossen, so wie wir es schon bei P. Tanner beobachtet haben. Stöcklein präsentierte sein Werk als Heilsgeschichte von Männern, die durch das Ertragen aller Beschwerlichkeiten die höchsten himmlischen Ehren verdient hatten. Auf diese Weise und durchaus im Einklang mit der katholischen Lehre gab Stöcklein dem Individuum herausragendes Gewicht, dem einzelnen Menschen, der zwischen Sünde, Buße und guten Werken sein ewiges Heil entscheidet. Ein Individualismus dieser Art mag typisch für den Geist des Humanismus und der Renaissance gewesen sein, dem auch die Gesellschaft Jesu entstammte. Doch Stöcklein ging noch ein Stück weiter. Noch bevor die ersten Denkmäler für die Toten der Kriege (inspiriert von der bürgerlichen Revolution und der Ausbildung der Nationalstaaten) errichtet wurden, widmete Stöcklein sein Buch keinem Herzog oder Grafen, auch nicht dem Papst oder dem Ordensgeneral, sondern den gemeinen Soldaten des universalen Kriegs gegen den Teufel. Er ehrt die unbekannten Alltagshelden des Missionswerks. Am Ende gestand Stöcklein, dass er die von ihm gesammelten Briefe auf Deutsch veröffentlichte, damit ihre Verbreitung in Deutschland die Jugend dazu motivierte, sich an der Mission zu beteiligen. Die dem gewöhnlichen Missionar erwiesene Ehre entsprach so dem Sinn und Zweck, den die Nationalstaaten mit der Ehrung der gefallenen Soldaten verfolgten: es galt die Zusammengehörigkeit und den Gruppengeist und damit die Bereitschaft zu stärken, sich am Kampf um das „Gemeinwohl“ zu beteiligen. Am Ende erweist sich das Individuum als Instrument eines Kollektives.

4.2 *Allgemeine Vorrede des Verfassers über dieses gantze Werck*

Ist der Zweck der Sammlung einmal klar, musste dafür gesorgt werden, dass sie auch gelesen wurde. Ganz offen instrumentalisierte Stöcklein dafür das exotische Potential der überseeischen Mission:

Niemand ist zwar unbekant / wie daß alles / was aus fremden Ländern nach Teutschland gebracht wird / unsern Vorwiz mehr als alle einheimische Seltsamkeiten kizle; Weil nemlich fast allen Menschen fremdes Brod besser zu schmäcken pfleget.[50]

[50] Stöcklein, Allgemeine Vorrede, in: Welt-Bott 1/1, o. S.

Auch bei Soutermans stößt man auf ähnliche Stellen, und die *Lettres édifiantes* folgten derselben Maxime, gleichzeitig gefallen wie belehren zu wollen, um so ein erzieherisches Ziel zu erreichen. Man war damit einer klassischen Ästhetik verpflichtet.[51] Doch anders als der etwas despektierliche Ton Soutermans über solche Neugierden und im Gegensatz zur Militanz des P. Tanner, gab sich Stöcklein recht entspannt, vielleicht in Folge des größeren Abstandes zu den religiösen Kriegen des 17. Jahrhunderts oder angesteckt durch den neuen Geschmack am Exotischen, wie es von *Tausend und einer Nacht* oder dem *Robinson Crusoe* geboten wurde. Er pries sein Werk den Lesern im Süden Deutschlands an, die nirgendwo koloniale Ambitionen verfolgten, sondern die weite Welt von ihren Häusern aus, zurückgelehnt im Lesestuhl und mittels einer angenehmen Lektüre kennen lernen wollten:

> [...] *doch gebührt dißfalls denen Welt-und Reis-Beschreibungen ohne Widerrede der Vorzug / angesehen ein neugieriger Leser / ohne einen Schritt zu thuen / zu Hauß sizend mit blosser Lesung dergleichen Schrifften die ganze Welt durchwandern / hiedurch aber ihm selbst eine bey nahe vollständige Erkantnuß des ganzen Erd-Kreyß ohne Gefahr / Unkösten oder Mühe erwerben kan.*[52]

Auch Ramusio und Hakluyt brachten vor, dass dank ihrer Arbeit andere sich häuslicheren Vergnügungen widmen konnten. Doch Ramusio scheint dabei eher an intellektuelle und wissenschaftliche Beschäftigungen gedacht zu haben,[53] und Hakluyt benutzte das

51 Reichler, Préface, 13-14. P. Keller schloss seine Einleitung zum 38. Teil des Welt-Botts mit der Einladung an den Leser: *Er bediene sich selber zu seiner Erbauung und Ergötzung* (Keller, Vorrede, in: Welt-Bott 5/38, 3).

52 Stöcklein, Allgemeine Vorrede, in: Welt-Bott 1/ 1, o. S.

53 *Ma che dico io del piacere che ne aranno li dotti e studiosi? Chiè colui che possa dubitare che ancor molti dei signori e principi non si abbiano a dilettare di cosi fatta lezione? Ai quali piú che ad alcuno altro appartiene il saper i secreti e particolarità della detta parte del mondo e tutti i siti delle regioni, provincie e città di quella, e le dependenzie che hanno l'uno dall'altro i signori e popoli che vi abitano. Perché, ancora che ne possino esser informati e instrutti da altri che abbino quei paesi trascorsi, gli scritti e ragionamenti de'quali essi leggendo e udendo hanno già fatto giudicio esser molto copiosi, son certisimo che, leggendo questo libro e considerando le cose in esso comprese e dichiarate, conosceranno quelle lor narrazioni a comparazione di questa esser brievi, manche e di poco momento, tanto serà il frutto ch'a piena satisfazione d'ogni lor desiderio ne trarranno i lettori* (Ramusio,

Argument mehr, um seine Leistung als Herausgeber hervorzuheben, als um damit die Leser anzulocken.[54] Stöcklein ließ dagegen keinen Zweifel daran bestehen, dass er sich mit seinem Werk das allgemeine deutschsprachige Publikum gewinnen wollte (und nicht – wie es etwa die *Lettres édifiantes,* zumindest in der Einleitung, erscheinen lassen – in erster Linie die Mitglieder seines Ordens).

Der allgemeine kulturelle Kontext seiner Zeit musste Stöcklein für sein Vorhaben durchaus günstig erscheinen. In der zweiten Hälfte des 17. Jahrhunderts hatte die Reiseliteratur sich ein immer breiteres Publikum gewonnen.[55] Allerdings ist auch zu beobachten, dass im deutschsprachigen Raum Schriftsteller erst ab den sechziger Jahren des 18. Jahrhunderts mit wirklich hohen Leserzahlen rechnen konnten.[56] Stöcklein wäre somit ein sehr frühes Beispiel oder ein Vorläufer einer solchen auf ein allgemeines Publikum zielenden Haltung, während Le Gobien noch für eine Gruppe gebildeter und informierter Freunde schrieb, in seinem Fall für seine Ordensbrüder. Stöcklein zählt dagegen alle des Lesens und Schreibens mächtige Berufsgruppen auf, von Theologen, Juristen, Apothekern, Philosophen, Mathematikern, Geographen und Lehrern der Ethik bis zu den Soldaten und Handwerkern, die alle aus der Lektüre der Jesuitenbriefe ihre nützlichen Lehren ziehen konnten. Sein Anliegen liegt in der Bildung des Volks und nicht so sehr in der Verbreitung der Wissenschaft. Tatsächlich befinden sich, abgesehen von den völkerkundlichen Notizen, im *Welt-Bott* kaum Informationen, die wissenschaftlich Neues geboten hätten, aber doch vieles, was dem breiten Publikum unbekannt war. Die Priester konnten zum Beispiel sehen, dass es auch unter den Angehörigen fremder Religionen nicht an Menschen wie den indischen Brahmanen fehlte, die ein Leben mit großer Disziplin und Entsagung führten. Man wäre verführt, hierin den Ausdruck jesuitischer Toleranz zu sehen, doch leider wird dieses Thema später mit einem ganz anderen Ton wieder aufgegriffen. Jedenfalls schien Stöcklein den außerordentlichen Gebräuchen der

---

All'eccellentiss. M. Ieronimo Fracastoro, in: Ramusio, Delle navigationi 1, o. S.).

54 *For the bringing of which unto this homely and rough-hewen shape, which here thou seest* [...] (Hakluyt, Preface to the Reader, in: Principal Navigations 1, o. S.).

55 S. Requemora, L'espace dans la littérature de voyages, in: Études Littéraires 34/1-2 (Québec 2002), 249-276.

56 Schramm, Widmung, 535-536.

Brahmanen keine geringere Wirkung auf die Aufmerksamkeit der Leser zuzuschreiben als den Opfern der Missionare.

Stöcklein ließ die Gelegenheit nicht verstreichen, nochmals auf die Entbehrungen und Verfolgungen hinzuweisen, die die Missionare litten. Er kritisierte zum Beispiel die holländischen Calvinisten, die nichts unversucht ließen, um das Bekehrungswerk der Jesuiten zu unterbinden; dagegen lobte er die Engländer, von denen man stets mit größter Höflichkeit behandelt wurde. Die Jesuiten ertrugen alle Schwierigkeiten und Erniedrigungen jedoch mit einzigartiger Geduld, den Sinn fest auf den Erfolg der Mission gerichtet. So lebten sie in Indien, um sich den Respekt der Bevölkerung zu gewinnen, in derselben Armut wie die verehrten Brahmanen. Doch die Patres – so fügte Stöcklein hinzu, und hier endete seine Toleranz und sein Respekt – lebten in echter Armut, während die Brahmanen alle Heuchler waren, *da sie indessen heimlich in der Schwelgerey / Unzucht / Knaben-Lieb und andern Lastern / wie die Säu in dem Sumpff sich herum welzen.*[57] Stöcklein wiederholte damit ein Urteil, das schon der Heilige Franz Xaver 1544 geäußert hatte.[58] Die Bereitschaft der Jesuiten zum Dialog mit anderen Religionen bewegte sich in recht engen Grenzen. Alle Anstrengungen zur Kommunikation dienten dem unverrückbaren Ziel, den anderen zu bekehren und so vor dem Höllenfeuer zu bewahren. Mit den Brahmanen und ihren Lustknaben war kein Dialog zu führen, man konnte sie nur zu retten versuchen.

Abschließend wandte sich Stöcklein der Frage zu, wo nun das Werk der Jesuiten am verdienstvollsten und heiligsten sei. Bezeichnenderweise sagte er in diesem Zusammenhang nichts über China, einst Schauplatz des prestigereichsten Missionsunternehmens. Doch das jesuitische Vorgehen im Reich der Mitte war mittlerweile innerkirchlich heftig umstritten und Gegenstand des so genannten Ritenstreits geworden, eine Polemik, in die er sich nicht verstricken lassen wollte.[59] Seine Rangliste wurde so von Indien vor Paraguay angeführt. Paraguay war vielleicht der Ort, wo es am besten gelungen war,

---

57 Stöcklein, Allgemeine Vorrede des Verfassers, in: Welt-Bott 1/1, o. S.

58 K. Schatz, Franz Xaver und die Herausforderung der nichtchristlichen Religionen, in: J. Meier (Hrsg.), Sendung – Eroberung – Begegnung. Franz Xaver, die Gesellschaft Jesu und die katholische Weltkirche im Zeitalter des Barock, Wiesbaden 2005, 99-117, hier 101.

59 So führte er selbst im *Vorbericht an den Leser* aus. Er folgte damit auch einer Aufforderung des Ordensgenerals Tamburini; Strasser, Case, 34.

eine ideale Gemeinschaft nach dem Vorbild der ersten Christen zu schaffen. Wegen der gewaltigen Schwierigkeiten, die die Missionare zu bewältigen hatten, gab Stöcklein jedoch Indien den Vorrang. Allerdings hatte die Erfahrung gezeigt, dass schon die Anwesenheit von Muslimen jeden Missionserfolg zu ersticken drohte. Daher musste man sich mit guten Gründen um den Bestand der Christenheit in Asien sorgen, während die amerikanische Mission als gefestigt gelten konnte. Auch Tanner hatte angemerkt, dass die Bewohner Amerikas, die im übrigen als *grausam wilde / und unter allen Menschen die unfreündlichste Leüte* beschrieb, *von dem Mahometischen Affter=Glauben unversehret geblieben.*[60] In Afrika ließen sich dagegen nur die geringsten Missionserfolge erwarten, denn es war beherrscht vom Islam und stand außerhalb der Kontrolle eines katholischen Königs, so dass es den sicheren Tod bedeute, den Kontinent als Missionar zu betreten (und den sicheren Tod zu suchen galt als sündig und nicht als heiliges Martyrium).[61]

4.3. *Vorbericht an den Leser*

Im letzten Teil seiner Einleitung erzählte Stöcklein von sich selbst. Das hatten andere schon vor ihm getan. Hakluyt schilderte ein Ereignis seiner Jugend, einen Besuch bei seinem Cousin desselben Namens, als Schlüsselerfahrung und Beginn seiner Leidenschaft:

> [...] *I found lying open vpon his boord certeine bookes of Cosmograhie, with an vniversall Mappe: he seeing me somewhat curious in the view therof, began to instruct my ignorance, by shewing me the diuision of the earth into three parts after the olde account, and then according to the latter, & better distribution, into more* [...] *From the Mappe he brought me to the Bible, and turning to 107. Psalme, directed mee to the 23 & 24 verses,*[62] *where I read, that they which go downe to the sea in ships, and occupy by the great waters, they see the works of the Lord, and his woonders in the deepe, &. Which words of the Prophet together with my cousins discourse (things of high and rare deleight to my yong nature) tooke in me so deepe an impression, that I constantly resolued* [...] *I would by*

---

60 Tanner, Die Societät in America, in: Gesellschaft JESU, o. S.

61 Ebd.

62 Die beiden Psalmen lauten: *Die mit Schiffen auf dem Meer fuhren und trieben ihren Handel in großen Wassern; die des Herrn Werke erfahren haben und seine Wunder im Meer*.

Gods assistance prosecute that knowledge and kinde of literature, the doores whereof [...] *were so happily opened before me.*[63]

Stöckleins Erfahrungen waren andere. Er berichtete, wie er während einer langen Krankheit begonnen hatte, die *Lettres édifiantes* zu lesen. Er fand dabei solche Inspiration, dass er begann, sie für das deutsche Publikum zu übersetzen.[64] Es besteht nun kein Grund, an der Erzählung Stöckleins zu zweifeln. Ihre Analogie mit der Geschichte des Hl. Ignatius ist jedoch evident. Letzterer hatte, von einer Kriegsverletzung zur Untätigkeit verurteilt und ohne seine bevorzugten Abenteuerromane, zu religiöser Literatur gegriffen, *una Vita Christi y un libro de la vida de los Santos en romance*, die ihn zu seiner geistlichen Karriere inspirierte.[65] Man kann hier erwähnen, dass Stöcklein möglicherweise am 31. Juli, dem Geburtstag des Hl. Ignatius, geboren wurde, was seine Identifikation mit dem Gründer seines Ordens gefördert haben mag.[66]

Nach seinem Berufungserlebnis erwies sich Stöckleins Projekt in erster Linie als harte Arbeit. Im Kontrast zum Genuss und dem Nutzen der Lektüre seiner Missionsberichte betonte Stöcklein die Mühen der Herausgabe. Es handelte sich um dieselben Schwierigkeiten, die auch alle anderen Herausgeber beklagten, manchmal in einem typischen Bescheidenheitsdiskurs, so Ramusio, der sich mit einer Herausforderung konfrontiert sah, die seinen schwachen Verstand zu übersteigen drohte.[67] Für Stöcklein begannen die Probleme damit, die Texte, die bei ihm einlangten, zu entziffern, korrekt zu transkribieren und aus dem Latein ins Deutsche zu übersetzen. Manche Kopien, die ihm zugeschickt wurden, waren so schlampig angefertigt worden, dass ihm schon das Lesen größte Mühen bereitete. Besonders knifflig erwies sich die Transkription der fremdlän-

---

63 Hakluyt, Dedication to the first edition [1589], in: Hakluyt, Principal Navigations 1, o. S.

64 Stöcklein, Vorbericht an den Leser, in: Welt-Bott 1/1, o. S.

65 I. de Loyola, Autobiografía de San Ignacio de Loyola (texto recogido por el P. Luis Gonçalves da Camara entre 1553 y 1555), cap. 1, § 5, Internet Ressource.

66 Auf die symbolische Bedeutung dieses Datums verwies auch P. Keller, in: Leben, Thaten, Reisen und Mission R. P. Josephi Stöcklein, in: Welt-Bott, Teil 29, 141.

67 *un debile e piccolo ingegno come è il mio* (Ramusio, All'eccellentiss. M. Ieronimo Fracastoro, in: Ramusio, Delle navigationi 1, o. S.).

dischen Toponyme.[68] Danach zwang die große Zahl eingegangener Berichte Stöcklein, eine Auswahl zu treffen:

> [...] *womit geschehen / daß (gleichwie vorher dern Frantzösischen Brieffen schon fünff-zehn Theil waren und der Teutsche / wo nicht mehr / eben so viel / mithin beyderseits zusammen über dreyßig Theil wurden ausgemacht haben) ich zwar alle mit Fleiß durchgelesen / aber in Betrachtung / daß in denen einen / wie in denen andern viel Sachen gar zu offt wiederholet werden / mit Auslassung dergleichen Uberflusses die meisten Brief mercklich verkürtzt / viel (die nichts Neues oder Unbekantes enthielten) völlig ausgelassen / mit dieser Behutsamkeit aber / ohne was wichtiges zu überspringen / das gantze Buch um drey Viertel enger zusammen gezogen / und aus zwey und dreyßig Theilen / so viel sonst wurden heraus kommen seyn / nur achte gemacht hab.*[69]

Stöcklein beklagte schließlich das Fehlen einer einheitlichen Standardisierung der deutschen Schriftsprache. Er suchte deshalb nach einer Ausdrucksform, die zwischen den verschiedenen Varianten des Hochdeutschen vermittelte, besonders zwischen den zwei wichtigsten, der sächsischen und der rheinischen. In Augsburg, so kommentiert er sarkastisch, hatten dann noch die Drucker Hand angelegt und nach Gutdünken einiges geändert, von den gewöhnlichen Druckfehlern ganz zu schweigen, ohne dass er das von Graz aus verhindern hätte können.

Für eine Person war die Arbeit des Herausgebers folglich gewaltig, doch sie wurde auch reich vergolten. Das hatte schon Hakluyt anerkannt:

> [...] *what restlesse nights, what painefull dayes, what heat, what cold I haue indured; how many long chargeable iourneys I haue trauailed ...; what expenses I haue not spared; and yet what faire opportunities of priuate game, preferment, and ease I haue neglected* [...] *Howbeit* [...] *the honour and benefit of this common weale wherein I liue and breathe, hath made all*

68 Vgl.: [...] *gli esemplari che mi son venuti alle mani erano estremamente guasti e scorretti.* (Ramusio, All'eccellentiss. M. Ieronimo Fracastoro, in: Ramusio, Delle navigationi 1, o. S.).

69 Stöcklein, Vorbericht an den Leser, in: Welt-Bott 1/1, o. S.

difficulties seeme easie, all paines and industries pleasant and all expenses of light an moment vnto me.[70]

5. Der Herausgeber als Missionar

Stöckleins Präsentation seines Werks stand in einer breiten literarischen Tradition. Darüber hinaus besaß sie aber eine symbolisch-religiöse Bedeutung. Auch die protestantischen Autoren gaben ihrem Tun eine religiöse Tiefe. Der Abenteurer Walter Raleigh etwa bezeichnete seine Expedition nach Guayana als „Pilgerfahrt", ein Begriff, der auch Purchas gefiel, und auf religiöse Elemente im Werk Hakluyts, ein ordinierter Priester, der die zweite Auflage der *Principal Navigations* als „Prediger" zeichnete,[71] wurde schon verwiesen. Doch war dem Engländer die Religion den Zielen der kolonialen Expansion untergeordnet. Bei Stöcklein scheinen mir jedoch Interpretationen, die in ihm einen Proponenten eines *imaginären* deutschen Kolonialismus sehen wollen,[72] den tiefen und vorrangig religiösen Sinn seines Werkes zu verkennen (was nicht ausschließt, dass seine Schriften wie die anderer Jesuiten in Deutschland tatsächlich zu einer „imperialen Sehnsucht" beigetragen haben, die sich erst im 19. Jahr-

70 Hakluyt, A preface to the Reader, in: Hakluyt, The Principal Navigations 1, o. S. – An anderer Stelle bemerkte er, dass er alle Mühen ertrüge *for the benefit and honour of my Country.*

71 "R. Hakluyt, preacher"; Hakluyt, Dedication to the second edition [1598], in: Hakluyt, Principal Navigations 1, o. S.

72 So die These von Strasser. Das stärkste Argument der Autorin ist der zeitliche Zusammenhang der Entstehung des *Welt-Botts* mit dem Erwerb der österreichischen Niederlande durch den Kaiser und der Gründung der *Kaiserlich-Ostindischen Kompagnie* im Jahre 1722 (Strasser, Case, 38). Auch die persönliche Nähe Stöckleins zu Prinz Eugen ließen einen solchen Bezug nicht unwahrscheinlich erscheinen. Jede Anlehnung der Jesuiten an eine imperiale Macht sollte aber in Bezug auf ihre religiösen Ziele interpretiert werden. Dasselbe gilt für die wissenschaftlichen Aktivitäten der Missionare, die z. B. von Dürr, Der „Neue Welt-Bott", und Zermeño, Entre el saber, sehr hervorgehoben werden. Wissenschaft war den Jesuiten kein Selbstzweck. Dürrs Interpretation des Frontispiz des *Welt-Bott* als symbolische Darstellung des Wissenstranfers *als zentrales Element christlicher Mission* (S. 448) würde ich eine Interpretation als Zusammenführung der Menscheit in der Einheit des Glaubens gegenüberstellen, die natürlich Wissenstransfer impliziert, aber vorrangig religiösen und nicht wissenschaftlichen Charakters war.

hundert unter den Vorzeichen des deutschen Nationalstaates ausleben konnte).

Stöcklein bemühte sich sehr, den Kontrast zwischen den Mühen des Missionars und den Annehmlichkeiten des Lesers narrativ herauszuarbeiten. Doch der Leser genoss die Lektüre nur dank der Anstrengungen des Herausgebers. Stöcklein verortet sich so als Vermittler zwischen den Missionaren und Autoren der Berichte und dem Leser, eine Rolle, die er unter persönlichen Opfern und Entbehrungen erfüllt. Seine Anstrengungen zielten nicht auf die Verbreitung der Religion in den exotischen Teilen der Welt, sondern vorrangig auf das Bewusstsein des Lesers. Diesen will er mit den gleichzeitig unterhaltsamen wie bildenden Attraktionen der Missionsberichte und dem Angebot, sie in aller Bequemlichkeit genießen zu können, verführen. Das Projekt Stöckleins ist manipulierend und, wenn man so will, missionarisch. Der Missionar interveniert in fremden Ländern zur Rettung der Ungläubigen; Stöcklein, der nicht in Übersee ist, interveniert zur Rettung des Lesers. Dieser muss die Lektion, die man ihn lehrt, nur annehmen. Die Lehre wurde nicht nur aus strategischen Gründen in angenehmer Form angeboten, sondern auch weil die Freude Bestandteil derselben sein sollte. Es handelte sich nicht um eine bedrückende Religion, die die Jesuiten anbieten wollten, sondern der Glaube sollte fröhlich und glücklich sein, so wie es Lodovico Antonio Muratori über die Christenheit in Paraguay postulierte.[73] Schon Ignatius von Loyola hatte über die Zeit seiner Berufung geschrieben, *que de unos pensamientos quedaba triste, y de otros alegre, y poco a poco viniendo a conocer la diversidad de los espíritus que se agitaban, el uno del demonio, y el otro de Dios.*[74]

Ignatius hatte, wie gesagt, das Glücksgefühl des Glaubens bei der Lektüre der *Vita Christi* und eines Buchs über das Leben der Heiligen entdeckt, und Ähnliches erlebte Stöcklein beim Lesen der *Lettres édifiantes* über die Abenteuer der Missionare. Jetzt wollte er dieselbe Erfahrung anderen vermitteln. Wie Ignatius gab Stöcklein sich nicht damit zufrieden, das Glück im Glauben gefunden zu haben, sondern er wollte es zum Heil seiner Nächsten verwenden und damit, zumindest potentiell, einen expansiven Prozess der Bekehrung auslösen.

---

73 L. A. Muratori, Il Cristianismo felice nelle Missioni de'Padri della Compagnia di Gesù nel Paraguai, t. 1-2., Venezia 1743-1749.

74 Loyola, Autobiografía, cap. 1, § 8.

Die Erzählungen von Stöcklein dienten, analog zum Heroismus der Märtyrer bei Tanner, als Vorbilder, wie auch Modelle weltlichen Heldentums bei der Erziehung der Fürsten und Könige der Barockzeit eingesetzt wurden. Die Leser sollten Inspiration in den Lebensgeschichten der Missionare finden, wie sie ja auch der Gründer der Gesellschaft Jesu in den Heiligenviten gefunden hatte:

> [...] *leyendo la vida de nuestro Señor y de los santos, se paraba a pensar, razonando consigo: ¿qué sería, si yo hiciese esto que hizo San Francisco, y esto que hizo Santo Domingo? y así discurría por muchas cosas que hallaba buenas, proponiéndose siempre a sí mismo cosas dificultosas y graves, las cuales cuando proponía, le parecía hallar en sí facilidad de ponerlas en obra* [...] *se le ofrecían los deseos de imitar los santos, no mirando más circunstancias que prometerse así con la gracia de Dios de hacerlo como ellos lo habían hecho.*[75]

Neben den Beispielen von Heldentum konnte auch die Beschreibung der Natur zum Wachsen des Glaubens beitragen, wie man in der Autobiographie des Hl. Ignatius lesen kann:

> *Y la mayor consolación que recibía era mirar el cielo y las estrellas, lo cual hacía muchas veces y por mucho espacio, porque con aquello sentía en sí un muy grande esfuerzo para servir a nuestro Señor.*[76]

Im frühen 17. Jahrhundert verwunderte sich P. Andrés Pérez de Ribas über den kunstvollen Nestbau der Webervögel,[77] und über hundert Jahre später erfreute sich P. Pfefferkorn während der langen Abende an den *unnachahmlichen Farben und Auszierungen* der Nachtfalter, die seine Lampe umkreisten und ihn mit Staunen und Dankbarkeit über *das unglaubliche Dasein des Schöpfers* erfüllten.[78] Solche Erlebnisse, erbauend im vollen Sinn des Wortes, gaben auch den wissenschaftlichen und naturgeschichtlichen Einstreuungen in den

75 Loyola, Autobiografía, cap. 1, § 7, § 9, Internet Ressource.

76 Ebd., cap. 1, § 11, Internet Ressource.

77 A. Pérez de Ribas, Historia de los triunfos de nuestra santa fe entre gentes las más bárbaras y fieras del Nuevo Orbe, Madrid 1645, Neuausgabe 1, México 1944, 124.

78 I. Pfefferkorn, Beschreibung der Landschaft Sonora samt anderen merkwürdigen Nachrichten von den inneren Theilen Neu-Spaniens und Reise aus Amerika bis in Deutschland, nebst einer Landkarte von Sonora 1, Köln 1794-1795, 382f.

Missionsberichten einen religiösen Sinn. Sie sollten – schriftlich vermittelt – einen geistigen Prozess in Gang setzen, wie es etwa P. Hazart erfahren hatte: *Als ich dieß gantze Welt=Wesen mit meinen Gedancken durchlauffe / und in reiffer Obsicht die darin waltende Allmacht Gotte uberlege* [...].[79]

Ebenso wie die Missionare die Botschaft der katholischen Kirche den Heiden der ganzen Welt übersetzen mussten, um den geistigen Vorgang der Bekehrung auszulösen, übersetzte Stöcklein die Texte der Missionare als Teile der universalen Heilsgeschichte für seine deutschen Zeitgenossen in eine allgemein verständliche Sprache.[80] Dazu musste er die komplexen Berichte kürzen, aufbereiten und in eine geeignete Sprache übersetzen. Fachbegriffe mussten erklärt werden. Sein Verzeichnis der Termini der Seefahrt oder die Erklärungen zur richtigen Aussprache spanischer Wörter ist durchaus analog zu den Bemühungen der Missionare zu sehen, die geeigneten Worte und Übersetzungen der Grundbegriffe der christlichen Lehre, wie Gott, Himmel, Hölle oder Jungfrau zu finden. Stöckleins Klage über die Teilung des Deutschen in zwei große Sprachblöcke und andere kleinere Varianten lässt sich als Metapher für die konfessionelle Teilung lesen, galt doch das sächsische Hochdeutsch als die Sprache Luthers und das rheinische eher als katholisch. Stöcklein betonte dagegen sein Bemühen, eine sprachliche Einheit herzustellen.[81] Er bedauerte die Hindernisse, die ihm in den Weg gelegt wurden, aber damit belegte er nur die Schwierigkeiten jeder religiösen Bekehrung. Eine solche ließ sich nicht mit einem Schlag erreichen. Selbst die Wandlung des Ignatius von Loyola vom sehr weltlichen Soldaten zum heiligen Diener Christi war ein allmählicher Prozess. Seine Lektüre erleuchtete ihn nicht plötzlich, sondern legten vielmehr ein Samenkorn, aus dem der Glauben sprießen sollte. So empfahl Stöcklein, seine Berichte entspannt zu lesen und kündigte erst einmal

---

79 Hazart, Vorrede deß Urhebers, in: Hazart – Soutermans, Kirchen-Geschichte 1, o. S.

80 Demgegenüber scheint mir die Interpretation Strassers (Case, 33), *Stöcklein's Welt-Bott thus was a textual instrument for forging an imaginary global community of German Christians unites inprayer and the common goal of conversion* die Idee des christlichen Universalismus national zu verengen.

81 Strasser, Case, 35 gibt diesem Thema eine nationale Interpretation (*a deeper desire for the synchronisation of national sentiments*).

einen zweiten Band seiner Sammlung an, in der Hoffung damit einen Berufungsprozess in Gang zu setzten.[82]

6. Abschließende Überlegungen

Obwohl man eine weite Verbreitung des *Welt-Botts* in den deutschen Bibliotheken nachweisen kann und der erste Band schon 1728 eine Neuauflage erlebte, scheint die Hoffnung Stöckleins (wie die anderer Autoren der Zeit), sich mit Büchern auf Deutsch eine wachsende Leserschaft erschließen zu können, sich nur zum Teil er-füllt zu haben. Denn während immer mehr Schriftsteller auf das Deutsche setzten, begann der Adel und die gebildete Schicht sich zunehmend dem Französischen zuzuwenden.[83] Das könnte erklären, warum der *Welt-Bott* selbst im deutschsprachigen Gebiet nie dasselbe Prestige besaß wie die *Lettres édifiantes*.[84] Doch Stöckleins Werk wurde an die verschiedenen Niederlassungen der Gesellschaft Jesu verteilt, wo es bei der Ausbildung der jungen Jesuiten verwendet wurde, von denen gerade im 18. Jahrhundert eine wachsende Zahl als Missionare in die weite Welt zog. Wahrscheinlich war es innerhalb der Organisation des Ordens, wo der *Welt-Bott* seine größte Wirkung tat.

War der *Welt-Bott* somit nur ein kurioses Produkt einer vormodernen Welt? Der Universalismus der katholischen Kirche war in der Zeit Stöckleins selbst in Europa nur mehr eine partikuläre Haltung, und trotz aller Anstrengungen der Missionare ließ er sich nie mehr als dominierende Weltsicht durchsetzen. Hakluyts nationalistischer

---

82 Vgl. Disselkamp, Barockheroismus, 54-82.

83 A. Martino, Barockpoesie, Publikum und Verbürgerlichung der literarischen Intelligenz, in: Internationales Archiv für Sozialgeschichte der deutschen Literatur 1 (1976), 107-145.

84 G. B. González, Die jesuitische Berichterstattung über die Neue Welt. Zur Verbreitungsgeschichte von Amerika-Nachrichten im Alten Reich am Beispiel der Briefe des Dominikus Mayer, in: J. Meier (Hrsg.), Sendung – Eroberung – Begegnung. Franz Xaver, die Gesellschaft Jesu und die katholische Weltkirche im Zeitalter des Barock, Wiesbaden 2005, 366-371; G. B. González, "Libros americanos, autores jesuitas y público alemán: la literatura jesuítica americana en el mercado de libros del siglo XVIII, in: K. Kohut – M. C. Torales (Hrsg.), Desde los confines de los imperios ibéricos. Los jesuitas de habla alemana en las misiones americanas, Madrid/Frankfurt 2007, 663-696.

Blickwinkel erwiese sich dagegen schnell als überaus erfolgreich und setzte sich spätestens im 19. Jahrhundert allgemein durch. Dem Glauben an die Wissenschaft eines Ramusio hat auch alle Skepsis wenig anhaben können, die die politischen, sozialen und ökologischen Folgen der wissenschaftlichen Eroberung der Erde mittlerweile erzeugt haben. Stöckleins Werk besaß dagegen eine religiöse Bedeutung, deren Wurzeln weit in der Zeit zurückreichen. Hier ist nicht der Platz, zu diskutieren, ob Kirche und Religion in unserer Zeit Aktualität haben könnte oder ob die westliche Modernität nicht zwangsweise mit der Entsakralisierung der Welt verbunden ist, wie Max Weber gemeint hat. Renate Dürr ist völlig zuzustimmen, dass sich etwa der Wissenschaftsoptimismus vieler Jesuiten kaum vom Fortschrittsglauben der Frühaufklärer unterschied.[85] Der Unterschied scheint eher in den Zielen zu liegen, die man mit dem Fortschritt der Wissenschaft zu ereichen hoffte. Historisch wäre es somit von größtem Interesse zu zeigen, inwiefern Werke wie der *Welt-Bott* und das missionarische Ideal der universalistischen Kirche zur Entstehung dieser Modernität, so wie sie sich heute präsentiert, beigetragen haben.

Worin konnte dieser Beitrag liegen? Zweifellos haben die Jesuiten zur wissenschaftlichen Erschließung der Welt durch Europa beigetragen. Faszinierend scheint vor dem Hintergrund des Individualismus der „Moderne" die Behandlung des Individuums und seine Verantwortlichkeit für seine Taten, wie wir anhand der Widmung Stöckleins an die ungenannten Missionare andeuten konnten. Dieses Menschenkonzept impliziert einen gewissen Egalitarismus, jenseits aller weltlichen Unterschiede von Rang, Stand und Status, die nicht in Frage gestellt werden. In Folge dieses tendenziell egozentrischen Individualismus stellten die Jesuiten am Ende die Selbsterfahrung des Missionars, seine Selbstrettung, über die Rettung des Nächsten. Deshalb bewertete Stöcklein auch die entbehrungsreiche indische Mission höher als die amerikanische, weil in der ersten die Jesuiten sich stärker beweisen konnten, während in der zweiten vor allem die Seelen anderer gerettet wurden.[86] Das Individuum ist auch das

---

85 Dürr, Der „Neue Welt-Bot", 466.

86 Vgl. J. J. Baegert, Nachrichten von der Amerikanischen Halbinsel Californien: mit einem zweyfachen Anhang falscher Nachrichten, Mannheim 1773, 184f.

Objekt der Bekehrung. Gerade aus dem Drang zur Bekehrung entwickelten die Missionare aller Orden ausgefeilte Kommunikationstechniken und Überzeugungsstrategien. Obwohl Stöckleins Einleitungstext als religiöse Parabel gelesen werden kann, nimmt sie mindestens in Teilen Methoden moderner Werbetexte vorweg. Stöcklein konstruiert einen emotionalisierten Diskurs, um sich „Kunden" zu gewinnen, so wie man heute Bilder von Stränden und Palmen einsetzt, um uns zum Kauf eines Stücks Seife zu veranlassen. Drittens muss das globale Weltverständnis betont werden, das sich Stöcklein und die Jesuiten im Allgemeinen, ausgehend vom Universalismus der Kirche, angeeignet hatten. Damit waren sie „Nationalisten" wie Hakluyt oder Purchas ein Stück voraus. Was sie alle gemeinsam hatten, war ihr hegemonistischer Anspruch, den man eurozentrisch nennen kann. Für P. Tanner ist, nicht zufällig und unabhängig vom Vormarsch des Protestantismus, Europa *der gantzen Welt vornehmster und bester Theil / nicht zwar ihrer Große halber / sondern ob wohl eingerichten Land- und Stadt-Rechten.*[87] Damit sind wir nun im Jahre 2010 angekommen.

Epilog

Im Herbst 1978 organisierte Günther Hamann, unterstützt von seinen damaligen Assistenten Johannes Dörflinger, Helmut Grössing und Marianne Klemun, eine Bücherschau mit Schätzen aus der Wiener Universitätsbibliothek zu seiner Vorlesung über Entdeckungsfahrten. Neben einer Faksimileausgabe der Sammlung des Ramusio beeindruckte mich schon damals der *Neue Welt-Bott.* Nicht ohne Nostalgie denke ich daran zurück und weiß nicht recht, was ich davon halten soll, dass ich mich mehr als dreißig Jahre später von Mexiko aus mit diesen Werken beschäftige. Einen geradezu niederschmetternden Eindruck hinterließ aber eine ebenfalls ausgestellte, in meiner Erinnerung vierbändige, jedenfalls monumentale Dissertation zum kartographischen Niederschlag der Expeditionen zur Suche nach einer Nordwestpassage. Zum ersten Mal wurde mir der Ernst der Existenz eines Wissenschaftlers bewusst, wenn auch noch nicht in allen ihren Tücken. Das kam erst später. Warum ich trotzdem bei

---

87 Tanner, Die Societät JESU in Europa, in: Gesellschaft JESU, o. S.

dieser Laufbahn geblieben bin? Johannes Dörflinger hat mit seiner uneitlen Berufsauffassung sicherlich dazu beigetragen.

Quellen- und Literaturverzeichnis

Baegert J. J., Nachrichten von der Amerikanischen Halbinsel Californien: mit einem zweyfachen Anhang falscher Nachrichten, Mannheim 1773.

Burghartz S. (Hrsg.), Inszenierte Welten. Die west- und ostindischen Reisen der Verleger de Bry, 1590-1630/Staging New Worlds. De Bry's Illustrated Travel Reports, 1590-1630, Basel 2004.

Burghartz S., Transformation und Polysemie. Zur Dynamik zwischen Bild, Text und Kontext in den „Americae" der de Bry, Internet Ressource, http://www.hist.net/debry/polysemie.pdf (3.1.2006).

Collani C. v., Stöcklein, Joseph Stöcklein, in: Traugott Bautz, Biographisch-Bibliographisches Kirchenlexikon 10 (1995): 1516-1517, http://www.bautz.de/bbkl/s/s4/stoecklein_j.shtml (19.8.2005).

Disselkamp M., Barockheroismus. Konzeption ›politischer‹ Größe in Literatur und Traktatistik des 17. Jahrhunderts, Tübingen 2002.

Downes P., Die Wahrnehmung des Anderen. Jesuitenmissionare und Indios im ‚Neuen Welt-Bott', in: Meier J. (Hrsg.): Sendung – Eroberung – Begegnung. Franz Xaver, die Gesellschaft Jesu und die katholische Weltkirche im Zeitalter des Barock, Wiesbaden 2005, 355-382.

Dürr R., Der „Neue Welt-Bott" als Markt der Informationen. Wissenstransfer als Moment jesuitischer Identitätsbildung, in: Zeitschrift für Historische Forschung 34 (2007), 441-466.

González G. B., Die jesuitische Berichterstattung über die Neue Welt. Zur Verbreitungsgeschichte von Amerika-Nachrichten im Alten Reich am Beispiel der Briefe des Dominikus Mayer, in: Meier J. (Hrsg.), Sendung – Eroberung – Begegnung. Franz Xaver, die Gesellschaft Jesu und die katholische Weltkirche im Zeitalter des Barock, Wiesbaden 2005, 366-371.

González G. B., Libros americanos, autores jesuitas y público alemán: la literatura jesuítica americana en el mercado de libros del siglo XVIII, in: Kohut K. – Torales M. C. (Hrsg.), Desde los confines de los imperios ibéricos. Los jesuitas de habla alemana en las misiones americanas, Madrid/Frankfurt 2007, 663-696.

Hakluyt R. (Hrsg.), Divers voyages touching the discouerie of America, and the Ilands adiacent vnto the same [...], London 1582.

Hakluyt R. (Hrsg.), The principal navigations, voiages, traffiqves and discoueries of the English nation [...], London 1598-1600.

Hazart C., Kirchen-Geschichte, Das ist: Catholisches Christenthum, durch die ganze Welt außgebreitet […], T. 1-3, Wien 1678-1701 (niederl. Antwerpen 1667-1671).

Lefèvre A. (Hrsg.), Montesquieu: Lettres persanes [1721], Paris 1873.

Le Gobien C. (u. a., Hrsg.), Lettres édifiantes et curieuses écrites des missions étrangères par quelques missionnaires de la Compagnie de Jésus, v. 1, Paris 1703, http://visualiseur.bnf.fr/Visualiseur?Destination =Gallica&O=NUMM-106074 (2.12.2005).

Loyola I. de, Autobiografía de san Ignacio de Loyola (texto recogido por el P. Luis Gonçalves da Camara entre 1553 y 1555), cap. 1, § 5, http://www.e-libro.net/E-libro-viejo/gratis/Loyola.pdf (29.11.2009).

Martino A., Barockpoesie, Publikum und Verbürgerlichung der literarischen Intelligenz, in: Internationales Archiv für Sozialgeschichte der deutschen Literatur 1 (1976), 107-145.

Maccoll A., The Construction of England as a Protestant 'British' Nation in the Sixteenth Century, in: Renaissance Studies 18/4 (2004), 582-608.

Menninger A., Die Macht der Augenzeugen. Neue Welt und Kannibalen-Mythos, 1492-1600, Stuttgart 1995.

Mitsi E., Nowhere is a Place: Travel Writing in Sixteenth-Century England, in: Literature Compass 2 (2005), 1-13.

Muratori L. A., Il Cristianismo felice nelle Missioni de'Padri della Compagnia di Gesù nel Paraguai, p. 1-2, Venezia 1743-1749.

Neuber W., Die frühen deutschen Reiseberichte aus der Neuen Welt. Fiktionalitätsverdacht und Beglaubigungsstrategien, in: König H. J. – Reinhard, W. – Wendt, R. (Hrsg.), Der europäische Beobachter außereuropäischer Kulturen. Zur Problematik der Wirklichkeitswahrnehmung, Berlin 1989, 43-64.

Pennington L., Samuel Purchase. His Reputation and the Uses of his Works, in: Pennington L. (Hrsg.), The Purchas-Handbook. Studies of the Life, Times and Writings of Samuel Purchas 1577-1626, London 1977, 5-27.

Pérez de Ribas A., Historia de los triunfos de nuestra santa fe entre gentes las más bárbaras y fieras del Nuevo Orbe, Madrid 1645, Neuausgabe, t. 1-3, México 1944.

Pfefferkorn I., Beschreibung der Landschaft Sonora samt anderen merkwürdigen Nachrichten von den inneren Theilen Neu-Spaniens und Reise aus Amerika bis in Deutschland, nebst einer Landkarte von Sonora, T. 1-2., Köln 1794-1795.

Purchas S., Hakluytus Posthumus, or Purchas His Pilgrimes, containing a History of the world, in Sea-voyages, and lande-Travells, by Englishmen and others [...], Some left written by Mr. Hakluyt, at his death more since added [...], v. 1-5, London 1625-1626.

Quinn D. B., Richard Hakluyt, Editor. A Study Introductory to the Facsimile edition of Richard Hakluyt's DIVERS VOYAGES (1582), to which is added a facsimile of A SHORT AND BRIEF NARRATION OF THE TWO NAVIGATIONS TO NEWE FRANCE, Translated by John Florio (1580), Amsterdam, 1967.

Ramusio G. B., Delle navigationi et viaggi, nel qual si contiene la descrittione dell' Africa et del paese del Prete Janni, con varii viaggi dal Mar Rosso a Calicut, et in sin all' isole Molucche [...], t. 1-3, Venezia 1550-1559.

Reichler C. (Hrsg.), Lettres édifiantes et curieuses des missions de l'Amérique méridionale par quelques missionaires de la compagnie de Jésus, Paris 1991.

Requemora S., L'espace dans la littératur de voyages, in: Études Littéraires 34/1-2 (Québec 2002), 249-276.

Saage R., Utopia und die drei Identitäten des Thomas Morus, in: Zeitschrift für Geschichtswissenschaft 53/12 (2005), 1077-1089.

Sauer S., Gottes streitbare Diener für Amerika. Missionsreisen im Spiegel der ersten Briefe niederländischer Jesuiten (1616-1618), Pfaffenweiler 1992.

Schatz K., Franz Xaver und die Herausforderung der nichtchristlichen Religionen, in: Meier J. (Hrsg.), Sendung – Eroberung – Begegnung. Franz Xaver, die Gesellschaft Jesu und die katholische Weltkirche im Zeitalter des Barock, Wiesbaden 2005, 99-117.

Schramm G., Widmung, Leser und Drama. Untersuchungen zu Form- und Funktionswandel der Buchwidmung im 17. und 18. Jahrhundert, Hamburg 2003.

Schurhammer G. – Wicki J. (Hrsg.), Epistolae S. Francisci Xaverii aliaque eius scripta. Nova editio ex integro refecta textibus, introductionibus, notis, appendicibus aucta, v. 1-2., Roma 1944-1945, erw. Neudruck 1996.

Sherman W. H., Bringing the World to England. The Politics of Translation in the Age of Hakluyt, in: Transactions of the Royal Historical Society 14 (2004), 199-207.

Sironval M., The Image of Sheherezade in French and English Editions of the Thousand and One Nights (Eighteenth–Nineteenth Centuries), in: Yuriko Yamanaka Y. – Nishio T. (Hrsg.), The Arabian Nights and Orientalism. Perspectives from East & West, London/New York 2006, 219-244.

Sommervogel C., Bibliothèque de la Compagnie de Jésus 7, Brüssel/Paris 1896.

Stöcklein J. (u. a., Hrsg.), Der Neue Welt-Bott. Allerhand so Lehr- als Geist=reiche Brief, Schriften und Reise=Beschreibungen [...], t. 1-5, Augsburg/Graz bzw. Wien 1726-1761.

Strasser U., A Case of Empire Envy? German Jesuits Meet an Asian Mystic in Spanish America, in: Journal of Global History 2 (2007), 23-40.

Tanner M., Die Gesellschaft JESU Bis zur Vergiessung ihres Blutes wider den Gotzendienst, Unglauben, und Laster, für Gott, den wahren Glauben, und Tugendten in allen vier Theilen der Welt streitend [...], Prag 1683 (latein.: Prag, 1675).

Wieckenberg E.-P., Johann Heinrich Voss und „Tausend und eine Nacht“, Würzburg 2002.

Zermeño G., Entre el saber y la edificación: una relación inestable, in: Zermeño G. (Hrsg.), Cartas edificantes y curiosas de algunos misioneros jesuitas del siglo XVIII. Travesías, itinerarios, testimonios, México 2006, 17-57.

Zum Autor

Bernd Hausberger, geb. 1960, Studium der Geschichte und Hispanistik an der Universität Wien, 1991-1993 Vertragsassistent bei Johannes Dörflinger am Institut der Geschichte an der Universität Wien, 1993-2006 zuerst wissenschaftlicher Mitarbeiter, dann wissenschaftlicher Assistent am Lateinamerika-Institut der Freien Universität Berlin und seit 2006 Professor am Centro de Estudios Históricos des Colegio de México in Mexiko-Stadt, Hauptforschungsbereiche: Kolonialgeschichte Lateinamerikas, Globalgeschichte und Geschichte im Film.

Kontakt: bhausberger@colmex.mx

Erich Woldan,
Die älteste Weltkarte mit dem „wirklichen“
Australien,
Wien 1981

Markus Heinz, Berlin

Versuch einer *umständlichen Historie* der Historie *der Land-Charten*

In Eberhard David Haubers Buch von 1724,[1] an dessen Titel derjenige dieser Skizze angelehnt ist, breitet der Autor das Gerüst einer ausführlichen Beschreibung aller Aspekte der Kartographie seiner Zeit und ihrer Geschichte systematisch vor dem Leser aus. Das nie in der geplanten vollständigen Fassung erschienene Werk ist eines der ersten, das sich explizit und ausführlich mit der Geschichte dieses Faches auseinander setzt, und somit einer der Grundsteine dieses Beitrags.

Aufbauend auf einer Analyse der zentralen Fragestellungen von Werken zur Geschichte der Kartographie sollen im Folgenden Strömungen der Forschungsgeschichte herausgearbeitet und chronologisch eingeordnet werden.[2] Dem Thema und Anlass des Bandes

1 E. D. Hauber, Versuch einer umständlichen Historie der Land-Charten, Ulm 1724.

2 Die Idee zu dieser Beschäftigung entstand bei der Vorbereitung einer Sitzung zur Geschichte der Kartographie auf dem Kartographentag in Oldenburg 2008 und in der vorbereitenden Auseinandersetzung mit aktuellen Strömungen in der Erforschung der Geschichte der Kartographie. Zentrale Arbeit zum Thema ist nach wie vor der Abschnitt von B. Harley, The Map and the Development of the History of Cartography, in: B. Harley – D. Woodward (Hrsg.), Cartography in prehistoric, ancient, and medieval Europe and the Mediterranean (The History of Cartography 1), Chicago 1987, 1-42, in dem sich auch detaillierte Hinweise auf weitere Werke, Autoren und teilweise auch weitere Strömungen finden lassen. Wertvolle Hinweise liefert auch T. Horst, Die Altkarte als Quelle für den Historiker – Die Geschichte der Kartographie

entsprechend wird hier vor allem auf die Einordnung der deutschen und österreichischen Kartographiegeschichte, die nicht zuletzt durch Protagonisten wie Erich Woldan oder Johannes Dörflinger geprägt wurde, Bezug genommen. Es ist festzustellen, dass die kartographiehistorische Forschung zu fast allen Zeiten relativ heterogen auftrat und somit die im Weiteren skizzierten Strömungen notwendiger Weise Vereinfachungen darstellen. Es erscheint dem Autor dennoch sinnvoll, die verschiedenen Einflüsse auf einen rückschauend konstruierten Hauptstrom der Kartographiegeschichte aufzuzeigen und damit eine erste Periodisierung zu skizzieren.

1. Einzelhinweise zur Geschichte der Kartographie

Bis gegen Ende des 17. Jahrhunderts ist keine systematische Auseinandersetzung mit der Geschichte der Kartographie erkennbar. An keiner Stelle scheint ein Autor diesem Aspekt auch nur ein ganzes Kapitel gewidmet zu haben. Die Hinweise auf alte Karten oder Kartenmacher der Vergangenheit waren vereinzelt und finden sich in Zusammenhängen, die nichts mit einer Geschichte der Kartographie als Überblick zu tun haben. In manchen Fällen wurde auf antike Vorbilder oder Vorläufer hingewiesen, um dem eigenen Tun eine höhere Relevanz zu verleihen (so wurde auf der Ebstorfer Weltkarte auf die „Vermessungen" durch Julius Caesar angespielt[3]). In anderen Fällen wurden aus historischem Interesse die Leistungen Einzelner hervorgehoben, wie beispielsweise auch später in den Biographien Johann Gabriel Doppelmayrs von 1730.[4] Wieder andere nahmen Bezug auf Vorgängerarbeiten, um die Vorzüge der eigenen Karte herauszustreichen.

---

als Historische Hilfswissenschaft, in: Archiv für Diplomatik, Schriftgeschichte, Siegel- und Wappenkunde 54 (2008), 309-377.

3 D. Woodward, Medieval Mappaemundi, in: B. Harley – D. Woodward (Hrsg.), Cartography in prehistoric, ancient, and medieval Europe and the Mediterranean (The History of Cartography 1), Chicago 1987, 309.

4 J. G. Doppelmayr, Historische Nachricht von den Nürnbergischen Mathematicis und Künstlern, Nürnberg 1730 (zum Beispiel die Ausführungen zu Behaim mit Nachzeichnung des Globus).

3. Karten als historische Quelle für die Geschichte der Geographie

Obwohl die Beschäftigung mit der Geographie des Ptolemäus am Beginn der Neuzeit, das Faksimilieren der *Tabula Peutingeriana* oder das Sammeln von Karten zur Entdeckungsgeschichte durch Richard Hakluyt bereits im 16. Jahrhundert die alte Karte als Quelle für einen historischen und oft auch rechtlich-politischen Zustand der Vergangenheit belegen (z. B. territoriale Ansprüche aus der Entdeckungsgeschichte), entstand hieraus wenig spezifische Literatur. Erst knapp nach 1800 trafen sich die Geschichtswissenschaft und besonders die aufstrebende und im Laufe des 19. Jahrhunderts institutionalisierte Geographie (vertreten durch Personen wie Alexander von Humboldt, Jomard oder Santarem) und etablierten eine positive Einschätzung der alten Karte als Quelle für die Entdeckungsgeschichte beziehungsweise das geographische Wissen bestimmter Epochen wie das des Mittelalters. Dies spiegelte sich – technisch möglich gemacht durch die Einführung der Lithographie – vor allem in der Herstellung von Faksimiledrucken alter Karten wieder. Alexander von Humboldt ließ beispielsweise 1853 die Juan de la Cosa Karte reproduzieren,[9] Jomard und Santarem publizierten jeweils ganze Sammlungen mittelalterlicher sowie entdeckungsgeschichtlicher Karten und markierten so den Beginn einer neuen Beschäftigung mit alten Karten. Für diese stand die Geschichte der Kartographie nicht im Zentrum des Interesses, sondern diente der quellenkritischen Bewertung der Karten für andere Fragestellungen. Dieser Forschungsstrang setzt sich bis in die Gegenwart fort, wo man ihn in Gestalt der historischen Geographie, aber auch nach wie vor in einer besonders engen Beziehung zwischen Kartographiegeschichte und Entdeckungs- respektive Expansionsgeschichte wieder antrifft. In Wien wurde diese Richtung vor allem durch Günther Hamann und teilweise Johannes Dörflinger, aber auch Erich Woldan und andere vertreten.

---

9 A. v. Humboldt, Ueber die ältesten Karten des Neuen Continents und den Namen Amerika. Mit einer genauen Abbildung des Behaim'schen Globus vom Jahr 1492 in 2 Planigloben nach seiner natürlichen Größe und 3 ältesten Karten von Amerika, in: F. W. Ghillany, Geschichte des Seefahrers Ritter Martin Behaim, Nürnberg 1853.

4. Regionale oder nationale Kartengeschichte

Der Beginn dieser Strömung ist aufgrund der zahlenmäßig stark anwachsenden Publikationen im 19. Jahrhundert schwer zeitlich oder personell fest zu machen. Selbstverständlich fielen der sich steigernde Nationalismus oder auch ein verstärktes regionales Bewusstsein in diese Periode und so waren Untersuchungen zur Geschichte der kartographischen Darstellung eines kleineren Raumes für diese gesellschaftspolitischen Strömungen als Legitimation nutzbringend. Ob sich hier, ebenso wie bei der Entstehung der Nationalatlanten, besonders jene Regionen hervortaten, die nach Selbstständigkeit strebten oder diese zu befestigen suchten,[10] bliebe noch zu untersuchen. Relativ früh verband sich dieser Forschungsstrang mit institutionellen aber auch privaten Sammlungen. Bis heute erfreuen sich entsprechende Ausstellungen oder Veröffentlichungen einer besonderen Publikumswirksamkeit. Dies soll in keinem Fall den oft hohen wissenschaftlichen Rang entsprechender Publikationen in Frage stellen, was an Werken wie *Descriptio Austriae,*[11] *Cartographia Bavariae*[12] oder *Berlin-Brandenburg im Kartenbild*[13] deutlich abzulesen ist. Die Grundfrage dieser Richtung orientiert sich – ähnlich der vorgenannten Strömung – allerdings ebenfalls stark an der Karte selbst sowie dem Karteninhalt und betrachtet somit das Umfeld der Karte nur unter dem Aspekt der Zusatzinformation.

---

10 I. Kretschmer, Zur Entwicklung thematischer Atlanten im 19. und 20. Jahrhundert, in: H. Wolff (Hrsg.), 400 Jahre Mercator - Vierhundert Jahre Atlas. Die ganze Welt zwischen zwei Buchdeckeln. Eine Geschichte der Atlanten (Bayerische Staatsbibliothek Ausstellungskataloge 65), Weißenhorn 1995, 233.

11 J. Dörflinger – R. Wagner – F. Wawrik, Descriptio Austriae. Österreich und seine Nachbarn im Kartenbild von der Spätantike bis ins 19. Jahrhundert, Wien 1977.

12 H. Wolff (Bearb.), Cartographia Bavariae. Bayern im Bild der Karte (BSB Ausstellungskataloge 44), Weisenhorn 1988.

13 W. Scharfe – H. Scheerschmidt (Hrsg.), Berlin-Brandenburg im Kartenbild. Wie haben uns die anderen gesehen? Wie haben wir uns selbst gesehen? (Staatsbibliothek zu Berlin – Ausstellungskataloge N. F. 42), Berlin 2000.

5. Disziplingeschichte

Ab dem Ende des 19. Jahrhunderts stellte sich für die sich langsam von der Geographie emanzipierende kartographische Wissenschaft die Frage nach der eigenen Entwicklung und damit nicht zuletzt nach der Legitimation der Eigenständigkeit. Hierbei rückte erstmals fast die gesamte Bandbreite kartographischer Betätigung direkt ins Blickfeld des historischen Interesses: Hintergründe, Personen, Techniken und natürlich weiterhin die Karten selbst, allerdings unter dem besonderen Blickwinkel vom Ende einer Entwicklung aus. So birgt dieser Forschungsstrang oftmals die Gefahr, zeitspezifische Phänomene aus heutiger Sicht als Nebenwege zu qualifizieren und das Forschungsfeld auf die jeweils aktuell relevanten Fragen der Kartographie einzuengen.

Nicht untypisch für die frühe Phase der Disziplingeschichte waren Biographien ‚Großer Kartographen' (z. B.: Behaim, Mercator), wobei die Darstellung der geschichtlichen Entwicklung anhand einzelner Biographien eine deutlich ältere Tradition aufweist. In den Zwanzigerjahren des 20. Jahrhunderts findet sich eine explizit auch die Geschichte betrachtende Grundhaltung in den Werken von Max Eckert.[14] Seit dieser Zeit hat die Geschichte der eigenen Disziplin in den grundlegenden Werken zur Kartographie einen festen Platz (z. B. Lexikon der Kartographie von Arnberger oder Lexikon der Kartographie und Geomatik).[15]

Nach dem ersten Weltkrieg entwickelte sich unter dem Eindruck der nun ihrerseits als Wissenschaft weitgehend etablierten Kartographie wiederum ein spezialisierter Forschungszweig, der sich explizit mit der Geschichte der Kartographie beschäftigte und sich dabei mehr und mehr von einer bloßen Erforschung des Fortschritts bis hin zur jeweils aktuellen Kartographie abwandte. Erstmals wurden

---

14 M. Eckert, Die Kartenwissenschaft. Forschungen und Grundlagen zu einer Kartographie als Wissenschaft, Berlin 1921 und 1925.

15 E. Arnberger (Hrsg.), Die Kartographie und ihre Randgebiete. Enzyklopädie, Wien 1975-1989; J. Bollmann – W. G. Koch (Hrsg.), Lexikon der Kartographie und Geomatik, Heidelberg 2001 und 2002. Als ein relativ rezenter Einzelbeitrag zur Disziplingeschichte sei beispielhaft ein Aufsatz von Ulrich Freitag erwähnt: U. Freitag, Von der Physiographik zur kartographischen Kommunikation. 100 Jahre wissenschaftliche Kartographie, in: Kartographische Nachrichten 58 (2008), 59-67.

Überblickswerke zur Geschichte der Kartographie geschrieben[16] und 1935 mit der Gründung der Zeitschrift *Imago Mundi* durch Leo Bagrow eine entscheidende Basis für eine dauerhafte Verselbständigung als eigener Forschungszweig gelegt.

6. „Klassische Kartographiegeschichte"

Ab den 1970er Jahren entwickelte sich aus diesen noch relativ lose verbundenen Strömungen unter dem Einfluss von Kommunikationstheorien in der Kartographie und von kulturgeschichtlichen Tendenzen in der Geschichtswissenschaft (die Karten als zeit- und kulturspezifische Artefakte zu betrachten) eine breitere Forschungslandschaft, die man aus heutiger Sicht als „klassische Kartographiegeschichte" bezeichnen könnte. Diese nähert sich der Kartographiegeschichte mit einem nochmals verbreiterten Ansatz und begreift alle Phänomene im Zusammenhang mit Kartographie und Karte als relevante zeitspezifische Ausprägungen. Sie hinterfragt auch die politischen, wirtschaftlichen, personellen, technischen und wissenschaftsgeschichtlichen Hintergründe und Implikationen ihrer Entstehung, ihrer Herstellung und Nutzung, respektive Wirkung. Sie adaptierte die Idee des kartographischen Informationsprozesses für die historische Dimension und wandelte sich von einer stark objekt- und personenbezogenen Kartengeschichte mehr und mehr zu einer prozessorientierten Kartographiegeschichte. Ein weiteres Charakteristikum ist das Bestreben, die Kartographie aller Kulturen in die Gesamtbetrachtung einzubeziehen. Zentrale Figuren dieser Strömung waren oder sind Brian Harley, Helen Wallis, David Woodward, Tony Campbell und die Zeitschrift *Imago Mundi* sowie die Herausgeber der *History of Cartography*.[17] Im deutschen Sprachraum vertraten in den letzten Jahren unter anderem Johannes Dörflinger und seine Schüler oder Wolfgang Scharfe mit der ‚Kommission Geschichte der Kartographie der deutschen Gesellschaft für Kartographie' diese Richtung. Markstein dieser Periode war das Lexikon zur Geschichte der Karto-

---

16 Vgl. L. Bagrow, Die Geschichte der Kartographie, Berlin 1951. Das Manuskript war jedoch bereits 1943 fertiggestellt worden.

17 B. Harley – D. Woodward – M. Edney (Hrsg.), The History of Cartography, Chicago 1987ff.

graphie,[18] das auch nach über zwanzig Jahren noch einen hervorragenden Einstieg in nahezu alle Fragen der Kartographiegeschichte bietet. Die Arbeit von Wolfgang Scharfe und der DACH-Arbeitsgruppe deutscher, österreichischer und schweizerischer Kartographiehistoriker manifestierte sich vor allem in den Kartographiehistorischen Colloquien, die seit 1982 in zweijährigem Rhythmus der aktuellen Forschung im deutschen Sprachraum bis heute eine Plattform bieten.

7. Verschiedene neue Strömungen in den Geisteswissenschaften

In den späten 1990er Jahren entstand in den Geistes- und Kulturwissenschaften unter dem Einfluß von Theorien, die mit Schlagworten wie *pictorial turn* oder *spatial turn* belegt sind, neues Interesse an alten Karten. Die Vertreterinnen und Vertreter dieser Richtungen gehören keiner fest gefügten oder organisierten Gruppe von kartographiehistorisch forschenden Personen an; es fanden jedoch in den letzten Jahren mehrere Tagungen mit spezifischer Ausrichtung auf die Geschichte der Kartographie statt, die zumindest einen losen personellen Zusammenhalt erkennen ließen. Die meisten Vertreterinnen und Vertreter gehören zur universitären Forschung und Lehre unterschiedlichster geistesgeschichtlicher Fachrichtungen.[19] In einer gelegentlich bewussten Abgrenzung gegen die bestehende Forschungsrichtung betrachten sie die Geschichte der Karto-

---

18 J. Dörflinger – I. Kretschmer – F. Wawrik (Hrsg.), Lexikon zur Geschichte der Kartographie von den Anfängen bis zum Ersten Weltkrieg (Die Kartographie und ihre Randgebiete C), Wien 1986.

19 Vgl. z. B. M. Stercken, Regionale Identität im spätmittelalterlichen Europa. Kartographische Zeugnisse, in: I. Baumgärtner – H. Kugler (Hrsg.), Europa im Weltbild des Mittelalters: Kartographische Konzepte, Berlin 2008, 279–300; U. Schneider, Kartenreisen – Reisekarten. Überlegungen zur Vermessung weißer Flecken in der Reisehistoriographie, in: S. Krimm – M. Sachse (Hrsg.), Die Welt erfahren - Zur Geschichtlichkeit von Wissen und Horizonten, München 2007, 128-147; T. Michalski, Medien der Beschreibung. Zum Verhältnis von Kartographie, Topographie und Landschaftsmalerei in der Frühen Neuzeit, in: J. Glauser – C. Kiening (Hrsg.), Text-Bild-Karte. Kartographie der Vormoderne (Reihe Litterae), Freiburg 2007, 319-349.

graphie vor mehr theoretischem Hintergrund und unter nochmals erweitertem Blickwinkel. Kernfragen sind die Rolle der Karte und der Kartographie bei der Konstituierung von Raumvorstellungen, das Verhältnis von Text und Karte oder die Bedeutung von Karten im Zusammenhang aller graphischen Darstellungen bzw. von wissenschaftlich-technischen „Bildern“. Stark betont wird die – nicht neue – Erkenntnis, dass Karten keine objektiven Darstellungen „realer“ Räume sind, sondern dass sie durch zweckbedingte Festlegungen und Kompromisse im Entstehungsprozess bis hin zu gezielter Fehlinformation überformte komplexe Werke sind, die erhebliches Machtpotential bezüglich der Vorstellungswelten ihrer Nutzer haben. Ohne quellenkritisches Hinterfragen all dieser Prozesse ließen sich keine verwertbaren Aussagen gewinnen. Damit werden nicht zuletzt die schwierigen Fragen nach der Nutzung und Wirkung von Karten (Schlagwort *Macht der Karten*) neu gestellt, die zwar klassische Themen der Kartographiegeschichte sind, aufgrund der unübersichtlichen Quellenlage aber kaum je erschöpfend bearbeitet wurden.

Schluss

Es bleibt zu hoffen, dass künftig eine stärkere Annäherung zwischen dem ausgeprägten Detailwissen, dem wissenschaftlich-praktisch fundierten Umgang mit alten Karten sowie der Kartographie allgemein und diesen neuen Ansätzen, die Kartographie in größeren Zusammenhängen zu begreifen, stattfinden wird. Es steht zu erwarten, dass die seit Jahrzehnten bestehenden Organisationsformen der klassischen Kartographiegeschichte auf lange Frist auch die stärker von methodischen Moden geprägten neuen Fragestellungen als selbstverständlichen Teil des Forschungsspektrums integrieren werden können. Auf den *International Conferences on the History of Cartography* ist dieses Ineinandergreifen bereits seit mehreren Jahren sichtbar – selbst wenn nicht immer alle Beiträge jeweils beide „Seiten“ zu überzeugen vermögen. Die Erforschung der Geschichte der Kartographie wird auch durch diese Ansätze vielseitiger, anregender und für ein breiteres Publikum attraktiver.

Quellen- und Literaturverzeichnis

Arnberger E. (Hrsg.), Die Kartographie und ihre Randgebiete. Enzyklopädie, Wien 1975-1989.

Bagrow L., Die Geschichte der Kartographie, Berlin 1951.

Bollmann, J. – Koch W. G. (Hrsg.), Lexikon der Kartographie und Geomatik, Heidelberg 2001 und 2002.

Dörflinger J. – Wagner R. – Wawrik F., Descriptio Austriae. Österreich und seine Nachbarn im Kartenbild von der Spätantike bis ins 19. Jahrhundert, Wien 1977.

Dörflinger J. – Kretschmer I. – Wawrik F. (Hrsg.), Lexikon zur Geschichte der Kartographie von den Anfängen bis zum Ersten Weltkrieg (Die Kartographie und ihre Randgebiete C), Wien 1986.

Doppelmayr J. G., Historische Nachricht von den Nürnbergischen Mathematicis und Künstlern, Nürnberg 1730.

Eckert M., Die Kartenwissenschaft. Forschungen und Grundlagen zu einer Kartographie als Wissenschaft, Berlin 1921 und 1925.

Freitag U., Von der Physiographik zur kartographischen Kommunikation. 100 Jahre wissenschaftliche Kartographie, in: Kartographische Nachrichten 58 (2008), 59-67.

Gottschling C., Versuch von einer Historie der Land-Charten, Halle 1711.

Gregorii J. G., Curieuse Gedancken von den vornehmsten und accuratesten alt- und neuen Land-Charten, Frankfurt 1713.

Harley B. - Woodward D. - Edney M. (Hrsg.), The History of Cartography, Chicago 1987ff.

Harley B., The Map and the Development of the History of Cartography, in: Harley B. – Woodward, D. (Hrsg.), Cartography in prehistoric, ancient, and medieval Europe and the Mediterranean (The History of Cartography 1), Chicago 1987, 1-42.

Hauber E. D., Versuch einer umständlichen Historie der Land-Charten, Ulm 1724.

Horst T., Die Altkarte als Quelle für den Historiker – Die Geschichte der Kartographie als Historische Hilfswissenschaft, in: Archiv für Diplomatik, Schriftgeschichte, Siegel- und Wappenkunde 54 (2008), 309-377.

Humboldt A. v., Ueber die ältesten Karten des Neuen Continents und den Namen Amerika. Mit einer genauen Abbildung des Behaim'schen Globus vom Jahr 1492 in 2 Planigloben nach seiner natürlichen Größe und 3 ältesten Karten von Amerika, in: Ghillany F. W., Geschichte des Seefahrers Ritter Martin Behaim, Nürnberg 1853.

Katalog sämmtlicher in dem K. K. Kriegsarchiv befindlichen Karten und Pläne, Wien 1859.

Kretschmer I., Zur Entwicklung thematischer Atlanten im 19. und 20. Jahrhundert, in: Wolff H. (Hrsg.), 400 Jahre Mercator – Vierhundert Jahre Atlas. Die ganze Welt zwischen zwei Buchdeckeln. Eine Geschichte der Atlanten (Bayerische Staatsbibliothek Ausstellungskataloge 65), Weißenhorn 1995, 231-264.

Michalski T., Medien der Beschreibung. Zum Verhältnis von Kartographie, Topographie und Landschaftsmalerei in der Frühen Neuzeit, in: Glauser J. – Kiening C. (Hrsg.), Text-Bild-Karte. Kartographie der Vormoderne (Reihe Litterae), Freiburg 2007, 319-349.

Nordenskiöld A. E., Facsimile-Atlas of the Early History of Cartography with reproductions of the most important maps printed in the 15th and 16th centuries, Stockholm 1889.

Scharfe W. – Scheerschmidt, H. (Hrsg.), Berlin-Brandenburg im Kartenbild. Wie haben uns die anderen gesehen? Wie haben wir uns selbst gesehen? (Staatsbibliothek zu Berlin – Ausstellungskataloge N. F. 42), Berlin 2000.

Schneider, U., Kartenreisen – Reisekarten. Überlegungen zur Vermessung weißer Flecken in der Reisehistoriographie, in: Krimm S. – Sachse M. (Hrsg.), Die Welt erfahren – Zur Geschichtlichkeit von Wissen und Horizonten, München 2007, 128-147.

Stercken M., Regionale Identität im spätmittelalterlichen Europa. Kartographische Zeugnisse, in: Baumgärtner I. – Kugler H. (Hrsg.), Europa im Weltbild des Mittelalters: Kartographische Konzepte, Berlin 2008, 279–300.

Wolff H. (Bearb.), Cartographia Bavariae. Bayern im Bild der Karte (BSB Ausstellungskataloge 44), Weisenhorn 1988.

Woodward D., Medieval Mappaemundi, in: Harley B. – Woodward D. (Hrsg.), Cartography in prehistoric, ancient, and medieval Europe and the Mediterranean (The History of Cartography 1), Chicago 1987, 286-370.

Zum Autor

Markus Heinz, Studium der Geschichte und Kunstgeschichte in Salzburg und Wien. Magisterarbeit und Dissertation bei Johannes Dörflinger. Dazu zwei Forschungsprojekte zur Geschichte des Verlages Homann in Nürnberg 1702-1848 als Assistent bei Johannes Dörflinger. Seit 2002 stellvertretender Leiter der Kartenabteilung der Staatsbibliothek zu Berlin und Leiter der Kommission Geschichte der Kartographie der Deutschen Gesellschaft für Kartographie.

Kontakt: Staatsbibliothek zu Berlin – Kartenabteilung, Potsdamer Str. 33, D – 10785 Berlin; Markus.Heinz@sbb.spk-berlin.de

Ferdinand von Hochstetter,
Neu-Seeland,
Stuttgart 1863

Gerhard Holzer, Wien

... i nga tini mea o to tatou nei motu ...[1]

Ferdinand von Hochstetter als Kartograph in Neuseeland

Als ich 1985 an der Akademie der Wissenschaften zu arbeiten begann, übernahm ich unter anderem auch die Aufgabe, die Sammlung des Wiener Geographiehistoriker Erich Woldan zu betreuen. Im Laufe der dreieinhalb Jahre, die mir vergönnt waren, mit Woldan gemeinsam an der Erstellung seines Kataloges zu arbeiten, entwickelte sich bald eine gegenseitige wissenschaftliche Vertrautheit und wohl auch Freundschaft. Wohl wissend, dass ich meine Diplomarbeit bei Prof. Hamann und Prof. Dörflinger über Ferdinand von Hochstetter als Neuseelandforscher geschrieben hatte, war es einer der ersten „Freundschaftsdienste" Woldans, mir mit Hilfe seiner exzellenten Kenntnis der internationalen Antiquariatsszene zu einem angemessenen Preis ein Exemplar der Erstausgabe von Hochstetters *Neu-Seeland* zu verschaffen. Es darf also nicht verwundern, dass ich als Beitrag für diesen ersten Sammelband zu Objekten aus der Sammlung Woldan auf mein wissenschaftliches Dreigestirn Hamann – Dörflinger – Hochstetter zurückgreife und mich nochmals mit diesem deutsch-österreichischen Geologen, diesmal aber hauptsächlich in seiner Rolle als Kartograph in Neuseeland, auseinandersetzen möchte.

Diese neuerliche Beschäftigung mit dem Thema Hochstetter erscheint mir umso angebrachter, da in der Zwischenzeit eine Reihe

---

1 [...] *laß ihn schauen die verschiedenen Dinge unserer* Insel [...]: aus einem Empfehlungsschreiben an Häuptling Pini te kore kore, Text abgedruckt in: F. Hochstetter, Neu-Seeland, Stuttgart 1863, 530f.

von neuen Quellen bzw. Quelleneditionen zugänglich geworden sind, die verschiedene Aspekte von Hochstetters frühen geologischen, aber eben auch kartographischen Arbeiten ergänzen bzw. in kritischem Licht erscheinen lassen.

Ankunft und Aufgabenbereiche

Als einer der wissenschaftlichen Teilnehmer an der Weltumsegelung der österreichischen Fregatte Novara war Hochstetter am 22. Dezember 1858 in Auckland an Land gegangen.

Mit der Beschreibung der Stadt, die für mehrere Monate das Zentrum seiner Tätigkeiten werden sollte und die er im Gegensatz zu seinen euphorischen Schilderungen in seinem Neuseelandwerk lediglich, als [...] e*in Sydney in kleinerem Maaßstab* [...] empfindet, endet der IX. und letzte erhaltene Band seiner wissenschaftlichen Tagebücher von der Weltumsegelung der Novara.[2] Wie es der Zufall wollte, begann damit nicht nur in Hochstetters Tagebüchern ein neuer Abschnitt: auch für ihn persönlich war damit ein neues Kapitel seines Lebens eröffnet worden.

Seine Ankunft in Neuseeland und seine Freistellung durch den Kommandanten der Weltumsegelung, Baron Bernhard von Wüllerstorf-Urbair, bedeuteten sowohl für ihn als auch für die Kolonialregierung Neuseelands einen enormen Glücksfall. Für Hochstetter ergab sich aus der Pionierarbeit, die er bei der geologischen Erforschung des Inneren der Nordinsel leistete, eine hervorragende Möglichkeit, sein Können unter Beweis zu stellen und darauf seine weitere wissenschaftliche Karriere aufzubauen. Andererseits erhielt die Provinzialregierung durch die geologisch-topographischen Aufnahmen des „Ausländers" endlich die notwendigen Daten, die sie für die wirtschaftliche Nutzung des Hinterlandes von Auckland – vorwiegend ging es dabei um Lagerstätten von Bodenschätzen, allen voran Kohle – und damit auch für den wirtschaftlichen Aufschwung der Hafenstadt im beginnenden Dampfschiffzeitalter benötigte.

Den Nutzen solcher Beurteilungen von Lagerstätten hatte die Provinzialregierung schon nach dem Gutachten Hochstetters über

2 F. Hochstetter, Reise um die Erde an Bord Sr. Maj. Fregatte Novara. IX, 53, Archiv der Österreichischen Akademie der Wissenschaften (AÖAW), Ferdinand von Hochstetter, Novara-Tagebücher.

die Kohlefelder im Drury und Hunua District erkannt, die dieser noch als Mitglied einer Exkursion der Novara-Wissenschaftler Ende Dezember 1858 erarbeitet hatte.[3] In diesem Zusammenhang steht daher auch das Ansuchen des Gouverneurs von Neuseeland, Thomas Gore Brown, an den Leiter der Novara-Expedition, [...] *giving permission to Dr. Hochstetter to extend his researches for a few months longer.*[4] Hochstetter selbst war sich des Risikos eines Scheiterns mit allen negativen Konsequenzen auf seine weitere Karriere durchaus bewusst und zögerte zunächst – sei es nun aus verhandlungstechnischen Gründen oder weil er wirklich Bedenken hatte –, auf das Angebot einzugehen. Dabei führte er seine Unkenntnis der Maori-Sprache, das weitgehend mit dichtem *bush* bewachsene Terrain und natürlich das Fehlen einer topographischen Karte als Gründe für seine Zurückhaltung an. Auf der anderen Seite war es aber gerade die Komplexität der geologisch-topographischen Aufnahme eines weitgehend von Vermessungsingenieuren unberührten Gebietes, das ihn im Landesinneren der Provinz Auckland erwartete, die den ehrgeizigen jungen Wissenschaftler reizte und ihn schließlich doch zur Annahme des Ansuchens bewegte. Betrachtet man seine weitere wissenschaftliche Laufbahn und seine Publikationstätigkeit,[5] wird schnell erkennbar, welch großen Stellenwert – ungefähr ein Drittel seiner Veröffentlichungen beschäftigt sich ganz oder zu einem beträchtlichen Teil mit Ergebnissen seines Neuseelandaufenthaltes – dabei seine erfolgreiche Expedition in das Zentralgebiet der Nordinsel hatte.

Doch zunächst blieb trotz aller Zusagen von Unterstützung durch die Provinzbehörden und eines gesunden Selbstbewusstseins ein ungutes Gefühl zurück, das Hochstetter in seiner Tagebuch-

---

3 F. Hochstetter, Report of a Geological Exploration on the Coalfield in the Drury and Hunua District, in the Province Auckland, in: The New Zealand Gazette 2 (1859), 9f.

4 K. v. Scherzer, Reise der Oesterreichischen Fregatte Novara um die Erde, in den Jahren 1857, 1858, 1859. Beschreibender Teil 3, Wien 1862, Beil. III B.

5 Vgl. dazu die Literaturliste in der Biographie von F. Heger, Ferdinand von Hochstetter, in: Mittheilungen der Geographischen Gesellschaft in Wien 27 (1884), 383-392.

eintragung vom 8. Jänner 1859 zum Ausdruck brachte, als er sich eingestand, dass er sich *Allein bei den Antipoden* fühlte.[6]

Vorbereitungen

Allzu lange hielt seine Unsicherheit nicht an, denn bereits Mitte Jänner 1859 brach er nach ersten Kurzexkursionen in die nördliche Umgebung Aucklands zu einer zweiwöchigen geologischen Untersuchung des Gebietes zwischen Manakau-Harbour und zur Waikato-Mündung auf. Damit hatte er sich zunächst aber nur in jene Gebiete vorgewagt, die bereits durch die Küstenaufnahmen der britischen Admiralität 1848 bis 1855 durch die H. M. S. Acheron und Pandora unter dem Kommando von John Lort Stokes (1812-1885) und Byron Drury (1815-1888) kartographisch erfasst worden waren.[7] Deshalb konnte er sich bei dieser Exkursion ganz auf die Geologie konzentrieren und die aus diesen Ergebnissen erarbeitete Karte vom *Isthmus von Auckland mit seinen erloschenen Vulkankegeln* [8] war einer der Grundsteine für seinen späteren Ruhm in der geologischen Erforschung Neuseelands, auch wenn es wegen einer von Charles Heaphy (1820-1881) knapp vorher veröffentlichten Karte desselben Gebietes einige Kontroversen gab.[9] Was Hochstetter in diesem Zusammenhang wohl am meisten störte, war, dass Heaphy die geologischen Ergebnisse des österreichischen Wissenschaftlers ohne dessen Wissen und auch ohne ihn zu erwähnen in seine Karte eingetragen

---

6 Für eine genauere Schilderung von Hochstetters Reisetätigkeit in Neuseeland vgl. G. Holzer, Ferdinand von Hochstetter, insbesondere als Neuseelandforscher, Wien 1983, 26-47.

7 C. A. Lawn, The Pioneer Land Surveyors of New Zealand, Wellington 1977, 65f.

8 DER ISTHMUS VON AUCKLAND | mit seinen erloschenen Vulkankegeln | von | D.r Ferdinand von Hochstetter | 1859, | Die Zeichnung u. geographische Grundlage (haupt- | sächlich nach den Aufnahmen von Stokes u. Drury) von A. Petermann. | Maassstab 1:120.000 / Novara Expedition, Neu-Seeland, Nordinsel / Petermann's geogr. Mittheil. / Jahrgang 1862 Tafel 6. / A. Welker del. / Lith. Anst. v. C. Hellfarth in Gotha. / GOTHA: JUSTUS PERTHES | 1862.

9 Die Angelegenheit wurde schon damals sehr ausführlich in der Öffentlichkeit behandelt. Vgl. dazu Dr. Hochstetter and Dr. Charles Heaphy, in: Daily Southern Cross 20/2217 (1864), 5.

und diese dann im *Quarterly Journal of the Geological Society of London* 1861 veröffentlicht hatte.[10]

Am südlichsten Punkt seiner Reise nahe den Waikato-Heads – der Einmündung des größten Flusses Neuseelands in die Tasmann-See – bekam er erstmals auch Kontakt zu Maoris, die nicht unter direktem Einfluss der *Pakehas*[11] standen. Sein von Anfang an gutes Verhältnis zu der indigenen Bevölkerung sollte sich als sehr entscheidend für das Gelingen der Expedition ins Landesinnere, insbesondere für den Bereich der topographischen Aufnahme erweisen.

Im Gegensatz zu den britischen *Provincial Surveyors*, deren Arbeit von den zu Recht misstrauischen Maoris immer wieder stark behindert wurde, da sie befürchteten, dass die Provinzregierung vermessenes Stammesland für Landspekulationen freigeben würde, konnte sich Hochstetter fast ungehindert unter der indigenen Bevölkerung Neuseelands bewegen; denn

> *Sie wußten, daß ich ein Fremder sei, der nur kurze Zeit im Lande bleibe, und halfen mir noch auf jedliche Weise mit, damit ich in meiner Heimath recht viel Schönes über ihr Land erzählen könne.*[12]

Eine Ansicht, die durchaus nicht als Einbildung des „Ausländers" abgetan werden kann, denn auch der langjährige Direktor des New Zealand Institute of Surveyors, C. A. Lawn, bestätigt dies in seinem Aufsatz über die Pioniere der neuseeländischen Landaufnahme:

> *Luckily for Hochstetter, the Maoris, who were suspicious of land speculators and who placed all sorts of obstacles in the way of the Provincial Surveyors, on learning that he was a foreigner spending only a short time in the country, gave him every assistance.*[13]

Die Sichtweise Hochstetters birgt natürlich eine gewisse Naivität in sich, da die Vermessungen [...] *unter thatkräftiger Mitwirkung Herrn Drummond Hay's* [...] – niemand anderes als der spätere District

---

[10] Hierzu sei lediglich auf Hochstetters eigene Kommentare verwiesen: F. v. Hochstetter – A. Petermann, Geologisch-topographischer Atlas von Neu-Seeland, Gotha 1864, 8.

[11] Maoribezeichnung für alle Neuseeländer europäischer Abstammung.

[12] F. v. Hochstetter, Neu-Seeland, 20f.

[13] Lawn, The Pioneer Land Surveyors of New Zealand, 96.

Surveyor George W. Hay (1826-1881)[14] – durchgeführt wurden und die Ergebnisse seiner Reise als Ganzes vereinbarungsgemäß der Provinzregierung zustanden. Die Folgen seiner kartographischen Aufnahmen zeigten sich schon wenige Jahre später bei der erfolgreichen britischen Invasion des Waikato-Territory 1863/1864. An diesen militärischen Operationen nahmen sowohl George W. Hay als auch der bereits erwähnte Landvermesser Charles Heaphy[15] teil. Die genaue Lokalisierung vieler Maori-Siedlungen in Hochstetters Karte erleichterte es dem Oberbefehlshaber der britischen Truppen, General Sir Duncan Cameron, wesentlich, geeignete Angriffspunkte für seine Invasionsstreitmacht zu finden, und seine Geländekenntnis ermöglichte es ihm, mehrmals Verteidigungslinien der Maoris zu umgehen.

Feldarbeiten

Hochstetter war sich der Probleme, die während der langen Reise durch unvermessenes Gebiet auf ihn warten würden, durchaus bewusst und er versuchte seine beschränkten Mittel möglichst optimal einzusetzen. Ganz neu war diese Problematik nicht für ihn, denn bereits zu Beginn seiner Karriere als Geologe war er bei der geologischen Landesaufnahme Böhmens weitgehend auf sich gestellt gewesen. Konnte er damals auf die Mittel der k. k. Geologische Reichsanstalt in Form von auf exakter Vermessung im Rahmen der Josephinischen und Franziszeischen Landesaufnahme beruhenden topographischen Karten und auf geeichten Instrumenten zurückgreifen, half ihm diesmal das *Provincial Survey Office* bei der Ausstattung. Von dieser Stelle bekam er allerdings keine auf systematischer Landvermessung basierenden topographischen Karten zur Verfügung gestellt, sondern lediglich Kartenskizzen im Maßstab 1 Zoll engl. = 1 Seemeile = ca. 1: 74.962, die allerdings kaum Terrainzeich-

14 Hay blieb auch in Neuseeland relativ wenig bekannt, doch handelst es sich trotz der scheinbar unterschiedlichen Namensansetzung wohl um dieselbe Person; zu seinen biographischen Daten vgl. Cambridge Museum – Biographies, Internet-Ressource.

15 Über Heaphy, dem auch als erstem Kolonialbriten das Victoria Cross verliehen wurde, finden sich wesentlich detailliertere Angaben u. a. in: The Dictionary of New Zealand Biography 1, Wellington 1990, 181-183.

nung sondern nur die Küstenumrisse und das Flussnetz enthielten. So mussten er und seine Helfer neben der geologischen Aufnahme im günstigsten Fall eben auch noch die Terrainzeichnung ergänzen.[16]

Da die spärlich vorhandenen Angaben und Eintragungen im Landesinneren aber nicht einmal auf einem Vermessungsnetz basierten, hatte er

> [...] *schon vom Beginn der Reise an ein Triangulierungssystem mittels Azimuthcompasses adoptirt, basirt auf die nautische Küstenaufnahme von Capitän Drury, und führte dieses unter thatkräftiger Mitwirkung Herrn Drummond Hay's durch von der Westküste nach der Ostküste.*[17]

Bezüglich der tatsächlich erreichten messtechnischen Qualität seiner Aufnahme blieb Hochstetter aber doch auf dem Boden der Realität und war sich der Unzulänglichkeit seiner Vermessungskapazitäten durchaus bewusst. Schließlich bestand seine diesbezügliche Ausrüstung in der Hauptsache aus dem bereits erwähn-

Abb. 1: Hochstetter bei Vermessungsarbeiten auf dem Mount Ngariha (aus: Hochstetter, Neu-Seeland, 216.)

16 F. v. Hochstetter, Neu-Seeland, 16.
17 Ebd., 20f.

ten *Azimuth- oder prismatischen Compass* sowie einem Barometer der Firma Kapeller in Wien aus dem Besitz der k. k. geologischen Reichsanstalt; Vergleichstabellen für die während der Reise gemachten barometrischen Höhenmessungen hatte das Royal Engineer's Observatorium in Auckland zur Verfügung gestellt.[18] Weiters stand ihm zur Triangulierung noch ein *Lamontscher magnetischer Reise-Theodolit* zur Verfügung, den ihm der Stuttgarter Physiker Johann Gottlieb Nörrenberg (1787-1861) überlassen hatte.[19]

Er erhob keinen Anspruch auf trigonometrische Genauigkeit, er sah sein Verdienst vielmehr darin, dass er eine […] *richtige Anschauung von den Fluß- und Bergsystemen des durchreisten Landes* […] erstellt hatte. Was ihn von vielen anderen Expeditionskartographen seiner Zeit aber durchaus positiv abhob, war sein stetes und bewusstes Bemühen, die korrekten topographischen Namen der indigenen Bevölkerung zu erfahren und in seine Karten zu übernehmen:

> *Es waren Häuptlinge selbst, welche zu allen merkwürdigen Punkten die Führer machten, mir mit größter Bereitwilligkeit alle Namen für Berge, Flüsse, Thäler und Seen gaben und nach ihrer Art die Geographie ihrer Gegend erklärten. Ich sammelte mit Sorgfalt alle Namen der Eingeborenen, und glaube manchen schönen und bezeichnenden Maorinamen der Vergessenheit entrissen zu haben.*

Doch nicht nur bei der Namensgebung, sondern auch beim Erfassen der topographischen Gegebenheiten waren ihm die Maoris, wie das folgende Beispiel zeigt, eine wertvolle Hilfe:

> *Meine Quelle war in Mangawhitikau der Häuptling Reihana te Huatare, ein dicker freundlicher Mann mit intelligentem Gesicht, der, als ich ihn um die Geographie der Umgegend befragte, mir die Karte auf dem Boden plastisch entwarf, indem er die Flüsse mit Holzstäbchen und die Berge mit Steinen bezeichnete.*[20]

Der Abstecher zur Westküste brachte die Expedition wieder in den Bereich der Küstenaufnahme von Stokes und Drury und so

---

18 Zum Einsatz seines Standart-Barometers und einer Liste der Messpunkte vgl.: F. v. Hochstetter, Geologie von Neuseeland, in: Reise der Novara, Geologischer Theil 1/1, Wien 1864, 187-192.

19 Kais. Akademie der Wissenschaften (Hrsg.), Bemerkungen und Anweisungen für die Naturforscher, Wien 1857, VI.

20 F. v. Hochstetter, Neu-Seeland, 203f.

konnte sich Hochstetter hier weitgehend auf die geologische Datenerfassung konzentrieren und seine Ergebnisse in eine vorhandene, topographisch nur geringfügig erweiterte Grundkarte einarbeiten. Daraus entstand ein zweites Detailblatt seiner Neuseelandkarten über die beiden großen Buchten Aotea- und Kawhia-Harbour.[21]

Der folgende Abschnitt der Reise vom Oberlauf des Waipa bis zum Lake Taupo war topographisch völliges Neuland und da er abseits der großen Flüsse lag, auch weitgehend weglos. Umso größeren Wert legte Hochstetter darauf, während und nach jeder Marschetappe seine Eindrücke und Ergebnisse zu Papier zu bringen:

> *Den Nachmittag* [d. 5.4.1859] *benützte ich, um von einer umweit Piopio am rechten Mokau-Ufer bei dem Dorfe M a i r o a gelegenen Anhöhe die Umgegend zu skizziern und in Karte zu bringen.*[22]

Um freies Sichtfeld für die systematische Vermessung aller markanten topographischen Punkte zu bekommen, verfiel er, wie diese Skizze von Julius Haast zeigt, auf originelle Lösungen:

Abb. 2: Hochstetter bei der Geländeaufnahme, nach einer kolorierten Bleistiftzeichnung von J. Haast (Alexander Turnbull Library, MS-0921-059)

---

21 DIE HÄFEN UND BUCHTEN | AOTEA UND KAWHIA, | Topographisch und geologisch untersucht | von | D.r Ferdinand von Hochstetter | 1859, | Die Zeichnung und geographische Grundlage (haupt- | sächlich nach den Aufnahmen von Stokes u. Drury) | von A. Petermann.

22 F. v. Hochstetter, Neu-Seeland, 206.

Bei den meisten anderen Gelegenheiten war das Besteigen eines geeigneten Vermessungspunktes allerdings mit wesentlich größeren zusätzlichen Strapazen verbunden:

In der Hoffnung, eine Aussicht auf die beiden Vulkankegel Tongariru und Ruapahu zu bekommen, denen wir uns jetzt bis auf 25 englische Meilen genähert haben mußten, und um für den Entwurf der Karte eine weitere Reihe von Beobachtungen auszuführen, bestieg ich am 10. April den Ngariha-Berg. Obwohl sich dieser nur 900 Fuß über die Thalsohle erhebt, und waldfrei ist, war die Besteigung doch mühsam, da wir durch mannshohes Farnkraut uns den Weg erst bahnen mußten. Wir kamen im Schweiße gebadet auf dem Gipfel an, sahen uns aber für die Mühe reichlich belohnt durch den Anblick, der sich uns hier darbot. Sogar die Eingeborenen von Katiaho, die uns begleiteten und noch nie diesen Berg bestiegen hatten, waren überrascht. Aller Augen waren nach einer Seite gerichtet. Da lag der Tongariru-Vulkan vor uns klar vom Fuß bis zum Gipfel.[23]

Bei seinen Arbeiten war er aber natürlich wie alle Expeditionskartographen besonders von den Launen der Natur und des Wetters abhängig. Wenn man daher zum Beispiel die dürftigen topographischen Angaben zwischen Puketapu und dem Südende des Taupo-Sees in seiner Reisekarte mit seinem Reisetagebuch vergleicht, zeigt sich sogleich die Ursache für diese Lücke: Dichter Busch und Regen machten Messungen und Beobachtungen einfach unmöglich.[24] Sein Triangulationsnetz wurde in diesen Gebieten entsprechend weitmaschig, und auch die genaue Lokalisierung der vermessenen Punkte litt bei diesem Mangel an Vergleichsmessungen. Fehlten markante Bodenerhebungen, so erschienen auch auf Hochstetters Karten die aus anderen wenig erforschten Regionen bekannten summarischen Angaben wie:

Vulkanisches Tafelland, 2000' hoch ohne hervorragende Kuppen aus trachytischen Felsen und mit dichten Wäldern besetzt, unerforscht.

Mit dem Erreichen des Lake Taupo hatte Hochstetter den schwierigsten Teil seiner Reise hinter sich gebracht. Nun bewegte er sich wieder in Gebieten, die auch schon durch zahlreiche andere

---

23 F. v. Hochstetter, Neu-Seeland, 215f.; vgl. auch Abb. nach 216.

24 Ebd., 219-221.

Reisende besucht worden waren. Überdies bestand durch das, was sich großartig *Overland Mail Track* auf Hochstetters Karte betitelte – in Wahrheit meist aber nur ein schmaler Maoripfad war, der Auckland über den Waikato und Lake Taupo mit Ahuriri (Napier) an der Ostküste Neuseelands verband – ein ständiger Kontakt zum Siedlungs- und Verwaltungszentrum im Norden.[25] Wie geringe Auswirkungen dieser Postweg an sich auf die topographische Erschließung einer Gegend hatte, zeigt sich u. a. daran, dass die oben als *unerforscht* bezeichnete Waldregion fast zur Gänze von dem *Overland Mail Track* durchquert wurde.

Hochstetters kleine Reisegruppe hielt sich nur kurz an diese Route und verließ den *Track* dann in nordöstlicher Richtung, um nach Überquerung des Waikato das Thermalgebiet um Rotorua zu erreichen. In diesem vulkanisch höchst aktiven Gebiet, das genau auf der geologischen Bruchlinie vom Tongario zu White Island lag, konzentrierte sich Hochstetter nun endlich wieder auf seine eigentlichen geologischen Forschungen. Trotzdem gewann er auch in den Regionen um den Lake Taupo und um Rotorua einiges neues topographisches Material, das er in seiner geologischen Version *Der Südliche Theil der Provinz-Auckland* in den beiden Nebenkarten *Der Taupo-See* und *Der See-Distrikt* verarbeitete.[26]

Als eigentliches kartographisches Hauptprodukt ist aber seine exakte Vermessung der Gegend um den *Rotomahana* oder *Warmen See* mit der weltbekannten rosa Sinterterrasse zu werten, die später als eigenes Kartenblatt sowohl in *Petermanns Mittheilungen* als auch in den beiden Fassungen des Atlas von Neuseeland veröffentlicht wurden.[27]

Mit seiner Rückkehr nach Auckland hatte Hochstetter seine topographische Erfassung und Vermessung von Teilen der Nordinsel Neuseelands weitgehend abgeschlossen.

---

25 C. A. Lawn, The Pioneer Land Surveyors of New Zealand, 96.

26 DER SÜDLICHE THEIL| DER | PROVINZ AUCKLAND | IN | NEU-SEELAND. | Zur Übersicht der Routen und Aufnahmen | von | D.r Ferdinand von Hochstetter | 1859. | Nach den Originalzeichnungen, Skizzen und Messungen | von Hochstetter's und den Englischen Küstenaufnahmen | unter Stokes und Drury zusammengestellt von | A. Petermann. / Nebenkarten: (1) DER | TAUPO-SEE, | im doppelten Maassstabe der Hauptkarte. (2) DER | SEE-DISTRIKT, | im doppelten Maassstabe der Hauptkarte

27 ROTOMAHANA ODER DER WARME SEE | mit seinen heissen Quellen, | aufgenommen von | D.r Ferdinand von Hochstetter | 1859.

Hatte er die Terrainverhältnisse immer gleich vor Ort skizziert, so brachte er so viel kartographisches

> *[…] Material von der Reise zurück, nach welchem ich noch in Auckland selbst eine topographische Karte der südlichen Theile der Provinz in großem Maßstabe entwarf, es der besseren Muße späterer Tage überlassend, diesen provisorischen Entwurf einer sorgfältigen Revision zu unterziehen […] Eine Copie meiner Originalkarte im Maßstab von zwei englischen Meilen = 1 Zoll blieb in Auckland für den Gebrauch der Regierung zurück,*[28] *eine zweite Copie erhielt Herr J. Arrowsmith in London zur Benützung für die Zusammenstellung einer großen Neu-Seelandkarte in sechs Blättern, welche derselbe zu bearbeiten im Sinne hatte, jedoch mit dem Bedeuten, daß die Karte, so wie sie in der in Auckland ausgeführten Copie in seine Hände kam, nur als provisorischer Entwurf meiner Beobachtungen zu betrachten sei.*[29]

Kleinere Exkursionen zu geologisch-montanistisch interessanten Gebieten, wie z. B. zu den Goldvorkommen auf der Coromandel Peninsula, wo Charles Heaphy zu dieser Zeit als *Comissioner of Gold Fields* tätig war, brachten keine nennenswerten topographisch-kartographischen Ergebnisse.

Was nun noch folgte, waren eine erste Aufarbeitung der gewonnenen Daten und ein abschließender Vortrag vor den Mitgliedern des *Mechanics Institute,*[30] der den ihm gebührenden Ruhm in der geologischen Forschung Neuseelands begründete. Für die unmittelbaren politisch-militärischen Pläne der Provinzregierung war es aber mindestens ebenso wichtig, dass er, wie bereits oben zitiert, eine Kopie seiner Originalkarte im Maßstab 1: 126.720 des von ihm bereisten Gebietes der Provinzregierung übergab (vgl. Abb. 3 u. Farbtafel S. 318). Damit stand dieser erstmals eine weitgehend exakte Planungskarte für die Invasion des unteren Waikatobeckens zur Verfügung.

---

28 [F. Hochstetter, Map of the southern part of the province of Auckland, Auckland 1859 (?)], Manuskripkarte, Alexandre Turnbull Library, Reference Number: MAPCOLL-R832.1CBA/1859/ACC.620.

29 F. v. Hochstetter, Neu-Seeland, 20f.

30 F. Hochstetter: Lecture on the Geology of the Province Auckland, New Zealand, in: Auckland Provincial Government Gazette 8/14 (1859), 83-100.

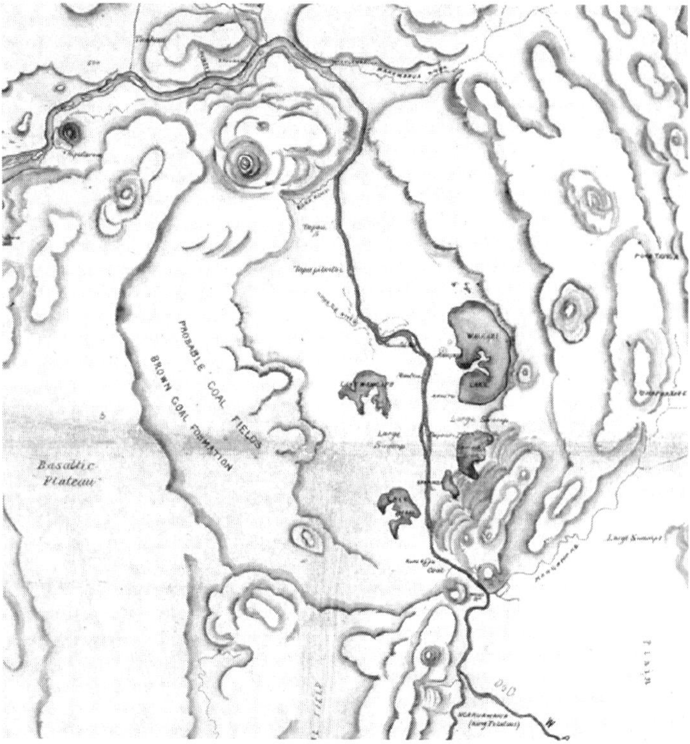

Abb. 3: Das Untere Waikatobecken, Ausschnitt aus:
[F. Hochstetter, Manuskriptkarte des südlichen Teils der Provinz Auckland]
(Alexandre Turnbull Library, MAPCOLL-R832.1CBA/1859/ACC.620)

Vom heutigen Standpunkt aus gesehen fiel die Aufenthaltszeit Hochstetters in Neuseeland mitten in eine Serie von kriegerischen Auseinandersetzungen zwischen den englischen Siedlern und den Maoris. Diese *Landwars* zogen sich über den Zeitraum von 1845 bis 1872 hin. Hochstetter hatte das Glück gehabt, seine Unternehmungen während einer friedlichen Phase durchführen zu können, doch hatten seine Forschungen, speziell die Vermessungsarbeiten, Auswirkungen auf sein neuseeländisches Umfeld, die er entweder nicht wahrhaben wollte oder bewusst in Kauf genommen hatte. Denn trotz Sympathien für die Maoris war er doch ein Kind seiner Zeit und seine Auffassung von der kulturell-geistigen Überlegenheit der Europäer für ihn selbstverständlich. Man muss ihm jedoch zugute halten, dass er nicht nur die scheinbar unabwendbare Tendenz des Aussterbens der Maoris bedauert, sondern versucht beide Seiten zu sehen – immerhin hatte er auf seiner Reise den 1. *Maori-King* Pōtatau Te Wherowhero (geb. um 1800, König von 1856-1860) in Ngaruawahia am Zusammenfluss von Waikato und Waipa getroffen – und auch gute Kontakte zu den neuseeländischen Behörden gepflegt. Aber das ändert nichts daran, dass er die expansive Kolonisierungspolitik der Briten als selbstverständlich und das Ziel der *Maori King Movement* (*Kīngitanga*) als Rebellion versteht. Aber wieder relativiert er im gleichen Atemzug seinen probritischen Standpunkt, indem er diese Rebellion auch als einen Akt der Notwehr sieht:

> *Es ist ein Versuch der eingeborenen Race sich los zu machen nicht bloß von der Herrschaft Englands, sondern sich los zu machen von allen Einflüssen einer Civilisation, die, seit der erste Europäer seinen Fuß auf Neu-Seeland gesetzt hatte, trotz Christenthum und höherer Gesittung, welche sie unter ein blutrünstiges Cannibalenvolk brachte, doch nur zersetzend und verderbenbringend auf dasselbe wirkte.*[31]

So gesehen hatte er das Vertrauen der Maoris, ihn als jemanden einzuordnen, der in seiner […] *Heimath recht viel Schönes über ihr Land erzählen könne* […], durchaus gerechtfertigt. Das änderte jedoch nichts daran, dass sich die britischen Behörden auch mit von ihm erbrachten Informationen im Kampf gegen die Maoris bedienten und danach durch den *New Zealand Settlements Act* von 1863 ca. 16.000 km^2 Land insbesondere auch in der rebellischen Waikato-Region konfiszierten.[32]

Quellenlage

Von diesen kommenden Ereignissen war jedoch im Winter 1859 in Auckland noch nicht wirklich viel zu spüren, und Hochstetter machte sich am 28. Juli 1859 auf, um auch die Südinsel Neuseelands kennen zu lernen.

In der von ihm besuchten Provinz Nelson herrschten allerdings völlig andere Voraussetzungen: Die Maoris hatten sich auf der

---

31 F. v. Hochstetter, Neu-Seeland, 499f.

32 M. King, The Penguin History of New Zealand, North Shore 2003, 212-216.

klimatisch wesentlich rauheren Südinsel kaum dauerhaft niedergelassen und so war das Gebiet um Nelson durchwegs von europäischen Auswanderern besiedelt worden. Zwar waren auch hier zunächst nur die Karten der Küstenaufnahme verfügbar, doch war abzusehen, dass der zuständige *Land Survey* hier innerhalb der nächsten Zeit ohne Probleme die Vermessungsarbeiten durchführen würde. Damit konnten sich Hochstetter und sein Freund und Reisegefährte Julius Haast auf die geologische Aufnahme einzelner besonders interessanter Punkte konzentrieren, was insofern für Hochstetter wichtig war, als sein geplanter Abreisetermin schnell näher rückte.

Auch die Quellenlage ist für diesen zweiten Teil von Hochstetters Neuseelandaufenthalt eine wesentlich andere: Denn im Gegensatz zu Hochstetters wissenschaftlichen Tagebüchern für die Zeit von seiner Ankunft in Auckland bis zum Beginn der Präsenz auf der Südinsel – diese vier Bände fielen weitgehend den Redaktionsarbeiten für die verschiedenen Publikationen „zum Opfer“ – blieb der Band für die letzte Phase des Nelsonaufenthaltes und für seine Heimreise erhalten.[33] Ergänzt wird diese Quelle noch durch Tagebücher einzelner Repräsentanten der Provinz Nelson wie z. B. Sir David Monroe (1813-1877), die Hochstetter auf seinen Exkursionen zum Teil begleiteten.[34]

Ähnliches gilt auch für die kartographische Quellenlage. Hochstetters Vorgehensweise von den Feldarbeiten in den von ihm in Nelson besuchten Gebieten wie Dun Mountain, Jenkin’s Hill, dem Aorere Valley bis zum kartographischen Endprodukt in Form der *Geologischen Übersichtskarte der Provinz Nelson*[35] lässt sich fast durchgehend mit Originalmaterial belegen. Denn im Gefolge des Teilverkaufs der Sammlung Infrid Schütz-Müller kamen im Wiener Dorotheum auch die zwei Blätter von Hochstetters Manuskriptkarte

---

33 F. Hochstetter, Neu-Seeland. 1859. Tagebuch Nro 5, Privatbesitz Albert Schedel, Wien.

34 C. A. Fleming, Dr. Hochstetter in Nelson. Extracts from the diary of Sir David Monroe, 1813-1877, in: New Zealand Journal of Geology and Geophysics 2/5 (1959), 954-963.

35 GEOLOGISCHE ÜBERSICHTSKARTE | DER | PROVINZ NELSON | (SÜD INSEL). | Nach den Aufnahmen | von | D.r Ferdinand von Hochstetter 1859 | und | Julius Haast 1860.

der Provinz Nelson zur Versteigerung, die von der ÖNB-Kartensammlung angekauft wurden.[36]

Anhand dieser beiden Blätter kann man die Arbeitsweise Hochstetters bei seinen Feldarbeiten sehr gut nachvollziehen: Er pauste die Küsten- und die Gewässerlinien auf dünnes Seidenpapier durch und klebte diese empfindlichen Blätter auf ein kräftigeres Trägerpapier bzw. Leinen. In diese in kleine Segmente gefaltete, robuste topographische Grundlage konnte er dann auch unter schwierigen Verhältnissen vor Ort die Ergebnisse seiner Untersuchungen eintragen.

Trotzdem war Hochstetter klar, dass er für den Entwurf einer geologischen Übersicht der Provinz mehr Material brauchen würde, als er in der kurzen ihm zur Verfügung stehenden Zeit sammeln konnte. In Hinblick auf seine weiteren Pläne war es daher ein Glücksfall, dass sich sein Freund Haast entschlossen hatte, in Neuseeland zu bleiben. Als „Schüler" Hochstetters – ohne weitere akademische Ausbildung – hatte dieser bei der geologischen Aufnahme des Wairau und Awatere Valley im Ostteil der Provinz den Beweis für seine Befähigung erbracht[37] und wurde zunächst als Provinzgeologe angestellt. So konnte er seinem Freund und Lehrer problemlos die noch fehlenden Informationen liefern, die für den Entwurf der Karte notwendig waren. Hochstetter würdigte diese Kooperation auch, indem er in einem Brief vom 18. Dezember 1861 schreibt:

> *Mein Vorschlag an Peterm. gieng nun dahin, als 6tes Blatt die Provinz Nelson unter der gemeinschaftlichen Firma Haast u. Hochstetter zu geben, u. zwar als topograph. geolog. Blatt. d. h. eine geolog. Karte mit vollst. Terrainzeichg. Sobald ich die Karten von Peterm. wieder zurück habe, werde ich Dein u. mein Stück an der geol. Karte in Uebereinstimmung bringen, zusammen einarbeiten die Formationen endgültig festsetzen u. Dir dann eine Copie im Kleinen schicken.*[38]

36 F. Hochstetter, [Neuseeland - Nordteil der Südinsel, 1859], [ca. 1:360 000], 1 Kt. auf 2 Bl., mehrfarb., Handz., Österreichische Nationalbibliothek – Kartensammlung, Sign.: K I 125435 Kar.

37 F. Hochstetter, Neu-Seeland. 1859. Tagebuch Nro 5, 37f., Privatbesitz Albert Schedel, Wien.

38 S. Nolden, The letters of Ferdinand von Hochstetter to Julius von Haast, Auckland 2007, 171.

Das war also der Beginn einer Zusammenarbeit, die trotz der großen Entfernung über zwanzig Jahre bis zum Tod Hochstetters 1885 funktionierte, da die beiden ihr ganzes Leben durch Briefe in engem Kontakt blieben.[39] Hochstetters Briefe blieben, soweit bekannt, zur Gänze erhalten und kamen mit Haasts Nachlass an die Alexander Turnbull Library, Wellington, wo sie von dem neuseeländischen Germanisten Sascha Nolden im Rahmen seiner Dissertation 2007 transkribiert, ins Englische übersetzt und ediert wurden.[40]

Am 2. Oktober 1859 verließ Hochstetter endgültig Neuseeland. Im Gepäck hatte er nicht nur eine ungeheure Menge an geologischen, botanischen, zoologischen und ethnologischen Fundstücken, sondern auch ein umfangreiches kartographisches Material, das noch auf Aufarbeitung wartete. Noch auf der Heimreise hatte er [...] *His Excellency Sir W. Denison K.C.B. Governor General of Australia Guverneur von Australien* [...] *meine Neuseelandkarte gezeigt* [...][41] und damit selbst dokumentiert, welchen Stellenwert er seiner kartographischen Tätigkeit in Neuseeland einräumte.

Hochstetters Aufarbeitung der Materialien

Bald nach seiner Rückkehr aus Neuseeland nahm Hochstetter Kontakt mit dem führenden geographischen Verlag des deutschen Sprachraumes *Justus Perthes* auf.

Der leitende Kartograph des Verlages, August Petermann (1822-1878), zeichnete in der Folge für fast alle kartographischen Produkte, an denen Hochstetter beteiligt war, verantwortlich.

> *Meinem hochverehrten Freund, Dr. A. Petermann in Gotha, habe ich für die Neubearbeitung und Revision aller meiner Originalskizzen und Beobachtungen zu danken,* [...].[42]

Eine Sonderstellung nimmt in diesen Produkten die Gesamtkarte von Neuseeland im Maßstab 1: 5 000 000 ein. Sie wurde zum Groß-

---

39 H. F. v. Haast, The life and times of Sir Julius von Haast, Wellington 1948, 28.

40 S. Nolden, The letters of Ferdinand von Hochstetter to Julius von Haast, XI.

41 F. Hochstetter, Neu-Seeland. 1859. Tagebuch Nro 5, 54, Privatbesitz Albert Schedel, Wien

42 F. v. Hochstetter, Neu-Seeland, 21.

teil von Petermann selbst – basierend auf britischen Vorlagen, insbesondere den verschiedenen Ausgaben der mehrblättrigen Neuseelandkarten von John Arrowsmith (1780-1873) – entworfen und von seinem Schüler Ludwig F. Friederichsen (1841-1915) bearbeitet. Obwohl Petermann selbst noch im Textteil der englischen Ausgabe des Atlas von Neuseeland 1864 den großen Anteil Hochstetters und Haasts an der geographischen Erschließung des Landes würdigt, indem er schreibt

> *Hochstetter's and Haast's observations have been used for the first time, will show at the first glance how much our former conception of New Zealand is enlarged and corrected.* [43]

scheint die Beteiligung der beiden an der Entstehung der Karte lediglich in jenem Blatt auf, das der deutschen Ausgabe von Hochstetters Neuseeland-Reisewerk von 1863 beigegeben ist.[44] Bei allen anderen der zahlreichen Varianten, sogar bei der deutschen und englischen Ausgabe, die mit geologisch-mineralogischen Eintragungen ergänzt wurden, fehlen die Namenserwähnungen völlig (vgl. Anhang 1, 1). Dabei besteht kein Zweifel, dass beide an der Entstehung der Karte federführend beteiligt waren. So schreibt Hochstetter seinem ehemaligen Reisegefährten und Freund Julius Haast (1822-1887), der zu dieser Zeit als Geologe und Museumsdirektor auf der Südinsel Neuseelands tätig war, z. B. in einem Brief vom 18. August 1861:

> *Zugleich aber bitte ich Dich recht sehr, da ich meinem Neuseelandwerk schon eine geolog. Uebersichtskarte über ganz Neuseeland beifügen möchte, mir nur in kleinem Maasstab natürlich die südliche Insel zu entwerfen, nur die Hauptgruppen von Formationen mit bes. Farben* [...].[45]

---

43 HOCHSTETTER, Ferdinand; PETERMANN, August: The Geology of New Zealand in Explanation of the Geographical and Topographical Atlas of New Zealand (Auckland: Delattre, 1864). 7.

44 NEU-SEELAND | (NEW ZEALAND) | IM MAASSSTABE 1:5.000.000. | Mit Benutzung der Aufnahmen von v. Hochstetter und Haast. | Von A. Petermann 1862. / Nebenkarte: DER ISTHMUS VON AUCKLAND, | (der Hauptstadt von Neu-Seeland) | im Maassstabe 1:500.000.

45 S. Nolden, The letters of Ferdinand von Hochstetter to Julius von Haast, 154.

Aber der geschäftstüchtige Petermann hatte mit der Produktion dieser Karte mehr vor, als sie lediglich als Kartenbeilage zu Publikationen Hochstetters zu verwenden. Vielmehr findet sich der identische Entwurf mit geringen Veränderungen auf der Südinsel auch Jahre später noch in *Stieler's Hand Atlas* als Blatt 50a Westaustralien / Tasmania / Neu-Seeland wieder.[46]

Aber Hochstetter konnte sich darüber kaum beklagen, Petermann ermöglichte ihm schließlich zahlreiche Publikationen in den *Mittheilungen aus Justus Perthes' Geographischer Anstalt über wichtige neue Erforschungen auf dem Gesammtgebiete der Geographie* – oder in der Kurzform *Petermann's Geographische Mittheilungen* –, der damals im deutschen Sprachraum führenden geographischen Fachzeitschrift. Da sich diese Zeitschrift besonders beim Bildungsbürgertum des 19. Jahrhunderts großer Beliebtheit erfreute, trug sie wesentlich zum gestiegenen Bekanntheitsgrad Hochstetters bei. Ein wesentlicher Bestandteil und eines der Erfolgsrezepte dieser Zeitschrift war die Ausstattung der oft mehrseitigen Artikel mit hervorragendem Kartenmaterial. Und so erschienen auch die Gesamtkarte von Neuseeland und drei der vier Spezialkarten, bearbeitet von einem weiteren Schüler August Petermanns, A. Welcker (gest. 1888), in zwei aufeinander folgenden Jahrgängen der Zeitschrift (siehe Anhang 1). Nur das Blatt über die Westküstenhäfen wurde erst im geologisch-topographischen Atlas publiziert.

Dies war für Hochstetter umso wichtiger, da er bei seiner Rückkehr nach Wien im April 1860 zahlreiche Ehrungen, aber keine Möglichkeit bekam, seine Forschungsergebnisse zu publizieren. So klagt er in einem früheren Brief an Haast vom 20. Oktober 1860:

> *So kam ich erst Ende Sept. hierher, u. mußte dann alsogleich meine Professur übernehmen. Ich habe Mineralogie u. Geognosie zu lesen, je 3 Stunden wöchentlich, u. bin unglücklich darüber, daß mir diese Vorlesungen d. h. die Vorbereitung dazu, da ich in meinem Leben nicht dozirt so viel Zeit wegnehmen, daß ich bis jetzt zu wissenschaftlichen Arbeiten gar nicht kam. So sieht es leider mit meinen Publikationen bis jetzt schlecht aus.*

---

46 A. Stieler (Hrsg.), Handatlas über alle Theile der Erde nach dem neuesten Zustande und über das Weltgebäude, Gotha 1870.

> *Der österr Staat hat kein Geld, u. denkt auch gar nicht an die wissenschaftl. Publikationen der Novara. Nur eine populäre Reisebeschreibung, von Scherzer bearbeitet ist bis jetzt im Druck. Um allerende doch etwas zur Welt zu bringen, habe ich mit Baron Cotta in Stuttgart privatim wegen meiner Reisen in Australien u. Neuseeland abgeschlossen, die Illustrationen werden bereits ausgeführt, desgleich, habe ich eben angefangen, an den Karten arbeiten zu lassen, vom Text aber noch keine Seite, der soll nun diesen Winter geschehen; das Werk soll im Allgemeinen wie die neueren englischen Reisewerke populär gehalten sein, aber dennoch alle naturwissenschaft. Resultate enthalten, u. gleichzeitig mit der deutschen Ausgabe soll eine englische erscheinen. Was Du mir noch geologisches mittheiltst, werde ich natürlich sowohl in Karten, wie in Text unter Deinem Namen aufnehmen.*[47]

Es war dem Heimkehrer also schnell klar geworden, dass die Habsburgische Monarchie nach dem verlorenen Krieg von 1859 zunächst andere Sorgen hatte, als die Ergebnisse der Novara-Reise, zu denen vereinbarungsgemäß auch die Ergebnisse von Hochstetters Neuseelandaufenthalt zählten, herauszugeben. Es lag deshalb nahe, sich in seiner württembergischen Heimat nach einem passenden Verlag umzusehen. In Person des Verlegers Johann Georg [IV.] Freiherrn von Cotta von Cottendorf (1796-1863), der den traditionsreichen Familienverlag seit 1832 führte, hatte er bald einen verlässlichen Partner gefunden. Aus dem bis heute in der Handschriftenabteilung des Deutsches Literaturarchivs, Marbach am Neckar erhalten gebliebenen *Verlags-Vertrag* (siehe Anhang 2) geht hervor, dass Hochstetter selbst für die beiden beigegebenen Karten bei Perthes zu sorgen hatte.

Neben der bereits beschriebenen Übersichtskarte von Neuseeland handelte es sich dabei um die Reisekarte des südlichen Teils der Provinz Auckland,[48] im Wesentlichen eine reduzierte Ausgabe der Karte, die Hochstetter der Provinzregierung in Auckland und

---

47 S. Nolden, The letters of Ferdinand von Hochstetter to Julius von Haast, 131.

48 DER SÜDLICHE THEIL| DER | PROVINZ AUCKLAND | IN | NEU-SEELAND. | Zur Übersicht der Routen und Aufnahmen | von | D.r Ferdinand von Hochstetter | 1859. | Nach den Originalzeichnungen, Skizzen und Messungen | von Hochstetter's und den Englischen Küstenaufnahmen | unter Stokes und Drury zusammengestellt von | A. Petermann.

Arrowsmith zur Verfügung gestellt hatte und in die er seine restlichen Forschungsergebnisse eingearbeitet hatte:

> […] *und so wie die Karte nun in reducirtem Maßstabe diesem Buche beigegeben ist, ist sie das Product seiner Arbeit und das endgültige Resultat meiner Beobachtungen.*[49]

Im Gegensatz zu den anderen neu angefertigten thematischen Spezialkarten Hochstetters, die zuerst in Form von einzelnen Artikeln mit topographisch-geologischen Kartenbeilagen in *Petermanns Mitteilungen* erschienen waren (vgl. Anhang 1), handelte es sich bei diesem Blatt zunächst nur um eine rein topographische Karte, in die die Reiseroute eingetragen worden war. Ein Bericht Haasts über die Reise selbst war bereits 1860, allerdings ohne Kartenmaterial, in *Petermanns Mittheilungen* veröffentlicht worden.[50] Alle späteren Ausgaben dieser Karte sollten die geologischen Verhältnisse des bereisten Gebietes darstellen. Sie beruhten auf exakt der gleichen Basiskarte inklusive der nicht kolorierten Reiseroute und waren lediglich mit Flächenfarben, Spezialsignaturen (u. a. für Vulkankrater und heiße Quellen), einer eigenen Legende und zwei Nebenkarten (Lake Taupo, Seen-Distrikt) erweitert (siehe Anhang 1).

Damit standen Hochstetter insgesamt sechs fertige Karten zur Verfügung, die nach dem Publikationsprogramm von August Petermann gemeinsam in einem topographisch-geologischen Atlas veröffentlicht wurden. Ursprünglich hatte Hochstetter, wie in einem Brief an Haast vom 18. August 1861 hervorgeht, geplant, einen zwölfblättrigen Atlas gemeinsam mit Haast herauszugeben:

> *Perthes in Gotha will meine Neuseelandkarten nur dann übernehmen u. herausgeben, wenn er dieselben als einen Neuseelandatlas besonders erscheinen lassen darf. Dies hat nun von Seiten der Novaraexped. u. von meiner Seite keinerlei Anstand, wohl aber bedarf er dazu meiner u. Deiner Unterstützung: Mein Vorschlag an ihn gieng nun dahin: Neuseelandatlas in 12 Blättern, 6 für die nördl. I. 6 für die südl. Insel, ausgegeben in 4 Lieferungen à 3 Blättern:*

---

49 F. v. Hochstetter, Neu-Seeland, 21.

50 J. Haast, Dr. Ferdinand Hochstetter's Reise durch die nördliche Insel Neu-Seelands, 5. März bis 24. Mai 1859, in: Petermanns Mittheilungen 6 (1860) 107-111.

1 Lieferung: Nro 1. der südl. Theil der Provinz Auckland – d. i. meine große Reisekarte, darauf in den Ecken auch besonders in gröserem Maaßstab: der Tauposee, u. Rotomahana folio 1/800,000

In Quart. Nro. 2 der Isthmus von Auckland, geologisch,

Nro. 3. Reisekarte für die Seegegend 1/200,000

Karten von Hochstetter.

2. Lieferung Nro 4 die nördl. Insel von Neuseeland 1/2,000,000, compilirt von Dr Petermann

Nro. 5 Südl. Theil der Provinz Auckland geologisch Kawhiahafen darauf besonders

Nro. 6. Geolog. Uebersichtskarte von ganz Neuseeland.

Die 6 weiteren Karten dann die Südl. Insel

Zunächst 2 Karten im Maaßstab 1/2,000,000 topogr.

dann topograph Detailkarte

u. geologische Karten bei der 3ten u. 4t. Liefg nun müsstest Du durch Einsendg des Nöthigen Materials ebenso als Mitarbeiter auftreten, wie ich bei der 1ten u. 2ten Liefg. Ueberlege Dir einmal die Sache u. schreibe mir darüber.[51]

Petermann war aber nicht davon zu überzeugen, dass eine derartige erweiterte Publikation in einem realistischen Zeitrahmen fertig werden würde, und beschränkte sich lieber auf die bereits in Vorbereitung befindlichen Karten, und so konnte Hochstetter in seinem Brief vom 18. April 1862 seinem Freund berichten:

An dem geolog. topograph. Atlas bei Perthes wird tüchtig gearbeitet: Derselbe wird vorderhand folgende 6 Blätter enthalten:

1) Geolog. Uebersichtskarte von ganz N. S., ich erwarte mit Schmerzen Deine Skizze der Südinsel

2) geol. topogr. Karte des südl. Theils der Prov. Auckld.

3) Isthmus v. Auckland

4) Aotea u. Kawhia

5) Rotomahana

6) Provinz Nelson von Dir u. mir.

sind einmal die Karten fertig, dann gehe ich an das offizielle geolog. Werk […].[52]

---

51 S. Nolden, The letters of Ferdinand von Hochstetter to Julius von Haast, 153.

52 Ebd., 185.

Auch in dieser reduzierten Form dauerte es noch bis Ende Sommer 1863, bis der Atlas in den Verkauf kam, aber Petermann sollte recht behalten: Der Absatz des Atlas lief speziell in Neuseeland wirklich gut. Allein hundert Exemplare gingen direkt nach Neuseeland, wo Hochstetters Freund und Mittelsmann in Auckland Dr. Carl F. Fischer (gest. 1893) den Vertrieb übernahm. Schon im Jahr darauf bestellte er weitere tausend Exemplare der geologischen Karten in einer englischen Variante, die er im Verlag Delattre herausgab und die er offenbar auch wirklich verkaufte. Den zu den Karten gehörenden Text übersetzte er und publizierte ihn im Gegensatz zur deutschen Originalausgabe, die den Text beigebunden hatte, in einem eigenen Bändchen.[53] Daran konnte auch die bereits oben erwähnte Hochstetter-Heaphy Kontroverse nicht ändern, aber Hochstetters Freude über den Erfolg war natürlich dadurch getrübt:

> *Die letzten N. S. Zeitungen haben mir Heaphy's Gemeinheiten gebracht unter der Ueberschrift Major Heaphy u. Dr. Hochst. – Lügen von Anfang bis Ende, positive Lügen. Ich habe alsbald eine Entgegnung geschrieben u. direkt an den Editor des New Zealanders eingeschickt. Die Sache ist mir übrigens höchst unangenehm, u. wenn ich eine Ahnung gehabt, daß Fischer den Text zum Atlas übersetzt, so hätte ich ihm geschrieben, er möge die Stelle gegen Heaphy unterdrücken. Denn vor Skandal mit Leuten seiner Art muß man sich hüten. Er scheut sich nicht zu lügen, ich habe bei den Holzschnitten u. Bildern s. Namen nicht erwähnt; u. doch ist jede Illustration, die von seiner Hand herrührt unter s. Namen gegeben. Ich habe seine Karte von Auckland u Nelson copirt!! er scheut sich nicht zu schreiben, er habe bei der Publik. der Karte von Auckland mich erwähnt, u. führt eine Anmerkung an, die Sir Ch. Lyell zu s. Aufsatze macht, um mein wissensch. Eigenthum gegen sein Stillschweigen in Schutz zu nehmen. Ich werde Dir meine Erwiderung in Abschr. mit nächster Post einsenden, hoffe aber, daß gegen die persönl. Verdächtigungen auch schon von Auckland aus durch Fischer geantwortet wurde.*[54]

[53] F. Hochstetter – A. Petermann, The Geology of New Zealand in Explanation of the Geographical and Topographical Atlas of New Zealand, Auckland 1864.

[54] S. Nolden, The letters of Ferdinand von Hochstetter to Julius von Haast, 247.

Während in Neuseeland die Kontorverse, wer denn nun von wem abgezeichnet hätte, über die Zeitungen ausgetragen wurde, machte sich Hochstetter daran, seine Texte für den wissenschaftlichen Teil der *Reise der österreichischen Fregatte Novara* fertig zu stellen. Die dafür benötigten Karten wurden mit geringfügigen Veränderungen beim Beschriftungstext aus *Petermanns Mittheilungen* bzw. dem *geologisch-topographischen Atlas* entnommen (siehe Anhang 1), was keinerlei rechtliche Probleme mit sich brachte, denn auch der Atlas war ja mit dem Zusatz *Aus den wissenschaftlichen Publicationen der Novara-Expedition* versehen worden.

Schlussbetrachtung

Mit der Publikation der drei Bände der Geologie des offiziellen wissenschaftlichen Novarawerkes und schließlich dem Erscheinen der englischen Ausgabe seines Reisewerkes,[55] das um die neuesten Forschungen Haasts in den Neuseeländischen Alpen erweitert worden war, hatte Hochstetter seine kartographischen Arbeiten betreffend Neuseeland beendet. Zwar schrieb er weiterhin kleinere Abhandlungen zu verschiedenen Neuseeland betreffenden Themen, wie etwa über Moas,[56] aber eigene Karten fügte er diesen Publikationen nicht bei. Auch für den Bildband über Rotomahana[57] des neuseeländischen Fotographen Daniel Louis Mundy (1826 od. 1827-1881), den Hochstetter auf der Wiener Weltausstellung 1873 kennen gelernt hatte, schrieb er zwar die Begleittexte für die Bilder, die beiliegende *Sketch Map to illustrate the southern Wonderland, Rotomahana, & the boiling springs of New Zealand* stammte allerdings nicht mehr von ihm. Sehr wohl fertigte er allerdings eine Reihe anderer Karten zu neuen Arbeitsbereichen an. Dabei blieb er eng mit dem Verlag Perthes verbunden, bei dem er z. B. eine Karte der Erdbeben-

---

55 Nach langwierigen Problemen bei der Übersetzung erschien endlich die englische Ausgabe: F. v. Hochstetter, New Zealand. Its physical Geography, Geology and Natural History, Stuttgart 1867.

56 Einer dieser angesprochenen Aufsätze war: F. v. Hochstetter, Die ausgestorbenen Riesenvögel von Neuseeland, in: Schriften des Vereins zur Verbreitung naturwissenschaftlicher Kenntnisse in Wien 1 (1860/1861), 215-246.

57 D. L. Mundy – F. v. Hochstetter, Rotomahana; and the Boiling Springs of New Zealand, London 1875.

flutwelle, die nach einem Beben in Peru im August 1868 quer über den Pazifik gelaufen war, und eine Karte des zentralen Balkan, die 1869 anlässlich einer Reise in das damalige Gebiet der europäischen Türkei entstanden war (vgl. Anhang 1).

Doch ungeachtet dessen, dass nach 1867 Hochstetter selbst keine Karten mit Neuseelandbezug veröffentlichte, obwohl er von Haast genügend Material und Informationen für eine Neuauflage einzelner Karten erhalten hatte, blieb sein großes Verdienst um die geologische Erschließung der Doppelinsel davon unberührt. Und so ist nicht verwunderlich, dass die zweite für die Geologie Neuseelands bedeutende Persönlichkeit, Sir James Hector (1834-1907), noch 1873 auf seiner geologischen Karte Neuseelands im Untertitel vermerkt: *Constructed from official Surveys and the Explorations of D.r F. von Hochstetter, D.r Julius Haast and Others.*[58] Eine Anerkennung von Verdiensten, die nicht nur bis heute anhält, sondern 2008 in der von Sascha Nolden ausgerichteten Gedächtnisausstellung in den Auckland City Libraries[59] neuerlich eine Würdigung erhielt. Der Titel der Ausstellung bringt es auf den Punkt, für die Neuseeländer bleibt er einfach: *Ferdinand von Hochstetter - Father of New Zealand Geology.*

Anhang

Anhang 1: Liste der Karten Hochstetters bzw. August Petermanns mit Neuseelandbezug

1) GESAMTKARTE VON NEUSEELAND

Variante A:

NEU-SEELAND | (NEW ZEALAND) | IM MAASSSTABE 1:5.000.000. | Von A. Petermann. / Petermann's geogr. Mittheil. / Jahrgang 1862 Tafel 14. / Bearbeitet von L. Friedrichsen. / Lith. Anst. v. C. Hellfarth in Gotha. / GOTHA: JUSTUS PERTHES | 1862.

Abmessungen: 19 x 32,5 cm, mehrfarb. Lithogr.

In: Mittheilungen aus Justus Perthes' geographischer Anstalt 8 (1862), 367-269, Tafel 14

58 J. Hector, Geological Sketch Map of New Zealand.

59 Auckland City Libraries – Ferdinand von Hochstetter, Internet-Ressource.

Variante B:

NEU-SEELAND | (NEW ZEALAND) | IM MAASSSTABE 1:5.000.000. | Zur Übersicht der Mineral-Befunde. | Von A. Petermann. / v. Hochstetter's & Petermann's Geol.-topogr. Atlas von Neu-Seeland. / Tafel 4. / Bearbeitet von L. Friedrichsen. / Druck v. C. Hellfarth in Gotha. / GOTHA: JUSTUS PERTHES | 1863.

Nebenkarte:

DER ISTHMUS VON AUCKLAND, | (der Hauptstadt von Neu-Seeland) | im Maassstabe 1:500.000

Abmessungen: 21 x 33 cm, mehrfarb. Lithogr.

In: (1) F. v. Hochstetter – A. Petermann, Topographisch-geologischer Atlas von Neuseeland.

(2) F. v. Hochstetter, Neu-Seeland.

(3) F. v. Hochstetter, Geologie von Neu-Seeland, in: Reise der Novara, Geolog. Theil 1/1, Tafel 1.

Variante C:

New-Zealand : Scale 1:5.000.000 / by A. Petermann. Drawn by L. Friederichsen.

Gotha: Justus Perthes, 1864 / No. 1.

Nebenkarte:

THE ISTHMUS OF AUCKLAND | (the Capital of New-Zealand) | Scale 1:500.000

Abmessungen: 21 x 33 cm, mehrfarb. Lithogr.

In: F. v. Hochstetter – A. Petermann, Geological and Topographical Atlas of New Zealand.

Variante D:

NEW ZEALAND | SCALE 1:5.000.000. | By A. Petermann. / Printed by C. Hellfarth Gotha. / GOTHA: JUSTUS PERTHES | 1867.

Nebenkarte:

THE ISTHMUS OF AUCKLAND | (the Capital of New-Zealand) | Scale 1:500.000

Abmessungen: 21 x 33 cm, mehrfarb. Lithogr.

In: F. v. Hochstetter, Its physical Geography, Geology and Natural History.

Variante E:

STIELER'S HAND-ATLAS N.° 50a. | Ausg. in 63 Bl. N.° LXII. / GOTHA: JUSTUS PERTHES | 1870. / Gestochen von Stier u. Stichart.

(1) WEST- | AUSTRALIEN. | MASSSTAB 1:5.000.000. | Von A. Petermann.

(2) TASMANIA. | früher genannt | VAN DIEMENS LAND | Im Maassstabe 1:5.000.000. | Von A. Petermann

(3) NEU-SEELAND | (NEW ZEALAND) | IM MAASSSTABE 1:5.000.000. | Von A. Petermann.

Nebenkarte:

DER ISTHMUS VON AUCKLAND, | (der Hauptstadt von Neu-Seeland) | im Maassstabe 1:500.000

In: A. Stieler (Hrsg.), Handatlas über alle Theile der Erde nach dem neuesten Zustande und über das Weltgebäude.

2) UMGEBUNG VON AUCKLAND

Variante A:

DER ISTHMUS VON AUCKLAND | mit seinen erloschenen Vulkankegeln | von | D.r Ferdinand von Hochstetter | 1859, | Die Zeichnung u. geographische Grundlage (haupt- | sächlich nach den Aufnahmen von Stokes u. Drury) von A. Petermann. | Maassstab 1:120.000 / Novara Expedition, Neu-Seeland, Nordinsel / Petermann's geogr. Mittheil. / Jahrgang 1862 Tafel 6. / A. Welker del. / Lith. Anst. v. C. Hellfarth in Gotha. / GOTHA: JUSTUS PERTHES | 1862.

Abmessungen: 19 x 24,5 cm, mehrfarb. Lithogr.

In: (1) Mittheilungen aus Justus Perthes' geographischer Anstalt 8 (1862), 81-83.; Tafel 6.

(2) F. v. Hochstetter – A. Petermann, Topographisch-geologischer Atlas von Neuseeland.

(3) F. v. Hochstetter, Geologie von Neu-Seeland, in: Reise der Novara, Geolog. Theil 1/1, Tafel 3.

Variante B:

THE ISTHMUS OF AUCKLAND | WITH ITS EXTINCT VOLCANOES | BY FERDINAND VON HOCHSTETTER | 1859 / the Drawing & geographical Foundation compiled principally from the Surveys of Stokes & Drury by A. Petermann. A. Welker del. / Gotha: Justus Perthes, 1865 / No. 3.

Abmessungen: 19 x 24,5 cm, mehrfarb. Lithogr.

In: F. v. Hochstetter – A. Petermann, Geological and Topographical Atlas of New Zealand.

3) SÜDTEIL DER PROVINZ AUCKLAND

Variante A:

DER SÜDLICHE THEIL| DER | PROVINZ AUCKLAND | IN | NEU-SEELAND. | Zur Übersicht der Routen und Aufnahmen | von | D.r Ferdinand von Hochstetter | 1859. | Nach den Originalzeichnungen, Skizzen und Messungen | von Hochstetter's und den Englischen Küstenaufnahmen |

unter Stokes und Drury zusammengestellt von | A. Petermann. / Maassstab 1:700.000 / Gezeichnet von A. Welker. / Lith. Anst. v. C. Hellfarth in Gotha. / GOTHA: JUSTUS PERTHES | 1863.

In: F. v. Hochstetter, Neu-Seeland.

Variante B:

DER SÜDLICHE THEIL| DER | PROVINZ AUCKLAND | IN | NEU-SEELAND. | Zur Übersicht der Routen und Aufnahmen | von | D.r Ferdinand von Hochstetter | 1859. | Nach den Originalzeichnungen, Skizzen und Messungen | von Hochstetter's und den Englischen Küstenaufnahmen | unter Stokes und Drury zusammengestellt von | A. Petermann. / Maassstab 1:700.000. / v. Hochstetter's & Petermann's Geol.-topogr. Atlas von Neu-Seeland. / Tafel 2. / Gezeichnet von A. Welker. / Lith. Anst. v. C. Hellfarth in Gotha. / GOTHA: JUSTUS PERTHES | 1863.

Nebenkarten:

(1) DER | TAUPO-SEE, | im doppelten Maassstabe der Hauptkarte.

(2) DER | SEE-DISTRIKT, | im doppelten Maassstabe der Hauptkarte.

Abmessungen: 33 x 41 cm, mehrfarb. Lithogr.

In: (1) F. v. Hochstetter – A. Petermann, Topographisch-geologischer Atlas von Neuseeland.

(2) F. v. Hochstetter, Geologie von Neu-Seeland, in: Reise der Novara, Geolog. Theil 1/1, Tafel 2.

Variante C:

THE SOUTHERN PART | OF THE | PROVINCE OF AUCKLAND | Explanatory of the Routes and Surveys | by | D.r Ferdinand von Hochstetter | 1859. | From the Original Drawings, Sketches and Measurements | by D.r von Hochstetter's and the Admiralty Surveys | by Stokes and Drury compiled by von | A. Petermann. | Scale 1:700.000. / Printed by C. Hellfarth Gotha. / GOTHA: JUSTUS PERTHES | 1867.

Abmessungen: 33 x 41 cm, mehrfarb. Lithogr.

In: F. v. Hochstetter, Its physical Geography, Geology and Natural History.

Variante D:

The Southern Part Of The Province Of Auckland, showing the Routes and Surveys by Ferdinand von Hochstetter / compiled by A. Petermann. Drawn by A. Welker / Gotha: Justus Perthes, 1865 / No. 2.

Nebenkarten:

(1) The Taupo-Lake.

(2) The Lake-District.

Abmessungen: 33 x 41 cm, mehrfarb. Lithogr.

In: F. v. Hochstetter – A. Petermann, Geological and Topographical Atlas of New Zealand.

4) ROTOMAHANA

Variante A:

ROTOMAHANA ODER DER WARME SEE | mit seinen heissen Quellen, | aufgenommen von | D.r Ferdinand von Hochstetter | 1859 | Maassstab 1:12:.000 / Jahrgang 1862 Tafel 10. / A. Petermann dir. A. Welker del. / Lith. Anst. v. C. Hellfarth in Gotha. / GOTHA: JUSTUS PERTHES | 1862.

Abmessungen: 19 x 24,5 cm, mehrfarb. Lithogr.

In: (1) Mittheilungen aus Justus Perthes' geographischer Anstalt 8 (1862), 263-266, Tafel 10

(2) F. v. Hochstetter – A. Petermann, Topographisch-geologischer Atlas von Neuseeland.

(3) F. v. Hochstetter, Neu-Seeland

(4) F. v. Hochstetter, Geologie von Neu-Seeland, in: Reise der Novara, Geolog. Theil 1/1, Tafel 5,

Variante B:

Rotomahana Or The Warm Lake and the hot Springs by Ferdinand von Hochstetter 1859 / A. Petermann dir. A. Welker del. / Gotha: Justus Perthes, 1865 / No. 5.

Abmessungen: 19 x 24,5 cm, mehrfarb. Lithogr.

In: F. v. Hochstetter – A. Petermann, Geological and Topographical Atlas of New Zealand.

5) WESTKÜSTEHÄFEN

Variante A:

DIE HÄFEN UND BUCHTEN | AOTEA UND KAWHIA, | Topographisch und geologisch untersucht | von | D.r Ferdinand von Hochstetter | 1859, | Die Zeichnung und geographische Grundlage (haupt- | sächlich nach den Aufnahmen von Stokes u. Drury) | von A. Petermann. | Maassstab 1:120.000 / v. Hochstetter's & Petermann's Geol.-topogr. Atlas von Neu-Seeland. / Tafel 4. / A. Welker del. / Lith. Anst. v. C. Hellfarth in Gotha. / GOTHA: JUSTUS PERTHES | 1863.

Abmessungen: 19 x 24,5 cm, mehrfarb. Lithogr.

In: (1) F. v. Hochstetter – A. Petermann, Topographisch-geologischer Atlas von Neuseeland.

(2) F. v. Hochstetter, Geologie von Neu-Seeland, in: Reise der Novara, Geolog. Theil 1/1, Tafel 4.

Variante B:

The Harbours And Bays Of Aotea And Kawhia, Topographically and geologically explored by Ferdinand von Hochstetter 1859 / Die Zeichnung u. geographische Grundlage (hauptsächlich nach den Aufnahmen von Stokes u. Drury) von A. Petermann. A. Welker del. / Gotha: Justus Perthes, 1865 / No. 4.

Abmessungen: 19 x 24,5 cm, mehrfarb. Lithogr.

In: F. v. Hochstetter – A. Petermann, Geological and Topographical Atlas of New Zealand.

6) NELSON

Variante A:

GEOLOGISCHE ÜBERSICHTSKARTE | DER | PROVINZ NELSON | (SÜD INSEL). | Nach den Aufnahmen | von | D.r Ferdinand von Hochstetter 1859 | und | Julius Haast 1860. | Maasstab 1:300.000. / Petermann's Geographische Mittheilungen / Jahrgang 1863 Tafel 1. / Petermann dir A. Welker del. / Lith. Anst. v. C. Hellfarth in Gotha. / GOTHA: JUSTUS PERTHES | 1862.

Abmessungen: 19 x 24,5 cm, mehrfarb. Lithogr.

In: (1) Mittheilungen aus Justus Perthes' geographischer Anstalt 9 (1863), 13-16; Tafel 1

(2) F. v. Hochstetter – A. Petermann, Topographisch-geologischer Atlas von Neuseeland.

(3) F. v. Hochstetter, Neu-Seeland.

(4) F. v. Hochstetter, Geologie von Neu-Seeland, in: Reise der Novara, Geolog. Theil 1/1, Tafel 6.

Variante B:

Geological Map Of The Province Of Nelson: by Ferdinand von Hochstetter 1859 and Julius Haast 1860 / A. Petermann dir. A. Welker del.. / Gotha: Justus Perthes, 1864 / No. 6.

Abmessungen: 19 x 24,5 cm, mehrfarb. Lithogr.

In: F. v. Hochstetter – A. Petermann, Geological and Topographical Atlas of New Zealand.

7) ATLAS VON NEUSEELAND

Deutsche Ausgabe:

GEOLOGISCH-TOPOGRAPHISCHER | ATLAS VON NEU-SEELAND | BEARBEITET | VON | D.R FERDINAND VON HOCHSTETTER UND D.R A. PETERMANN, | SECHS KARTEN | HAUPTSÄCHLICH GEBIETE DER PROVINZEN AUCKLAND UND

NELSON UMFASSEND | MIT KURZEN ERLÄUTERUNGEN. | AUS DEN WISSENSCHAFTLICHEN PUBLIKATIONEN DER NOVARA-EXPEDITION. | GOTHA: JUSTUS PERTHES, 1863.

Abmessungen: 19 x 24,5 cm, mehrfarb. Lithogr.

Englische Ausgabe:

GEOGRAPHICAL AND TOPOGRAPHICAL | ATLAS OF NEW ZEALAND | BY | DR. F. VON HOCHSTETTER AND DR. A. PETERMANN. | SIX MAPS | OF THE PROVINCES OF AUCKLAND AND NELSON. AUCKLAND, T. DELATTRE, QUEEN STREET. | 1864.

Mit Textband:

THE | GEOLOGY OF NEW ZEALAND: | IN EXPLANATION OF THE | GEOGRAPHICAL AND TOPOGRAPHICAL | ATLAS OF NEW ZEALAND | BY | DR. F. VON HOCHSTETTER AND DR. A. PETERMANN. | FROM THE SCIENTIFIC PUBLICATIONS OF THE | NOVARA EXPEDITION. | TRANSLATED BY DR. C. F. FISCHER. | ALSO, | LECTURES BY DR. F. HOCHSTETTER | DELIVERED IN NEW ZEALAND.| AUCKLAND, T. DELATTRE, QUEEN STREET. | 1864.

Abmessungen: 19 x 24,5 cm, mehrfarb. Lithogr., Textbd.: 14 x 21 cm

7) WEITERE KARTEN AUS DEM VERLAG JUSTUS PERTHES

Die durch das Erdbeben in Peru, 13. August 1868 | erzeugte | ERDBEBEN-FLUTH | im | Pacifischen Ocean, 13.-16. August 1868 | dargestellt | von Ferdinand von Hochstetter. | Äquatorial Maassstab 1: 70.000.000. / GOTHA: JUSTUS PERTHES. | 1869. / Autogr. v. Fr. Hanemann. / Druck v. C. Hellfarth. / Petermann's Geographische Mittheilungen, Jahrgang 1872, Tafel 12.

Abmessungen: 25 x 18 cm, mehrfarb. Lithogr.

In: Mittheilungen aus Justus Perthes' geographischer Anstalt 15 (1869), 222-226; Tafel 12.

ORIGINALKARTE | DER | CENTRAL-TÜRKEI | NACH AUFNAHMEN VOM JAHRE 1869 | entworfen von | Ferdinand von Hochstetter | 1870. | Maaßstab 1:420.000. / GOTHA: JUSTUS PERTHES. | 1872. / Druck v. C. Hellfarth. / Petermann's Geographische Mittheilungen, Jahrgang 1872, Tafel 1.

Abmessungen: 48 x 36 cm, mehrfarb. Lithogr.

In: Mittheilungen aus Justus Perthes' geographischer Anstalt 18 (1872), 1-4; Tafel 12.

8) KLEINERE KARTENSKIZZEN IM TEXT

(1) Te Ika a Maui | der Fisch des Maui. (F. Hochstetter, Neu-Seeland, 30).
Te Ika a Maui. (F. Hochstetter, New-Zealand, 203).

(2) Der Isthmus von Auckland. (F. Hochstetter, Neu-Seeland, 81)

(3) Waitomokia, ein erloschener Vulkankegel am Manukau Hafen, südlich von | Auckland. / E. Ade x. A. (Karte mit Ansicht, F. Hochstetter, Neu-Seeland, zw. 88/89)

Waitomokia Krater am Manakauhafen | Südlich von Onehunga. / E. Ade x. A. (Karte mit Ansicht, F. Hochstetter, Geologie von Neu-Seeland, 180).

Waitomokia, an extinct volcano on Manakau Harbour, | South of Onehunga. / E. Ade x. A. F. (Karte mit Ansicht, F. Hochstetter, New-Zealand, zw. 232/233).

(4) Mount Wellington oder Maunga Rei bei Auckland (Kartenskizze mit geolog. Profil, F. Hochstetter, Neu-Seeland, 93 und F. Hochstetter, Geologie von Neu-Seeland, 175)

Mount Wellington or Maunga Rei near Auckland (F. Hochstetter, New-Zealand, 237).

(5) Maori-Skizze des Rotokakahi. (F. Hochstetter, Neu-Seeland, 265)

Maori sketch of the | Rotokakahi. (F. Hochstetter, New-Zealand, 404).

(6) Skizze des Waikato bei Aniwhaniwha. (F. Hochstetter, Neu-Seeland, 311 und F. Hochstetter, Geologie von Neu-Seeland, 60)

Sketch of the Waikato near Aniwhaniwha. (F. Hochstetter, New-Zealand, 450).

(7) Die Geröllbank am Hafen von Nelson. ((F. Hochstetter, Neu-Seeland, 328)

Die Geröllbank (Boulder-Bank) am Hafen von Nelson (nach einer Skizze von Haast, (F. Hochstetter, Geologie von Neu-Seeland, 230).

The boulder-bank, Nelson Harbour. (F. Hochstetter, New-Zealand, 471).

(8) Die Gletscher am Mount Cook, | nach einer Skizze von Julius Haast 1862. (kleine Skizze im Text, F. Hochstetter, Neu-Seeland, 347).

The Glaciers | on Mount Cook and Mount Tasman | from a sketch | by Dr. Julius Haast 1866. / Printed by . F. Köke, Vienna. (ganzseitige kolorierte Karte, F. Hochstetter, New-Zealand, zw. 498/499).

(9) Höhlen mit Moa-Knochen im Aorere-Thal. (nach einer Skizze von Haast, F. Hochstetter, Neu-Seeland, 450).

Höhlen mit Moa-Knochen im Aorere-Thale (Provinz Nelson). (nach einer Skizze von Haast, F. Hochstetter, Geologie von Neu-Seeland, 242).

Caves with Moa bones in the Aorere Valley. (nach einer Skizze von Haast, F. Hochstetter, New-Zealand, 185).

(10) Mangere oder Mount Elliot. (F. Hochstetter, Geologie von Neu-Seeland, 182).

(11) Der Kufperbergbau am Dun Mountain. (nach einer Skizze von Haast, F. Hochstetter, Geologie von Neu-Seeland, 222).

Anhang 2: Vertragsverlag zwischen Hochstetter und dem Verlag Cotta, Stuttgart vom 29. Jänner 1862

Verlags-Vertrag | zwischen | Herrn Professor Dr. von Hochstetter in Wien | und der | J. G. Cotta'schen Buchhandlung in Stuttgart, | sowie deren beiderseitigen Rechtsnachfolgern. | §1., | Herrn Professor Dr. von Hochstetter übergibt der J. G. Cotta'schen Buchhandlung das Verlagsrecht eines Werkes über Neu Seeland betitelt: | Neu-Seeland von Dr. Ferdinand v. Hochstetter | §2. | Dasselbe wird an Text ungefähr 400 bis 500 Seiten im Maßstabe des Novara Werkes enthalten, und in Format und Druck nach dem von der J. G. Cotta'schen Buchhandlung Herrn Professor von Hochstetter vorgelegten Probeblatt hergestellt, auch mit | 6 Farbenstahlstichen | 6 großen Holzschnitten in 8° | 50 bis 60 Kleineren Holzschnitten (zum Einschalten in den Text); und | 2 Karten: 1, Uebersichts Karte von ganz Neu Seeland | 2, Topograph. Karte des südlichen Theils der Provinz Auckland | ausgestattet werden. Die Anfertigung der Karten geschieht bei Perthes in Gotha, und wird von Herrn Professor von Hochstetter alles hierauf Bezügliche angeordnet. Die Vollendung und Ausgabe | | des Werkes soll spätestens im Herbste laufenden Jahres 1862 erfolgen. | §3., | Die erste Auflage des Werkes soll zweitausend Exemplare stark abgezogen werden; der Herr Verfasser erhält hiefür als Honorar die Summe von: | f 2000.~ Zweitausend Gulden im f 24 fuss, zahlbar nach Vollendung des Druckes, nebst 25 Freiexemplaren. | §4., | Beim Nöthigwerden einer zweiten Auflage, die von Herrn Professor von Hochstetter, nach dessen Ausspruch, vielfacher Aenderungen erfahren wird, soll neue Vereinbarung zwischen beiden Contrahenten getroffen werden. | §5., | Herrn Professor von Hochstetter bleibt das Eigenthumsrecht für Uebersetzungen seines Werkes über Neu Seeland in fremde Sprachen vorbehalten. Zur Vervollständigung einer solchen Ausgabe liefert die J. G. Cotta'sche Buchhandlung Niederschläge der Holzstöcke, sowie Abdrücke der Farbenstahlstiche in nöthiger Anzahl mit englischer Unterschrift, gegen einen von derselben zu bestimmenden Preis. | Wien den | Stuttgart d. 29. Januar 1862 | Prof. Dr. F. v. Hochstetter

Cotta Vertrag, Cotta-Archiv (Stiftung der Stuttgarter Zeitung), Handschriftenabteilung, Deutsches Literaturarchiv, Marbach am Neckar.

Die Transkription des Verlagstextes wurde der Dissertation des neuseeländischen Germanisten Sascha Nolden entnommen, dem ich an dieser Stelle besonders für seine wirklich hervorragende Arbeit und für alle guten Ratschläge, die er mir dank seines exzellenten Quellen- und Hintergrundwissens zu Ferdinand von Hochstetter gegeben hat, danken möchte.

Quellen- und Literaturverzeichnis

Akademie der Wissenschaften in Wien (Hrsg.), Bemerkungen und Anweisungen für die Naturforscher, welche die Expedition von Sr. K. K. Apost. Maj. Fregatte „Novara" unter dem Commando des Herrn Obersten Bernhard v. Wüllerstorf – Urbai begleiten, Wien 1857.

Auckland City Libraries (Hrsg.), Ferdinand von Hochstetter Exhibition. Father of New Zealand Geology, Internet Ressource, http://0-www.aucklandcity.govt.nz.www.elgar.govt.nz/dbtw-wpd/virt-exhib/hochstetter/index.html (4.9.2009).

Cambridge Museum – Biographies, Internet Ressource, http://home.xtra.co.nz/hosts/Cambridge_Museum/Biographies/biosH.htm (4.9. 2009).

Daily Southern Cross, August 29th, 1864 20/2217(1864).

The Dictionary of New Zealand Biography 1769-1869, v. 1, Wellington 1990.

Fleming C. A., Dr. Hochstetter in Nelson (Extracts from the diary of Sir David Monroe, 1813-1877, in: New Zealand Journal of Geology and Geophysics 2/5 (1959), 954-963

Haast H. F. v., The life and times of Sir Julius von Haast. Explorer, Geologist, Museum Builder, Wellington 1948.

Hector J., Geological Sketch Map of New Zealand. Constructed from official Surveys and the Explorations of D.r F. von Hochstetter, D.r Julius Haast and Others, Wellington 1873.

Heger F., Ferdinand von Hochstetter, in: Mittheilungen der Geographischen Gesellschaft in Wien 27 (1884), 345-392.

Hochstetter F. v., Geologie von Neu-Seeland, in: Reise der österreichischen Fregatte Novara um die Erde in den Jahren 1857, 1858, 1859, unter dem Befehlen des Commodore B. v. Wüllerstorf-Urbair, Geologischer Theil 1/1, Wien 1864.

Hochstetter F. v. – Petermann A., Geologisch-topographischer Atlas von Neu-Seeland, Gotha 1864.

Hochstetter F. v. – Petermann A., The Geology of New Zealand in Explanation of the Geographical and Topographical Atlas of New Zealand, Auckland 1864

F. Hochstetter, [Map of the southern part of the province of Auckland, Auckland 1859 (?)], Manuskripkarte, Alexandre Turnbull Library, Reference Number: MAPCOLL-R832.1CBA/1859/ACC.620.

Hochstetter F. v., [Neuseeland - Nordteil der Südinsel] s. l. 1859, [ca. 1:360 000], 1 Kt. auf 2 Bl., mehrfarb., Handz., ÖNB, Kartensammlung, Sign.: K I 125435 Kar.

Hochstetter F. v., Neu-Seeland, Stuttgart 1863.

Hochstetter F. v., Neu-Seeland. 1859. Tagebuch Nro 5, Privatbesitz Albert Schedel, Wien.

Hochstetter F. v., New Zealand. Its physical Geography, Geology and Natural History with special Reference to the Results of the Government Expeditions in Provinces of Auckland and Nelson, Stuttgart 1867.

Hochstetter F., Lecture on the Geology of the Province Auckland, New Zealand, in: Auckland Provincial Government Gazette, July 8th, 1859 8/14 (1859), 83-100

Hochstetter F., Reise um die Erde an Bord Sr. Maj. Fregatte Novara. Archiv der Österreichischen Akademie der Wissenschaften, AÖAW, Ferdinand von Hochstetter, Novara-Tagebücher.

Hochstetter F. v. – Petermann A., Geographical and Topographical Atlas of New Zealand. Six maps of the provinces of Auckland and Nelson, Auckland 1864.

Hochstetter F. v. – Petermann A. – Fischer C. F., The Geology of New Zealand: in Explanation of the Geographical and Topographical Atlas of New Zealand. From the Scientific Publications of the Novara Expedition. Translated by Dr. C. F. Fischer. Also, Lectures by Dr. F. Hochstetter delivered in New Zealand, Auckland 1864.

Hochstetter F., Report of a Geological Exploration on the Coalfield in the Drury and Hunua District, in the Province Auckland, in: The New Zealand Gazette January 1st, 1859 2 (1859).

Holzer G., Ferdinand von Hochstetter, insbesondere als Neuseelandforscher, Dipl.-Arb., Wien 1983.

King M., The Penguin History of New Zealand, North Shore 2003.

Lawn C. A., The Pioneer Land Surveyors of New Zealand, Wellington 1977, Internet Ressource, http://www.surveyors.org.nz/Documents/parts%201_3_THE%20PIONEER%20LAND%20SURVEYORS%20OF%20NEW%20ZEALAND.pdf (4.9. 2009).

Mundy, D. L. – Hochstetter, F. v., Rotomahana; and the Boiling Springs of New Zealand. A photographic series of sixteen views, London 1875.

Nolden S., The letters of Ferdinand von Hochstetter to Julius von Haast, Diss., Auckland 2007.

Petermann A. (Hrsg.), Mittheilungen aus Justus Perthes' geographischer Anstalt über wichtige neue Erforschungen auf dem Gesammtgebiete der Geographie 6-18 (1860-1872).

Scherzer K. v., Reise der Oesterreichischen Fregatte Novara um die Erde, in den Jahren 1857, 1858, 1859, unter dem Befehlen des Commodore B. v. Wüllerstorf-Urbair. Beschreibender Teil, Bd. 3, Wien 1862.

Stieler A. (Hrsg.), Handatlas über alle Theile der Erde nach dem neuesten Zustande und über das Weltgebäude, Gotha 1870.

Zum Autor

Gerhard Holzer, geb. 1958, Studium der Geschichte und Geographie an der Universität Wien mit Schwerpunkt Wissenschaftsgeschichte, seit 1985 Mitarbeiter der Kommission für Geschichte der Naturwissenschaften, Mathematik und Medizin der Österreichischen Akademie der Wissenschaften, Leiter der Sammlung Woldan, Hauptforschungsbereiche: Geschichte der Kartographie und Geographie.

Kontakt: Österreiche Akademie der Wissenschaften, Dr.-Ignaz-Seipel-Platz 2, A-1010 Wien; gerhard.holzer@oeaw.ac.at

George Adams,
Globus hunc Terrestem Omnes hactenus
exploratos terrarum tractus,
London ca. 1765

Thomas Horst, München

Die Globen der Sammlung Woldan: Erdkugeldarstellungen als Quelle für die Geschichtswissenschaft*

Die Darstellung der Welt in Form von Erd- bzw. Himmelskugeln war bereits in der Antike bekannt; jedoch haben sich erst ab dem universaler ausgerichteten, humanistischen Zeitalter terrestrische Globen erhalten, wobei dem bekannten Manuskriptglobus von Martin Behaim (ältester erhaltener Erdglobus von 1492) und den Globensegmenten des Martin Waldseemüller (1507, mit der ersten Bezeichnung des amerikanischen Kontinents) eine besondere Rolle zukommt. Mit der kopernikanischen Wende, die das geozentrische Weltbild des Ptolemäus durch ein heliozentrisches ersetzte, entwickelte sich zugleich eine regelrechte Blütezeit in der serienmäßigen Herstellung von Globenpaaren in Form eines Erd- und Himmelsglobus, die in keiner barocken Bibliothek fehlen durften.

Diese Globen waren jedoch keineswegs nur Prestige- und Statussymbole, sondern auch Lehr- und Studienobjekte und somit wissen-

* Dieser globenkundliche Beitrag ist im Rahmen eines an der Universität der Bundeswehr in München angesiedelten Postdoktorandenprojektes zur *Interdisziplinären Erforschung der in Bayern verwahrten Erdgloben unter Berücksichtigung der Kulturgeschichte* unter der Betreuung von Prof. Dr.-Ing. Kurt Brunner entstanden. Der Autor dankt insbesondere der Dr. Meyer-Struckmann-Stiftung für die Finanzierung des Vorhabens, das zum Ziel haben wird, sämtliche in Bayern verwahrte Globen bis ca. 1850 vergleichend unter Berücksichtigung von kulturgeschichtlichen Aspekten zu analysieren.

schaftliche Instrumente[1], die zumeist neben dem abgebildeten Raum auch die Zeit mittels eines am Horizontalkreis befindlichen Kalenders und eines seitlich des Zenits angebrachten Stundenrings eindrucksvoll erfahrbar machten. Außerdem dienten sie astronomischen Zwecken, indem sie den Stand der Sonne zu einer bestimmten Jahreszeit visualisieren konnten.

Gemeinsam ist all diesen kugelförmigen Modellen der Welt, dass sie im Gegensatz zur Karte die Erde als dreidimensionales Medium unverzerrt, also der Realität entsprechend, wiedergeben. Zugleich verdeutlichen diese vielfach in Bibliotheken, Museen und Privatsammlungen verwahrten Kulturgüter das Weltbild einer vergangenen Zeit. Sie können deshalb in der Geschichtswissenschaft als vorzügliche, jedoch bisher nur am Rande beachtete Quellen herangezogen werden.

Die Globenkunde (Globographie) stellt – was wenig bekannt ist – ein eigenständig agierendes Forschungsgebiet dar, das sehr eng mit der Geschichte der Kartographie und der Entdeckungsgeschichte verbunden ist, jedoch im Gegensatz zu diesen Nachbardisziplinen stets „die Welt in Händen" hält und diese im Sinne einer Makrogeschichte untersucht. Von Bedeutung sind hierfür neben den Forschungen des italienischen Kartographiehistorikers Matteo Fiorini[2] insbesondere die wegweisenden Studien des amerikanischen Historikers Edward Luther Stevenson (1859-1944), der 1921 ein zweibändiges Kompendium zur Globenkunde vorlegte.[3]

Im deutschsprachigen Bereich sind die Einführungen von Oswald Muris (zusammen mit Gert Saarmann)[4] und von Alois Fauser[5] aus den 1960er Jahren die bislang einzigen Überblickswerke geblieben.

---

1 Die Anschaffung eines Globus war zu jeder Zeit sehr kostspielig und die richtige Handhabung dieser Geräte erforderte eine gewisse Bildung.

2 S. Günther, Erd- und Himmelsgloben. Ihre Geschichte und Konstruktion nach dem Italienischen frei bearbeitet, Leipzig 1895.

3 E. L. Stevenson, Terrestrial and celestial globes: Their history and construction …, 2 vol., London 1921.

4 O. Muris und G. Saarmann, Der Globus im Wandel der Zeiten. Eine Geschichte der Globen, Berlin und Beutelsbach bei Stuttgart 1961.

5 A. Fauser, Die Welt in Händen. Kurze Kulturgeschichte des Globus, Stuttgart 1967.

Globenforschung in Wien

Es ist an dieser Stelle vor allem auf die rege Globenforschung in Österreich hinzuweisen: Denn bereits Ende der 1940er Jahre hatte der Wiener Globensammler Robert Haardt (1884-1962)[6] in seiner Privatwohnung (damals Gußhausstraße 20) ein Globenmuseum eingerichtet, in dem er seine eigene Sammlung sowie aus öffentlichem Besitz stammende Objekte zur Schau stellte.[7] Für dieses Museum gestaltete kein Geringerer als der seit seiner Jugend an Geographica interessierte, 1948 bereits vorzeitig pensionierte Ministerialsekretär Erich Woldan (1901-1989) mehrere kleinere Ausstellungen und verfasste dazu auch die entsprechenden Erläuterungen.[8]

Der Privatsammler Erich Woldan zählt neben Robert Haardt auch zu den Mitbegründern des 1952 in Wien initiierten *Coronelli-Weltbundes der Globenfreunde* (ab 1978: *Internationale Coronelli-Gesellschaft für Globenkunde*, benannt nach dem venezianischen Minoritenpater Vincenzo Maria Coronelli, der 1684 mit der *Accademia cosmografica degli argonauti* die erste geographische Gesellschaft der Welt gegründet

---

6 Zu Haardt vgl. E. Bernleithner, Das Lebenswerk von Dipl.-Ing. Robert Haardt, in: Der Globusfreund 15/16 (1967), 123f.

7 Diese Globensammlung bildete den Grundstock des weltweit einzigartigen Globenmuseums in der Österreichischen Nationalbibliothek in Wien, das seit 2005 im Palais Mollard in Wien untergebracht ist und dort über 500 Objekte (Erd- und Himmelsgloben, Globen verschiedener Planeten sowie den Globen verwandte Instrumente wie Armillarsphären, Planetarien, Tellurien und Lunarien) verwahrt. – Bereits 1953 wurde durch das Österreichische Unterrichtsministerium die Einrichtung einer staatlichen Globensammlung (organisatorisch und räumlich angeschlossen an die Kartensammlung der ÖNB) beschlossen, die ab 1956 der Öffentlichkeit präsentiert werden konnte, vgl. J. Mokre, Das Globenmuseum der Österreichischen Nationalbibliothek, hrsg. von P. E. Allmayer-Beck, Wien 2005, 53-56.

8 R. Haardt, The Contribution of the «Coronelli-World League of Friends of the Globe». To the Development of the International Catalogue of Early Globes, in: Actes du VIIIe Congrès International d'Histoire des Sciences, Florence-Milan 3-9 Septembre 1956, Florenz 1958, vol. I, 468-472, hier: 469; I. Kretschmer, Professor Dr. Erich Woldan, 1901-1989. Vierzig Jahre ehrenamtlicher Bibliothekar der Österreichischen Geographischen Gesellschaft, in: Mitt. d. Österr. Geogr. Ges. 132 (1990), 342-345, hier: 343.

hatte). Diese Fachgesellschaft vereinigt Wissenschaftler, Sammler, Museumskuratoren, Restauratoren und Händler, die Globen besitzen bzw. verwahren, und hat es sich zur Aufgabe gemacht, alte Erd- und Himmelsgloben, Armillarsphären und Planetarien wissenschaftlich zu erforschen und die Ergebnisse dieser Forschungen in der Zeitschrift *Der Globusfreund/Globe Studies* zu publizieren, die mittlerweile (2009) in 56 Bänden vorliegt. Es ist ein Hauptanliegen dieser Fachdisziplin, die in zahlreichen Sammlungen verwahrten Erd- und Himmelsgloben und verwandte Instrumente der Astronomie (wie Globusuhren etc.) mittels ausführlicher (Ausstellungs-) Kataloge zu inventarisieren[9] sowie das Schrifttum zur Globenkunde[10] zusammenzustellen.

Dabei reicht das Spektrum von Einzelstudien zu Erd- und Himmelsgloben und dazugehörigen Globensegmenten bis hin zu biographischen Forschungen über ihre Hersteller bzw. Auftraggeber.[11] Die Ergebnisse dieser Studien werden in den von der Internationalen Coronelli-Gesellschaft veranstalteten Symposien (zuletzt in Venedig, 2007), die von internationalen Fachleuten und

---

9 Hierzu seien folgende Publikationen in Auswahl genannt: P. E. Allmayer-Beck (Hrsg.), Modelle der Welt. Erd- und Himmelsgloben. Kulturerbe aus österreichischen Sammlungen, Wien 1997; E. Dekker, Catalogue of Orbs, Spheres and Globes (Cataloghi di raccolte scientifiche 5; Istituto e Museo di Storia della Scienza Firenze), Firenze 2004; Dies., Globes at Greenwich: A catalogue of the globes and armillary spheres in the National Maritime Museum, Greenwich, Oxford 1999; W. Dolz (Bearb.), Erd- und Himmelsgloben, hg. vom Staatlichen Mathematischen Salon (Dresden, Zwinger), Dresden 1994; K. Kratzsch, Alte Globen, hrsg. von den Nationalen Forschungs- und Gedenkstätten der Klassischen Deutschen Literatur in Weimar, Weimar 1984; P. van der Krogt, Globi Neerlandici. The Production of Globes in the Low Countries, Utrecht 1993; G. und L. Zögner, Globen in Berlin, in: L. Zögner (Hrsg.), Kartographische Sammlungen in Berlin. Geschichte, Standorte, Informationen (Kartensammlung und Kartendokumentation 12), Gotha 2001, 69-90.

10 W. Bonacker, Das Schrifttum zur Globenkunde, Leiden 1960. – Nachträge in: Der Globusfreund 10 (1961), 29-34; 11 (1962), 151-157; 12 (1963), 55-60.

11 Historische Quellen wie Bilder, Enzyklopädien und Verlagsanzeigen können hierzu wertvolle Hinweise geben, vgl.: E. Dekker, The Scientific Instruments in Holbeins Ambassadors: A Re-Examination, in: Journal of the Warburg and Courtauld Institutes 62 (1999), 93-125; J. Mokre, Rund um den Globus. Über Erd- und Himmelsgloben und ihre Darstellungen, hrsg. von P. E. Allmayer-Beck, Wien 2009; F. Wawrik, Der Globus in den großen Enzyklopädien der Aufklärung. Darstellungen bis zum Jahr 1765, in: Der Globusfreund 47/48 (1999), 191-218.

wissenschaftlich Interessierten besucht werden, zur Diskussion gestellt. Dabei werden auch Themen wie die Herstellung, Pflege und Restaurierung von Globen, aber auch Fragen ihrer Verwendung behandelt.

Trotzdem harren noch immer viele Globen ihrer wissenschaftlichen Erforschung. Sie werden von Historikern (mit einigen Ausnahmen, wie dem Behaim-Globus von 1492)[12] bisher eher selten als Quellen herangezogen. Dies mag unter anderem auch daran liegen, dass Globen in der Regel als sperrig und der Umgang mit ihnen als sehr mühsam empfunden wird, weshalb es im multimedialen Zeitalter auch einer neuen Zugangsform bedarf (sog. digitale Globen).[13]

Der internationale Bestand alter Globen

Bereits 1957 hatte Robert Haardt als „Aufgabe der Gegenwart" die Katalogisierung von alten Globen gefordert.[14] In der Folgezeit wurde von der wachsenden Coronelli-Gesellschaft damit begonnen, den internationalen Globenbestand in entsprechenden Listen zu erfassen.[15] Diese Listen führen stichpunktartig den Typ des jewei-

---

12 J. K. Willers (Bearb.), Focus Behaim Globus. Ausstellungskatalog des Germanischen Nationalmuseums Nürnberg, 2 Bde., Nürnberg 1992. – Neuere Studien hierzu auch in einem Sonderheft der Zeitschrift Norica (Berichte und Themen aus dem Stadtarchiv Nürnberg 3), Nürnberg 2007.

13 Hinzu kommt noch, dass photographische Abbildungen in der Regel nur eine Seite eines Globus zeigen können. Inzwischen werden aber neueste Technologien angewendet, um alte Globen der Forschung zugänglich zu machen, vgl. zur Thematik der digitalen Globen insbesondere: F. Hruby – I. Plank – A. Riedl, Das interaktive 3D-Faksimile des Erdglobus von G. Mercator (1541), in: Cartographica Helvetica 33 (2006), 11-15; F. Hruby, Der digitale Globus – Begriff und Bedeutung für die Geographie, in: K. Kriz – W. Kainz – A. Riedl (Hrsg.), Geokommunikation im Umfeld der Geographie. (Wiener Schriften zur Geographie und Kartographie 19), Wien 2009, 154-160; H. Tomberger, Digitale Erfassung alter Globen. Ein Überblick über die aktuellen Technologien, in: Der Globusfreund 53/54 (2007), 176-188.

14 R. Haardt, Katalogisierung alter Globen, eine Aufgabe der Gegenwart, in: Festschrift zur Hundertjahrfeier der Geographischen Gesellschaft in Wien 1856-1956, Wien 1957, 528-535.

15 R. Haardt, Liste alter Globen in Österreich, in: Der Globusfreund 3 (1954), 7-36; A. Fauser, Alte Globen in Bayern, in: Der Globusfreund 11 (1962), 53-64; B. Olszewicz, Die alten Globen in Polen (Vorläufige Mitteilung), in: Ebd.,

ligen Kulturgutes (Erdglobus, Himmelsglobus oder Armillarsphäre bzw. Globusuhr), dessen Autor bzw. Fertiger (oder Auftraggeber), den Standort und den Durchmesser des Objektes samt entsprechenden Anmerkungen auf. Bis heute ist die Forderung nach einer internationalen Liste alter Globen jedoch ein Desiderat geblieben, wenngleich die diesbezüglichen Bemühungen von Peter van der Krogt (für die Niederlande),[16] von Ib Rønne Kejlbo (für Dänemark),[17] von Alois Fauser (für Bayern),[18] sowie von der Globensammlerin Heide Wohlschläger unter Mitwirkung von Prof. Rudolf Schmidt (für Österreich)[19] sehr zu würdigen sind. In jüngster Zeit wurde zudem der weltweite Globenbestand bis in das 16. Jahrhundert mit 181 Erd- und Himmelsgloben von Elly Dekker[20] vorbildlich in einer Liste zusammengetragen, weshalb nun eine systematische, vergleichende und quellenkritische Auswertung dieser Globen von Seiten der

---

85; E. O. Bratt, Provisorische Liste alter Globen in Schweden (Stand: 1. Januar 1965), in: Der Globusfreund 14 (1965), 28-35; K. Kuchar, Verzeichnis alter Globen in der Tschechoslowakei, in: Der Globusfreund 15/16 (1967), 261f.; R. Kejlbo, Provisorische Liste alter Globen in Dänemark, in: Der Globusfreund 17 (1968), 22-27; G. Duprat, Les globes terrestres et célestes en France, in: Der Globusfreund 21/23 (1973), 198-225; T. Petrovna Matvejeva, Alte Globen in der Sowjetunion, in: Ebd., 226-233; Z. von Ambrus-Fallenbüchl, Geschichte und Liste der Ungarischen Globen, in: Ebd., 243-254; B. Besevliev, Alte Bulgarische Globen und Landkarten bis zum Jahre 1878, in: Der Globusfreund 25/27 (1978), 259-263; P. van der Krogt, List of old globes in the Netherlands, in: Der Globusfreund 31/32 (1983), 78-106; A. Estácio dos Reis, List of old globes in Portugal, in: Der Globusfreund 35/37 (1987), 249-254; T. E. Rössaak, List of old globes in Norway, in: Ebd., 255-258.

16 P. van der Krogt, Old Globes in the Netherlands: A Catalog of Terrestrial and Celestial Globes Made Prior to 1850 and Preserved in Dutch Collections, Utrecht 1984.

17 I. R. Kejlbo, Rare globes. A cultural-historical exposition of selected terrestrial and celestial globes made before 1850, especially connected with Denmark, Copenhagen 1995.

18 A. Fauser, Ältere Erd- und Himmelsgloben in Bayern, Stuttgart 1964; Ders., Ältere Erd- und Himmelsgloben in Bayern (Nachträge zu dem 1964 erschienenen Buch), in: Der Globusfreund 31/32 (1983/1984), 107-128.

19 H. Wohlschläger unter Mitwirkung von R. Schmidt, Globen in Österreich, in: P. E. Allmayer-Beck (Hrsg.), Modelle der Welt, 320-379, insbes. 360f. (Liste der in der Sammlung Woldan befindlichen Globen).

20 E. Dekker, Globes in Renaissance Europe, in: D. Woodward (Ed.), The History of Cartography 3/1: Cartography in the European Renaissance, Chicago 2007, 135-173.

Geschichtswissenschaft im Rahmen des *Spatial Turn* erfolgen kann. Hierbei werden vor allem private und staatliche Sammlungen von Globen eine immer wichtigere Rolle spielen.

Die Globen der Sammlung Woldan – ein Inventar

Der gelehrte Kartensammler und Mäzen Prof. Dr. Erich Woldan, der sich selbst auch wissenschaftlich mit der Globographie auseinandergesetzt hat,[21] hatte seit seiner Jugend (ab ca. 1916) mit außergewöhnlicher Sachkenntnis eine systematische Sammlung in einer auf Geographica spezialisierten Privatbibliothek zusammengetragen, die nach seinem letzten Willen im Jahre 1989 an die Österreichische Akademie der Wissenschaften übertragen wurde.[22] Diese wohl bedeutendste Privatsammlung von geographischen Büchern und Karten in Mitteleuropa enthält neben wertvollen Cartographica, die in diesem Buch erstmals international dargestellt werden, auch zwanzig Erdgloben vom frühen 17. bis in das frühe 20. Jahrhundert,[23] die für die Geschichtswissenschaft einen optimalen Überblick über das sich verändernde Weltbild der Renaissance vermitteln können: Denn diese Globen bieten – sofern man sich kritisch mit ihnen in einer kartogaphiehistorischen, interdisziplinär angelegten Analyse auseinandersetzt – reichhaltige Hinweise auf die Vorstellung unserer Welt in vergangenen Zeiten und sollten deshalb vermehrt als Quelle für die Kultur- und Entdeckungsgeschichte herangezogen werden.[24] Sie

---

21 E. Woldan, Die Darstellung Australiens und der Antarktis auf den deutschen Globen des 17. und 18. Jahrhunderts, in: Der Globusfreund 7 (1958), 22-25; Ders., Der Erdglobus des Gemma Frisius, in: Unica Austriaca. Notring-Jahrbuch 1960, 23-25.

22 J. Dörflinger, Seltenes Mäzenatentum in der Kartographie. Sammlung Erich Woldan an die Österreichische Akademie der Wissenschaften, in: Kartographische Nachrichten 40 (1990), 187-189.

23 J. Dörflinger, Erich Woldan's Collection of Maps, Atlases and Globes, in: Imago Mundi 42 (1990), 117f.; Ders., In Memoriam Erich Woldan 1901-1989, in: Der Globusfreund 38/39 (1990), 195-202, hier: 202.

24 Vgl. hierzu beispielsweise: J. Dörflinger, Forschungsreisen auf Globen – das Beispiel James Cook, in: Lothar Zögner (Hrsg.), Die Welt in Händen. Globus und Karte als Modell von Erde und Raum (Staatsbibliothek Preußischer Kulturbesitz, Ausstellungskatalog 37), Berlin 1989, 113-119; T. Horst, Der Niederschlag von Entdeckungsreisen auf Globen des frühen 16. Jahrhunderts, in: Der Globusfreund 55/56 (2009), 23-38. – Als wichtige

dokumentieren nicht nur die „Entschleierung der Erde“[25] mit der Kolonialisierung Nordamerikas in der frühen Neuzeit[26] (und somit die humanistisch geprägte Sicht auf die neu entdeckte Welt),[27] sondern sie verdeutlichen auch die zunehmende Erkundung des afrikanischen und australischen Kontinents, was bisher in der Forschung zumeist vernachlässigt worden ist. So können spezifische Details erkannt werden, die wichtige Ver-bindungen und gegenseitige Einflüsse zwischen der Mikro- und Makrogeschichte aufzeigen. Dies soll im Folgenden ausführlicher untersucht werden.

---

biographische Lexika zur Erforschung der Erde sind folgende Werke zu nennen: W. Krämer, Die Entdeckung und Erforschung der Erde: Mit einem Abc der grossen Entdecker und Forscher, Leipzig ²1953; F. Salentiny, Das Lexikon der Seefahrer und Entdecker mit 134 Bilddokumenten und 10 Karten, Tübingen und Basel 1974; E. Schmitt, Dokumente zur Geschichte der europäischen Expansion, 8 Bde., München 1984-2008.

25 W. Behrmann, Die Entschleierung der Erde (Frankfurter Geographische Hefte 16), Frankfurt 1948.

26 Als Überblick vgl. U. Bitterli, Die Entdeckung Amerikas. Von Kolumbus bis Alexander von Humboldt, München ⁴1999; M. Häberlein, Britisch-Nordamerika von den ersten Koloniegründungen bis zum Ende des Siebenjährigen Krieges (1607-1763), in: P. Gassert – M. Häberlein – M. Wala (Hrsg.), Kleine Geschichte der USA, Stuttgart 2007, 15-102; A. Johnson, America explored. A cartographical history of the exploration of North America, New York 1974; L. Zögner (Hrsg.), Amerika im Kartenbild. Karten und Dokumente zur Entwicklung der Vereinigten Staaten von Amerika, Berlin 1976.

27 Vgl. beispielsweise W. Reinhard (Hrsg.), Humanismus und Neue Welt (Mitteilungen der Kommission für Humanismusforschung 15), Mannheim 1987.

Chronologische Liste der Globen der Sammlung Woldan[28]

Jahr	Autor	Art	Ø cm	Inv.nummer GL-V(L)
1601	Jodocus Hondius d. Ä.	EG	21	WE 54
ca. 1675	Isaac II. Habrecht	EG	21	WE 102
ca. 1675	Isaac II. Habrecht	HG	21	WE 103
1688	Vincenzo Coronelli	EG-Seg.	110	WE 115 (1-26)
1696	Vincenzo Coronelli	EG-Seg.	48	WE 122 (1-13)
1696	Vincenzo Coronelli	HG-Seg.	48	WE 123 (1-13)
1705	Georg Christoph Eimmart	EG	30	WE 130
< 1731	Matthäus Seutter	EG	21	WE 169
1750	Johann Gabriel Doppelmayr	EG	20	WE 165
1756	J. Ph. Andreae/ J. C. Berndt	EG	13,5	WE 172
ca. 1765	George Adams Senior	EG	46	WE 183
1791	Johann Gabriel Doppelmayr	EG	20	WE 200
1840	Charles Dien Junior	EG	30	WE 288
1840	Josef Jüttner/Franz Lettany	EG	32	WE 289
1844	F. L. Schönninger/F. Elekes	EG	16,5	WE 300
ca. 1855	Johann Adam Bühler	EG	14,5	WE 369
1865	Dietrich Reimer-Verlag	EG	34	WE 396
1867	Franz Schöninger/L. Hugl	EG	31,5	WE 404
vor 1870	Jan Felkl	EG	8,6	WE 411
1872	Franz Schönninger	EG	16,5	WE 417
>1875	Felkl & Sohn (Verlag)	EG	15,8	WE 422
>1875	Schotte & Co/Th. Mettke	EG	31	WE 408
1876	Felkl & Sohn/O. Delitsch	EG	31,86	WE 430
1927-31	Freytag & Berndt A.G.	EG	32	WE 497

Nicht in der Sammlung Woldan, jedoch im Besitz der ÖAW:

1849	J. Riedl von Leuenstern	MoG	23,5	AÖAW B089

28 Einen umfassenden Überblick über die Cartographica der Sammlung Woldan gibt der von G. Holzer zusammengestellte Bestandskatalog (ÖAW, Phil.-Hist. Klasse, Sitzungsberichte 628), Wien 1995. Darin wurden auch die globographischen Kulturgüter erstmals knapp beschrieben.

Erich Woldan (1901-1989) in seiner Privatbibliothek im Dezember 1986 (entnommen aus: Der Globusfreund 38/39 (1990), S. 196).

Entdeckungsgeschichtlicher Erkenntnisgewinn: Eine kulturgeschichtliche Annäherung

Die Globen der Sammlung Woldan sind als bedeutende bildliche Überreste der frühen Neuzeit auch von kulturhistorischem Interesse. So zeigt der älteste in der Sammlung befindliche Globus des Jodocus HONDIUS d. Ä. bereits das 1585 vom englischen Seefahrer John Davis (1550-1605)[29] befahrene *Fretum Davis* (Tafel S. 262). Die breite Meerenge zwischen der kanadischen Baffininsel und Grönland wurde erstmals in der Neuzeit im Zuge der imperialen Interessen der englischen Königin Elisabeth I. bei der Suche nach einer Nordwestpassage von John Davis im Jahr 1585 bis zu einer nördlichen Breite von fast 73° befahren.[30] Diese Erkenntnis hat sich auch in zeitgenössischen Karten niedergeschlagen: So zeigt etwa die zweite Nordpolarkarte *Septentrionalium Terrarum descriptio* des Gerhard Mercator von 1595 (im posthum erschienenen *Atlas sive Cosmographicum*)[31] wie auch die von Hondius selbst um 1597 in Amsterdam herausgegebene *Christian Knight-Karte* unter dem Titel *Typus Totius Orbis Terrarum* die nach Davis benannte Meeresstraße (zwischen Grönland und einer nordamerikanischen Landmasse liegend).[32] Auf dem Globus von 1601 ist sie oberhalb der einhundert Jahre zuvor von Gaspar Cortereal (*Terra de Cortereal*) erreichten Küste von Labrador (*Terra de Laborador*) abgebildet.

29 A. H. Markham, The voyages and works of John Davis the navigator (Hakluyt Society 59), London 1880 (ND 1970).

30 John Davis, der bei einer Ostindien-Expedition am 29. Dezember 1605 in Indonesien getötet wurde, veröffentlichte auch mehrere Buchwerke über die Navigationsmethoden seiner Zeit. Nach ihm ist auch der sog. Davisquadrant zur Bestimmung der geographischen Breite benannt worden, vgl. C. Markham, A life of John Davis, the navigator, 1550–1605: discoverer of Davis Straits, New York u. London 1889.

31 Vgl. das Exemplar in der Kartensammlung der ÖNB Wien, E 15.706-C/D (Mercator 393.698-D. Kart), abgebildet bei G. Sammet, Die Welt der Karten: Historische und moderne Kartographie im Dialog, Gütersloh und München 2008, 50.

32 Somit können Globen auch als Proxydaten für die Klimageschichte verwendet werden, vgl. K. Brunner, Kartographie und Geoinformation. Karten als Klimazeugen, in: Mitteilungen der Österreichischen Geographischen Gesellschaft 147 (2005), 237-264, hier: 244-249.

Die Toponomie des amerikanischen Kontinents auf dem Hondius-Globus erinnert insgesamt an die Karte *Americae Sive Novi Orbis, Nova Descriptio*[33] im *Theatrum Orbis Terrarum* des Abraham Ortelius von 1570. Bei der Darstellung der pazifischen Küste in Nordamerika wird hier zudem auf die von Sir Francis Drake (1540-1596)[34] für England im Jahre 1579 beanspruchte Region *Nov*(a) *Albion* (New Albion) hingewiesen. Das *Anian Fretum* im äußersten amerikanischen Nordwesten trennt Asien von Amerika in zwei geographisch selbständige Erdteile. Dieser Name verweist möglicherweise auf Annam, eine chinesische Provinz (heute in Nordvietnam), die auch in einer 1559 bei Giovanni Battista Ramusio erschienenen Ausgabe von Marco Polos Reisebericht erwähnt wird. Die sog. Straße von Anián ist neben einem *Regnum Anianum* und der Region *Quivira* oftmals auf kartographischen Produkten (erstmals auf einer Weltkarte des Giacomo Gastaldi von ca. 1561)[35] zu finden; ihre wirkliche Existenz wurde jedoch erst im 18. Jahrhundert durch Vitus Bering und James Cook nachgewiesen.[36]

Sehr deutlich ist auf dem Globus von Hondius auch die – bei der Suche nach einer Nordostpassage von Europa zum Pazifik von Willem Barentsz (um 1550-1597) erforschte[37] – russische Doppel-

---

[33] D. Reinhartz, The Americas Revealed in the Theatrum, in: M. P. R. van den Broecke (Hrsg.), Abraham Ortelius and the First Atlas. Essays commemorating the quadricentennial of his death 1598 – 1998, 't Goy-Houten 1998, 209-220.

[34] J. Paige, Sir Francis Drake, circumnavigator of the globe and privateer for Queen Elizabeth, New York 2003; P. Whitfield, Sir Francis Drake, New York 2004.

[35] M. Milanesi, Arsarot oder Anian? Identität und Unterscheidung zwischen Asien und der neuen Welt in der Kartographie des 16. Jahrhunderts (1500-1570), in: A. Prosperi – W. Reinhard (Hrsg.), Die Neue Welt im Bewusstsein der Italiener und Deutschen des 16. Jahrhunderts (Schriften des Italienisch-Deutschen Historischen Instituts in Trient 6), Berlin 1993, 15-67, hier: 37f. und Abb. 37.

[36] Vgl. Behrmann, Entschleierung der Erde, 37: Die Anianstraße wurde 1648 auch von Deschnew bzw. 1741 von Bering befahren.

[37] F. Abdelouahab, L'aventure des poles. Entdecker im ewigen Eis: fünf Jahrhunderte Polarreisen in Reisetagebüchern, Kehl 2006, 37-39; H. J. Doeksen (Hrsg.), Proceedings International Willem Barentsz Symposium 1997. Conference on exploration and exploitation of the European Arctic Region in past, present and future, Maritime Institute Willem Barentsz, Terschelling, The Netherlands 1988.

insel Novaja Semlja dargestellt (*Nova Zembla detecta 1596*). Östlich davon findet sich ein Hinweis auf das Eismeer (*Congelat(um) Mare*) mit der Inschrift *Septentrionalis ab hac parte omnino incognitae sunt* (Tafel S. 264 oben), was indirekt Hinweise für die Geschichte des Klimas dieser Region geben könnte.[38] Insgesamt beherrscht der riesige, unbekannte Südkontinent (hier: *Terra Incognita Australis*), der bereits auf den Globen des Johannes Schöner von 1515 abgebildet war, die Erdkugel, auf der zudem noch folgende kleinere pazifische Inseln ihre Erwähnung finden: *I. de Paxaros* (bei einer Länge von 220°, zwischen Japan und dem nordamerikanischen Kontinent),[39] *Roca Partida* (die kleinste der zu Mexiko gehörenden Revillagigedo-Inseln, 1569 gesichtet) und *I. de Anublada* (westlich der Region *Nova Hispania* bei einer Länge von 270°, heute: San Benedicto).[40]

Im Atlantik hingegen erscheinen folgende Inseln für die heute sich an Universitäten etablierende Atlantische Geschichte (v.a. auch für die Wirtschaftsgeschichte)[41] von besonderem Interesse: *Island* und die Phantominsel *Frisland*;[42] *I. de Bermuda* (Bermudasinseln, von den Spaniern entdeckt, aber erst 1609 kolonisiert), *I. Fernando de Laronno* (das ist die östlich von Brasilien liegende Inselgruppe Fernando de Noronha, die bereits von Vespucci erwähnt wurde), *I. de Fernando Pó*

---

38 Zu dieser Thematik vgl. K. Brunner, Nordeuropa-Darstellungen in den Ptolemäus-Ausgaben am Ende des 15. Jahrhunderts, in: W. Scharfe (Hrsg.), 6. Kartographiehistorisches Colloquium Berlin 1992, Berlin 1994, 19-29.

39 Es könnte sich dabei um eine bei den Erkundungsfahrten des italienischen Seefahrers Juan Gaetano (Mitte 16. Jahrhundert) gefundene Insel handeln. Ob er Hawaii entdeckt hat, ist bis heute umstritten, vgl. B. Judd, Voyages to Hawaii before 1860. A record, based on historial narratives in the libraries of the Hawaiian Mission Children's Soc. and The Hawaiian Historical Soc., extended to march 1860. Enl. and ed. by Helen Yonge Lind, Honolulu 1974; G. L. Fitzpatrick, The early mapping of Hawaii, London u. a. 1987.

40 Diese unbewohnte Vulkaninsel war erstmals 1533 vom spanischen Entdecker Hernando de Grijalva gesichtet worden und hatte von Ruy López de Villalobos 1542 ihren Namen erhalten, vgl. C. Varela, El viaje de don Ruy López de Villalobos a las islas del Poniente 1542-1548 (Letterature e culture dell'America Latina 7), Milano 1983.

41 C. Schnurmann, Europa trifft Amerika. Atlantische Wirtschaft in der frühen Neuzeit 1492-1783, Frankfurt am Main 1998.

42 D. S. Johnson, Fata Morgana der Meere. Die verschwundenen Inseln des Atlantiks. Aus dem Amerikanischen von A. Menneke, München u. Zürich 1999, Kapitel VI: Frisland, 99-124.

im Golf von Guinea (heute: Bioko/Äquartialguinea),[43] *I. de Tristan da Cunha* (westlich des Kaps der Guten Hoffnung)[44] und schließlich *I. dog°. al(vare)s* (Gough-Insel).[45] Außerdem finden die patagonischen Riesen oberhalb der Magellanstraße (Mann und Frau mit der Inschrift *Imago viri et feminae circa Fretum Magellani,* s. Tafel S. 264 unten), die bereits vor ihrer Entdeckung auf Karten und den Globen Johannes Schöners abgebildet war,[46] sowie die chinesische Mauer ihre visuelle Berücksichtigung.

Der Erdglobus des Isaak HABRECHT ist auf den ersten Blick als reiner Nachstich der Hondius-Segmente von 1601 zu erkennen, was vor allem beim Vergleich des kartographischen Bildes und der Legenden zu ersehen ist. Jedoch finden sich hier auch bemerkenswerte Unterschiede:

- Die bei Hondius vorkommenden Orte *Lucach*, *Beach Regnum*, *Petan I.* und *Iava Minor* (nach Mercator, oberhalb der Titelkartusche) werden nun nicht mehr bezeichnet, dafür westlich der Kartusche ein *Terra d'Eintracht* (Westaustralien).
- Neue Ergebnisse der Entdeckungen im Norden Amerikas werden hinzugefügt, womit vor allem auf die vierte Entdeckungsreise des

43 Diese vulkanische Insel wurde 1472 vom portugiesischen Seefahrer Fernão do Pó, dessen Namen sie bis 1973 trug, im Auftrag König Alfonsos V. von Portugal entdeckt. 1778 wurde die Insel von Portugal an Spanien abgetreten und war von 1827 bis 1843 an das britische Königreich verpachtet, vgl. M. Liniger-Goumaz, Historical dictionary of Equatorial Guinea, Lanham/Md. 2000.

44 Diese Insel wurde im März 1506 vom portugiesischen Admiral Tristão da Cunha entdeckt, der sie nach sich selbst benannte, vgl. Obras de Tristão da Cunha, 2 Bde., Rio de Janeiro 1979.

45 Diese vulkanische Insel wurde im frühen 16. Jahrhundert vom portugiesischen Seefahrer Gonçalo Álvares entdeckt und wurde später nach dem britischen Kapitän Charles Gough benannt, der sie 1731 wiederentdeckte.

46 R. Henning, The Representation on Maps of the Magalhães Straits before Their Discovery, in: Imago Mundi 5 (1948), 32-37; F. v. Wieser, Magalhães-Strasse und Austral-Kontinent auf den Globen des Johannes Schöner: Beiträge zur Geschichte der Erdkunde im XVI. Jahrhundert, Innsbruck 1881 (Nachdruck Amsterdam 1967).

Henry Hudson (ca. 1565-1611)[47] und die nach ihm benannte Hudsonbay Bezug genommen wird (*Magni huius Oceani primus Detector ille fuit Mr. Hudson*). Hudson begab sich wie sein Zeitgenosse William Baffin (1584-1622) auf die Suche nach einer Nordwestpassage vom Atlantik in den Pazifik. In der Hudsonbay saß sein Schiff jedoch während des kalten Winters 1610/1611 im Eis fest (*Hudson winter*), was auf dem Rückweg zu einer Meuterei führte, bei der er bei Tagesanbruch des 23. Juni 1611 in einem kleinen Boot ausgesetzt wurde und für immer verschollen blieb.

Zudem findet sich der Sankt Lorenz Strom (*R. S. Laurens*) auf dem Habrecht-Globus abgebildet. Darüber hinaus verweist die oberhalb der Region *Montagnata* gestochene Inschrift *Prince Henrii The Kinges Fort Worsnam* auf den Prinzen von Wales, Henry Frederick (1594-1612)[48] aus dem Hause Stuart, dessen früher Tod eine Katastrophe für das englische Königshaus bedeutete.

- Eine umfangreiche Legende unter *Nova Guinea* weist außerdem (wohl beeinflusst durch den spanischen Entdecker Íñigo Ortiz de Retez, der 1545 hier landete) auf die Ähnlichkeit der Insel mit der Küste des afrikanischen Guinea hin: *Sic dicta, quod ejus littora locor(um) facies Guineæ Africanæ multum sint similia*[49] (vgl. Farbtafel S. 300 unten).
- Auffällig ist auch die Verwendung eines anderen Fabelwesens im Pazifik westlich des südamerikanischen Kontinents, an dessen Südspitze weiterhin die zum hypothetischen Südkontinent gehörige *Terra del Fogo* (jedoch ohne Bezeichnung des Cabo Espíritu Santo) neben der Magellanstraße (*Fretum Magellanicum*) liegt. Zusätzlich findet sich hier mit der Ergänzung *Fret. Le Maire* die nach Jacob Le-Maire (1585-1616) benannte Straße am öst-lichsten Ausläufer des argentinischen Teils von Feuerland vermerkt. Der niederländische Seefahrer Le Maire hatte diese Straße und den südlichsten Punkt Südamerikas, das Kap Hoorn (benannt nach dem Geburtsort des mitreisenden Seefahrers Willem Cornelisz

---

[47] P. Edwards, Last voyages: Cavendish, Hudson, Ralegh: the original narratives, Oxford 1988; D. S. Johnson, Charting the sea of darkness: the four voyages of Henry Hudson, Camden, Me. 1993.

[48] Henry Frederik war der Sohn des Königs Jakob VI. und seiner Gemahlin Anna von Dänemark. Er starb vermutlich an Typhus und wurde in Westminster Abbey beigesetzt.

[49] Im Jahre 1623 kartographierte Jan Carstenszoon im Auftrag der Niederländischen Ostindien-Kompanie große Teile der Küste Neuguineas.

Schouten), im Januar des Jahres 1616 entdeckt.[50] Diese Route war nicht nur weniger gefährlich als die Magellanstraße, sondern hatte zudem den Vorteil, dass die Händler hier keinen Tribut an die Ostindische Kompanie entrichten mussten.

Auch die Globensegmente des Vincenzo CORONELLI verzeichnen in italienischer Sprache die Schiffsrouten der Reise von Le Maire und Schouten (1613-1617). Zudem sind hier neue, nach der Entdeckung Australiens (*Nuova Hollanda*) und Tasmaniens (*La Terra di Antonio Diemens fu scoperta li 24 Novembre del 1642 da Abel Tazman Hollandese*, vgl. Tafel S. 268)[51] gewonnene geographische Kenntnisse verwendet worden. Außerdem ist die gesamte Reiseroute (Hin- und Rückreise) des französischen Gesandten Alexandre Chevalier de Chaumont eingetragen (*Viaggio da Brest à Siam nell'anno 1685*). Ziemlich genau wurde der 1541 von Hernando de Soto entdeckte Mississippi (*Michissipi*) kartiert, dessen Mündung erstmals vom Europäer René-Robert Cavelier, Sieur de la Salle, 1683 aufgefunden wurde.[52] Kalifornien ist nun als Insel dargestellt. Eine genaue Untersuchung aller kartographischen Parerga (beispielsweise die südlich von Neuseeland vermerkten Antipoden von Venedig[53]) und Legenden, die

---

50 Bei der Weiterfahrt durch den Pazifik wurden u. a. die Tonga-Inseln entdeckt, die jedoch auf dem Globus noch keine Berücksichtigung gefunden haben. Auf Java wurden Le Maire und Schouten verhaftet, da man ihnen vorwarf, das Monopol der Ostindischen Kompanie verletzt zu haben; man glaubte ihnen die Geschichte der neu entdeckten, alternativen Seestraße nicht. Jakob Le Maire starb – in Ketten gelegt – auf der Rückfahrt; sein Vater erhielt nach einem Prozess jedoch eine Entschädigung, vgl. J. van Spilbergen, Oost ende West-Indische spieghel, Amsterdam 1621 (http://www.ub.uni-bielefeld.de/diglib/2005/lemaire, aufger. am 3.12.2009).

51 Der niederländische Seefahrer Abel Janszoon Tasman (1603-1659) unternahm für die Ostindien-Kompanie, die sich die Erschließung neuer Handelgebiete zum Ziel machte, mehrere Expeditionen und entdeckte 1642 das nach ihm benannte Tasmanien, das er jedoch noch nicht als Insel erkannt hatte. Zeitgenössisch wurde dieses Eiland nach Antonio van Diemen (1593-1645), dem damaligen Generalgouverneur von niederländisch-Ostindien, bezeichnet, vgl. A. J. Tasman, Entdeckung Neuseelands, Tasmaniens und der Tonga- und der Fidschi-Inseln, 1642-1644, Stuttgart 1985.

52 H.-O. Meissner, Louisiana für meinen König. Die Abenteuer des Robert La Salle (Die Abenteuer der Weltentdeckung 3), Stuttgart 1966.

53 *Questo preciso luogo segnato coll'Asterismo è Antipodo alla Città di Venetia, Patria dell'Autore della presente Opera.*

besonders für die Entdeckungs-, jedoch auch für die Umweltgeschichte (Erdbeben von Lima, 1688[54]) interessante Aspekte liefern können, steht jedoch noch aus.

Auf der nur wenige Jahre später in Nürnberg von Georg Christoph EIMMART (Farbtafel S. 302 oben) gefertigten Erdkugel sind vor allem die Erforschungen der Ostindienkompanien im Pazifik auf den neuesten Stand gebracht worden, wenngleich die 1700 vom dreimaligen Weltumsegler William Dampier (1651-1715) entdeckte und nach ihm benannte Schifffahrtsstraße bei Neuguinea noch keine Berücksichtigung fand.[55] Auch die im Süden und Osten noch nicht endeckten Küsten Australiens werden für die Phantasie des Globenbetrachters offengelassen. Dafür finden wir hier mit den *Navigationi Oliverii* die Reiseroute des ersten Niederländers (Olivier van Noort, 1558-1627), der die Welt umsegelte, punktiert eingetragen. Der westliche Teil Nordamerikas war damals aber immer noch ein recht unbekanntes Land (*Terra incognita sive Terra Esonis*) und der Gedanke von Kalifornien als Insel wurde weitergeführt. Jedoch ist nunmehr interessanterweise der hypothetische Südkontinent verschwunden.

Diese geographische Berichtigung ist auf vielen Globen des 18. Jahrhunderts zu erkennen. Am weitesten verbreitet waren damals im deutschsprachigen Bereich die Globen des Nürnberger Mathematikers Johann Gabriel DOPPELMAYR, von dem sich zwei Erdgloben in der Sammlung Woldan erhalten haben (Farbtafel S. 304). Der ältere von beiden, eine Auflage von 1750, zeigt im Gegensatz zu den bisher untersuchten Globen nun auch die Schifffahrtsroute des Dampier (*Navigatio Dampieri 1688*). Australien wird weiterhin mit

---

54 Man vgl. beispielsweise die Kartusche im Pazifik vor Peru: *La Cità di Lima capitale del Perù fù sommersa del Tremuotonel i688* – Das Erdbeben, das über 5000 Todesopfer forderte, fand am 10. Oktober 1688 statt, vgl. J. F. Seyfahrt, Allgemeine Geschichte der Erdbeben, Frankfurt und Leipzig 1756, 67f.

55 Hier ist eine Verbindung zum niederländischen Globenhersteller Gerhard Valk zu erkennen, vgl. P. van der Krogt, 122, sowie F. Wawrik und H. Hühnel, Das Globenmuseum der Österreichischen Nationalbibliothek, in: Der Globusfreund 42 (1994), 3-188, hier: 61f.; H. Walz (Hrsg.), Freibeuter 1683-1691. Das abenteuerliche Tagebuch eines Weltumseglers und Piraten, Darmstadt 1997.

offener Ostküste kartiert (aber ohne die Insel Tasmanien). Auch hatte sich mittlerweile die Vorstellung einer kalifornischen Halbinsel durchgesetzt und der in den 1680er Jahren erreichte *Rio Colorado* findet sich in *AMERICA SEPTENTRIONALIS* ebenso verzeichnet, wenngleich Alaska noch unbekannt war. Dafür wurde die 1728 von Vitus Bering (1681-1741) erforschte Ostküste Asiens mit der Tschuktschenhalbinsel sorgfältig dargestellt; die zweitgrößte japanische Insel Hokkaidō (früher: Jesso) wurde aber mit Kamtschatka gleichgesetzt (*Kamtzadalia alias Terra jedso*).
Die Globen Doppelmayrs waren die genauesten ihrer Zeit, verzeichnen aber auch Phantominseln wie die angeblich 1670 vom niederländischen Seefahrer John Lindestz Lindeman gesichtete Insel Saxemberg im Südatlantik (*I. Saxenburg*).[56]
Das zweite Doppelmayr-Exemplar von 1791 weist folgende geographische Änderungen auf:

- Die australische Küste ist nun als abgeschlossener Kontinent kartiert worden und die südöstliche Ecke von Australien bildet Tasmanien.
- Erstmals gelangten auch die neuesten Entdeckungsreisen des britischen Seefahrers James Cook (1728-1779) zur Darstellung (vgl. beispielsweise den Hinweis auf seine zweite Südseereise unterhalb der Magellanstraße: *Cap. Cook 3. Ian. 1775*).[57]

Insgesamt ist zu konstatieren, dass die Globen Doppelmayrs gemäß des rational aufgeklärten Zeitalters sachlich und nüchtern erscheinen und auf jegliches kartographisches Beiwerk verzichten.

Für das 18. Jahrhundert, in dem man sich über die genauen Ausmaße der einzelnen Kontinente noch recht unsicher war, sind

---

[56] R. H. Ramsay, No Longer on the Map. Discovering Places That Never Were, New York 1972, hier: 109f. – Weitere noch nicht identifizierte Orte auf dem Globus sind etwa die *Nachtigal I.* (unterhalb Madagaskars, evtl. die Kerguelen?) oder die *Terra detecta per David Angliam 1680* (westlich vom *Mare chilense*, evtl. die Osterinsel?).

[57] J. Frings (u. a.), James Cook und die Entdeckung der Südsee. Katalog der Ausstellung in der Kunst- und Ausstellungshalle der Bundesrepublik Deutschland, Bonn, 28. August 2009 bis 28. Februar 2010; Museum für Völkerkunde, Wien, 10. Mai bis 13. September 2010; Historisches Museum Bern, 7. Oktober 2010 bis 13. Februar 2011, München 2009.

ansonsten noch folgende Globen in der Sammlung Woldan bemerkenswert:

- Ein Unikat stellt die Erdkugel des Kaiserlichen Geographen Matthäus SEUTTER aus der Zeit vor 1731 dar (Farbtafel S. 303). Es handelt sich dabei um eine geringfügig veränderte Neuauflage von dessen um 1710 gestochenen Segmenten für einen Erdglobus, die insgesamt an das globographische Bild von Eimmart erinnern. Jedoch findet sich hier wieder die Vorstellung eines imaginären Südkontinents (*Promontorium Terrae Australis*, darunter: *Australis Incognita Magellan*) und Neuguinea wird als Halbinsel Australiens kartiert. Die Westküste Tasmaniens, das sich wie das 1642 von Abel Tasman entdeckte Neuseeland zu weit östlich befindet, ist imaginär; die Nordküste bleibt offen. Die Phantominsel Saxemberg fehlt. Auffallend ist zudem der Hinweis auf die Nilquellen in Afrika: *Nili originem cum Recentioribus Geographis* […] *huc opposuismus.*
- Die Quellflüsse des Nils treten auch auf dem Erdglobus von Johann Philipp ANDREAE (Neuauflage 1756, vgl. Farbtafel S. 305 oben) auf; der von Johann Christoph Bernhard gefertigte Kupferstich weist nicht nur im Horizontring, sondern insgesamt im Kartenbild überraschende Ähnlichkeiten zum Seutter-Erdglobus von 1710 auf (*Nova Hollandia* verbunden mit Neuguinea und offen gelassener Süd- bzw. Ostküste, vgl. Farbtafel S. 305 unten; Symbolisierung des Südkontinents in der Südsee).
- Der in London um 1765 von George ADAMS Senior gefertigte Erdglobus hingegen vermerkt mit der Inschrift *Ansons Return* die Weltreise des britischen Admirals George Anson (1697-1762), der auf seiner Erdumsegelung in den Jahren 1740-1744 den Auftrag erhalten hatte, im Rahmen des österreichischen Erbfolgekriegs die spanischen Niederlassungen in Übersee zu zerstören.[58] Im *OCEANUS GLACIALE* ist hier die Küste der Bouvetinsel im Südatlantik zu erkennen, die erstmals 1739 vom französischen Forschungsreisenden Jean-Baptiste Charles Bouvet de Lozier (1705-1786) gesichtet wurde, der sie für ein nördliches Kap der *Terra Australis* hielt (*I. de la Circumcision detecta 1739*). Außerdem ist der australische Kontinent *Hollandia Nova* zwar mit seinen Küsten

58 Darüber berichtet u. a. ein vom Marineprediger Richard Walter verfasster Bericht unter dem Titel *Voyage round the world*, London 1748, vgl. G. Williams, Der letzte Pirat der britischen Krone. Captain Anson und der Fluch des Meeres, Berlin 2000.

einigermaßen richtig kartiert, jedoch fälschlicherweise im Süden mit der nach der niederländischen Provinz Diemen genannten *Terra de Diemen* (Tasmanien) verbunden worden. Dafür wurde nun der Küstenverlauf von *Kamzatka* im Vergleich zu den Doppelmayr-Globen wesentlich verbessert.

Wenn wir im Folgenden bei unserer globographischen Analyse bis in die Mitte des 19. Jahrhunderts vordringen, so fällt grundsätzlich auf, dass die Globen des aufgeklärten Zeitalters zunehmend ihren Horizontring ablegen und vom ehemals rein wissenschaftlichen Instrument zu einem in Serienproduktion hergestellten Gebrauchsgegenstand degradiert werden. Die früher sehr zahlreichen weißen Flecken des Erdballs, die oftmals mit Phantasiewesen ausgefüllt worden waren, begannen nun langsam zu verschwinden. Der Schein trügt aber: Denn zumeist waren lediglich die Umrisse der anderen Kontinente bekannt, während man über das Landesinnere noch sehr wenig wusste. Ein Präzedenzbeispiel stellt hierbei der afrikanische Kontinent dar, dessen Kartenbild entscheidende Hinweise auf die Datierung von Erdgloben geben kann.[59]

So vermerkt der nach den neuesten Entdeckungen mit französischen Legenden versehene Erdglobus des Charles DIEN von 1840 (Farbtafel S. 307) in Südafrika neben der Kalahariwüste noch hohe, unbekannte Gegenden (*Contrée élevée et fort peu connue* [sic!]) und in Zentralafrika ein Plateau, über dessen Höhe sich der Autor selbst nicht ganz sicher ist (*Plateau de Dembo d'une environ 2000 metres?* – vermutlich, weil darunter ein portugiesischer [*Marché des Portugais*] und kein französischer Handelsplatz gekennzeichnet ist).

Auch der von einem französischen Seeadmiral im Jahre 1772 bereits entdeckte Kerguelen-Archipel, eine subantarktische Inselgruppe im südlichen Indischen Ozean, wird nun kartographisch vermerkt (*I. Kerguelen ou de la Désolation C. Louis*).[60] Der

59 R. Schmidt, Zur Datierung der Erdgloben nach dem Kartenbild Afrikas (Unter Verwendung der Ausarbeitung von Prof. Dr. H. Wels), in: Der Globusfreund 30 (1982), 88-97.

60 Der französische Seefahrer Yves Joseph de Kerguelen de Trémarec (1734-1797) hatte auf seiner Rückreise von Ostindien diese Inselgruppe im Februar 1772 entdeckt und wurde daraufhin von König Ludwig XV. beauftragt, den unbekannten Südkontinent zu finden und Beziehungen mit dessen vermeintlichen Bewohnern aufzunehmen. Diese Theorie wurde aber mit den

nordamerikanische Kontinent ist mit dem russischen Territorium (*Amérique Russe*: Westküste Alaskas mit den *Escimaux*, entdeckungsgeschichtlich auch bedeutsam die Vermerke *Ent. de Kotzebue* und *C. Krusenstern*[61]), das sich vom amerikanisch-kanadischem Gebiet abgrenzt, dargestellt worden. Südamerika wird erstaunlich genau kartiert.[62] Auch die Erforschung des nunmehr als *Australie* bezeichneten Kontinents war inzwischen vorangekommen, wie die kartographische Darstellung der Flüsse Darling und Lachlan sowie von *Port Jackson* (Hafen von Sydney) mit vorgelagerter Lord-Howe-Insel beweist.[63] Südlich von *ELLE ZEELANDE* (Neuseeland mit der Südinsel *Tarai-Pounammon* und der Nordinsel *Ikas-Na-Mawi*) sind jetzt zudem die Campbell- und Auckland-Inseln,[64] die Chatham- und Bounty-

---

Reisen von James Cook endgültig widerlegt. Cook schreibt in seinem Logbuch von 1776 über den Kerguelen-Archipel, der nun auch erstmals kartographiert wurde: *I could have very properly called the island Desolation Island to signalise its sterility, but in order not to deprive M. de Kerguelen of the glory of having discovered it, I have called it Kerguelen Land.* – Die Inseln waren im 19. Jh. für Wal- und Robbenfänger (Trangewinnung!) eine wichtige Einnahmequelle.

61 Dieses Kap wurde von einem russischen Seefahrer mit deutscher Herkunft, Otto von Kotzebue (1788-1846), im Sommer 1816 entdeckt und nach dem deutsch-baltischen Admiral Adam Johann Baron von Krusenstern (1770-1846), der im Dienste des Zaren die Welt umsegelte, benannt.

62 Dies ist vor allem erkennbar, wenn man den Blick auf die kleinsten Orte richtet: So ist sehr deutlich der Madeira, ein Nebenfluss des Amazonasstromes, mit der 1756 von Jesuiten gegründeten Stadt *Borba Villa* (Borba) auf dem Globus eingetragen. Zur Ethnologie dieses Gebiets vgl. meinen Bericht: Bayern am Amazonas. Der Münchener Thomas Horst traf im Urwald den katholischen Pfarrer Pedro Maria Gawlik – einen früheren Absolventen der Ludwig-Maximilians-Universität, in: Literatur in Bayern 77 (September 2004), 62-64.

63 Diese Insel wurde erstmals im Februar 1788 von Leutnant Henry Lidgbird Ball (um 1756-1818) gesichtet und nach dem englischen Flottenadmiral Lord Richard Howe (1726-1799) benannt. Die ersten permanenten Siedler ließen sich aber erst ab 1833 hier nieder, vgl. V. Parsons, Henry Lidgbird Ball, in: D. Pike, Australian Dictionary of Biography, Melbourne (u. a.) 1966, Bd. 1; D. Syrett, Admiral Lord Howe: A Biography, Annapolis/Md. 2006.

64 Die subantarktischen Auckland-Inseln waren – wie archäologische Funde beweisen – bereits im 13. Jahrhundert von Polynesien aus besiedelt worden. Sie wurden erstmals von Europäern im Jahr 1806 entdeckt und nach dem englischen Diplomaten und Schriftsteller William Eden, dem ersten Baron Auckland (1745-1814) benannt. Zur Zeit der Globusherstellung waren die Inseln auch für Frankreich von Interesse, wie der Aufenthalt des Seefahrers Jules-Sébastien-César Dumont d'Urville (1790-1842) beweist.

Inseln,[65] die unbewohnten Antipodeninseln (eine Felsinselgruppe vulkanischen Ursprungs, 1800 entdeckt), sowie die *Antipodes de Paris* auf dem Globus verzeichnet. Nördlich von Neuseeland sind außerdem die Norfolk-Insel (in dieser Zeit britische Sträflingskolonie) und die Kermadec-Inseln (*Raoul*[66] und *Curtis*) zu sehen.

Ein sehr ähnliches, aber dennoch in Einzelheiten divergierendes Bild stellt der vom k. k. Hauptmann der österreichischen Artillerie, Joseph JÜTTNER, zusammen mit Leutnant Franz LETTANY gefertigte Erdglobus aus demselben Jahr dar (Farbtafel S. 308). Hier wird erneut auf *Unbekannte Länder* in Zentralafrika Bezug genommen und auch die ostafrikanischen Gebirgslandschaften in schematischer Rautenzeichnung erscheinen noch mehr als ungewiss (*Wahrscheinliche Stelle des Hochgebirges Kumri*). Bei den darüberliegenden Mondbergen, die bereits in der Ulmer Ptolemäus-Ausgabe von 1482 verzeichnet sind, findet sich hier zudem ein Hinweis auf *Kupferminen* westlich von Abbysinien (Äthiopien). Zudem sind folgende Inschriften von entdeckungsgeschichtlichem Interesse:

- Südlich der Insel Madagaskar finden sich Sandbänke und Felsriffe mit entsprechenden Jahreszahlen vermerkt (*Sandbank (1811)*, *Union Felsen 1812*, *Hagus (Untife 1815)*). Bei der bis heute unbewohnten Amsterdam-Insel im südlichen Indischen Ozean erscheint zudem der Zusatz *vulkanisch*.
- Südöstlich von NEU HOLLAND *oder Australien* ist im südlichen Pazifischen Ozean die zufällig im Jahr 1810 entdeckte Macquarieinsel[67] mit den Felsgruppen Judge und Clerk sowie Bishop und

65 Die Bounty-Inseln wurden am 8. September 1788 vom britischen Seeoffizier William Bligh (1754-1817) entdeckt und nach seinem Schiff benannt, das durch die Meuterei auf der Bounty berühmt wurde, vgl. H. Homann (Hrsg.), Meuterei auf der Bounty. Die Piratenjagd der Pandora, Augsburg 2005. (Aufzeichnungen des Dr. George Hamilton 1787-1792).

66 Die Raoulinsel war bereits in den 1780er Jahren von einer britischen Expedition entdeckt worden, die sie als *Sunday Island* taufte. 1793 landete hier der bretagnische Seefahrer Jean-Michel Huon de Kermadec (1748-1793, nach ihm ist die Inselgruppe benannt). Im 19. und frühen 20. Jahrhundert wurden immer wieder Versuche unternommen, Raoul zu besiedeln. Diese blieben aber erfolglos.

67 Erstmals kartographiert wurde diese Insel durch Fabian Gottlieb Thaddeus von Bellingshausen (1778-1852), der im Auftrag des russischen Zaren Alexander I. die Insel erforschte. Derselbe hatte auch Expeditionsreisen im Südlichen Eismeer durchgeführt, die jedoch auf dem Erdglobus von Jüttner nicht weiter behandelt werden, vgl. F. G. v. Bellinghausen, Forschungs-

Clerk (*Judg Clerk v. Judge Neuentdekte Maquarie In. Bischof u. Clerk B.*) eingetragen. Zudem sind hier auch einige Sandbänke sowie zaghafte Umrisse von *Enderbys Land von Cap. Biscee 1831-32 entdeckt*[68] vermerkt.

- Das *Kiteguen Land* (Alaska, mit dem Kap *C. George Simpson*[69]) wird mit einer punktierten Demarkationslinie von Kanada abgegrenzt.[70] Der auf dem Globus eingetragene Mackenzie-Fluss erinnert an Sir Alexander MacKenzie (1764-1821), der 1789 von Montreal aus zu einer Expedition in den Nordwesten Kanadas aufgebrochen war und dabei 1793 den Pazifik erreicht hatte.[71]
- Mehrere, auch kleinere Indianerstämme in Nord- und Südamerika finden sich verzeichnet (beispielsweise: *Aldea dos Mundrucus*[72] in Amazonien).
- Im Südatlantik springen vor allem die mythische Insel Sachsenburg, das Eiland *Columbus (1809)*, sowie die Gough-Insel (diesmal mit der Jahreszahl 1788) und ganz im Süden die Bouvetinsel (*In. Bouvet od. Circumcision von Schwan und Otter gesehen im Jahr 1808*)[73] ins Auge. Auffallend sind hier zudem die subantarktischen Insel-

---

fahrten im Südlichen Eismeer 1819-1821. Reisebericht in deutscher Sprache auf Grund des russischen Originalwerks, Leipzig 1902.

68 Diese, nach einem Gründer einer Walfang-Firma, Samuel Enderby Junior, benannte Insel wurde 1806 entdeckt. Sie gehört zur subantarktischen Auckland-Inselgruppe.

69 Sir George Simpson (1787-1860) war von 1821 bis 1856 Gouverneur in Kanada für die Hudson's Bay Company, vgl. A. S. Morton, Sir George Simpson, overseas governor of the Hudson's Bay Company; a pen picture of a man of action, Toronto 1944.

70 Auf Jüttners Globussegmenten von 1839 findet sich hier verzeichnet: *Grenze zw. d. russischen und britischen America nach d. Tactate von 1825.*

71 D. Hayes, First crossing: Alexander MacKenzie, his expedition across North America, and the opening of the continent, Seattle 2001.

72 Die mitten im tropischen Urwald liegende Stadt Manaus wurde damals noch als *Fort Rio Negro* bezeichnet.

73 Am 6. Oktober 1808 sahen die Walfänger James Lindsay und Thomas Hopper mit ihren Schiffen *Swan* und *Otter* dieses ferne Eiland im Südatlantik. Sie konnten aber aufgrund von Packeis nicht an Land gehen. Zwar wurde die Insel bereits 1739 gesichtet, aber auch James Cook konnte sie nicht auffinden. Eine erste Landung glückte erst 1822, aber bereits in den 1840er Jahren konnte man das Eiland nicht wieder finden, ehe es 1898 von der deutschen Valdivia-Expedition wiederentdeckt wurde, vgl. G. Engelmann, Die Bouvet-Insel. Mit vier Karten, in: Geographische Zeitschrift 46 (1940), Nr. 11, 408-424.

gruppen *Neu-Süd-Schetland*, *Neu Orkneis* und *Neu-Georgien*.

Jüttner hatte 1839 einen Erdglobus (Ø 63 cm) bei der Wiener Firma von Franz Leopold Schöninger (1790-1877) herausgebracht. In diesem Verlag arbeitete auch der Militärkartograph Franz VON ELEKES, dessen Segmente (bereits mittels Lithographie hergestellt, vgl. Farbtafel S. 309) folgende Besonderheiten aufweisen:

- Oberhalb der Marquesas-Inseln findet sich eine *mit Wald bedeckte niedrige Ins. 1814 von Capt. Walker entdeckt* (vgl. Tafel S. 286).
- In Afrika werden die *Neger Länder* mit dem *Aethiopisches Hochland*, der *Sklavenküste* im Westen und zahlreichen, noch unbekannten Gegenden verzeichnet.
- In Australien wird auf die englische Statthalterschaft Bezug genommen (*NEUSÜDWALES Englische Statthaltersch. Sidney Covo*).

Der australische Kontinent ist auch für die Datierung eines Erdglobus von Johann Adam BÜHLER (Farbtafel S. 310) hilfreich, der in der Abel-Klinger'schen Kunsthandlung in Nürnberg herausgegeben wurde. Denn die unabhängige Kolonie Queensland, die 1859 ausgerufen wurde, findet sich hier noch nicht dargestellt; sehr wohl aber die Kolonie Victoria, die sich 1851 von New South Wales separiert hatte.

Somit gelangen wir bei unserem Überblick in die 2. Hälfte des 19. Jahrhunderts, dessen gründliche globographische Aufarbeitung noch aussteht.[74] Hier spielten folgende Globenhersteller, von denen mindestens je ein Exemplar in der Sammlung Woldan vorhanden ist, eine wichtige Rolle:

- Carl ADAMI, dessen Globen im Berliner Verlag von Dietrich REIMER ab 1852 verlegt wurden (Farbtafel S. 311 oben). Sie wurden von keinem Geringeren als dem Kartographen Heinrich Kiepert (1818-1899) überarbeitet und zeigen ein detailgenaues, zeitgenössisches Erdbild. So finden sich auf dem Exemplar in der Sammlung Woldan von 1865 beispielsweise die in dieser Zeit

[74] Erste Ansätze hierzu bei H. Wohlschläger, Globen der neueren Zeit, in: Allmayer-Beck (Hrsg.), Modelle der Welt, 199-223. In der Staatsbibliothek Berlin sind zudem in einer virtuellen Ausstellung die Globen der Berliner Produktion von 1801 bis 1912 erschlossen worden, vgl. http://staatsbibliothek-berlin.de/kartenabteilung/bestaende/globen-sammlung.htm (aufger. am 21. Dezember 2009).

unabhängigen Buren-Republiken in Südafrika (*Oranje Republik*; *Transvaal Republik*, vgl. Farbtafel S. 311 unten) eingezeichnet und auch die Grenzen der Vereinigten Staaten von Amerika, die nach dem Krieg gegen Mexiko (1846-1848) einen erheblichen Flächengewinn im Südwesten zu verzeichnen hatten.

- Das Wiener Verlagshaus SCHÖNINGER, von dem neben dem Erdglobus von Jüttner noch zwei weitere Exemplare aus der Zeit von ca. 1867 und 1872 in der Sammlung Woldan verwahrt werden (Farbtafel S. 312 oben). Dies ist verhältnismäßig wenig, wenn man bedenkt, dass um 1870 etwa 10.000 Globen in diesem Verlag hergestellt worden sein sollen. Der von Leopold HUGL um 1867 entworfene Globus zeigt den schon wesentlich realitätsnäheren antarktischen Kontinent sowie mit dem *Kilimandjaro* (Tafel S. 290) die höchste Erhebung in Afrika.[75] In diesem Verlag erschien auch der nicht im Bestand der Sammlung Woldan befindliche, jedoch der Österreichischen Akademie der Wissenschaften gehörige Mondglobus, der 1849 von Josef RIEDL VON LEUENSTERN gefertigt wurde. Dieser Globus (Farbtafel S. 312 unten) zeigt nur die Hälfte des Mondes, wie dies auch durch eine zeitgenössische Beschreibung von Franz Locher im Jahr 1856 berichtet wird: *Bei seinem Umlaufe um die Erde wendet der Mond uns stets dieselbe Seite zu; seine andere Hälfte hat noch kein menschliches Auge je gesehen.*[76]
- Vom bedeutendsten Globenhersteller der Österreichisch-Ungarischen Monarchie des 19. Jahrhunderts,[77] Jan FELKL (1817-1887), haben sich gleich drei Erdgloben unterschiedlicher Größe (Ø 8,6 cm, 16,8 cm und 31,86 cm) in der Sammlung Woldan erhalten (Farbtafel S. 313). Für die Datierung dieser Globen spielen neben den Kartuschen unter anderem auch Verlagsanzeigen[78] sowie das Kartenbild des afrikanischen,[79] jedoch auch

---

75 Ab 1860 wurden mehrere Expeditionen zum Kilimandscharo durchgeführt. Diese hatten die Erstellung von topographischen Karten zum Ziel, vgl. K. Brunner, Frühe Karten des Kilimandscharo – Ein Beitrag zur Expeditionskartographie, in: Cartographica Helvetica 30 (2004), 3-9.

76 F. Locher, Inductionsglobus oder Globus zur praktischen Einführung in den mathematisch-geographischen Unterricht, Ellwangen 1856, 10.

77 Vgl. J. Dörflinger, Printed Austian Globes (18th to early 20th century), in: Der Globusfreund 35-37 (1987), 191-210.

78 J. Dörflinger, Verlagsanzeigen als Quelle zur Geschichte der Globen, in: Der Globusfreund 33/34 (1985), 81-98.

australischen Kontinents eine wichtige Rolle, auch wenn die Abbildung der Länder zumeist nicht nach den neuesten Entdeckungen geschah.[80]

- Inzwischen waren auch Expeditionen zur Bestimmung des magnetischen Südpols aufgebrochen. Dies ist deutlich auf dem vom Berliner Verlagshaus Ernst SCHOTTE & CO. herausgegebenen Erdglobus des Theodor METTKE aus der Zeit nach 1875 ablesbar, verzeichnet dieser doch das nach Charles Wilkes (1798-1877), dem Leiter der *United States Exploring Expedition* (1838-1842) benannte *Wilkes Land* in der Antarktis.[81] Die eigentliche Eroberung der Antarktis begann aber erst mit dem Sechsten Internationalen Geographischen Kongress, der 1895 in London stattfand.

Somit gelangen wir mit dem jüngsten, in der Sammlung Woldan vorhandenen Erdglobus in das 20. Jahrhundert und wollen mit diesem – einem beim Wiener Verlag FREYTAG & BERNDT zwischen 1927 und 1931 gefertigten Schulglobus, der auch physikalische Gegebenheiten der Erdkugel aufzeigt – unseren Rundgang durch die hier vorgelegte, über dreihundert Jahre umfassende Globengeschichte beenden. Im Nordpolarmeer ist nun der bei der Österreichisch-Ungarischen Nordpolexpedition[82] unter Oberleutnant Julius Peyer und dem Marineoffizier Carl Weyprecht im Jahre 1873 entdeckte Archipel nördlich der großen Doppelinsel Nowaja Semlja vermerkt worden (*Franz Josefs Land*). Deutlich finden sich zudem die ehemaligen deutschen Kolonien eingetragen, die auch für die

79 R. Schmidt, Zur Datierung von Erdgloben nach dem Kartenbild Afrikas (unter Verwendung der Ausarbeitung von Prof. Dr. H. Wels), in: Der Globusfreund 30 (1982), 88-97.

80 So zeigt die aufgrund der Kartusche *Felkl & Sohn Roztok bei Prag* sicher nach 1875 gefertigte Erdkugel mit der Inventarnummer GL-V(L): WE 422 beispielsweise den australischen Kontinent in seiner politischen Gliederung des Jahres 1862, weshalb hier besondere Vorsicht bei Datierungsfragen vonnöten ist. Leider hat sich das Archiv des Verlags nicht erhalten.

81 Vgl. F. Abdelouahab, L'aventure des poles, 139-145.

82 Diesbezüglich sei besonders auf den Beitrag von Marianne Klemun, Österreichisch-Ungarische Nordpolexpedition (1872 – 1874): Euphorie der Gemeinsamkeit und die Positionierung der Akademie der Wissenschaften (Wien) im Vorfeld der Expeditionsplanung (in diesem Sammelband, S. 359-381) verwiesen.

Kolonialkartographie eine wichtige Rolle spielen:[83] Togo wird als *kerndeutsch* bezeichnet; bei Kamerun, Deutsch-Südwestafrika (Namibia)[84], Deutsch-Ostafrika (Tansania)[85] und den ehemaligen deutschen Kolonien in der Südsee begnügt sich der Kartenautor mit dem Verweis *Mandatgebiet des Völkerbundes, ehemals deutscher Besitz*. Diese und weitere Hinweise könnten für die Raumforschung des 20. Jahrhunderts von Bedeutung sein, sind aber in größerem Umfang noch nicht untersucht worden.

Zusammenfassung

Im vorliegenden Beitrag wurden erstmals 20 in der Sammlung Woldan verwahrte Erd- und Himmelsgloben sowie entsprechende Globensegmente vergleichend analysiert. Dabei konnte festgestellt werden, dass diese Globen, die uns das Weltbild vom 17. bis zum 20. Jahrhundert vor Augen führen, Kulturgüter ersten Ranges sind, denn sie dokumentieren augenscheinlich die Entdeckung der Welt. Dies ist seltsamerweise in der neueren Geschichtsforschung bisher (trotz der Hinwendung zum *Pictorial Turn*) nur am Rande betrachtet worden. Jedoch konnten in diesem Beitrag lediglich einige Tendenzen aufgezeigt werden, die man im Detail noch genauer untersuchen sollte. Denn die Globenkunde ist ein weites Feld, das es erst noch von den Historischen Grundwissenschaften zu erschließen gilt. Dies ist aufgrund der Tatsache, dass Globographie in der universitären Lehre nicht berücksichtigt wird, aber nicht ganz einfach. Jedoch könnten die auf den Globen visualisierten Theorien eines riesigen

83 Vgl. R. Hafeneder, Deutsche Kolonialkartographie 1884-1919, Diss. Universität der Bundeswehr München 2008.

84 Vgl. hierzu auch J. Moser, Untersuchungen zur Kartographiegeschichte von Namibia. Die Entwicklung des Karten- und Vermessungswesens von den Anfängen bis zur Unabhängigkeit 1990, Diss., Dresden 2007.

85 Zur Kartographie dieses Gebiets vgl. I. J. Demhardt, Die Kartographie des Kaiserlichen Schutzgebiets Deutsch-Ostafrika, in: Cartographica Helvetica 30 (2004), 11-21, sowie allgemein: K. Brunner, Karten Ostafrikas um die Jahrhundertwende. Ein Beitrag zur Kolonialkartographie, in: U. Lindgren (Hrsg.), Kartographie und Staat. Interdisziplinäre Beiträge zur Kartographiegeschichte (Algorismus. Studien zur Geschichte der Mathematik und der Naturwissenschaften 3), München 1990, 47-53.

Südkontinents und die Suche nach der sog. Nordwestpassage[86] für die frühe Neuzeit von Interesse sein.

Im Anhang zu dieser Studie wird versucht, dem Historiker eine detaillierte Übersicht über alle in der Sammlung Woldan befindlichen globographischen Produkte an die Hand zu geben, um diese Globen der Forschung und einer breiteren Öffentlichkeit bekannt zu machen. Jeder Globenhersteller wird hier deshalb zunächst in einer biographischen Skizze vorgestellt. Anschließend werden die Globen nach einem vom Autor dieses Beitrags entwickelten System untersucht. Dabei finden folgende Punkte ihre Berücksichtigung: die Inventarnummer des Objektes, die Maße und das Gestell (dazugehörig der Horizont- und Meridianring sowie der Stundenkreis und Stundenzeiger), der Erhaltungszustand, die Segmente, die auf dem Globus befindlichen Längen- und Breitenkreise (samt Nullmeridian) sowie eine Transkription der Hauptkartuschen. Abschließende Bemerkungen sowie weiterführende Literatur sollen insbesondere dazu dienen, es jedem an diesem Thema Interessierten zu ermöglichen, weiterzuforschen. Denn nur im interdisziplinären Dialog können Ergebnisse erzielt werden, die für die Globographie von Bedeutung sind. Dieser Beitrag soll deshalb den Auftakt für weitere diesbezügliche Untersuchungen bilden.

86 Grundlegend hierzu ist nach wie vor J. Dörflinger, Nordwestpassage-Theorien von 1731-1823, 2 Bde., Wien 1969.

Erdglobus des Jodocus Hondius d. Ä. (Antwerpen, 1601)

Der flämische Kartograph, Kupferstecher und Verleger Jodocus Hondius d.Ä. (1563-1612) war als Protestant 1584 nach England geflohen, um den religiösen Wirren in Flandern im Zuge des Freiheitskampfes der Niederlande zu entkommen. Dort arbeitete er u. a. als Kupferstecher an den von Molyneux herausgegebenen Globen. 1593 übersiedelte er nach Amsterdam und entfaltete dort eine rege kartographische Verlagstätigkeit, was sich auch darin zeigt, dass Jan Vermeer in seinen Gemälden *Der Astronom* und *Der Geograph* (1668/1669) jeweils einen Hondius-Erdglobus verewigte. 1604 kaufte Hondius die Druckplatten des Weltatlas des Gerhard Mercator, den er in erweiterter Ausgabe ab 1606 in mehreren Auflagen (sog. Mercator-Hondius-Atlas) veröffentlichte. Nach seinem Tode wurde der Verlag zunächst von seiner Witwe, dann schließlich von den Söhnen Jodocus d. J. und Henricus (1597-1651) sowie von seinem Schwiegersohn, dem Verleger Joannes Janssonius (1588-1664), weitergeführt.

1) Inventarnummer: GL-V(L): WE 54.
2) Maße und Gestell: Ø = 21 cm; U = 66 cm; h = 32,5 cm; b = 30 cm. Auf vier Füßen (spitz auslaufend) und runder Bodenplatte ruhender Erdglobus, am unteren Haltegestell etwas ausgebrochen.
3) Horizontring: teilweise abgeblättert, ursprünglich farbig und nur noch zum Teil erkennbar (Windrichtungen in niederländischer Sprache, lateinische Monatsnamen, Heiligenkalender, Sternzeichen).
4) Meridianring: jeweils viermal in 0-90° eingeteilt, teilweise oxidiert.
5) Stundenkreis und -zeiger: fehlen.
6) Erhaltungszustand: Insgesamt sehr restaurierungsbedürftig: Horizontalkreis zweimal gebrochen und etwas verblasst; Delle bei Peru; zudem zwischen Afrika und Südamerika mehrere Bruchstellen; Riss von Rußland bis zur arabischen Halbinsel.
7) Segmente: 12 durchgehende Kupferstiche (ohne Polarkalotten).
8) Längen- und Breitenkreise: Der Nullmeridian verläuft durch die Azoren; die Wendekreise sind mit „TROPICUS CANCRI” bzw. „TROPICUS CAPRICORNI” bezeichnet, die Ekliptik und der Äquator sind zu jeweils 10° eingeteilt.
9) Kartuschen: zwischen 150° und 210° Länge erstreckt sich eine ausführliche Legende beim Südkontinent:

I. Hondius Lectori S. In locorum longitudine hactenus mirifice peccatum eßse omnibus hydrographiæ peritis satis sup(er)q(ue) constat: Regiones enim ferè omnes descriptæ sunt prout naucleri in suis navigationibus directionem duorum locorum ab uno loco ad alterum invenerunt; idque nulla habita ratione loci tertij, vel deviationis acus [nauticæ], vel etiam directorij nautici, quo indifferenter [utuntur], quamvis in uno, non aeque ac in alio chalybeu illa acus ponatur, et a vero Septentrione magis vel minus divertatur, pro usu loci in quo directoria fabricata sunt, un[de] [ne]ceßario longitudo locorum distorta est.
Multi hos errores frustra conati [sunt] emendare per polares stellas, alij per Lunæ cursum: alij certius per eclipses: Verum hoc opus, ille labor, quis enim in tanta locorum multitudine eclipses observabit? At cum jam tandem per variationem aut deviationem aecus nauticæ (ut vocant) locorum longitudo inveniatur, opere: proetium me facturum putavi si in hoc globo regiones omnes (saltem quarum longitudo fiam cognita est) suis quasq. veris longitudinis gradibus delineavero, quamvis id non exigui [laboris] fuerit. Longitudine [incepi]mus non ab Insulis fortunatis, ut Ptoleme(us), sed ab ijs quae açores vocantur, quod acus nautica ibi recte in Septentrionem vergat. Vale. Anno 1601.

Die Widmungskartusche oberhalb des nordamerikanischen Kontinents ist an das Statthalterpaar der spanischen Niederlande gerichtet; Isabella Clara Eugenia de Austria y Valois (1566-1633) hatte zur Zeit der Globusherstellung (am 18. April 1599) den österreichischen Erzherzog Albrecht VII. (1559-1621) geheiratet (vgl. Farbtafel S. 299 unten): *SERENISSIMIS PRINCIPIBUS ALBERT[O ET ISABE]LLAE CLA. AUSTRIACIS BRABANTIAE DUCIBUS, ETC. D.D. Jodoc Hondius auctor et Joan Baptista friendt Antwerpior(um).*

Das Fretum Davis auf dem Erdglobus von Hondius.

Bemerkungen:
Bei dem Erdglobus (Ø 21 cm) von 1601 handelt es sich um ein seltenes Exemplar, das Hondius in Kooperation mit dem Antwerpener Verleger Joan Baptista Vriendt veröffentlichte (man vgl. die Kartusche). 1615 wurden die Segmente dieses Globus von Giuseppe de Rossi in Rom nachgestochen.

Auffallend ist der riesige Südkontinent (*Terra Australis Incognita*, jedoch nicht verbunden mit *Novo Guinea*). Die einzelnen Landstriche sind sorgfältig farbig unterschieden (Ortssignaturen mittels roter Punkte); in Maulwurfshügelmanier werden Gebirge dargestellt (vgl. *Serra Nevada*). Schiffsabbildungen sowie Meeresungeheuer finden sich als für diese Zeit nicht unübliches kartographisches Beiwerk. Australien fehlt naturgemäß noch.

Weiterführende Literatur

A. Fauser, Ältere Erd- und Himmelsgloben in Bayern, 99; P. van der Krogt, Old Globes in the Netherlands, 148-154, Ders., The Globes of Hondius. A most important pair of globes showing the results of the earliest Dutch exploration voyages to the East Indies (Antiquariaat Forum «Utrecht»: Catalogue 98), Utrecht 1991; Ders., Globenproduktion in den Niederlanden und ihre Auswirkung auf Europa, 1525-1650, in: Der Globusfreund 49/50 (2001/2002), 49-65; M. Pelletier, Les globes terrestre et céleste de Jodocus Hondius, 1600, in: Bulletin de la Bibliothèque Nationale 5 (1980), Heft 3, 99-103; A. de Smet, Jodocus Hondius, continuateur de Mercator, Bruxelles 1963; E. L. Stevenson, Terrestrial and celestial globes, Bd. 2, 2-19; H. Wohlschläger unter Mitwirkung von R. Schmidt, Globenhersteller aus aller Welt in: Allmayer-Beck, Modelle der Welt, 236-319, hier: 273f.

Details auf dem Erdglobus von Hondius: Polgebiet (oben)
und patagonische Riesen in Südamerika (unten).

Globenpaar des Isaak Habrecht (Nürnberg, um 1675)

Isaak II. Habrecht (1589-1633) entstammte einer bedeutenden Uhrmacherfamilie. Sein Vater, Isaak I. Habrecht, arbeitete 1574 zusammen mit seinem Onkel Josuah Habrecht unter der Leitung des Dasypodius an der zweiten Uhr des Straßburger Münsters. Zudem fertigte er 1594 eine astronomische Uhr mit einem Himmelsglobus, die sich heute auf Schloss Rosenborg in Kopenhagen befindet. Von seinem Sohn Isaak II., der Arzt und Physiker war, hat sich ein Globenpaar von 1621 erhalten. Die diesbezüglichen Segmente wurden vom Straßburger Kupferstecher Jakob van der Heyden (1573-1645) gestochen. Bei dem vorliegenden Globenpaar handelt es sich jedoch um eine durch den Nürnberger Kupferstecher und Kunsthändler Johann Christoph Weigel (1654-1725) geschaffene Neuauflage der Habrecht-Globen (ohne Aktualisierung des dargestellten Weltbildes). Habrecht schrieb 1628 auch einen Traktat zum *Planiglobium Coeleste et Terrestre*, der von Johann Christoph Sturm überarbeitet wurde und in deutscher Fassung bei Paulus Fürst 1666 erschien.

1) Inventarnummer: GL-V(L): WE 102 und 103.
2) Maße und Gestell: Ø = 21 cm; U = 66 cm; h = ca. 32,5 cm; b = 30 cm. Auf vier rund auslaufenden, Füßen ruhender, gut verankerter Globus (sehr schwere Holzkugel!), darunter eine dreiteilig zusammengesteckte Bodenplatte mit Ø = 18 cm.
3) Horizontring: erstaunlich gut erhalten, führt die Windrichtungen in niederländischer (grün) und lateinischer Sprache (gelb), die Monatsnamen nach dem Gregorianischen Kalender und dem *Calendario antiquum* (rot), die Heiligentage (grün), sowie die Tierkreiszeichen in lateinischer Sprache auf.
4) Meridianring: viermal in 0-90° eingeteilt, teilweise abgenutzt.
5) Stundenkreis und -zeiger: beides vorhanden (Stundenzeiger zweimal in XII römischen Stunden); beim Himmelsglobus hat sich nur der Stundenkreis erhalten.
6) Erhaltungszustand: Horizontalkreise relativ gut erhalten; beim Erdglobus mehrere Löcher in Asien und Riss bis in den Ozean hinein; zudem ein Riss nördlich der Straße von Anian und im nördlichen Europa der Länge nach.

7) Segmente: 12 durchgehende Segmente (vom Nord- zum Südpol) im Kupferstich.
8) Längen- und Breitenkreise: Der Äquator und die Ekliptik sind in 10° eingeteilt, der *Meridianus Primus* verläuft durch die Insel Corvo (Azoren).
9) Kartuschen:
Per Illustri et Generosiss(imo) Dn^o^ Dn^o^ EBERHARDO Dijnastæ in Rappolstein, Hohenac[cio] et Gero[l]tzeccio ad Vogasinum Divi Mathiæ II Imp. nec non Sereniss(imo) Maximiliani Archiducis Austriæ P.M. Camerario et Citeriorum Ordinu[m] Provincialium Præsidi Magnifico : ex antigua Ducu(m) Spoleti familia oriundo : Domino meo Clementiss(im)^o^: Triplicem hunc globum : Caelestem sc(ilicet): con vexum et concavum vt et hunc terrestrem novissimæ editionis et correctionis D.D.D. Isaacus Habrecht Phil. et med. d. Argentinensis.
Versus Articum Polum ulter[ior tran]sgressus hacten(us) non dabatur Herculis licet Davis Angli labore id exanilatum fuerit sicut et circa Antarcticum, fretum noviter à Guilielmo Schout detectum, La Maire nuncupatum extremus adhuc navigationum est terminus. Quamvis nullus dibitem [sic!], *maxima totius orbis magnalia sub polis delitescere, quoru(m) detectionem forte an summus Deus suo tempore reservat. Typis Jacobi ab Heyden ARGENTINÆ.*

Bemerkungen:
Vorbild des Kartenbildes waren die Segmente des Globenpaares des Jodocus Hondius d. Ä. von 1601. Beim Erdglobus (Farbtafel S. 300) erscheinen vor allem die handgemalten, farblich differenzierten Grenzen von Interesse. Der Himmelsglobus (Farbtafel S. 301) hingegen zeigt die unterschiedlichen Sternbilder, weist aber auch einige Novae und Kometenbahnen auf. In der Kartusche des Himmelsglobus von Habrecht findet sich zudem ein Hinweis auf Joh. Christoph Weigel; die Sterne sind nach ihren Größen bis zur siebten Klasse geordnet.

Weiterführende Literatur

M. Bauer, Christoph Weigel (1654–1725), Kupferstecher und Kunsthändler in Augsburg und Nürnberg (Archiv für Geschichte des Buchwesens 23 (1983), Sp. 693-1186; P. van der Krogt, Old Globe in the Netherlands, 137-140; E. L. Stevenson, Terrestrial and celestial globes 2, 49-52; F. Wawrik – H. Hühnel, Das Globenmuseum der Österreichischen Nationalbibliothek, in: Der Globusfreund 42 (1994), 38f. (zum HG von 1621); H. Wohlschläger unter Mitwirkung von R. Schmidt, Globenhersteller aus aller Welt in: Allmayer-Beck, Modelle der Welt, 236-319, hier: 269.

Globensegmente des Vincenzo Coronelli (Venedig, 1688 und 1696)

Der Franziskanermönch P. Vincenzo Maria Coronelli (1650-1718) ist zweifelsohne der bekannteste Globenhersteller der frühen Neuzeit. Von besonderer Bedeutung sind hierbei seine riesigen Manuskriptgloben von Marly (1681-1683 gefertigt für König Ludwig XIV., mit einem Ø von 384 cm), die heute in der Bibliothèque National in Paris bestaunt werden können. Jedoch soll Coronelli bereits 1678 als Sekretär des Herzogs Ranuccio II. Farnese ein handgezeichnetes Globenpaar angefertigt haben. Nach seiner Rückkehr nach Venedig gründete er 1684 mit der *Accademia cosmografica degli argonauti* die erste geographische Gesellschaft der Welt. Im folgenden Jahr wurde er zum Kosmographen der Republik Venedig ernannt und war vor allem kartographisch (*Atlante Veneto* 1691; *Corso geografico universale* 1692-1694), aber auch enzyklopädisch tätig (seine *Biblioteca universale sacra-profana* von 1701-1706 blieb aber unvollendet). Ab 1701 war Coronelli zudem Generalminister der Minoriten. Er wurde jedoch von Papst Clemens XI. aufgrund interner Intrigen abgesetzt. 1717 wurde er von Kaiser Karl VI. zum *Commissario Perpetuo del Danubio* berufen, um Pläne für die Donauregulierung zu erarbeiten.
Von besonderer Bedeutung ist sein *Libro dei Globi* von 1693, in dem er die Meridianstreifen einiger von ihm gedruckter Globen (so auch im Ø von 48 cm) abbildet.

In der Sammlung Woldan befinden sich die folgenden Globensegmente:

- Inventarnummer: K-V(Bl): WE 115 (1-26): Auf 26 Blättern (b = 26 cm; h = 45,5 cm) Globenstreifen für den Erdglobus mit Ø = 110 cm (Venedig 1688), jedoch unvollständig: Es sind zwölf nördliche und elf südliche Segmente sowie drei Polkappen (Blatt 13, 14 und 26) vorhanden. Somit ist die Nordpolarkappe (Blatt 26: mit Ø = 35,5 cm auf einer Doppelseite farbig angelegt) doppelt vorhanden, während die Segmente mit dem Kap Hoorn und die zwölf Äquatorialsegmente fehlen. Zwischen dem 240. und dem 270. Längengrad findet sich folgende Kartusche:
  IL GENIO della Virtù raccamandò all'Eternità. Il Nome di CESARE, Cardinale Eminentissimo D'ESTREES, Duca, e Pari di Franchia, mentre fece elaborare Per LUDOVICO IL MAGNO, Dal P. Coronelli Due Gran

Globi, l'Idea dè quail hà poi epilogata in questi Per L'ACCADEMIA COSMOGRAFICA DEGLI ARGONAUTI l'Anno MDCLXXXVIII in Venezia.

- Inventarnummer: K-V (BI): WE 122 (1-13) und HG K-V (Bl): WE 123 (1-13): Hierbei handelt es sich um die vollständig erhaltenen Segmente eines Globenpaares im Ø von 48 cm (Venedig, 1696) auf jeweils 13 Blättern (h = 33 cm, b = 12,5 cm; Blatt 13: 2 Polkappen mit Ø von 8 cm), die auch im Libro dei Globi abgebildet sind (vgl. beispielsweise UB Wien: Signatur III 35.125). Die Hauptkartusche befindet sich unterhalb der *Terra D'Antonio Diemens* (Tasmanien): *Pater Magister Vincentius Coronelli Min: Conv: S. Francisci Serenissimae Venetorum Reipublicae Cosmographus MDCLXXXXVI.* Hier fehlt jedoch die sonst übliche Widmung an den Grafen Aloysio Pauluccio (man vgl. beispielsweise den schlecht erhaltenen, aber aufgrund seiner Konstruktion [hohles Blechgestell!] interessanten Erdglobus selber Größe in der Studiensammlung des Globenmuseums in der ÖNB, Signatur: Gl. 14, auf dem die Widmung noch teilweise zu erkennen ist).

Weiterführende Literatur

E. Armao, Vincenzo Coronelli. Cenni Sull'Uomo e la sua vita. Catalogo ragionato delle sue opere lettere – Fonti Biographiche – Indici. Con disegni nel testo e tavole fuori testo (Biblioteca di Bibliografia Italiana XVII), Firenze 1944; M. L. Bonelli – L. Luzio, Catalogo dei globi antichi conservati in Italia, (Istituto e Museo di Storia della Scienza «Firenze» Biblioteca), Bd. 2: I Globi di Vincenzo Coronelli, Firenze 1960; V. Coronelli: Libro die Globi, Venice 1693 (1701). With an introduction by Dr. H. Wallis (Theatrum Orbis Terrarum, Series of Atlases in Facsimile, Fourth Series, Volume V), Amsterdam (u. a.) 1969; H. Hassinger, Vincenzo Coronelli, ein Kartograph an der Wende des 17. und 18. Jahrhunderts, in: Mitt. der Geogr. Gesellschaft Wien 92 (1950), 200-202, hier: 201; G. Holzer, Die Globen des Vincenzo Coronelli. Bildmaterial und Anleitung zum Nachbau eines Erd- und Himmelsglobus, CD-Rom, Wien 2009; M. Milanesi, Le Globe terrestre de Coronelli, in: La Revue 37 (2002), 34-44; F. Nicolini di Marzio, Vincenzo Coronelli (1650 – Venezia – 1718). Epitome storica veneziana nel culto ambivalente della loro identità. Memorie e risonanze, Napoli 2005 (v. a. zum venezianischen Umfeld); H. Richard, Les Globes de Coronelli, Bibliothèque nationale de France, Paris 2006 (viele farbige Abb. des Globenpaares von Marly).

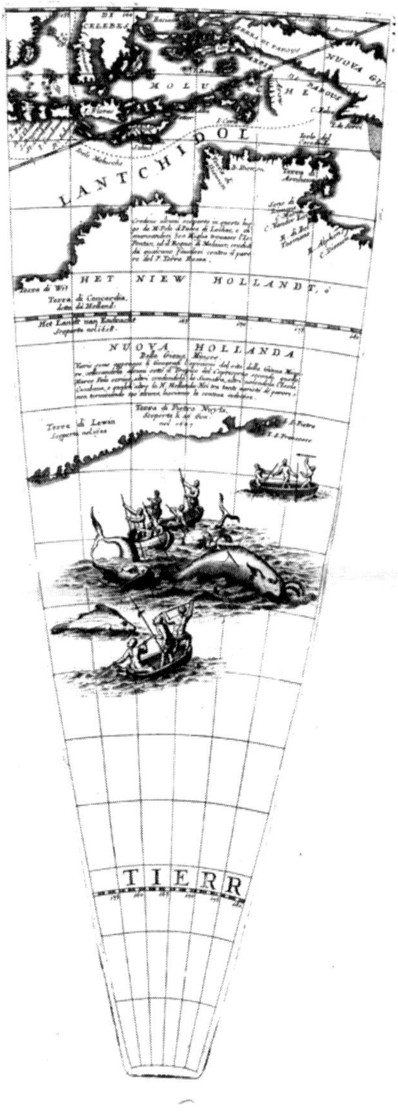

Globensegment 6 des Erdglobus von Coronelli mit Ø von 48 cm: Australien

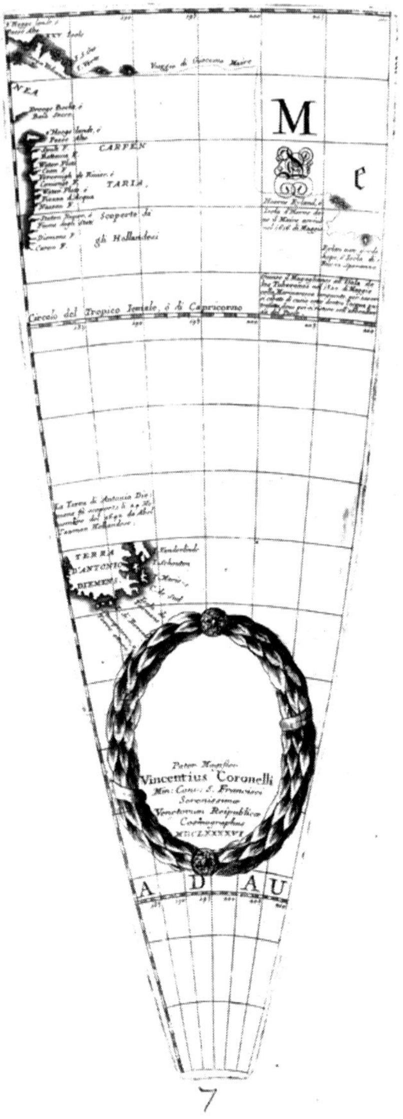

Globensegment 7 des Erdglobus von Coronelli mit Ø von 48 cm: unausgefüllte Kartusche und Diemens-Land (Tasmanien)

Erdglobus des Georg Christoph Eimmart (Nürnberg, 1705)

Georg Christoph Eimmart d. J. (1638-1705) stellt für die Astronomie der Reichsstadt Nürnberg aufgrund der von ihm im Herbst 1678 auf der Vestnertorbastei der Nürnberger Burg errichteten Sternwarte, der ersten Volkssternwarte der Neuzeit in Europa, einen regelrechten Höhepunkt dar. Der Kartograph und Astronom wurde am 22. August 1638 als Sohn des aus Königsberg stammenden gleichnamigen Vaters Georg Christoph d. Ä. († 1658), der zu dieser Zeit als Kupferstecher in Regensburg tätig war, geboren. Nach dem Besuch des Regensburger *Gymnasium Poeticum* begab sich Eimmart 1655 an die Universität Jena, wo ihn vor allem der Mathematiker Erhard Weigel (1625-1699), mit dem ihn eine lebenslange Freundschaft verband, beeinflusste. Ende 1658 kehrte er nach Regensburg zurück und folgte 1660 seiner Schwester Regina Christina (1636-1708) nach Nürnberg. Diese hatte den Kupferstecher Jacob von Sandart (1630-1708) geheiratet, der dort einen kleinen Kunstverlag betrieb. Eimmart nutzte die Gunst der Stunde, in diesem Umfeld als Karten- und Globenhersteller tätig zu werden, was ihm auch gelang: So war er neben Joachim von Sandrart (1606-1688) an der 1662 gegründeten Malerakademie beteiligt, der er von 1699 bis kurz vor seinem Tode auch als Direktor vorstand. 1667 fertigte er eine Gesamtansicht der Stadt Nürnberg an, der weitere kartographische Werke folgen sollten. Seine Globen, von denen sich auch Segmente erhalten haben, sind äußerst selten. Nachdem diese jedoch auf sein Todesjahr datieren, ist es fraglich, inwieweit er selbst daran noch beteiligt war. Hierzu könnte eventuell sein in St. Petersburg verwahrter Nachlass größere Aufschlüsse geben. Seine Tochter, die Astronomin Clara Maria Eimmart (1676-1707), spielt für die Selenographie, die kartographische Erfassung der Mondoberfläche, eine bedeutende Rolle.

1) Inventarnummer: Gl-V(L): WE 130.
2) Maße und Gestell: Ø = 30 cm; U = 94 cm; h = ca. 49,5 cm; b = ca. 45 cm. Sehr wuchtiges, vierfüßiges Holzgestell.
3) Horizontring: relativ gut erhalten und abnehmbar; mit auf Pappe aufgeklebtem Kupferstich (Winde, Kalender mit Monaten und Heiligentagen, Sternzeichen und Gradeinteilungen zeigend).

4) Meridianring: aus Messing (mit Gradzählung alle 5°).
5) Stundenkreis und -zeiger: nicht mehr vorhanden.
6) Erhaltungszustand: ordentlich, aber teilweise abgeblättert.
7) Segmente: 12 (teilweise nicht ganz richtig zusammengesetzt) und zwei Polarkalotten.
8) Längen- und Breitenkreise: Auffällig und wohl einmalig ist, dass der Nullmeridian und die Ekliptik im pazifischem Ozean östlich der Marianen (*ILHAS DE LADRONES*) beginnt (alle 10° bezeichnet).
9) Kartuschen: westlich von Südamerika als Hauptkartusche (vgl. Farbtafel S. 302 unten): *L. F. Cum Geographica Orbis Terrarum Descriptio secundum long. et lat. non nisi vel per peregrinationes maritimas vel Observationes Coelestes emendatior in dies prodeat, istud autem per experimenta propria (quoad exiguam saltem partem) perfecisse, è mille vix uni contingat; Oportuit nos Recentiorum accuratissimis Observationibus insistere, et quatenus cum Veritate congruant vel discrepent, exactiori Tuo judicio relinquest. Nos eadem Loca, bona fide, nihil immutantes, prout ab Authoribus novissimis accepimus, Usui Tuo exhibemus. Norimbergæ apud G.C. Eimmartum. A(nn)° Chr(isti) 1705.*

Bemerkungen:
Von diesem Globus haben sich in der Kartensammlung der ÖNB (Inv. Nr. FKB 273-1;4;5-7) entsprechende Segmente erhalten. Im Jahre 1958 wurde bei Nebehay in Wien zudem ein Faksimile angefertigt, wovon sich ein Belegexemplar in der Studiensammlung des Globenmuseums befindet (Signatur: GL 79 Kar (V 4 E 2)).

Weiterführende Literatur
H. Gaab, Zur Geschichte der Eimmart-Sternwarte. „Nürnberg [...] der beste ort pro studio Astronomiae in gantz Deutschland“, in: Regiomontanusbote 18 (2005), 5-70; H. Gaab (Hrsg.), Astronomie auf der Kaiserburg Nürnberg. Zur Geschichte der ersten Volkssternwarte: Georg Christoph Eimmart, Lauf a. d. Pegnitz 2007; R. Stoyan, Die Nürnberger Mondkarten 1: Die Mondkarte von Georg Christoph Eimmart (1638-1705) und Maria Clara Eimmart (1676-1707), in: Regiomontanusbote 14 (2001), 29-39; F. Wawrik – H. Hühnel, Das Globenmuseum der Österreichischen Nationalbibliothek, in: Der Globusfreund 42 (1994), 3-188, hier: 61-64; H. Wohlschläger unter Mitwirkung von R. Schmidt, Globenhersteller aus aller Welt in: Allmayer-Beck, Modelle der Welt, 236-319, hier: 260.

Erdglobus des Matthäus Seutter (Augsburg vor 1731)

Der Augsburger Kupferstecher Matthäus Seutter (1678-1757) stammte väterlicherseits aus einer Goldschmiede- und Stecherfamilie und war in der ersten Hälfte des 18. Jahrhunderts einer der bedeutendsten und produktivsten Landkartenverleger des deutschsprachigen Raumes. Er war Schüler und späterer Konkurrent von Johann Baptist Homann (1664-1724) und gründete um 1707 in Augsburg seine eigene kartographische Offizin, in der er sehr erfolgreich Karten, Atlanten sowie Globen verlegte. Seutter erleichterte die Herstellung von Globen, indem er die Segmente am Äquator durchschnitt und außerdem einen vereinfachten Stich der Polarkalotte einführte, indem er aus der Kreisscheibe einen Sektor herausschnitt. Die Kreisscheibe für die Pole beträgt deshalb bei den Seutter-Globen nur 350°. Der hier vorgestellte Erdglobus zeigt (im Unterschied zu seinen anderen erhaltenen Erdgloben) eine Besonderheit im geänderten Umriss der australischen Küste. Dieser Globus ist in die Zeit vor Seutters Ernennung zum Kaiserlichen Geographen von 1731/1732 zu datieren.
Seine Firma wurde ab 1762 von Johann Michael Probst und Seutters Schwiegersohn Tobias Konrad Lotter (1717-1777) weitergeführt und später von dessen Sohn Matthäus Albrecht Lotter (1741-1810), von dem sich ebenso Kupferstiche von Globensegmenten erhalten haben, übernommen.

1) Inventarnummer: GL-V(L): WE 169.
2) Maße und Gestell: Ø = 21 cm; U = 66 cm; h = 33 cm; b = 26 cm. Das Gestell ist demjenigen der Doppelmayr-Globen sehr ähnlich (jedoch ohne runde Bodenplatte und etwas verzierter, vgl. Farbtafel S. 303 oben).
3) Horizontring: Der aufgeklebte, kolorierte Stich zeigt von außen nach innen die Himmelsrichtungen, die Monate, die Sonntagsbuchstaben (zum Teil mit Heiligentagen), die Sternbilder sowie die Gradeinteilung.
4) Meridianring: Umlaufender Meridianreif mit eingravierter Skala der komplementären Breitengrade (0-90°).
5) Stundenkreis und -zeiger: fehlen.
6) Erhaltungszustand: gut, der Globus war ursprünglich farbig

angelegt (nur noch in Ansätzen erkennbar).

7) Segmente: 12 (am Äquator geteilt) und zwei Polkappen ab 66° (mit handschriftlichen Ergänzungen bzw. Verbesserungen vgl. *Mare Chilense*).
8) Längen- und Breitenkreise: Deutlich sind der Äquator, die Wende- und Polarkreise und die Ekliptik mittels verstärkter Linien hervorgehoben. Der *MERIDIANUS PRIMUS* verläuft durch die Kanarischen Inseln; der Breitengradabstand beträgt je 10°.
9) Kartuschen: östlich von *NOVA SELANDIA*: *GLOBUS TERRESTRIS juxta recentissimas observation. et navigationes peritissimor. Geograph. Accuratissme delineat[us] cura et sumtibus MATTH. SEUTTERI. Chalcogr: August.* Westlich der kalifornischen Halbinsel: *Hoc est illud punctum quod inter tot Gentes ferro et Igne dividitur. Senec.* (Dieses Zitat ist dem Vorwort des ersten Buches der *Naturales Quaestiones* (62-65 n. Chr.) des römischen Philosophen Seneca d. J. entnommen).

Bemerkungen (an dieser Stelle sei Herrn Dipl.-Geogr. Michael Ritter, München herzlichst für seine Hilfe gedankt):
In der Privatsammlung von Prof. R. Schmidt (Wien) haben sich die Segmente der Globenstreifen von ca. 1710 erhalten. Von dieser Ausgabe befinden sich auch zwei aufgezogene, kolorierte Exemplare im Globenmuseum der Österreichischen Nationalbibliothek (Inv. Nr. 57 und 96). Waldgebiete sind darauf mittels kleiner Bäume gekennzeichnet. Ob Seutter wirklich ein Globenpaar im Ø von 160 cm angefertigt hat (wie Stevenson, Terrestrial and celestial globes, Bd. 2, 155 schreibt), erscheint mehr als fraglich, vor allem, weil diesbezügliche Quellen fehlen.

Weiterführende Literatur

M. Ritter, Augsburger Landkartenproduktion im 18. Jahrhundert. Die Verleger Seutter, Lotter und Probst, in: J. R. Paas (Hrsg.), Augsburg, die Bilderfabrik Europas. Essays zur Augsburger Druckgraphik der Frühen Neuzeit, Augsburg 2001, 53–162; Ders., Seutter, Probst and Lotter: An Eighteenth-Century Map Publishing House in Germany, in: Imago Mundi 53 (2001), 130–135; Ders.: Rund um den Globus, in: Internationale Coronelli-Gesellschaft für Globenkunde, News 2002, 43f. (zu den Kupferstichsegmenten der Globen von Matthäus Albrecht Lotter); P. Plaßmeyer, Weltenglanz. Der Mathematisch-Physikalische Salon Dresden zu Gast im Maximilianmuseum Augsburg [Ausstellung im Maximilianmuseum Augsburg, 20. November 2009 bis 14. Februar 2010], München 2009, 62-65.

Zwei Erdgloben des Johann Gabriel Doppelmayr (Nürnberg, 1750 und 1791)

Der universal gebildete Nürnberger Johann Gabriel Doppelmayr (1677-1750) war nach seinem Studium der Jurisprudenz, Mathematik und Physik zunächst auf den damals üblichen Kavaliertouren in den Niederlanden und England unterwegs, wo er sich wohl Anregungen zum Globenbau holte. 1704 wurde er als Professor der Mathematik an das Ägidiengymnasium seiner Heimatstadt berufen; ab 1710 war er zudem als Nachfolger des Georg Christoph Eimmart für die Sternwarte auf der Kaiserburg verantwortlich. In dieser Zeit nahm er auch mit dem Verlagshaus Homann Kontakt auf und war aufgrund seines handwerklichen Geschicks als Kartograph (vgl. seinen *Atlas Novus Coelestis* von 1742) und Globenmacher (ab 1728) tätig. Von seinen zahlreichen Schriften hat vor allem die Übersetzung eines Buches des französischen Globen- und Instrumentenbauers Nicolas Bion (1652-1733) unter dem Titel *Neu-eröffnete mathematische Werck-Schule* (mehrere Auflagen, erstmal Nürnberg 1712 – darin findet sich auch ein Kapitel über die Verwendung von Globen) sowie seine *Historische Nachricht von den nürnbergischen Mathematicis und Künstlern* (Nürnberg 1730) bleibenden Wert. Seine Globen im Ø von 10, 20 und 32 cm zählen zu den am weitesten verbreiteten und zugleich genauesten der deutschen Produktion im 18. Jahrhundert, weshalb nicht verwundert, dass dieselben noch ein halbes Jahrhundert nach seinem Tode, den er wohl in Folge eines physikalischen Unfalls erlitt, zu bekommen waren. Der Nürnberger Feinmechaniker Johann Georg Puschner (1706-1754) stellte die Kupferstiche der Segmente her und fertigte vermutlich auch die Kugeln und die Armierung. Die späteren Auflagen seiner Globen wurden von Wolfgang Paul Jenig († 1805) und von Johann Bernhard Bauer (1752-1839) gefertigt.

1) Inventarnummer: GL-V(L): WE 165 und 200.
2) Maße und Gestell: Ø = 20 cm; U = 63 cm; h = ca. 33 cm; b = ca. 27 cm. Das vierfüßige Holzgestell der beiden Globen wird von gedrehten Säulenfüßen und einer runden Bodenplatte getragen.
3) Horizontring: aus Holz mit aufgeklebten Kupferstichen, achteckig, jedoch teilweise abgeblättert (auf dem Exemplar von 1750), zeigend die Windrichtungen in deutscher Sprache, in lateinischer Sprache hingegen die Monatsnamen nach dem gregorianischen

und julianischen Kalender (mit Heiligentagen) sowie astronomische Termina wie das Azimut und den Tierkreis (Zodiak).

4) Meridianring: vorzüglich erhaltene Messingringe, auf dem Exemplar von 1750 mit dem Zusatz: *Jenig fecit*; auf dem Exemplar von 1791 mit Angabe *No. 360*.

5) Stundenkreis und -zeiger: fehlen; auf dem Exemplar von 1791 hat sich aber der Stundenzeiger erhalten.

6) Erhaltungszustand: soweit gut, die Farben aber sehr ausgebleicht.

7) Segmente: 12 Segmente vom *Polus Arcticus* bis zum *Polus Antarcticus*.

8) Längen- und Breitenkreise: Dargestellt sind der *ÆQUATOR*, die *Ecliptica* sowie die Wendekreise (*Tropicus Cancri*, *Tropicus Capricorni*), ebenso der *Circulus Polaris Arcticus*. Der Anfangsmeridian ist durch die Kanaren-Insel Ferro gelegt (vgl. die entsprechende Inschrift im Südpazifik: *Meridianus primus per Insulam Fer, quæ inter Canarias occidentalissima, ductus est, a quo Parisiensis 20. Gradibus, Noribergensis autem 28. Grad. 40. Minutis distat*).

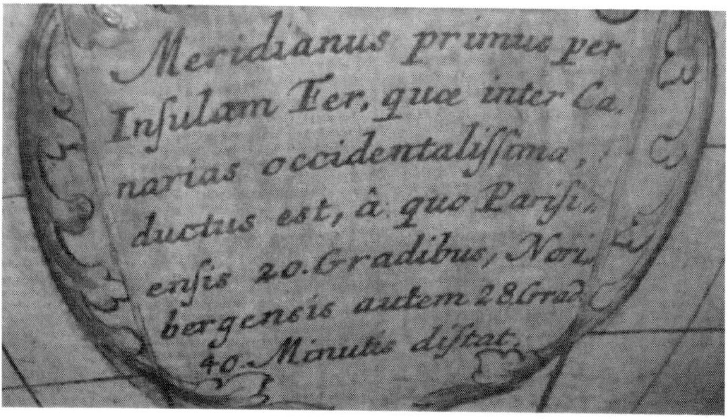

9) Kartuschen:

Auf dem Exemplar von 1750: *GLOBUS TERRESTRIS NOVUS Loca Terræ insigniora sec. praestant: Astron. et Georgr. observationes sistens opera IOH. GABR. DOPPELMAIERI M.P.P. concinne traditus à Ioh. Georg. Puschnero Chalcographo Norib. A.C. 1750.* (Jahreszahl handschriftlich ergänzt?).

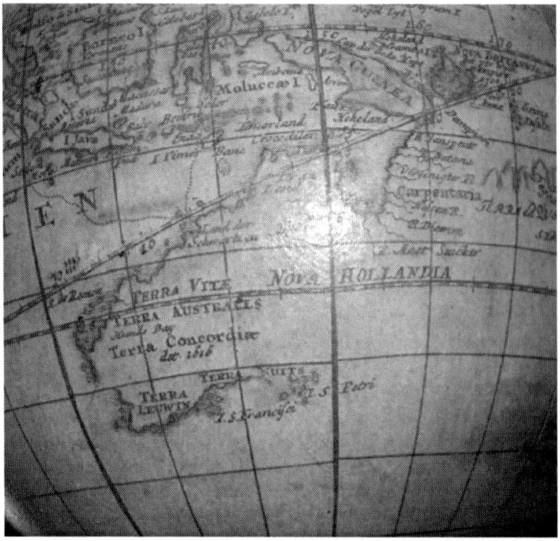

Auf dem Doppelmayr-Globus von 1750 ist die australische Ostküste noch nicht erkundet und mit Neuguiea als Halbinsel verbunden (oben). Sehr zahlreich wurden hier Entdeckungsreisen eingetragen, so auch (unten) die Schifffahrtsroute des dreimaligen Weltumseglers William Dampier (1651-1715).

Auf dem Exemplar von 1791 zwischen 180° und 210° Länge im Nordpazifik östlich von Japan: *GLOBUS TERRESTRIS, in quo locorum insigniorum situs terraeque facies, secundum praecipuat observationes celebrimorum nostri aevi Astronomorum et Geographorum, adiectis etiam novissimis tam Septentrionalium quam Australium terrarum prius incognitarum descriptionibus exhibentur olim opera IOH. GAB. DOPPELMAIERI M.P.P. concinnatus á I.G. Puschnero. Nunc audem* [sic!] *denuo editus W.P. Ienig Noriberge 1791* (Jahreszahl überklebt).

Bemerkungen:
Beide Globen (Farbtafel S. 304) haben denselben Durchmesser. Siedlungen sind mittels Ortsringen, einzelne Länder und Territorien auch farbig gekennzeichnet; auffallend sind zudem die schwarzen und roten Reiserouten der Entdeckungsreisen, von denen auf dem Exemplar von 1791 insbesondere die drei Cookreisen zu erwähnen sind. Erobererportäts sind aber nur auf den Doppelmayr-Globen mit einem Ø von 32 cm zu finden. Von Interesse ist außerdem die zeitgenössische Bezeichnung der Meere: *MARE DEL ZUR MARE PAZIFICUM* (Pazifik), *OCEANUS ORIENATALIS sive MAR DE INDIA* und *MARE ZANGEBARIAE* (Indischer Ozean mit dem Meer von Sansibar), *MARE DEL NORT* (Atlantischer Ozean und das Karibische Meer). Für die übrigen Teile des Atlantiks stehen die folgenden Namen: *MARE CANADENSE SIVE NOVAE FRANCIAE*, *MARE VIRGINAE*, *MARE BRASILICUM*, *MARE PARAGUAYA*, *MARE GUINEÆ* und *OCEANUS AETHIOPICUS SIVE MERIDIONALIS*.

Weiterführende Literatur

Anonyme Nachricht über kürzlich gefertigte Doppelmayrgloben, in: Neue Zeitung von gelehrten Sachen 1728, 690-692; Anonymer Nachruf auf Johann Gabriel Doppelmayr, in: Hamburgische Berichte von den neuesten gelehrten Sachen 20 (1751), 102-104, 116-120 und 134-136; A. Fauser, Ältere Erd- und Himmelsgloben in Bayern, 80-89; W. Dolz (Bearb.), Erd- und Himmelsgloben, 38-41 und 47-53; H. Gaab, Johann Gabriel Doppelmayr (1677-1750), in: Beiträge zur Astronomiegeschichte 4 (2001), 46-99; K. Kuhn, Professor Johann Gabriel Doppelmayr 1671-1750. Das erste Opfer eines elektrischen Unfalls, in: Die Stimme Frankens 26 (1960), 87-89; P. van der Krogt, Old Globe in the Netherlands, 111-117; E. L. Stevenson, Terrestrial and celestial globes 2, 158-160.

Erdglobus des Johann Philipp Andreae, gefertigt vom Nürnberger Kupferstecher Johann Christoph Berndt (Nürnberg 1756)

Johann Philipp Andreae (1700-1760) war der Sohn des aus Württemberg stammenden evangelischen Pfarrers Johann Ludwig Andreae (1667-1725), der sich ab 1711 auf die Herausgabe von Globenpaaren spezialisiert hatte. Von seinem Vater haben sich allein in Bayern über 20 Globen im Ø von 48 cm erhalten. Seine Himmelsgloben tragen ein Geleitwort des Rektors des Nürnberger Ägidiengymnasiums, Samuel Faber (1657-1716); die meisten der Kartuschen der Erdgloben sind aber unausgefüllt geblieben (vgl. das Exemplar in den Museen der Stadt Landshut; für die Untersuchung dieses Globus in situ sei Herrn Max Tewes, M.A., herzlichst gedankt). Johann Ludwig Andreae entwickelte zudem eine Methode, um das Sternbild maßstabgetreu auf einer Himmelskugel abbilden zu können. Von ihm stammen auch zwei globographische Schriften (1718 und 1724).
Sein Sohn Johann Philipp Andreae brachte ab 1718 – wohl zunächst unter Anleitung seines Vaters – in Nürnberg (noch als Student der Mathematik) Globenpaare von unterschiedlicher Größe heraus (sein größtes Globenpaar im Ø von 95 cm wird im Salzburger Museum Carolino Augusteum verwahrt), produzierte aber auch Sonnenuhren, ein astronomisches Kartenspiel (1719) und Landkarten. 1733 wurde er als Verfasser einer Schmähschrift gegen den Nürnberger Rat verhaftet und zu lebenslänglicher Zuchthausstrafe verurteilt. Er konnte jedoch fliehen und gründete im nahegelegenen Schwabach in der Markgrafschaft Brandenburg-Ansbach eine *Fabrik für Leonische Gold- und Silberwaren* sowie eine Globenfabrik, mit der er aber bald Konkurs anmelden musste.
Sein Mitarbeiter war der Nürnberger Mechanikus Johann Christoph Berndt (1707-1798), der den vorliegenden Globus der Sammlung Woldan als Neuauflage des Exemplars von 1726 mit nur geringfügigen Verbesserungen und einer Kartusche in französischer Sprache gestochen hat; darin wird auf *la plus nouvelle maniere* (sic!) des französischen Mathematikers und Instrumentenbauers Jean Pigeon (1654-ca. 1739) Bezug genommen.
Die Kupferplatten von Andreae wurden in überarbeiteter und deutlich verbesserter Form erneut gegen Ende des 18. Jahrhunderts

von Johann Georg Klinger (1764-1806) und Johann Bernhard Bauer (1752-1839) für die Herstellung von Globenpaaren verwendet.

1) Inventarnummer: Gl-V(L): WE 172.
2) Maße und Gestell: Ø = 13,5 cm; U = 42,5 cm; h = 25 cm; b = 19 cm; massives, vierfüßiges Holzgestell auf zwei Querstreben und runder Bodenplatte sitzend (vgl. Farbtafel S. 305).
3) Horizontring: teilweise verblasst, zeigt wie beim Erdglobus von Seutter die Himmelsrichtungen, die Sonntagsbuchstaben, das Tagedatum, die Monate (inkludiert die Heiligentage) und die Tierkreiszeichen, sowie die Gradeinteilung.
4) Meridianring: einseitig, aus Messing mit einer Skala der Breitengrade (viermal 0°-90°).
5) Stundenkreis und -zeiger: fehlen.
6) Erhaltungszustand: gut, aber etwas verblasst.
7) Segmente: 12 und zwei Polkappen.
8) Längen- und Breitenkreise: Der *Meridianus Primus* verläuft durch die kanarische Insel Ferro; neben dem Äquator, den Wendekreisen und der fein gestrichelt ausgeführten Ekliptik fallen der *Colurus Aequinoctiorum* und der *Colorus Solstitiorum* auf.
9) Kartuschen: unter dem Südmeer: *Ce Globe de la terre est desiné le plus nouvelle maniere de Iean Pigeon, Mathematicien dedié à S.A.R. Mg.r Le Duc d'Orleans, il est à trouver chez de Iean. chretop. Berndt Mecanique, et Graveur, à Nürenberg l'An 1756.*

Weiterführende Literatur

J. L. Andreae, Mathematische und historische Beschreibung des gantzen Welt-Gebäudes zum nutzlichen Gebrauch zweyer auf eine neue Art verfertigten Himmels- und Erd-Kugeln, Nürnberg 1718; J. P. Andreae, Neu inventirtes, belehrendes und ergötzendes astronomisches Karten Spiel: Das ist kunstrichtige Abbildung aller Gestirn des gantzen Firmamentes. Nürnberg 1719; T. Hampe, Johann Philipp Andreae und das Medaillen-Pasquill auf den Nürnberger Rat vom Jahre 1731, in: Mitteilungen des Vereins für Geschichte der Stadt Nürnberg 22 (1918), 244-279, hier: 260-263; R. Schmidt, Andreae – Pigeon: Ein Problem der Zuschreibung (anhand von 14 cm Himmelsgloben), in: Der Globusfreund 28/29 (1980/1981), 23-42; H. Wohlschläger unter Mitwirkung von R. Schmidt, Globenhersteller aus aller Welt, in: Allmayer-Beck, Modelle der Welt, 236-319, hier: 240, 245 und 296.

Erdglobus des George Adams Senior (London, ca. 1765)

Der 1748 zum *Mathematical Instrument Maker to his Majesty's Office of Ordnenance* ernannte Brite George Adams Senior (1709-1772) hatte sich ab 1734 auf die Herausgabe von mathematischen und optischen Instrumenten (darunter auch Globenpaare in verschiedenen Größen) spezialisiert und diesbezüglich mehrere Traktate verfasst (z. B. *A Treatise Describing and Explaining the Construction and Use of new Celestial and Terrestrial Globes*, 1766). Sein Wirken wurde von seinen Söhnen George Adams Junior (1750-1810) und Dudley (1762-1801) erfolgreich weitergeführt; die in der Londoner Fleet Street beheimatete Firma musste jedoch im Jahr 1817 Konkurs anmelden.

1) Inventarnummer: GL-V(L): WE 183.
2) Maße und Gestell: Ø = 46 cm; U = 144 cm; h = 86 cm; b = 64 cm. Das Holzgestell besteht aus einem dreigeteilten, gedrechselten Holzfuß, in dem eine signierte Kompassrose eingefügt ist und der auf vier Querbalken den Horizontring hält, worin der Globus ruht (Farbtafel S. 306).
3) Horizontring: zeigt neben den Tierkreiszeichen auch die Gradeinteilungen und einen ausführlichen Kalender.
4) Meridianring: auf der Vorderseite die übliche Breitenskala (viermal 0°-90°), auf der Rückseite hingegen sind die Tage eines jeden Monats in Bezug auf die Missweisung der Sonne angegeben.
5) Stundenkreis und -zeiger: vorzüglich vorhanden.
6) Erhaltungszustand: insgesamt gut, aber etwas fleckig.
7) Segmente: zwei Sets von 12 Segmenten (durchgehend, vom Äquator bis zum Nord- bzw. Südpol verlaufend).
8) Längen- und Breitenkreise: Äquator, Ekliptik und punktierte, waagrechte Linien zwischen den beiden Wendekreisen. Der Nullmeridian geht durch London.
9) Kartuschen: Eine Widmung findet sich im Pazifik zwischen 165° und 180° östlich von London:
   Britanniarum Regi Augustissimo GEORGIO TERTIO Scientiarum Cultori pariter et Præsidio Globus hunc Terrestrem Omnes hactenus exploratos terrarum tractus, Ad Observationes Navigantium Itinerantium et Astronomorum recentiores accuratissime descriptos exhibentem, Grati animi et pietatis monumentum, D.D.Q. Omni cultu et officio devinctissimus G.

Adams. Londini apud G. ADAMS artificem Reguim [sic!] *in vico Fleet-street.*

Bemerkungen:

Im Atlantik sind die Passatwinde mittels Pfeilen in Richtung Amerika eingezeichnet worden. Mehrere Inseln weisen zudem noch recht ungenaue Küstenlinien auf (vgl. *Terra detecta 1680* wie bei der Neuauflage von Doppelmayr von 1791).

Weiterführende Literatur

J. R. Millburn, Adams of Fleet Street, Instrument Makers to King George III, Aldershot (u. a.) 2000; H. Wallis, Thematische Elemente auf frühen englischen Globen (1592-1900), in: Der Globusfreund 15/16 (1966/1967), 75-85, hier: 79; P. van der Krogt, Old Globes in the Netherlands, 35-44; E. L. Stevenson, Terrestrial and celestial globes 2, 184-187.

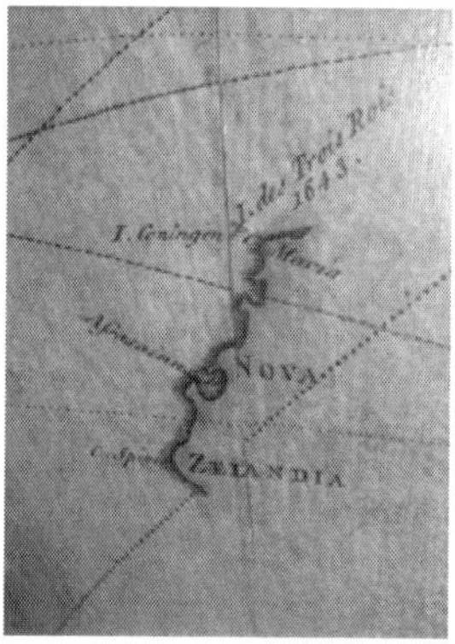

Der Globus von Adams zeigt nicht nur die Passatwinde (oben), sondern auch die noch offene Ostküste Neuseelands (rechts).

Erdglobus des Charles Dien Junior (Paris 1840)

Der Pariser Astronom und Kosmograph Charles Dien Junior (1809-1870) war nicht nur publizistisch tätig (vgl. seinen *Atlas Céleste*, Paris 1851), sondern fertigte zunächst für Félix Delamarche und später unter eigenem Namen Globen an. Ab 1840 arbeitete er mit dem Ungarn Nagy Károly zusammen und brachte Himmelsgloben in ungarischer Sprache heraus.

1) Inventarnummer: GL-V(L): WE 288.
2) Maße und Gestell: Ø = 30 cm; U = 94 cm; h = 86 cm; b = 35 cm. Massive runde Bodenplatte; kräftiger dicker Holzfuß, darauf ruhend die Erdkugel.
3) Horizontring: vorhanden, aber sehr reduziert (nur Gradeinteilung und Himmelsrichtungen in französischer Sprache).
4) Meridianring: aus Messing, zeigend die Breitengrade (viermal 0-90°).
5) Stundenkreis und –zeiger: Nur Stundenkreis (zweimal Stunde 1-12) vorhanden.
6) Erhaltungszustand: gut, jedoch sind Segmente bei den Philippinen abgebröckelt, weshalb hier ein Blick auf den Gipsuntergrund möglich ist.
7) Segmente: 12 (pro Segment dreimal unterteilt), durchgehend.
8) Längen- und Breitenkreise: Äquator und Wendekreise wie üblich. Der Nullmeridian verläuft nun durch Greenwich, die Ekliptik wurde außerdem mit Monatsnamen versehen, was folgendermaßen erläutert ist: *Nota. Les Signes et Degrés du Zodiaque comme les jours des mois gravés à l'Ecliptique donnent immédiatement et plus exactement la declinaison du Soleil, qu'en la cherchant sur l'horizon où ils deviennent alors inutiles.*
9) Kartuschen: südlich von Ceylon: *GLOBE Terrestre redigé d'après les découvertes les plus récentes par CH. DIEN, Paris 1840 Rue Hautef[...] 13.*

Bemerkungen:
Die *Antipodes de Paris* finden sich in der Südsee (vgl. Farbtafel S. 307).

Weiterführende Literatur

L. Bartha, The first educational globes in Hungarian language, in: Der Globusfreund 43/44 (1995), 237-256.

Erdglobus des Josef Jüttner (Prag 1840)

Der Deutschböhme Josef Jüttner (1775-1848) war Soldat und gehörte dem k. k. Bombardierkorps an, dessen Kommandant er ab 1842 war. Der in einer im Kriegsarchiv liegenden Individualbeschreibung als *Artillerist und Mathematiker sehr gebildet* geschilderte Jüttner nahm mehrere trigonometrische Vermessungen (u. a. für die Franziszeische Landesaufnahme) vor und fertigte auch entsprechende Karten an (Umgebungskarte von Prag, 1811-1815). Ab 1822 war er mit der Herstellung von Globen beschäftigt, die er in seiner *Anleitung zum Gebrauch der Erdkugel* (Prag 1822, 2. Auflage: Wien 1838) beschrieb. Mit ihm beginnt der Übergang zur serienmäßigen Globenherstellung in Österreich. Ein Mitarbeiter an seinen Globen war der k. k. Leutnant Franz Lettany (1793-1863), von dem sich ein selbst gefertigter Globus in der Globensammlung von Prof. Schmidt erhalten hat.

1) Inventarnummer: GL-V(L): WE 289.
2) Maße und Gestell: Ø = 32 cm; U = 100 cm; h = ca. 54 cm; b = ca. 38 cm. Das robuste Holzgestell besteht aus einem dreigeteilten Bogen, darauf ruhend ein gedrechselter Standfuß, der ursprünglich viergeteilt war (jedoch fehlt ein Fuß, weshalb der gesamte Globus schief steht und restauriert werden sollte).
3) Horizontring: nur noch fragmentarisch erhalten, da das aufgeklebte Papier sehr ausgebleicht ist. Enthält außer der Gradskala auch figurale Tierkreiszeichen, eine reduzierte Tages- und Monatsteilung sowie Windrichtungen (vgl. das Salzburger Exemplar).
4) Meridianring: aus Messing (die Breitengrade 0°-90° zeigend).
5) Stundenkreis und -zeiger: fehlen.
6) Erhaltungszustand: sehr schlecht und restaurierungsbedürftig, weil ein tragender Fuß fehlt. Großes Loch in den Segmenten beim Südpazifik.
7) Segmente: 18 und zwei Polarkappen.
8) Längen- und Breitenkreise: Äquator mit Ekliptik und den beiden Wendekreisen. Nullmeridian: westlich von Porto Santo.
9) Kartuschen: westlich von den *Gallopagos-Inseln od(er) Schildkröten-I(nseln)*, unterhalb des Äquators (vgl. Farbtafel S. 308): *ERDKUGEL von 1. Wiener Schuh im Durchmesser, nach Astron: Bestimmungen entworfen*

und gezeichnet, von Jos: Jüttner Hauptm: und Franz Lettany Lieut. der k.k. Oest: Artillerie im Jahre 1822 zu Prag: Berichtiget und vermehrt von Oberstlieutenant Jos: Jüttner im Jahre 1840.

Bemerkungen:
In der Universitätsbibliothek Wien haben sich die Segmente der Erdgloben von Jüttner aus den Jahren 1822 (Ø 31 cm, Signatur: III-300 476) und 1839 (Ø 62 cm, Signatur: III-300 477) erhalten, außerdem auch die beiden Polarkappen und der Horizontring. Diese Exemplare (in einem Atlas zusammengefasst) gehörten u. a. zu seinen *Anweisungen für den Gebrauch eines neu verfertigten Globus* (Prag 1822). Sie weisen aber untereinander auch erhebliche Unterschiede auf. Gebirgsformationen werden sorgfältig in schematischer Rautenzeichnung ausgeführt. Vorbild ist wohl Stielers Hand-Atlas.
In der Universitätsbibliothek Salzburg (Magazin der Sondersammlungen) wird ein besser erhaltenes Globenpaar gleicher Größe (beide Ø 31 cm, HG von 1824 und EG von 1840) verwahrt. Dabei zeigt insbesondere die *Him(m)els-Kugel auf welcher der Stand der Gestirne für den 1. Jänner 1850 vorgestellet ist*, kulturgeschichtliche Details: Denn es werden hier neben 8200 Sternen, die hauptsächlich dem Sternverzeichnis von Johann Elert Bode (1747-1826) entnommen worden sind, auch zeitgenössische Sternbilder dargestellt, die heute nicht zu den von der Internationalen Astronomischen Union festgelegten 88 anerkannten Sternbildern gehören. Diese sind bei Jüttner nach Erfindungen des Industriezeitalters benannt. So findet sich beispielsweise eine *Electrisier-Maschine*, eine *Luftpumpe* oder das Teleskop abgebildet, mit dem der deutsch-britische Astronom Friedrich Wilhelm Herschel (1738-1822) im Jahre 1781 den Planeten Uranus entdeckt hatte.

Weiterführende Literatur

H. Hühnel, Österreichische Globenhersteller. Ein biographischer Streifzug durch die Jahrhunderte, in: Allmayer-Beck, Modelle der Welt, 89-105, insbes. 93-97; O. Regele, Die Globen des Josef Jüttner (1775-1848) und des Franz Ritter von Hauslab (1798-1883), in: Der Globusfreund 22 (1953), 16-22; R. Schmidt, Franz Lettany, k. k. Offizier im Kaiserreich Österreich und Globushersteller, in: Informationen 8 (1984), 3-8; H. Wohlschläger, Die Globensammlung Rudolf Schmidt, in: Der Globusfreund 42 (1994), hier: 245f. und Tafel 24.

Erdglobus des Franz von Elekes (Wien 1844)

Der Militärkartograph Franz von Elekes war Mitarbeiter von Franz Leopold Schönninger und brachte in diesem Verlag im Jahre 1831 einen Globus im Ø von 12 cm, 1844 hingegen im Ø von 16,5 cm heraus.

1) Inventarnummer: GL-V(L): WE 300.
2) Maße und Gestell: Ø = 16,5 cm; U = 52 cm; h = 35 cm; b = 18 cm. Einfüßiges Holzgestell auf runder Bodenplatte.
3) Horizontring: bereits aus der Mode gekommen.
4) Meridianring: aus Messing (einseitig von 0° bis 90° verlaufend).
5) Stundenkreis und -zeiger: abhanden gekommen.
6) Erhaltungszustand: relativ gut.
7) Segmente: 18 durchgehende Segmente (Lithographien).
8) Längen- und Breitenkreise: Der Nullmeridian verläuft durch die Kanareninsel Ferro.
9) Kartuschen: westl. von Neuholland: *die ERDE. Nach den besten Ortsbestimmungen und Carten bearbeitet, WIEN. Bey F.L. Schönninger bürgl. Buchbinder. 1844. Entworf. u. Lith. Fr. von Elekes.*

Bemerkungen:
Der amerikanische Kontinent ist in grüner, Europa und Afrika hingegen in roter Farbe dargestellt. Der Globus der Sammlung Woldan (vgl. Farbtafel S. 309) findet sich erstmals auf der Umschlagseite von Globusfreund 13 (1964) abgebildet.

Weiterführende Literatur

H. Wohlschläger unter Mitwirkung von R. Schmidt, Globenhersteller aus aller Welt, in: Allmayer-Beck, Modelle der Welt, 236-319, hier: 260.

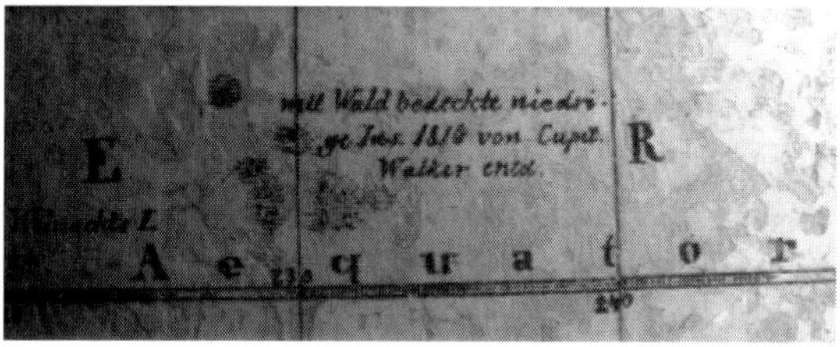

Hinweise auf eine von Capt. Walker im Jahre 1814 entdeckte Insel.

Erdglobus des Johann Adam Bühler (Nürnberg, ca. 1855)

Der Nürnberger Graveur Johann Adam Bühler (1813-1870) stellte ab 1836 Reise- und Verkehrskarten sowie Globensegmente im Stahlstich her, die in der zweiten Hälfte des 19. Jahrhunderts in der Abel-Klinger'schen Kunsthandlung in Nürnberg vertrieben wurden.

1) Inventarnummer: GL-V(L): WE 370.
2) Maße und Gestell: Ø = 14,5 cm; U = 46 cm; h = 30 cm; b = 14,5 cm. Das einfüßige, gedrechselte Holzgestell steht auf einer runden Bodenplatte.
3) Horizontring: aus der Mode gekommen.
4) Meridianring: dito.
5) Stundenkreis und -zeiger: waren niemals vorhanden.
6) Erhaltungszustand: oben fehlt die Spitze, ansonsten gut.
7) Segmente: 12 und Polkappen; Stahlstich.
8) Längen- und Breitenkreise: Äquator von 10° zu 10° geteilt, Ekliptik und Wendekreise; Meridian durch die Insel Ferro.
9) Kartuschen: westlich von Australien (vgl. Farbtafel S. 310): *DIE ERDE nach den neuesten Entdeckungen. NÜRNBERG Verlag von C. Abel Klinger Gez. u. gest. v. J.A. Bühler.*

Bemerkungen:
Die Segmente von ca. 1845 für einen von Bühler für Johann Georg Klinger gestochenen Erdglobus, der *zum Gebrauche in Schulen* vorgesehen war, sind in der Bayerischen Staatsbibliothek unter der Signatur Mapp, I 31 zu finden. Die Kontinente werden nun deutlich in einer fetten, gesperrt ausgeführten Majuskelschrift voneinander abgegrenzt; Stammesverbände werden in kursiver Schrift angegeben. Das Woldan-Exemplar ist ein grober, veränderter Nachstich dieser Segmente und auf die Zeit nach 1852 zu datieren, denn erst in diesem Jahr wurde Carl Casimir Abel Teilhaber der Kunsthandlung.

Weiterführende Literatur

A. Fauser, Lebensdaten von Nürnberger Globenmachern vom späteren 18. bis zur Mitte des 19. Jahrhunderts, in: Der Globusfreund 31/32 (1983), 123-128, hier: 125f.; H. Wohlschläger, Die Globensammlung Rudolf Schmidt, in: Der Globusfreund 42 (1994), 89-362, hier: 240-242.

Erdglobus des Carl Adami aus dem Dietrich Reimer-Verlag (Berlin 1865)

Der heute noch existierende Verlag von Dietrich Reimer (1818-1899) wurde in Berlin 1845 gegründet und hatte 1852 die Globusproduktion der Firma Adami & Co. übernommen. Carl Ludwig Adami (1802-1874) war Garnisonslehrer in Potsdam und hatte ab 1838 Globen (darunter auch Reliefgloben) im Kupferstichverfahren wie auch in Lithographie hergestellt. Er war für Reimer als Kartograph tätig. Seine Globen wurden von Heinrich Kiepert (1818-1899) neu bearbeitet, der seit 1865 auch eigene Zeichnungen für Erdgloben anfertigte. Die Globusproduktion dieses Verlags wurde bis ca. 1960 weitergeführt.

1) Inventarnummer: GL-V(L): WE 396.
2) Maße und Gestell: Ø = 34 cm; U = 106 cm; h = 51 cm; b = 33 cm. Das Holzgestell besteht aus einer runden Bodenplatte, darauf ruhend ein schmaler Holzfuß, der die Erdkugel trägt.
3) Horizontring: ist in dieser Zeit bereits unüblich.
4) Meridianring: dito.
5) Stundenkreis und -zeiger: fehlen.
6) Erhaltungszustand: gut, aber wackelige Angelegenheit.
7) Segmente: 12 Farblithographien (jeweils dreigeteilt).
8) Längen- und Breitenkreise: Der Nullmeridian geht durch die Kanarischen Inseln.
9) Kartuschen: westlich von Nordamerika: *C. ADAMI'S ERDGLOBUS neu bearbeitet und gezeichnet von H. KIEPERT. BERLIN D. REIMER 1865. Stich v. J. Sulzer * Druck v. F. Barth.*

Bemerkungen:
Auffällig ist die Einzeichnung von Meeresströmungen. Die verschiedenenen Länder sind in fetter Majuskelschrift und auch farblich hervorgehoben (vgl. Farbtafel S. 311).

Weiterführende Literatur

J. Espenhorst – G. R. Crossmann, Petermann's Planet: a guide to German handatlases and their siblings throughout the world 1800-1950, 2 Bde., Schwerte 2003 und 2008; L. Zögner (Hrsg.), Antike Welten – Neue Regionen. Heinrich Kiepert 1818-1899 (Staatsbibliothek zu Berlin – Preußischer Kultur-besitz, Ausstellungskataloge, N.F. 33), Berlin 1999.

Zwei Erdgloben des Wiener Verlagshauses Schöninger (Wien, ca. 1867 und 1872)

Das in Wien befindliche Verlagshaus Schöninger spielt für die Globengeschichte Österreichs im 19. Jahrhundert eine zentrale Rolle. Der Buchbinder Franz Leopold Schöninger (1790-1877) führte das Geschäft ab 1815 und hatte Globen verschiedener Größen im Angebot. Seine Söhne Joseph (1819-1882) und Franz Xaver (Schönninger, 1820-1897) ließen die Globenerzeugung regelrecht florieren; bei der Wiener Weltausstellung von 1873 erhielt Franz für einen astronomischen Globus (Armillarsphäre) sogar eine Verdienst-Medaille.

Leopold Hugl (1822-1887) stammte aus Niederösterreich. Er war Lehrer und später Direktor einer Volks-Bürgerschule. Seine Globen (vgl. Farbtafel S. 312) erschienen bei Franz Xaver Schönninger.

1) Inventarnummer: GL-V(L): WE 404 und 417 (alte Signatur: 41 EW).
2) Maße und Gestell: Globus von ca. 1867: Ø = 31,5 cm; U = 98 cm; h = 62,5 cm; b = 26 cm. Das Holzgestell besteht aus einer runden Bodenplatte mit einem darauf stehenden, gedrechselten Fuß; Erdkugel von 1872: Ø = 16,5 cm; U = 52 cm; h = 42 cm; b = 21 cm. Das Gestell ist in barockem Stil gehalten, dreifüßig; unten war sicher einmal ein Kompass eingefügt.
3) Horizontring: Nur beim Exemplar von 1872 erhalten. Dieser zeigt (etwas abgeblättert) neben den Sternzeichen und der Graduierung auch die Himmelsrichtungen samt einem Kalender (bei dem Heiligentage nicht mehr von Interesse sind).
4) Meridianring: den Globus umfassend, mit Angabe der Breitengrade (viermal 0°-90°, beim Exemplar von ca. 1867 nur zweimal).
5) Stundenkreis und -zeiger: Lediglich der Stundenkreis des Exemplars von ca. 1867 hat sich erhalten. Darauf sind die Stunden zweimal folgendermaßen angegeben: 3-6-9-12.
6) Erhaltungszustand: gut bis mäßig (Exemplar von 1872), weil dieses sehr verblasst ist und der Stab zur Gestellhalterung fehlt, teilweise auch abgerieben (v. a. unten).
7) Segmente: 18 (durchgehend, ca. 1867) bzw. 12 (Exemplar von 1872) und jeweils zwei dazugehörige Polkappen.

8) Längen- und Breitenkreise: Äquator, Wendekreise und Ekliptik. Der Nullmeridian geht durch die Kanarischen Inseln (Ferro).
9) Kartuschen:
   Auf dem älteren Exemplar findet sich südwestlich von Australien bei den Kerguelen-Inseln: *ERDGLOBUS entworfen und herausgegeben von L. HUGL. Bei F. Schöninger in WIEN.*
   Auf dem jüngeren Exemplar hingegen unterhalb von Capstadt: *Erdglobus nach den neuesten Quellen bearbeitet von F. Schönninger 1872 Globus Fabrik Wien VI. Bez. Mariahilferstr. 111* (heute befindet sich dort ein Bekleidungsgeschäft).

Bemerkungen:
Darstellung der Meeresströmungen mittels feiner weißer Linien; Trennung von Gebirgsgruppen mittels brauner Schraffen. Länderbezeichnungen werden vernachlässigt, weil es sich sich hier um einen thematischen, physikalischen Globus handelt.

Weiterführende Literatur
Das Exemplar von 1872 findet sich abgebildet auf der Umschlagsseite von Globusfreund 13 (1964), vgl. dazu: Hadmar Wolfdietrich Schandl, Die Stellung der Buchbinderfamilie Schöninger in der Geschichte der Wiener Globenerzeugung, in: Der Globusfreund 13 (1964), 49-63; H. Wohlschläger unter Mitwirkung von R. Schmidt, Globenhersteller aus aller Welt, in: Allmayer-Beck, Modelle der Welt, 236-319, hier: 274 und 305.

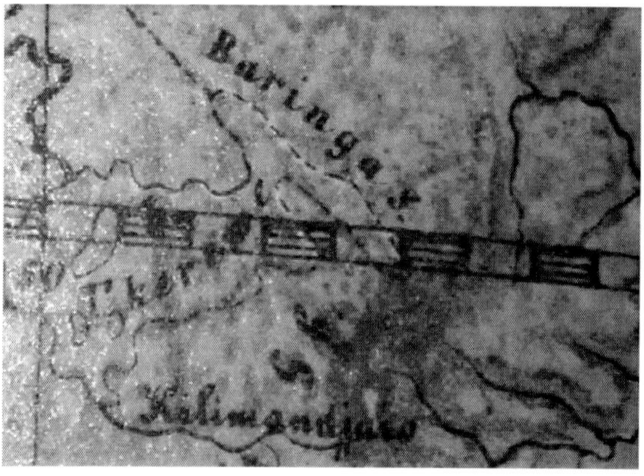

Abbildung der höchsten Erhebung Afrikas: der Kilimandscharo auf dem Globus von Leopold Hugl gefertigten Exemplar von ca. 1867

Drei Erdgloben verschiedener Größe der Firma Jan Felkl (Prag vor 1870 bzw. Rostok bei Prag nach 1875 und 1876)

Der Böhme Jan Felkl (1817-1887) war wohl der bedeutendste Globenproduzent der Österreichisch-ungarischen Monarchie. Er hatte um 1850 die Werkstatt des Prager Kupferstechers und Kartographen Václav (Wácslaw) Merklas (1809-1866) übernommen und bereits bei der Herstellung von Globen (in deutschen und tschechischsprachigen Ausgaben) mitgewirkt. Ab den 1860er Jahren bot er zudem Tellurien, Lunarien, Planetarien und Induktionsgloben (Kugeln mit schwarzer Oberfläche, auf denen man mit Kreide beispielsweise die Kontinente zeichnen konnte) an und fertigte seine Globen auch in anderen Sprachen (u. a. auf Ungarisch, Russisch, Polnisch, Niederländisch oder Englisch). Insgesamt sind von ihm Globen in 17 verschiedenen Sprachen überliefert, was ihm internationale Anerkennung einbrachte. In dieser Zeit erhielt der Globus einen höheren Stellenwert im Schulunterricht, weshalb die Nachfrage nach derartigen Lehrmitteln sprunghaft anstieg. 1870 verlegte Felkl sein prosperierendes Unternehmen nach Rostok (Roztok) bei Prag, wo er auch entsprechende Anleitungen zur Globenkunde herausgab (vgl. beispielsweise: *Der Globus und seine Anwendung in Schule und Haus. Ein unentbehrliches Hilfsbuch für Lehrer und Laien* von 1877). 1875 wurde sein jüngster Sohn Krystof Zikmund (1855-1894) Teilhaber dieser Globusfabrik. Ab diesem Zeitpunkt lautete der Firmenname *J. Felkl & Sohn*, was für die Datierung seiner überwiegend undatierten Globen von Belang ist. Nach dem Tode Krystofs übernahm sein Bruder Ferdinand Felkl (1846-1925) die Leitung der Firma, die noch zwei Generationen weitergeführt wurde. 1950 wurde das Unternehmen stillgelegt. Bis vor vierzig Jahren waren die früheren Fabrikgebäude in Rostok bei Prag noch zu sehen.

Der bedeutendste Autor der Felkl-Globen war der sächsische Geograph Otto Delitsch (1821-1882), der 1865 promovierte und 1875 zum außerordentlichen Professor für Geographie an der Universität Leipzig ernannt worden war. Er war ab ca. 1860 für den Entwurf von Kartensegmenten für die Erdgloben der Firma Felkl zuständig und verwendete dabei zwei bis drei Höhenschichten, verbunden mit der damals üblichen Schraffenmanier.

1) Inventarnummer: GL-V(L): WE 411 (alte Signatur 2 RS, 4 EW), GL-V(L): WE 422 (auf der Unterseite des Gestells findet sich die alte Inv. Nr. 12 EW) und GL-V(L): WE 430.
2) Maße und Gestell: Ø = 8,6 cm; U = 27 cm; h = 22 cm; b = 9 cm.–
Ø = 15,8 cm; U = 50 cm; h = 34 cm; b = 15 cm. –
Ø = 31,86 cm; U =101 cm; h = 62 cm; b = 29 cm.
Das Holzgestell dieser Globen besteht einheitlich aus einer runden Bodenplatte, auf der sich ein einfüßiger Holzfuß mit Verbindungsstab befindet, der die Erdkugel trägt (vgl. Farbtafel S. 313).
3) Horizontring: zu dieser Zeit bereits unüblich.
4) Meridianring: aus der Mode gekommen.
5) Stundenkreis und -zeiger: dito.
6) Erhaltungszustand: mäßig bis gut, teilweise etwas verblasst.
7) Segmente: 12 durchgehende, kolorierte Segmente.
8) Längen- und Breitenkreise: Der Nullmeridian verläuft westlich von Madeira.
9) Kartuschen: östl. von den Gesellschaftsinseln: *Die Erde Prag bei J. Felkl* bzw. auf dem Exemplar von nach 1875 (*Die Erde Herausgegeben J. Felkl & Sohn in Roztok b. Prag*) und 1876 (*DER GLOBUS Bearbeitet von Prof. Otto Delitsch in Leipzig. FABRIK J. FELKL & SOHN Rostok bei Prag. 1876*).

Weiterführende Literatur:

J. Dörflinger, Jan Felkl (1817-1887): Zum 100. Todestag eines altösterreichischen Globenfabrikanten, in: Kartographische Nachrichten 37 (1987), 182f.; F. G. Hahn, Otto Delitsch und seine Bedeutung für die Erdkunde, in: Das Ausland 55 (1882), 841-844; L. Mucha, Der tschechische Globenmacher Jan Felkl, in: Der Globusfreund 33/34 (1985/86), 122-142; Ders., Felkls Söhne Siegmund und Ferdinand Felkl, in: Der Globusfreund 38/39 (1990/91), S. 97-104; Ders., Die Globen von Otto Delitsch, in: Der Globusfreund 15/16 (1966/1967), 243-247; K. Patay, Ein kombinierter Erd-Himmels-Globus der Firma J. Felkl in der Széchényi Nationalbibliothek in Budapest, in: Der Globusfreund 33/34 (1985/86), 117-121; R. Schmidt, Jan Felkl – Ein Rostoker Globenmacher. Ausstellung in Rostok bei Prag, 24. September 1992 bis 15. November 1992, in: Informationen 19 (1993), 13 f.; H. Wohlschläger – J. Dörflinger, Österreichische und deutsche Globenhersteller der zweiten Hälfte des 19. und des beginnenden 20. Jahrhunderts, in: Der Globusfreund 30 (1982): Katalog zur Ausstellung „Globen aus Urgroßvaters Zeit. J. Felkl und seine Zeitgenossen“, 9-40, v. a. 12-20.

Erdglobus des Theodor Mettke (Berlin, nach 1875)

Theodor Mettke war Inhaber einer lithographischen Anstalt in Berlin und trat als Stecher auf den Globen der Verlagsanstalt von Ernst Schotte & Co. besonders hervor. Dieser Verlag bestand ab 1855 in Berlin und bot Erdgloben in 15 verschiedenen Sprachen in einfacher Ausfertigung oder mit vollständiger Armierung (Meridianring, Stundenring und Kompass) an; jedoch waren auch Himmelsgloben und Tellurien Bestandteil des Firmensortiments. Somit reiht sich der Verlag unter den bedeutendsten Globenherstellern der 2. Hälfte des 19. Jahrhunderts ein.

1) Inventarnummer: GL-V(L): WE 408.
2) Maße und Gestell: Ø = 31 cm; U = 97 cm; h = 66 cm; b = 37 cm. Das Gestell besteht aus einem dreifüßig gedrechselten, gezwirbelten Holzfuß, auf dem der den Horizontring tragende Messingring ruht.
3) Horizontring: aus Holz, verzeichnet die Himmelsrichtungen, einen Kalender, die Gradeinteilungen sowie die Tierkreiszeichen.
4) Meridianring: Messing (etwas fleckig) mit Gradeinteilung 0°-90°.
5) Stundenkreis und -zeiger: nur der Stundenkreis (aus Holz) hat sich erhalten, darin Stundeneinteilung von 1-12.
6) Erhaltungszustand: ganz passabel, aber teilweise abgerieben.
7) Segmente: 12 und zwei Polkappen.
8) Längen- und Breitenkreise: Äquator, Ekliptik und Wendekreise. Der Nullmeridian geht durch die Insel Ferro.
9) Kartuschen: südwestlich von Australien: *Die* ERDKUGEL *nach den neuesten Quellen bearbeitet. Geographisch. artist. Anstalt* E. SCHOTTE *&* C^O^. BERLIN *gez. u. gest. v. Th. Mettke, Lith. Anst. Berlin.*

Bemerkungen:
In roter Farbe werden die britischen Besitzungen in Übersee vermerkt. Dass es sich um einen Schulglobus handelt, der auch die physischen Gegebenheiten aufzeigen sollte, belegen die eingetragenen Meeresströmungen (vgl. Farbtafel S. 314).

Weiterführende Literatur

H. Wohlschläger unter Mitwirkung von R. Schmidt, Globenhersteller aus aller Welt, in: Allmayer-Beck, Modelle der Welt, 236-319, hier: 289 und 305.

Schulglobus vom Verlag Freytag und Berndt A.G. (Wien, um 1930)

Die Firma Freytag & Berndt wurde 1879 gegründet und produzierte ab 1894 Lehrmittel (wie Landkarten, Atlanten und Wandtafeln) für den Unterricht. Von 1907 bis 1941 wurden in diesem Verlag auch physische und politische Erdgloben in verschiedenen Sprachen und unterschiedlicher Größe hergestellt, ebenso Himmelsgloben, Induktionsgloben und Tellurien. Für den Kartenentwurf dieser Globen war ab 1908 der Kartographieprofessor Johann Georg Rothaug (1850-1924) zuständig.

1) Inventarnummer: GL-V(L): WE 497.
2) Maße und Gestell: Ø = 32 cm, U = 101 cm; h = 69 cm; b = 30 cm. Das Gestell besteht aus einer runden Bodenplatte, darauf ein Holzfuß, und einem kleinen Kompaß im Bodengestell.
3) Horizontring: bereits aus der Mode gekommen.
4) Meridianring: als Halbmeridianring (zweimal zu je 90°) ausgeführt.
5) Stundenkreis und -zeiger: Der Kreis ist in Ansätzen vorhanden.
6) Erhaltungszustand: gut.
7) Segmente: 24 durchgehende Segmente im Farbdruck (unterteilt).
8) Längen- und Breitenkreise: Nullmeridian von Greenwich. Außer den üblichen Angaben sind die nördliche bzw. südliche Palmengrenze sowie die Baum-, Getreide- und Treibeisgrenze auf dem Globus eingetragen.
9) Kartuschen: Westlich der Galapagosinseln: *WIENER SCHULGLOBUS. Große Ausgabe Maßstab 1: 40,000.000. Kartogr. Anstalt G. FREYTAG & BERNDT A.G. WIEN* (vgl. Farbtafel S. 315).

Bemerkungen:

Die Datierung dieses Globus kann auf den Zeitraum zwischen dem 1. Februar 1927 und dem 12. Juni 1931 eingegrenzt werden, weil nur in dieser Zeit das hier abgebildete, australische Territorium Central Australia (*Mittel-Australien*, geschaffen durch den North Australia Act von 1926) bestand. Ein Globus desselben Durchmessers findet sich im Globenmuseum der ÖNB unter der Signatur Gl. 107.

Weiterführende Literatur

H. Wohlschläger unter Mitwirkung von R. Schmidt, Globenhersteller aus aller Welt, in: Allmayer-Beck, Modelle der Welt, 236-319, hier: 263.

Nicht in der Sammlung Woldan, sondern im AÖAW:

Mondglobus der Firma Franz Schönninger (Wien 1849)

Josef Riedl von Leuenstern (1786-1856) hatte 1814 das 1801 gegründete Kunst- und Industrie Comptoir, dessen Name er in *J. Riedl's Kunsthandlung* änderte, übernommen. Er war der Lehrer von Franz Xaver Schönninger und arbeitete in späteren Jahren für dessen Verlag in Wien, für den er 1849 mit dem vorliegenden Globus (Farbtafel S. 312) einen ersten serienmäßig gefertigten Mondglobus nach der *Mappa Selenographica* des Dorpater Astronomen Johann Heinrich von Mädler (1794-1874) schuf. Ähnliche Mondhalbkugeln sind uns schon früher überliefert, so beispielsweise eine von Hofrätin Witte in Hannover in Relief ausgeführte Darstellung, über die Mädler 1840 berichtet.

1) Inventarnummer: B 89.
2) Maße und Gestell: Ø = 23,5 cm; h = 42 cm; b = 22 cm. Auf einem gedrechselten Fuß stehender Mondglobus.
3) Horizontring: fehlt.
4) Meridianring: dito.
5) Stundenkreis und -zeiger: waren niemals vorhanden.
6) Erhaltungszustand: gut.
7) Segmente: 12 Segmente, jedoch sechs davon ohne Darstellung (Mond nur einseitig zu sehen!) sowie die beiden Polarkalotten.
8) Längen- und Breitenkreise: Bemerkenswert ist der Maßstab (5,2 cm) mit *Wiener Meilen zu 4000 Klaffter.*
9) Kartuschen: *Der MOND nach der orthograpischen Karte von Beer und Mädler SPHÄRISCH DARGESTELLT VON RIEDL LEUENSTERN, Wien 1849. Gestochen von H. Mansfeld. Zu finden b. F. Schönninger.*

Bemerkungen: Von kulturgeschichtlichem Interesse sind zudem die Bezeichnungen *Mare Humorum*, *Serenitatis, Tranquilitatis* und *Nectaris.* – Im Kriegsarchiv in Wien befinden sich unter der Signatur A I 12 außerdem die Segmente für diesen Mondglobus.

Weiterführende Literatur

Schreiben des Herrn J. H. Mädler an den Herausgeber, Hildesheim 1839, in: Astronomische Nachrichten 17 (1840), Sp. 29f.

Epilog

Als ich vor einigen Jahren, im Sommersemester 2002 (noch als Student) während eines Auslandssemesters in Wien ein Seminar über *Alte Landkarten* besucht habe, wurde ich vom Veranstalter dieses Seminars, Prof. Dr. Johannes Dörflinger, nicht nur für die Thematik der Geschichte der Kartographie (der ich bis heute treu geblieben bin) begeistert, sondern auch tatkräftig unterstützt. Insbesondere sind mir dabei eine Exkursionen anhand von alten Stadtplänen innerhalb Wiens sowie die Vorträge der Seminarteilnehmer in der Kartensammlung der Österreichischen Nationabibliothek in bester Erinnerung geblieben.
Jedoch konnte ich damals auch für ein Referat über die Münchener Stadtpläne entsprechendes Material in der *Sammlung Woldan* auffinden, wo ich bereits als Student herzlichst aufgenommen und betreut worden bin. Als Initiator dieses Sammelbandes, der mit Hilfe von Frau Prof. Dr. Christine Harrauer als dritter Band der *Edition Woldan* erscheinen kann, möchte ich die Gelegenheit wahrnehmen, meinem Mentor Johannes Dörflinger herzlich zu danken. Zusätzlich will ich entsprechende Dankesworte auch an Frau Dr. Petra Svatek und den langjährigen Leiter der Sammlung Woldan, Herrn Mag. Gerhard Holzer, richten, denn ohne die Hilfe dieser Mitherausgeber wäre die hier vorgelegte Studie nicht möglich gewesen.

Zum Autor

Thomas Horst, geb. 1980, Studium der Geschichte und Ethnologie an der Ludwig-Maximilans-Universität München, 2002 Auslands-Semester in Wien bei Prof. Dr. Johannes Dörflinger. 2003 und 2005 ethnologische Feldforschung bei den Nachkommen der Mundurukú-Indianer im Amazonasgebiet. 2008 Promotion (über die älteren Manuskriptkarten Altbayerns). Seit 2005 als Wissenschaftler bei Prof. Dr.-Ing. Kurt Brunner (Kartographie und Topographie) an der Universität der Bundeswehr München tätig.
Projekte als Postdoktorand: Erforschung der in Bayern verwahrten Globen; frühneuzeitliche Kartographie- und Entdeckungsgeschichte.
Kontakt: Eberwurzstraße 93, 80935 München, Deutschland;
thomashorst@gmx.net

Tafeln / Plates

zu Band 1

Ältester Globus in der Sammlung Woldan:
Der Erdglobus des Jodocus Hondius d. Ä. von 1601

Erdglobus des Isaak II. Habrecht von ca. 1675 mit Stundenkreis und Horizontring (oben) sowie Detailausschnitt (Inschrift bei *Nova Guinea*)

Himmelsglobus des Isaak II. Habrecht von ca. 1675 mit Stundenkreis und Horizontring (oben) sowie Kartusche beim Sternbild *Pegasus*

Erdglobus des Georg Christoph Eimmart von 1705:
Gesamtansicht und Hauptkartusche

Erdglobus des Matthäus Seutter von 1750:
Gesamtansicht und Hauptkartusche

Die beiden Doppelmayr-Erdgloben in der Sammlung Woldan:
Titelkartusche des Exemplars von 1750 (oben) und der Globus von 1791 (unten)

Erdglobus des Johann Philipp Andreae, gefertigt von Johann Christoph Berndt (Nürnberg, 1756): Gesamtansicht und Detail: Australien

Erdglobus desGeorge Adams Senior von 1765
mit einem in das Bodengestell eingefügten Kompass

Der Erdglobus des Charles Dien Jun. von 1840 zeigt
östlich von der Bounty-Insel die *Antipodes de Paris*

Dem Erdglobus des Josef Jüttner fehlt ein Fuß (oben); er wurde zusammen mit Franz Lettany im Jahr angefertigt (s. Kartusche unten)

Erdglobus des Franz von Elekes, Wien 1844

Erdglobus des Johann Adam Bühler, Nürnberg ca. 1855

Der in Berlin im Dietrich Reimer-Verlag erschienene Erdglobus
des Carl Adami zeigt die Buren-Republiken in Südafrika im Jahr 1865

Globenerzeugnisse des Wiener Verlagshauses Schöninger: Erdglobus des Leopold Hugl (um 1867, oben links) und weiteres Exemplar von 1872 (oben rechts) sowie Mondglobus des Josef Riedl von Leuenstern (1849, unten)

Globenerzeugnisse der Prager Firma Jan Felkl vor 1870 und nach 1875 (oben) sowie von 1876 (unten)

Erdglobus des Theodor Mettke, Berlin nach 1875

Wiener Schulglobus, erschienen um 1930 im Wiener Verlag Freytag & Berndt A.G.

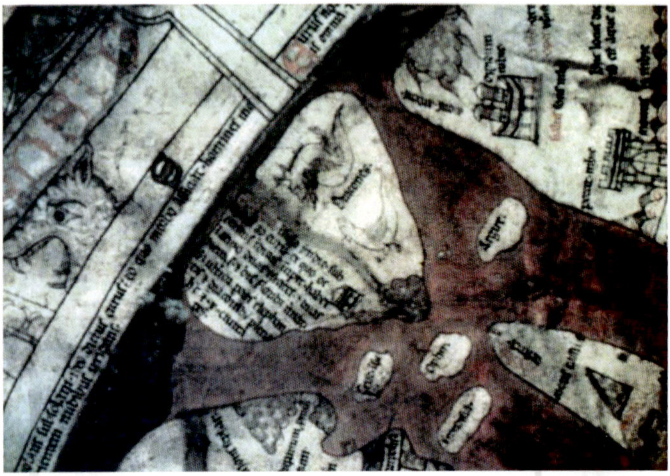

a) Detail from the Hereford World Map
Courtesy of the Cathedral Library, The cathedral, Hereford

b) Detail from Fra Mauro's World Map (Venice, 1459), The Marciana Library.
Courtesy of the National Commission for UNESCO, Venice

a) Detail of Lukas Brandis, Rudimentum novitiorum, Lübeck, 1475.
Courtesy of the Incunabula Collection, Austrian National Library, Vienna

b) Detail of the Walsperger map, 1448.
Biblioteca Apostolica Vaticana, Rome

c) Detail of Leardus World Map (1448)
American Geographical Society Library – Maps, 050 A-1452

[F. Hochstetter, Manuskriptkarte des südlichen Teiles der Provinz Auckland, 1859 (?)], Alexandre Turnbull Library

G. HOLZER – T. HORST – P. SVATEK (Hrsg.)

DIE LEIDENSCHAFT DES SAMMELNS

Streifzüge durch die Sammlung Woldan

ÖSTERREICHISCHE AKADEMIE DER WISSENSCHAFTEN
PHILOSOPHISCH-HISTORISCHE KLASSE

EDITION WOLDAN

Band 3/2

Herausgegeben von Christine Harrauer

ÖSTERREICHISCHE AKADEMIE DER WISSENSCHAFTEN
PHILOSOPHISCH-HISTORISCHE KLASSE

GERHARD HOLZER – THOMAS HORST –
PETRA SVATEK (Hrsg.)

Die Leidenschaft des Sammelns

Streifzüge durch die Sammlung Woldan

Verlag der
Österreichischen Akademie
der Wissenschaften

ÖAW

Wien 2010

Vorgelegt von w. M. Hermann Hunger in der Sitzung am 18. Juni 2010

Logo der Edition Woldan:
Detail eines Holzschnitts vom Titelblatt der Cosmographie
des Gemma Frisius (Antwerpen 1530).

Umschlagbild:
Die Wohnung des Wiener Sammlers Erich Woldan
(Ausschnitt aus einem Foto im Archiv der Sammlung Woldan)

Umschlaggestaltung:
Reisenbauer / Wallisch / Holzer

Die verwendete Papiersorte ist aus chlorfrei gebleichtem Zellstoff hergestellt,
frei von säurebildenden Bestandteilen und alterungsbeständig.

ISBN 978-3-7001-6894-2

Druck und Bindung: Prime Rate kft., Budapest

http://hw.oeaw.ac.at/6894-2
http://verlag.oeaw.ac.at

Franz Xaver Embel,
Schilderungen der Gebirgsgegenden
um den Schneeberg in Oesterreich,
Wien 1803

Helga Hühnel, Wien

Erzherzogliche Gipfelstürmer

Handschriftliche Reisenotate Erzherzog Rainers über Expeditionen auf den Schneeberg und in angrenzende Gegenden Anno 1802 und 1805

Als ich bei Recherchen in der Handschriftensammlung der Österreichischen Nationalbibliothek auf Reiseaufzeichnungen Erzherzog Rainers über eine Schneeberg-Exkursion stieß, war ich persönlich sofort interessiert, da dieser höchste Berg Niederösterreichs schon immer eine magische Anziehungskraft auf mich ausgeübte hatte.

Er liegt nicht weit von Wien entfernt, bietet bei klarem Wetter einen majestätischen Anblick und lockt mit seinen wunderschönen Wanderwegen und Klettersteigen, seiner vielseitigen Flora und dem großartigen Gipfelpanorama die Naturliebhaber seit Jahrhunderten an. Trotzdem er nur knapp über 2000 m (Klosterwappen 2076 m) hoch ist, ähnelt er einem Dreitausender vor allem in klimatischer Hinsicht, und die hier oft auftretenden raschen Wetteränderungen wurden schon vielen Menschen zum Verhängnis.

Wann trat der Schneeberg bewusst ins Blickfeld von Naturliebhabern und wer waren die Ersten, die ihn bestiegen und zu welchem Zwecke? Wann erschienen die ersten topographischen Beschreibungen über diesen Berg? Und vor allem war ich nun neugierig geworden, welche Erlebnisse Erzherzog (EH) Rainer und seine Brüder im Schneeberggebiet erfuhren, was sie interessierte, welche Nutzen sie aus dieser Unternehmung zogen.

Am 1. August des Jahres 1802 fuhr eine neunköpfige Reisegruppe[1] mit Dienerschaft um 11 Uhr Nachts von Schönbrunn auf der gewöhnlichen Poststraße nach Neunkirchen, Richtung Schneeberg. Es handelte sich dabei um die Erzherzöge Johann, Rainer und Anton, die Obersthofmeister Nimptsch[2] und Spork[3], den Leibarzt Störck[4], den kaiserlichen Hofgärtner Bredemayer[5], den Direktor des k. k. Naturalienkabinetts Abbé Stütz[6] und den späteren Kammermaler Johann Kniep.

Den Grund dieser Unternehmung nennt EH Johann in einem Brief an den Salzburger Naturforscher Carl Erenbert Moll: *Mein Hauptzweck war, den grossen Nebenzweig der hohen Alpenkette zu bereisen, und zu untersuchen, nämlich vom Schneeberge bei Neustadt bis an den hohen Schneeberg*[7]*, der die 3fache Gränze Oberösterreich's Salzburg's und Steiermark's bildet.*[8] Weiters schildert hier Johann ziemlich genau den Reiseverlauf und den Tagesablauf; seine handschriftlichen Aufzeichnungen darüber sind leider in den Nachkriegswirren des 2. Weltkrieges verloren gegangen. Hingegen existieren in der Handschriftensammlung der Österreichischen Nationalbibliothek die Reisenotizen EH Rainers, des nachmaligen Vizekönigs von Lombardo-Venetien. Die Katalogisate Cod. Ser. n. 2023 (fol. 1-63) und Cod. Ser. n. 2026 (fol.1-17) beinhalten die handschriftlichen, überaus detailreichen und genauen Aufzeichnungen zweier Schneeberg-Exkursionen von 1802 und 1805, die im Folgenden behandelt werden.

---

1 Vgl. Bericht in der Wiener Zeitung Nr. 3097 vom 25. August 1802.

2 Josef Graf Nimptsch (1763-1838), General-Major; nach dem Luneviller Frieden (1801) wurde der Graf zum Obersthofmeister EH Johanns ernannt und blieb es bis zu seinem Tode 1838.

3 Johann Rudolf Graf von Spork (1755-1806), 1793 General-Major; 1797 Feldmarschall - Lieutenant; 1802 zum Obersthofmeister von EH Anton bestimmt.

4 Anton Freiherr von Störck (1731-1803), ein österreichischer Mediziner und Hochschullehrer. 1760 Leibarzt Maria Theresias und Promedicus der Erblande.

5 Franz Bredemeyer (1758-1839) unternahm einige botanische Sammelreisen nach Amerika, seit 1793 Hofgärtner.

6 Abbé Andreas Stütz (1747-1806).

7 Der Großglockner, der höchste Gipfel der Ostalpen (3798m) war am 28.7. 1800 erstmals bestiegen worden.

8 C. E. v. Moll, Mittheilungen aus seinem Briefwechsel 2, Augsburg 1830, 454.

Die Prinzen des Hauses Habsburg waren natürlich nicht die Ersten, die der Faszination dieses Berges und seiner Umgebung erlagen. Schon im 16. Jahrhundert durchwanderte einer der größten Naturforscher seiner Zeit Carolus Clusius (Charles d'Escluse 1526-1609) auf seinen botanischen Streifzügen u. a. Schneeberg und Schneealpe (1574, 1575, 1576).[9] Der Wechsel, Gahns, Schneeberg und die Preiner Alpe werden in seinem Werk als vorzügliche Punkte seiner botanischen Forschungen bezeichnet.[10] Er besuchte dabei, wenn möglich, dieselben Gegenden, um zu untersuchen, ob jedes Jahr die gleichen Pflanzen an den notierten Standorten anzutreffen seien.

Der nächste berühmte Botaniker, von dem wir explizit wissen, dass er im Schneeberggebiet viele Pflanzen sammelte, war Nikolaus Joseph Freiherr von Jacquin (1727-1817). Jacquin wurde vom Leibarzt Maria Theresias, Gerard van Swieten, nach Wien geholt und kurz darauf zum Leiter einer Expedition in die Karibik (1754-1759) bestimmt, die zum Auftrag hatte, schöne und wohlriechende Pflanzen für den botanischen Garten in Schönbrunn zu besorgen. Nach dieser Reise, die seinen Bekanntheitsgrad enorm gesteigert hatte, stand er mit den bekanntesten Botanikern Europas in regem Briefaustausch. Und so schreibt ihm z. B. Carl von Linné, er solle vorsichtig sein und nicht soviel in den Alpen herumspazieren, sonst würde er noch krank werden und sterben, bevor er seine Resultate von seiner Westindienreise publiziert hat.[11] 1761 besuchte Jacquin zum ersten Mal den Schneeberg und war von dessen Blumenreichtum so entzückt, dass er in seiner *Enumeratio*[12] schreibt: *Flora hat in diesem Absturze die seltensten Gewächse wie in einem Schaukasten ihren Gönnern hinterlegt.* Bei seinen Schneebergwanderungen fand er im Jagdhaus der Grafen Hoyos am Gahns öfters Unterkunft.[13] 1773

9 Vgl. W. Kubitschek, Exkursionen des Carolus Clusius während seines Wiener Aufenthaltes, in: Jahrbuch für Altertumskunde 6, Wien 1913, 213.

10 Die Ergebnisse dieser Exkursionen veröffentlichte er in zwei berühmten Werken: Rariorum stirpium per Pannoniam et Austriam observatarum historia und Rariorum plantarum historia, Antwerpiae 1583 und 1601.

11 Vgl. Carl Linnaeus an Nicolaus Joseph Jacquin, 13. Dezember 1761, Letter L3015. The Linnaean correspondence, linnaeus.c18.net.

12 N. J. Jacquin, Enumeratio Stirpium Plerarumque, Quae Sponte Crescunt In AgroVindobonensi, Montibusque confinibus, Wien 1762.

13 K. Kollar, Schneeberg – Rax – Schneealpe, Reichenau [2]1997, 42.

veröffentlicht Jacquin den ersten Band seiner *Florae Austriacae.* Das Titelblatt ziert ein kolorierter Kupferstich, der die Schneeberglandschaft des Gahns darstellt.

Zu Jacquins späteren Schülern zählte auch Thaddäus Haenke (1761-1816), der als Mitreisender der Malaspina-Weltreise und dann in Bolivien Hervorragendes für die Wissenschaften leistete. In den Jahren 1786 und 1788 bereiste er Nieder- und Oberösterreich, mehrmals den Schneeberg.[14] Die sehr ergiebigen Resultate dieser botanischen Ausflüge beschrieb er in zwei gehaltvollen Aufsätzen in Jacquins *Collectanea ad botanicam etc.*[15] Eine Anekdote schildert, dass Jacquin, der Haenke sehr schätzte und förderte, sich immer wieder nach dessen Verbleib erkundigte: *Haenke, hieß es dann, habe einen Ausflug auf den Schneeberg gemacht; und gewöhnlich trat er wenige Tage darauf mit verbranntem Gesicht und bestaubten Kleidern, plötzlich in Jacquins Zimmer und langte aus einer Schachtel die Pflanzen hervor, welcher der Gegenstand des letzten Streites gewesen.*[16] Bei der Landbevölkerung in der Schneeberggegend war Haenke schon lange unter dem Namen *der kühne Steiger* bekannt und selten fand er einen Bergführer, weil kaum einer mit ihm Schritt zu halten vermochte und alle sich weigerten, ihm bei seinen tollkühnen Klettertouren zu folgen.

Drei große Botaniker nur als Beispiel für die frühe wissenschaftliche Erforschung des Schneeberggebiets.

Die älteste Beschreibung des Schneebergs finden wir in der Topographie von Niederösterreich, die Friedrich Weiskern im Jahre 1769 verfasste, wo zu lesen ist: [...] *der Schneeberg U.W.W. ein großer hoher Berg, hinter Neunkirchen, rechts der Strasse bey Glocknitz und dem Sömmering nordwärts gelegen, im Gebiete des Grafen von Hoyos zu Gutenstein. Er raget über alle umliegende hohe Gebirge so weit hervor, dass man ihn 8 Meilen davon, zu Wien, bey heiterem Wetter genau sehen kann. Ohngeachtet des ewigen Schnees, von dem er den Namen trägt, und welcher denselben auf allen Seiten bedecket, befinden sich zuoberst auf demselben die schönsten grünen, mit*

---

14 In den *Reliqiae Haenkeanae* befindet sich eine Biographie über Haenke, verfasst von Sternberg. Darin heißt es auch, dass Haenke zusammen mit seinem Bruder Aloys, der als Kavallerie Offizier in Oedenburg stationiert war, viele botanische Wanderungen, u. a. im Juli 1787 auf den Schneeberg unternommen hat.

15 N. J. Jacquin, Nicolai Josephi Jacquin Collectanea ad botanicam, chemiam, et historiam naturalem spectantia 2, Wien 1788, 3-96.

16 J. W. Ridler, Taschenbuch für die vaterländische Geschichte, Wien 1814, 192.

den besten Kräutern versehenen Alben; wohin die benachbarten Landleute aus Österreich und Steiermark im Mond Julio ihre Pferde und Rinder zu treiben, und solche bis in den September daselbst auf der Weide zu lassen pflegen [...].[17]

1799 veröffentlicht der damalige Professor an der theresianischen Ritterakademie Josef Anton Schultes sein Werk, *Über Reisen im Vaterlande zur Aufnahme der vaterländischen NaturGeschichte, an die adeliche Jugend* [...], worin er befürchtet, dass der benachbarte Schneeberg bald keine der vegetabilischen Schätze mehr vorzeigen wird können, die *Jacquin auf demselben fand, seit jährlich Botaniker in kleinen Caravanen dahin ziehen, um nicht blos Exemplare sondern Körbe voll Samen für ihre Hortulos herabzuholen.*[18] Er selbst unternahm bis 1802 sechs „Fußreisen" zum Schneeberg, sammelte seine Aufzeichnungen und um seine Landsleute zum Reisen in die Umgebung zu überreden, gab er 1802 einen der ersten Schneebergführer heraus.[19]

Franz Xaver Embel hatte schon 1801 seine *Fußreise von Wien über den Schneeberg. Wien, 1801* veröffentlicht, worin man jedoch Gebiets- und Lokalbezeichnungen, wie z. B. Gahns, Kaltwasser etc. vermisst. In seinem Vorwort heißt es: *Ich wünsche nicht, dass man diesen Aufsatz in der Erwartung zur Hand nimmt, darinn eine Beschreibung zu finden, welche den Gelehrten interessirt. Einestheils besitze ich weder die ausgedehnten Kenntnisse zu solch einem Werke, noch erübrigte ich von meinen Berufs= und häuslichen Arbeiten so viele Zeit, um etwas zu sammeln, das nur mich selbst ganz befriediget hätte; – Anderntheils habe ich auch anfänglich diesen Aufsatz nicht für ein gelehrtes, kritisches Publikum, sondern bloß für meine Freunde geschrieben, die, so hoffte ich, mir die Nachlässigkeiten im Vortrage verzeihen, und keine überspannten Erwartungen hegen würden* [...].

1803 folgen dann seine*: Schilderungen der Gebirgs-Gegenden um den Schneeberg in Oesterreich,* worin er bekrittelt, dass durch viele Reisebeschreibung ferne Länder und Weltteile bekannter sind, dass viele Menschen besser über die Südseeinseln und das Innere von Afrika unterrichtet sind, als über die nähere Umgebung ihrer Heimat: [...] *dass bisher nur wenige die Schönheiten und Wunder der Natur zu ahnden schienen, mit denen sich unser Vaterland schon in den so nahen Gegenden des*

---

17 F. Weiskern, Topographie von Niederösterreich, Wien 1769, 156.

18 J. A. Schultes, Über Reisen im Vaterlande zur Aufnahme der vaterländischen NaturGeschichte, an die adeliche Jugend, in der k. k. Theresian. Ritterakademie bey Gelegenheit des Endes des II. Jahr-Cursus nach ihrer Wiedererrichtung, Wien 1799, 75.

19 Die zweite vermehrte und verbesserte Auflage folgte 1807.

Schneeberges auszeichnet. Nur einigen gelehrten Männern, deren Schriften nur wieder Gelehrte lesen, sind sie bekannt, und von Malern werden sie besucht, deren Zeichnungen und Beobachtungen verborgen blieben […]. *In der gegenwärtigen Schrift habe ich mich bemüht, eine getreue lebhafte Schilderung der uns nahen Gebirgsgegenden um den Schneeberg zu geben. Ich wollte es versuchen, jene Leser, die warmes Gefühl für die Schönheiten der Natur haben, in diese wirklich schönen und merkwürdigen Gegenden zu führen, und sie dort auf alles aufmerksam zu machen, was Kopf und Herz angenehm beschäftigen und vergnügen kann.*[20]

In der Aufklärung beginnt die geistige Elite sich also vermehrt der Natur zuzuwenden, und im Biedermeier entdeckten dann die Künstler, vor allem die Maler und die Bürger den Reiz der Landschaften ihrer Heimat. Sie fuhren hinaus in den Wienerwald, das Alpenvorland, aber auch bis in hochalpine Bereiche.

Die hier besprochenen Exkursionen fanden noch in der Anfangszeit der allgemeinen Bergeuphorie statt, als noch nicht alles hinaus in die Natur drängte.

Rückschauend erinnerte sich EH Johann: *Zu Anfang des 19. Jahrhunderts waren unsere Gebirge vollkommen unbekannt; es gab keinen Touristen; ich war der erste, welcher von dem österreichischen Schneeberge aus die steiermärkischen Alpen sehend, mich dahin wandte* […] *und so nach und nach die Alpen* […] *kennen lernte.*[21] Damit betont Johann seinen Anspruch, vieles initiiert, aus eigenem Bemühen erkundet zu haben und nicht nur auf vorhandenen Pfaden einem Führer gefolgt zu sein. Diese Lebenseinstellung lässt sich bei ihm auch in anderen Bereichen nach verfolgen. Später hat er den Ausbau von Klettersteigen und den Bau und die Instandhaltung von Hütten, als Übernachtungsmöglichkeit für Wanderer und Bergsteiger, zu einem wichtigen Anliegen gemacht.

Um zu verstehen, warum die Erzherzöge so naturverbunden waren und solche Wanderfahrten zu ihrem jährlichen Reiserepertoire gehörten, ist ein kurzer Exkurs über ihre Erziehung notwendig.

---

20 F. X. Embel, Schilderungen der Gebirgs Gegenden um den Schneeberg in Oesterreich, Wien 1803, Vorerinnerung.

21 Zit. nach H. Lambauer, Mit Erzherzog Johann in den Bergen, in: Erzherzog Johann von Österreich 2, Graz 1982, 391.

Von EH Rainer ist im Gegensatz zu seinen berühmteren Brüdern, dem späteren Kaiser Franz II. (I.), EH Johann und EH Carl, biographisch sehr wenig bekannt.

Er wurde als 14. Kind (10. Sohn) des damaligen Großherzogs der Toskana und späteren Kaisers Leopold II. und der Maria Ludovica, Tochter des Königs Carl III. von Spanien, am 30. 9. 1783 in Pisa geboren. Wie seine Brüder erhielt er eine sorgfältige wissenschaftliche Ausbildung, in welcher der gründliche Unterricht in staatswissenschaftlichen und militärischen Fächern die Hauptsache bildete. Als Leopold nach dem plötzlichen Tod Josephs II. Kaiser wurde, musste er sich sofort nach Wien begeben, während seine jüngeren Kinder Anton, Johann, Rainer, Ludwig und Rudolf noch bis Mai 1790 in Florenz zurückblieben. Rainers erste große Reise war die Übersiedlung nach Wien, die ihn als Siebenjährigen zusammen mit den jüngeren Brüdern in zwei großen Reisewägen quer durch die Alpen in einer 15tägigen Fahrt nach Wien führte. Als 1792 innerhalb von fünf Monaten Leopold und seine Gemahlin verstarben, übernahm Franz II. (I.) die Oberaufsicht über die Erziehung seiner fünf jüngeren Brüder. Es existiert heute noch ein handgeschriebenes Erziehungskonzept des Kaisers mit eigenhändig eingetragenen Korrekturen aus der Zeit um 1793. Er bestimmte einen neuen Erzieher, in diesem Fall den General Baron Franz Alois Haager[22] zum Ajo (Hofmeister), für den er genaue Erziehungsgrundsätze festlegte. Dass ihm neben der Charakterfestigung auch die Herzensbildung ein großes Anliegen war, zeigt folgender Ausschnitt: […] *so muß des Ajo größte Sorge und Befliessenheit seyn gleich von der ersten Jugend Ihre Herzen zu bilden.* […] *so wäre es höchst schädlich, wenn die Herzen der Erzherzöge zu weich und zu sehr empfindsam gebildet, so wäre es eben schädlich, wenn solche zu hart und zu unempfindsam ausfallen, und gerathen sollten. Gute Beyspiele aus der Geschichte, denen Erzherzogen zeitlich das gute lieben, das nützliche davon erkennen zu machen, Selben entgegen wieder das schädliche eines harten,*

22 Franz Alois Haager von Altensteig (1722-1812) hatte noch unter Prinz Eugen gedient, war im Siebenjährigen Krieg schwer verwundet worden und aus dem aktiven Dienst ausgetreten. 1792 wurde er zum Feldmarschallleutnant befördert und anschließend Erzieher der fünf jüngeren Brüder des Kaisers.

unempfindsamen Herzens vorzustellen, und die üblen Folgen davon begreiflich zu machen, wird und kann von erwünschter Wirkung sein.[23]

In Wien bekam Rainer Vinzenz Stingel als Instruktor zugeteilt, der 1807 Kustos an der k. k. Hofbibliothek wurde.[24]

Das Erziehungsprinzip der Erzherzöge war die gewissenhafte Vorbereitung auf ihre jeweiligen Aufgaben. 1796 beschrieb der damals 14jährige EH Johann bereits seine Reise nach Mariazell, unter Zuziehung der bekannten Schriften darüber. Wie dadurch seine Beobachtungsgabe geschärft und auf alle, auch scheinbar bedeutungslose Dinge gelenkt wurde, schildert er selbst in seinem Tagebuch mit folgenden Worten: *Während der Reise suchte ich mir alles, was die Gegenden, in die ich gelangte, betraf, zu verschaffen und zu benützen;* [...] *nach allen genauen Besichtigungen jedes Gegenstandes blieb mir nur wenig Zeit, um alles, wenn auch nur mit Schlagwörtern, aufzuzeichnen. In späteren Reisen pflegte ich entweder während dem Fahren oder bey irgend einem Stillstand auf einzelne kleine Blätter Papier mit Reißbley das Gesehene, vorzüglich aber, was sich so leicht und schnell vergißt, aufzumerken; dieses habe ich in der Folge durch mein ganzes Leben geübt* [...],[25] und genauso hielt es wahrscheinlich EH Rainer. Geprägt durch das Gedankengut der Aufklärung, dem großen Interesse an den Naturwissenschaften, vor allem an der Botanik, war die Erziehung der Prinzen darauf ausgerichtet, sie schon in frühester Jugend mit allen Bereichen des praktischen Lebens in unmittelbare Berührung zu bringen, um so ihr Interesse für die wirtschaftlichen und sozialen Verhältnisse möglichst früh zu wecken.

Wenn wir nun die Berichte Rainers und Johanns und den des sie begleitenden Abbé Stütz, der in seinem posthum erschienenen *Mineralogischen Taschenbuch* diese Reise ebenfalls beschreibt, vergleichen, so ergibt sich ein ziemlich abgerundetes Bild dieser kleinen Inlandexpedition.

Voller Dankbarkeit schreibt Stütz: [...] *Diese Reise verdiente schon wegen der da vorkommenden merkwürdigen Gegenstände meine volle Aufmerk-*

---

23 HH StA, Familienakten 55, f. 83ff. zitiert nach B. Binder-Krieglstein (Hrsg.), Erzherzog Johann Gedächtnisausstellung, Graz 1959, 49.

24 Vgl J. Weichinger, Vincenz Stingel 1761-1815, in: Biblos 31, Wien 1982, 140-154, und V. Theiss, Leben und Wirken Erzherzog Johanns 1, Graz 1960, 42 u. 60.

25 Zit. nach V. v. Geramb, Erzherzog Johanns Bedeutung für die steirische Volkskunde, in: Das steiermärkische Landesmuseum Joanneum und seine Sammlungen 1811-1911, Graz 1911, 41.

samkeit; aber unvergesslich werden mir dabey die gnädige Herablassung der vortrefflichen Brüder unsers glorwürdigsten Monarchen, ihre Begierde, der Natur nachzuspüren, ihre gründliche Sachkenntnis, ihre reifen treffenden Beurtheilungen bleiben.[26]

Schneebergexpeditionen

Der erzherzogliche Konvoi fuhr also weiter von Neunkirchen über St. Johann und Sieding, besuchte das Schloss Stixenstein des Grafen Hoyos und gelangte in das enge Sierningtal: *An dem Bache liegt eine Reihe von Sägemühlen und Eisenhämmern ganz gedrängt so daß diese Gegend so wild als sie ist, dadurch ungemein belebt wird; das Traurige derselben verlohr sich auf einmal, als sich dieses Thal in dasselbe fruchtbare und angenehme Buchbergerthal erweiterte, an dessen Eingange das Dorf gleiches Nahmens liegt* [...].[27]

Höchstwahrscheinlich hatte man sich auf die Schneeberg-Reise sorgfältig vorbereitet, vielleicht auch die oben angeführten entsprechenden Werke Embels[28] und Schultes'[29], die ersten touristischen Führer dieser Gegend, studiert.

Puchberg mit seiner alten Burg und das aus einigen armseligen Hütten bestehende Schneebergdörfchen wurden inspiziert und *Nachmittags ging unser Zug, der mit allerley Geräthschaften und Nothwendigkeiten versehen, und aus 60 Personen, worunter mehrere Weiber, bestand, die theils mehr oder weniger schlecht beritten oder unberitten waren, mithin einer Karawane ziemlich ähnlich war, sogleich aufwärts einen sehr felsigten steilen Berg hinan; allmählich drangen wir in einen dichten sehr alten Buchenwald, der aber doch an manchen Orten durch das viele Kohlenbrennen sehr gelitten hatte, als uns auf einmal eine Blöße mit der angenehmen Aussicht auf das kleine Rohrbach überraschte; dieses Dörfchen liegt an einem Teiche in einer engen ganz mit Wald bewachsenen Bergschlucht, einsam und melancholisch; der Weg ging über den*

---

26 A. Stütz, Mineralogisches Taschenbuch. Enthaltend eine Oryctographie von Unterösterreich zum Gebrauche reisender Mineralogen. Hrsg. von J. G. Megerle v. Mühlfeld, Wien und Triest 1807, 116.

27 ÖNB/HAD Cod. Ser. n. 2023, fol. 3.

28 F. X. Embel, Fußreise von Wien über den Schneeberg, Wien 1801.

29 J. A. Schultes, Ausflüge nach dem Schneeberge in Unterösterreich. Ein Taschenbuch auf Reisen nach demselben, Wien 1802. Beide Reiseführer sind in der Sammlung Woldan vorhanden.

Hengst so fort bis zum Kaltwasser, einer aus dem Waxriegel, der ein Theil des Schneebergs ist, entspringenden Quelle, die die letzte, äusserst kalt und folglich für die Wanderer auf dem Schneeberg sehr gefährlich ist [...].[30]

Nach einem vierstündigen Aufstieg übernachtete man auf dem Pürschhof, einem hölzernen Jagdhaus am Gahns, welches von dem jagdbegeisterten Grafen Hoyos errichtet worden war:[31] [...] *er besteht in einem hölzernen Haus mit 6 Zimmern, worinnen man zwar Bequemlichkeit, aber nicht hinlänglichen Schutz gegen die gewöhnliche Nachtkälte findet. Sie sind mit uralten vergoldeten Ledertapeten ausspalirt, und mit einigen Tischen und Bänken versehen; hier wohnt beständig ein Jäger, und hier ist auch das gewöhnliche Nachtlager der Botaniker, die die hier herumliegenden an Subalpinen reichen Wiesen und den Schneeberg bereisen* [...].[32]

Immer wieder finden sich forst- und viehwirtschaftliche Betrachtungen in der Reiseerzählung eingestreut. Rainer ist gegen die Ochsenhaltung auf dem Schneeberg. Er meint, dass allein durch den großen [Butter-]Schmalzbedarf der Bevölkerung Wiens die Milchwirtschaft weitaus mehr Berechtigung hätte. Die Ochsenwirtschaft hinterließ jedoch in der lokalen Topographie ihre Spuren, wie *Ochsenhütte*, *Ochsenboden* etc.

Auch die Kohlenmeiler erfahren eine erzherzogliche Analyse, die sehr wohl ihre Schwachstellen hervorhebt. *Die Bauern brennen aus ihren jüngeren Bäumen Kohlen; die älteren lassen sie stehen, bis sie modernd vom Wind umgeworfen werden und dann ganz hinmodern; der Wind trägt hernach die Erde davon, so daß der Wald auf immer vernichtet wird, daher wären die vielen kahlen Flecke in den Waldungen, welche das Übel immer vergrößern; denn so wie der Wind einen größeren Spielraum bekömmt, so reißt er lange Strecken alter ohnehin fast nur auf Felsen stehender Bäume nieder, die unbenutzt verfaulen. Die Bäume besonders die Tannen werden ehe man sie zu Kohlen brennt, geschält und die Klafter Rinde den Gerbern zu 30-40 Groschen verkauft. Die Kohlen werden in kleinen liegenden Meilern gebrannt, welches viel zweckmäßiger als die Ungrische Verfahrensart in stehenden großen Meilern zu brennen ist,* [...] *ich*

---

30 ÖNB/HAD Cod. Ser. n. 2023, fol. 6,7.

31 Vgl. K. Leeder, Geschichte des Hauses Hoyos in Österreich 2, Wien 1914, 431: *Nichts ging aber dem Grafen über das Vergnügen der Jagd, und er widmete sich demselben vorzugsweise auf dem Gahns. Dort hatte er auf einer Lichtung, welche einen Ausblick auf den nahen Schneeberg und weithin auf die Berge gegen N bietet, ein hölzernes Wohnhaus erbaut und zum Teil mit alten, wahrscheinlich dem Schloße Stixenstein entnommenen Tapeten ausgeschmückt.*

32 ÖNB/HAD Cod. Ser. n. 2023, fol. 9.

bemerkte keine Erzeugung von Theer und Holzessig; Beyde gehen daher verlohren [...].[33]

Von der Bodenwiese, wo man botanisierte, ging es zurück zum Kaltwasser, über den Waxriegel Richtung Hochschneeberg: [...] *sonderbar war es daß von unsern Trägern, nachdem wir einem jeden nur ein Glas Wein geben liessen, dennoch während dem Marsche viele so trunken wurden, daß sie unterwegs liegen bleiben mußten.*[34] Im Felsen gab es immer wieder Nester voller Hornissen, worunter die Pferde sehr zu leiden hatten. Das Pferd des Abbé Stütz wurde derart gestochen, dass es den Reiter samt Sattel und Zaumzeug abwarf. Da dieser Zwischenfall ohne Schaden abging, durften die übrigen Expeditionsteilnehmer herzlich lachen.

Unter den am Ochsenboden, der sich zwischen Waxriegel und Gipfel befindet, wachsenden Alpenpflanzen werden folgende genannt: *Gentiana* (Enzian), *Soldanella* (Alpenglöckchen), *Achillea* (Scharfgarbe), *Campanula* (Glockenblume), *Erica*, *Saxifragia* (Steinbrech), etc. Man schlug hier die Zelte auf und machte drei große Feuer aus Krummholz, das durch seinen hohen Harzgehalt sehr hell brennt und einen guten Geruch verbreitet. *Es ward bald Abend; es erhob sich ein sehr kalter Wind vom Gipfel, wälzte dichte Nebelwolken auf uns herab, und sogleich ward es Nacht um uns; wir lagerten uns am hellen Feuer, nahmen unser Nachtmahl ein, und tranken geschmolzenen Schnee aus den Löchern; unsere Karawane that ein gleiches und ließ sich den eben angekommenen Wein recht sehr schmecken; Einer suchte den Andern aufzuheitern, und alles fing an lustig zu werden, welches uns nicht wenig ergötzte. Ein munterer Geselle setzte sich auf das Dach der Hütte und blies auf einer langen Pfeife allerley Tänze. Nachdem wir uns recht durchgewärmt hatten, legten wir uns in den Zelten schlafen; allein der feuchte Boden, die durchdringende Kälte, der heftige Wind, welcher die Zelte umzustürzen drohte, und der Lärm der bey den Feuern wachenden Bauern ließ uns nur wenige Stunden schlafen.*[35] Am nächsten Tag bekam man aber anstatt des Gipfels nur eine Nebelwand zu sehen. Man wartete bis 11 Uhr vormittags, der Nebel wurde jedoch immer dichter und so tastete man sich zum Gipfel weiter. Statt mit einer Aussicht, die nur von der vom *Montblanc oder Aetna* übertroffen wird, belohnt zu werden, zog die Gesellschaft *ganz nieder geschlagen und vor*

---

33 ÖNB/HAD Cod. Ser. n. 2023, fol. 11.
34 ÖNB/HAD Cod. Ser. n. 2023, fol. 13.
35 ÖNB/HAD Cod. Ser. n. 2023, fol. 16.

Kälte und Nässe gleich fieberhaft schauernd weiter; über den Kuhschneeberg und der Tränkwiese ging es nach Schwarzau. Man bemerkte unterwegs immer wieder Wald schädigende Angewohnheiten: Die Bauern zündeten in ihren Wäldern das Gras und Reisig an, um dadurch die Bäume auszudörren, die sie zum Kohlenbrennen brauchten; den Boden düngten sie mit Asche, bauten ihn nach einem Jahr an; dann zogen sie weiter und verfuhren mit einem andern Flecken auf gleiche Weise, dadurch wurde der Wald vernichtet; denn auf so einem abgesengten Flecke wuchs nur mehr schlechtes Gras und die Gefahr eines Waldbrandes war bei einfallendem Wind sehr groß.

Dann ging es weiter nach Schwarzau und durch das Höllental, das von der Schwarza durchflossen wird. Rainer notierte, dass der Forellenfang für die Herrschaft sehr einträglich ist und die Fische nach Baden zum Verkauf geschickt werden. Danach kam man ins Naßtal, wo das Holz bis Reichenau geschwemmt wurde. Das imposante Schauspiel wird hier genau geschildert. Man besuchte das große Gebirgsbauernhaus des Holzmeisters Hübmer,[36] wo die Gesellschaft mit schmackhaften *Griesklösern* bewirtet wurde. *Die hier in der ganzen Gegend angesiedelten Holzknechte und Holzhacker sind meistens aus dem Salzkammergute; sie sind starke gesunde sehr arbeitsame und rechtschaffene Leute; sie bewohnen hölzerne in den Wäldern zerstreute Häuschen; sie nähren sich fast blos von sehr fett zubereiteten Mehlspeisen; betrachten sich als ein abgesondertes Völkchen, und heyrathen blos Mädchen aus ihrer Mitte.*[37] Ein zweites Huebmersches Haus dient als Schule für die lutherischen Kinder und *besitzt einen Bücherschrank voll der neuesten Schulwerke.* Interessant ist hier, dass der katholische Erzherzog diese lutherische Gemeinde sehr positiv beurteilt. Völlig tolerant merkt er das Bemühen um die Verbesserungen der sozialen Verhältnisse dieser evangelischen Gemeinde aus ihren eigenen Reihen an. Eine handgezeichnete Kartenskizze, die dem Bericht beiliegt, zeigt die Huebmerische Niederlassung eingetragen, was wiederum die Hochschätzung für den

---

36 Georg Huebmer (* 1755 in Gosau – † 1833 in Naßwald), Schwemmmeister und Sozialreformer. Huebmer baute die Holzbringungsanlagen (Schwemmkanal) im Rax–Schneeberg–Gebiet. Über den Wiener Neustädter Kanal gelangte das Holz bis nach Wien. Im Volksmund war er als „Raxkönig“ bekannt. Er bemühte sich um die Verbesserung der sozialen Verhältnisse der evangelischen Holzknechte in Naßwald und Umgebung.

37 ÖNB/HAD Cod. Ser. n. 2023, fol. 26, 27.

fortschrittlichen Schwemmmeister widerspiegelt. Auf Pferden ging es dann zum Naßjoch, welches die Grenze zur Steiermark bildet. Hier warteten schon Führer und Gamsjäger und geleiteten den Tross auf die Schneealpe, wo sich damals um die 60 Schwaig- (Senn-) und Alpenhütten befanden. Die Mitglieder der kaiserlichen Familie wurden mit Böllerschüssen empfangen, die Schwaigerinnen überbrachten im Sonntagskleid Blumensträuße und kleine Schafkäsestücke. Bergwerksbeamte, Bergleute und andere Bewohner Neubergs waren zum Willkommen erschienen.

Man übernachtete hier und Rainer beschrieb das Innere der Hütten und Ställe. Tagwache war um halb vier, da die Brüder den Sonnenaufgang sehen wollten. So weckte Rainer seinen Bruder Anton und Abbé Stütz, und nach einer Stunde Anstieg war der Gipfel und damit eine außerordentliche Aussicht erreicht. *Nun sahen wir ein prächtiges und zugleich gewiß seltenes Schauspiel, die Thäler waren alle mit Nebel angefüllt, so daß alles einem Ocean glich; die hervorragenden Gipfel und Felsenrücken so vielen größern und kleinern Inseln in demselben glichen; bey Tagesanbruch war dessen Farbe grau, so wie es heller wurde hob sich der Nebel etwas, und als die Sonne gerade über den Gipfel des Schneeberges emporstieg, überzog den Nebel nach und nach ein Rosenroth, welches immer lebhafter ward; bald setzte ein sanfter Wind dieselben in eine allgemeine Bewegung, und nun glich er einem wogenden Meere; die Nebelwolken wälzten sich über einander an den verschiedenen Bergen fort; der Schneeberg sah wie ein feuerspeyender Vulkan aus, weil die Sonne gerade auf seinem Gipfel zu stehen schien; wir wurden mit Bewunderung erfüllt, aus der uns nur der von allen Seiten anstürmende Nebel, welcher uns ganz zu durchnetzen drohte, reissen konnte.*[38]

Nach einer erfolglosen Gamsjagd wurden aufgelassene Brauneisenstein-Gruben besichtigt. Abends unterhielten sich die Bauern mit Spielen, Tanz und Musik. *Die Viehwirtschaft ist hier ein wichtiger Erwerbszweig. Die Kühe werden mittags und abends gemolken, die Milch hat einen angenehmen Geruch von den Alpenkräutern, ist aber nicht sehr fett. Von abfallenden Resten werden die Schweine gemästet; zum Unterstreuen bedienen sich die Schwaigerinnen des Aconitum Napellus,*[39] *welches in großer Menge um die Schwaighütten fortkömmt, und da es damals gerade in der Blüte stand, die vom schönsten Blau ist, einen prächtigen Anblick gewährt, der einem unbe-*

38 ÖNB/HAD Cod. Ser. n. 2023, fol. 34.
39 gemeiner Eisenhut.

nommen bleibt, weil sie von den Kühen vermieden werden.[40] Ob dem EH die Giftigkeit dieser Pflanze bekannt war, geht aus dem Text nicht eindeutig hervor, anderseits bemerkte er sehr wohl, dass der Eisenhut von den Kühen verschont blieb. Weiter ging es dann nach Neuberg, wo sie Quartier im ehemaligen Kloster bezogen. Wieder wurden die Mitglieder der kaiserlichen Familie von Beamten und Bergleuten festlich empfangen. Bergwerke (Witherit und Spateisenstein) und Schmelzen und Zerrenhämmer werden besichtigt, die Vorteile der steirischen Hochöfen und die Nachteile der hier verwendeten Floßöfen vermerkt.

Danach fuhr man zur neu angelegten Gewehrlauffabrik[41] und der Kugelhütte, wo die neuen Produkte gleich ausprobiert wurden. Eine ehemalige Kaserne bot Unterkunft für die Arbeiter. Hier zeigt sich wiederum die soziale Ader des Erzherzogs, er ist sehr besorgt um die Gesundheit der Werktätigen: *Die Kaserne ist auch sehr klein, und 44 derselben liegen in 7 kleinen Stuben zusammengedrängt; Da nun die Krankheiten wegen der winterlichen Feuchtigkeit häufig sind, und hier kein Spital ist, so ist der Ansteckung bey Krankheiten gar nicht vorgebeugt.*[42]

Zusammen mit Abbé Stütz sammelt Rainer Mineralien. Aber auch die Kunstgeschichte findet ihren Platz in der Niederschrift. Die gotische Kirche von Neuberg und ihr holzgeschnitzter Flügelaltar aus dem 14. Jahrhundert werden einer eingehenden Betrachtung unterzogen. Der Neuberger Dom war unter Joseph II. säkularisiert worden: […] *man hatte nun daraus ein Magazin gemacht, daher waren alle Ornamente und die für eine so lange Zeit so gut erhaltene Freskomalereyen, ganz nach dem damaligen Geschmack ohngefähr wie jetzt die Malereyen der Chinesen, zum Theile vom Staube unkenntlich gemacht. Im alten Refektorium sahen wir die alten Altäre, die man bey der Aufhebung des Klosters aus der Kirche nahm; sie sind alle sehr alt; mit Flügelthüren versehen; gehen aber hier völlig zugrunde, da sie doch als schöne Alterthümer eine vorzügliche Sorge verdienten; es fehlen schon fast auf jedem einige Statuen oder Thürflügel.*[43]

---

40 ÖNB/HAD Cod. Ser. n. 2023, fol. 40.

41 1798/99 errichtete man in Lanau eine große Rohr- und Gewehrfabrik mit einem eigenen Holzrechen an der Mürz, um die Brennstoffversorgung zu sichern; 1801 wurden bereits 5577 Läufe für Infanteriegewehre, für 749 Husarenkarabiner und für 209 Pistolen hergestellt. Vgl. O. Pickl, Geschichte des Ortes und des Klosters Neuberg an der Mürz, Neuberg 1966, 306.

42 ÖNB/HAD Cod. Ser. n. 2023, fol. 47.

43 ÖNB/HAD Cod. Ser. n. 2023, fol. 50.

Danach folgt die Schilderung einer Bauernhochzeit mit Brautentführung. Man fährt weiter die Mürz entlang bis zum Wasserfall am *toten Weibl* und kehrt über Mürzsteg nach Neuberg zurück.

Am nächsten Tag besichtigen sie den Hochofen in Veitsch, dann das Alaunwerk bei Krieglach, wo sie vom Eigentümer, Baron Riegersperg, geführt werden. Das Gefolge wird hier entlassen, und man fährt heimwärts die Poststrasse über den Semmering, Maria Schutz, wo die Gipsbrüche und -brennereien besichtigt werden, nach Schottwien. Dort wird im Posthause soupiert und um halb sieben Uhr morgens ist man wieder in Schönbrunn.

Ein Jahr später musste man die Schneebergtour wegen Schlechtwettereinbruchs absagen. Aber 1805 war es dann wieder soweit.

Bei anhaltendem Schönwetter startete man zu dritt[44] abends am 25. Juni und war in den Morgenstunden in Neunkirchen. Dann ging es an der Ruine Stixenstein vorbei nach Puchberg, wo man Träger und Reitpferde bestellt hatte, und dort vom Schneebergdörfl den Hengst hinauf zum Schneeberg. Rainer beschreibt den blühenden Frauenschuh, der hier in Mengen wächst, und die zarte Röte des Seidelbasts. Die Brüder nehmen diesmal einen schwierigeren Aufstieg und treffen sich dann mit ihrem Gefolge bei der Ochsenhütte. Sie teilen sich hier auf, um die blühenden Alpenpflanzen zu sammeln, sind jedoch etwas enttäuscht, da die Vegetation oben noch nicht so weit fortgeschritten war wie in der Ebene. Trotzdem wird alles genau aufgelistet und unterschiedliche Höhenpunkte mit dem Barometer gemessen.[45]

Abends werden dann die von den Bauern schon vorbereiteten Holzstösse angezündet, um die versprochenen Feuersignale bis Wien senden zu können: *Während dieser Streifereyen, da es schon gegen Abend ging, eilten wir auf den Waxriegel, um von da und bey voller Nacht, wie wir es in Wien versprochen hatten, Feuer-Signale zu geben. Nebel zogen sich nun schon aus den Thälern herauf; der Gipfel des Berges war bald in dichten Wolken umhüllt, bald schien wieder die Sonne aus dessen mit Schnee bedeckten Scheitel. Leichte Nebel überzogen uns Augenblicke weise und bildeten völlige Nacht um*

---

44 Die Brüder werden nicht namentlich genannt.

45 Am Ende dieses Berichtes ist ein Verzeichnis der gesammelten Pflanzen, sowie eine Tabelle mit Barometermessungen zu Höhenbestimmungen angefügt.

uns; Bald verschwand sie wieder gänzlich, welches einen sonderbaren Effekt gewährte; so erreichten wir nach und nach in der Dämmerung den Waxriegel, wo die Bauern schon einen großen Stoß von Krummholz aufgerichtet hatten. Eine herrliche Aussicht über alle Vorgebirge Österreichs, über die Ebene Ungarns, und im Hintergrund über zahlreiche Alpenketten breitete sich vor unsern Augen aus […].[46] Es regnete und man beeilte sich, zur Ochsenhütte zurückzukehren. *Nun fing erst das wahre Elend an: da die Hütte sehr klein ist* […]*; wir dreye legten uns auf die schmale herum gehende hölzerne Bank; drey legten sich zusammengekrümmt auf den Boden, und die übrigen mußten die ganze Nacht einen heftigen Regen aushalten. Die unbequeme Lage, der durch die Ritzen hineintropfende Regen und die Kälte erlaubte uns nicht zu schlafen, jedoch machte es uns manchmal sehr Lachen. Nachdem es endlich Tag geworden war, sahen wir daß es nicht mehr regnete, wir gingen daher zum Feuer uns zu wärmen.*

Am nächsten Tag beschloss ein Teil, den bequemeren Weg nach Reichenau zu nehmen, die anderen, darunter Rainer, wollten unbedingt den Gipfel besteigen. Eiskalter Wind, Nässe und Nebel verhinderten leider den erwarteten herrlichen Rundblick. Man ging dann über den Kuhschneeberg hinunter nach Schwarzau, dann durch das gefürchtete Höllental: *Hier ereilte uns die Nacht, wobey es zu regnen anfing, wir beschloßen jedoch getrost unsern Weg fortzusetzen: immer enger wurde nun das Thal, meistens zog sich der nur mit Mühe nun zu findende Weg im dichten Wald; zahlreiche Brücken brachten uns bald auf die eine, bald auf die andere Seite, wo wir immer tief unter uns das Wasser rauschen hörten; bey der vorletzten Brücke zeichnete sich eine besonders schöne Felsengruppe aus, die kühn über den Fluß hängt und ganz mit Moos bewachsen ist. Lange tappten wir so im Finstern herum, es wollte gar nicht endigen, bis wir von Ferne Fackeln erblickten, die man uns von Reichenau entgegenschickte. Unser Muth der schon zu sinken anfing, erhob sich mächtig bey diesem Anblicke und bald hatten wir einige Bauern erreicht, die uns mit Fackeln versehen zum weitern Wegweiser dienten; noch lange jedoch zog sich der Weg durch den Wald bis wir den Kaiserbrunnen erreichten. In einer großen Höhle die die linke Bergwand des Schneeberges bildet; quillt aus einem ziemlich großen senkrecht in die Tiefe gehenden Loche, ein starker Bach hervor; mit Heftigkeit drängt er sich zwischen den Felsen durch, und stürzt sich in die Schwarza. Über diese ist ein großes hölzernes Haus gebauet, welches an dem Orte wo das Wasser hervorquillt, eine Öffnung im Fußboden hat. Diese Quelle wird der Kaiserbrunnen genannt: weil vorzeiten dieses Wasser wegen seiner vorzüglichen Reinheit zum Gebrauche des Kaisers nach Wien geführt wurde; es*

---

46 ÖNB/HAD Cod. Ser. n. 2026, fol. 6.

ist äußerst kalt, besonders rein und enthält manchmal auch Forellen, die von trefflichen Geschmacke sind. Nachdem wir die Quelle besucht und uns zugleich über das darüber gebaute Haus geärgert hatten, welches alle Untersuchung hindert, so setzten wir uns in die hier befindlichen Wägen, welche am Anfang des Weges auf uns warteten und setzten unsern Weg weiter fort.[47]

Die Nacht verbrachten sie im Schloss Reichenau und am nächsten Tag fuhr man über Payerbach nach Gloggnitz. Auf dem Weg nach Schottwien besuchte man noch die k. k. Smalten-Fabrik[48] bei Schlögelmühle, die eine detailreiche Beschreibung erhielt: *Sie ist sehr ansehnlich: zuerst ging es zu der Gifthütte, die nun eben nicht im Gange war, sie ist in einem ansehnlichen Gebäude, der Ofen ist sehr gros; er besteht aus einem sehr flachen niedrigen Gewölbe mit einem Roste, von diesem ziehet sich der Rauchfang in mehrere Kanäle, und endlich in einige große hinten angebaute Kammern; das rote Kobalt-Erz wird nun mit Holz darinnen eingetragen und stark geröstet, dadurch wird der darinnen enthaltene Arsenik verflüchtiget: der heftige Luftzug treibt nun diesen Dunst, der sonst für die Arbeiter tödlich wäre – nachdem der Ofen von vorne verschlossen worden, durch die mancherley gekrümmten Kanäle in die Kammer, an deren Wände nun sich der Arsenik-Kalk als ein graues Pulver ansezt; ist nun das Feuer im Ofen ausgegangen, so werden nach einigen Tagen die Zugfenster, die bey jeder Operation genau verschlossen werden, geöffnet, und der Arsenik in Gestalt eines grauen Pulvers gesammelt. Der Ofen ist ganz neu und so zweckmäßig gebauet, daß die Arbeiter wenig dabey leiden; jedoch erreicht keiner ein hohes Alter. Durch diese Operazion wird das Erz befreiet und derselbe gewonnen* […].

Nun besuchten wir die Fabrik selbst: die Schmalte wird wie bekannt aus Kobalt, Kiesel und Potasche gemacht […]. *In drey Brennöfen wird nun die Schmalte erzeugt, sie sind ganz wie die Glasöfen gebauet, mitten ist das Feuer, und auf jeder mit einer Öffnung versehenen Seite, sind drey Tiegel; sie werden von feuerfesten Thon, der von Joachimsthal kömmt, hier erzeugt und gebauet; sie dauern doch von 3 bis 10 Monaten, wenn immer gebrannt wird* […]. *Dieses dunkelblaue Kobaltglas kömmt nun in ein Pochwerk*[49] *mit 6 Schießern wo es fein gepocht wird.* […] *nun kommt es in die Mühlen, deren hier 14 sind, dort wird es zwischen zwey großen Mühlsteinen die horizontal wie bey den*

---

47 ÖNB/HAD Cod. Ser. n. 2026, fol. 10.

48 Blaufarbenfabrik.

49 Ein Pochwerk war eine dem Zerkleinern von Erzen dienende Maschine. Über eine Welle mit eingelassenen Zapfen wurden senkrechte Balken (Schießer), an deren Ende sich eiserne Pochschuhe befanden, kurz angehoben und auf das zu pochende Material fallengelassen.

Mahlmühlen laufen, auf welche immer Wasser fließt, welches das fein gemahlene wegspület, gemahlen, dann in einem Schlemmherde bis in den Sumpf geschlemmt, dieses wird nach den verschiedenen Gattungen abgetheilt, auf lange Bretter in stark geheizten Stuben getrocknet und endlich zum Verkauf in kleine Fäßchen gepackt. Es werden 11 Gattungen von Schmalte, die nachdem sie feiner – auch blässer in der Farbe ist – die alle Abstufungen vom höchsten Azurblau bis zum hellen Grau durchlaufen und von vorzüglicher Schönheit sind, erzeugt. Es werden jährlich 5000 Zentner davon gewonnen; der Absatz ist reissend davon, meistens ins teutsche Reich und Niederlande. Nächstdem wird noch aus dem Reste in den Tiegeln schöner blauer Streusand erzeugt; das übrige Metall mit Nuzen verkauft und zwey Gattungen weissen Arsenik gewonnen, der auch von vorzüglicher Güte ist. Diese Fabrik die sehr zweckmäßig eingerichtet und in der besten Ordnung ist, trägt jährlich 12.000 Gulden […].[50]

Das Mittagessen fand in Neunkirchen statt. Abends war man wieder in Schönbrunn.

Fazit

Das Erwandern eines Gebietes bedeutete für die Erzherzöge zugleich ein „Erschauen“, ein Kennenlernen, ein Erfahren, ein Bewusstwerden der jeweiligen Region. Nicht nur die Schönheit der Landschaft wurde von ihnen aufgesogen, Forst- und Landwirtschaft, Bergbau, beginnende Industrialisierung, soziale Zustände, Kunstgeschichte alles wurde wachsam beobachtet, niedergeschrieben und immer wieder reflektiert. Man fuhr nicht gemütlich in der Kutsche durch die Gegend, obwohl diese und Pferde immer wieder für bestimmte Streckenabschnitte zur Verfügung standen, man nahm viele Unannehmlichkeiten auf sich: schlechtes Wetter, schlechte Steige, elende Unterkünfte, karge Kost, um ein unverfälschtes Bild der Region und ihrer Bewohner erhalten zu können.

Der zur Dokumentation mitgeführte Kammermaler hatte den Auftrag, den jeweiligen Status quo festzuhalten. Als Hofbediensteter arbeitete der Künstler gegen monatliches Entgelt und Naturalleistungen für seinen Auftraggeber und ersetzte die heutige Digitalkamera.

---

50 ÖNB/HAD Cod. Ser. n. 2026, fol. 14.

Wie wir hier lesen können, war Rainer offen für die Schönheiten der Natur. Seine dargestellten Momentaufnahmen, z. B. einer Wettersituation oder seine Landschaftsschilderungen, sind sehr poetisch, ausdrucksstark und zeugen von einer großen Sensibilisierung gegenüber seiner Umwelt. Merkbar ist das erwachende wissenschaftliche Interesse an den ökonomischen Ressourcen im Zusammenhang mit einem erwachenden Lokalpatriotismus.

Die Abhängigkeit der Industrialisierung von der Holz- und Wasserwirtschaft ist in den Berichten spürbar. Kahlschläge, Verschwendung von Holz und Misswirtschaft im forstlichen Betrieb werden angemerkt. Holzschwemmen, Holzriesen, Eisenhämmer, Rechen, Floßöfen, Hochöfen, Alaunwerk, Smalten-Fabrik, alles wird genau beschrieben und zeugt von EH Rainers großem Interesse an der Technologie. Mit dem stetigen Ansteigen der Nachfrage nach Eisen und Stahl sowie durch veränderte Verarbeitungsverfahren trat ein fühlbarer Mangel an Holzkohle auf. Auch kristallisierte sich Wien mit seiner rasch wachsenden Einwohnerzahl als größter Holzverbraucher des Landes heraus. Es gab daher zahlreiche Versuche, althergebrachten Brennstoff durch Mineralkohle zu ersetzen. Alte aufgelassene Bergwerke wurden neu unter- und besucht.

Andreas Stütz, ein Teilnehmer an dieser Exkursion, schrieb enttäuscht, dass er geglaubt habe, der Schneeberg und seine Umgebung sei reich an geologischen Vorkommen, stattdessen bestehe der Schneeberg nur aus Kalkstein.[51] Außerdem lobt er EH Rainer wegen dessen mineralogischer Kompetenz.

Die Botanik nimmt natürlich einen breiten Raum in den Notizen ein. Immer wieder wird das Pflanzensammeln erwähnt und die lateinischen Namen der hier vorkommenden Blumen notiert. Franz Bredemayer, der kaiserliche Hofgärtner, begleitete die kaiserliche Gruppe, und die Wiener Zeitung, die 1802 über diese Exkursion berichtet, meldet, dass deren Teilnehmer eine *ganze Wagenladung für die Botanik merkwürdiger Kräuter sammeln, und nach Wien liefern ließen.* An seinen Bericht der 2. Schneebergtour 1805 fügte Rainer ein Verzeichnis der gesammelten Pflanzen an.

Am Ende des – mit noch weiteren Exkursionsberichten Rainers zusammengebundenen – Handschriftenkonvolutes befinden sich auch zwei handgezeichnete unsignierte Karten zu oben beschriebe-

51 A. Stütz, Mineralogisches Taschenbuch, 118.

nen Berichten. Die eineist die *Karte der Gegend vom Schneeberge*, die andere zeigt mit einer leeren Titelkartusche das Gebiet um Neuberg zwischen Schneealpe, Mürzzuschlag und der Veitschalpe. Es handelt sich um lavierte Federzeichnungen mit Betonung der Orographie; das Flusssystem ist blau eingetragen, die Ansiedlungen, auch einzelne Gehöfte rot und die Wege braun gekennzeichnet.[52] Die Vermutung liegt nahe, dass EH Rainer diese erstaunlich genauen kartographischen Skizzen selbst angefertigt hat, da sie bei seinen persönlichen Aufzeichnungen liegen, was wiederum das kaiserliche Erziehungsprinzip, bei genauer Beobachtung vieler Details den Überblick nicht zu verlieren, unterstreichen würde. Die Gesamtschau über ein Gebiet hatte für die Erzherzöge den gleichen Rang wie die kleinräumliche Erforschung, Erkundung.

Offene Augen für soziale Zustände der Landbevölkerung sind immer wieder in den Reisenotizen spürbar. Leider gibt es kaum persönliche Anmerkungen Rainers, z.B. über das Verhältnis zu seinen Brüdern während der Reise, über die verwendeten Gerätschaften, über die Reisekleidung usf. Schultes hingegen gibt in seinem Führer folgende Bekleidungstipps: […] *die beste Kleidung ist ein leichter Rock von Tuch oder ein Reitjäckchen und ein Mantel. Letzterer ist, zumahl bey Fussreisen, beynahe unentbehrlich: er dient nicht bloss zum Schutze bey den durchdringenden oft plötzlich einfallenden Nebelregen auf den Alpen, sondern auch Nachts zur Decke auf dem Stroh, das das gewöhnliche Nachtlager in den Dörfern ist. Und in dieser Hinsicht taugen die hochgepriesenen Mäntel von Wachstaffet oder Wachsleinwand nichts.*[53]

Alljährlich unternahm Rainer Ausflüge und Reisen in den Alpen-, Sudeten- oder Karpatenraum, um Natur und Menschen in verschiedenen Teilen der Monarchie durch Autopsie kennen zu lernen. Viele von ihm exakt und flüssig abgefasste Exkursionsberichte beweisen neben intensiver Neigung zu naturwissenschaftlichen Studien ein frühzeitig eminentes Verständnis für wirtschaftlich-sozialen Fragen, wie einige der hier angeführten Beispiele dies anschaulich vermitteln.

---

52 Farbtafeln S. 549 und 550: ÖNB/HAD Cod. Ser. n. 2023 Anhang 1 und 2.

53 J. A. Schultes, Ausflüge nach dem Schneeberge in Unterösterreich, Wien ^{2}1807, XIV.

Quellen- und Literaturverzeichnis

Binder-Krieglstein B. (Hrsg.), Erzherzog Johann Gedächtnisausstellung, Graz 1959.

Embel F. X., Fußreise von Wien über den Schneeberg, Wien 1801.

Embel F. X., Schilderungen der Gebirgsgegenden um den Schneeberg in Oesterreich, Wien 1803.

Festschrift anlässlich der 400jährigen Wiederkehr der wissenschaftlichen Tätigkeit von Carolus Clusius im pannonischen Raum, Eisenstadt 1973 (Burgenländische Forschungen, Sonderheft 5).

Geramb V. v., Erzherzog Johanns Bedeutung für die steirische Volkskunde, in: Das steiermärkische Landesmuseum Joanneum und seine Sammlungen 1811-1911, Graz 1911, 39-44.

Hoffmann R., Wissenstransfer durch Netzwerkbildung. Karl Erenbert von Moll und die Anfänge der wissenschaftlichen Landeskunde im Erzstift Salzburg, in: Scheutz M. – Schmale W.- Stefanová D. (Hrsg.), Orte des Wissens (Jahrbuch der Österreichischen Gesellschaft zur Erforschung des 18. Jahrhunderts 18/19), Bochum 2004, 135-151.

Kempf H., Die Flora des Schneeberges in Nieder-Oesterreich, Wien 1882.

Klingenstein G. (Hrsg.), Erzherzog Johann von Österreich, Graz 1982.

Kubitschek W, Exkursion des Carolus Clusius während seines Wiener Aufenthaltes, in: Jahrbuch für Altertumskunde, Wien 1913, 212-216.

Lambauer H., Erzherzog Johann als Bergsteiger und Naturfreund. Sonderausstellung, Bad Aussee 1982.

Leeder K., Geschichte des Hauses Hoyos, Bd. 1-2 , Wien 1914.

Leeder K., Johann Ernest Graf von Hoyos-Sprinzenstein, Wien 1902.

Mittermüller F., Holzkohle für Innerberg. Zur Brennstoffversorgung eines Reviers vom 16. bis zum 19. Jahrhundert, Diplomarbeit, Graz 1994.

Moll C. E. v., Mittheilungen aus seinem Briefwechsel. Prodromus seiner Selbstbiographie, Abt. 1-4, Augsburg 1829-1835.

Neilreich A., Geschichte der Botanik in Nieder-Österreich, in: Verhandlungen des zoologisch-botanischen Vereins in Wien 5, Wien 1855.

Österreichische Industriegeschichte GmbH, Linz (Hrsg.), Österreichische Industriegeschichte 1700 bis 1848. Die vorhandene Chance, Bd. 1, Wien 2003.

Pickl O., Geschichte des Ortes und Klosters Neuberg an der Mürz, Neuberg 1966.

Riegler J. (Hrsg.), Erzherzog Johann – Mensch und Mythos (Veröffentlichungen des Steiermärkischen Landesarchivs 37), Graz 2009.

Schultes J. A., Ausflüge nach dem Schneeberge in Unterösterreich. Ein Taschenbuch auf Reisen nach demselben, Wien 1802 (Nachdruck 1982).

Schultes J. A., Ausflüge nach dem Schneeberge in Unterösterreich. Ein Taschenbuch auf Reisen nach demselben. Zweyte vermehrte, verbesserte, mit Kupfern verschönerte Auflage, Wien 1807.
Schultes J. A., Über Reisen im Vaterlande zur Aufnahme der vaterländischen NaturGeschichte, an die adeliche Jugend, in der k. k. Theresian. Ritterakademie bey Gelegenheit des Endes des II. Jahr-Cursus nach ihrer Wiedererrichtung, Wien 1799.
Silberstein A., Land und Leute im Naßwalde, Wien 1868.
Stütz A., Mineralogisches Taschenbuch. Enthaltend eine Oryctographie von Unterösterreich zum Gebrauche reisender Mineralogen. Hrsg. von J. G. Megerle v. Mühlfeld, Wien und Triest 1807.
Theiß V., Leben und Wirken Erzherzog Johanns 1: Kindheit und Jugend (1782-1805) (Forschungen zur geschichtlichen Landeskunde der Steiermark 17), Graz 1960.
Weichinger J., Vincenz Stingel 1761-1815, in: Biblos 31, Wien 1982, 140-154.
Wertheimer E. (Hrsg.), Zwei Denkschriften Erzherzog Rainers aus den Jahren 1808 und 1809, Wien 1892.

Zur Autorin

Helga Hühnel, Studium der Geschichte und Volkskunde an der Universität Wien, 1981 Promotion, Studienassistentin bei Prof. Günther Hamann, wissenschaftliche Bibliothekarin, stellvertretende Leiterin der Kartensammlung der Österreichischen Nationalbibliothek, zahlreiche Publikationen zur Geschichte der Kartographie und zur Geschichte der Reisen sowie Mitarbeit an einschlägigen Ausstellungen.

Kontakt: Österreichischen Nationalbibliothek – Kartensammlung, Josefsplatz 1, A-1015 Wien; helga.huehnel@onb.ac.at

Julius Payer,
Die österreichisch-ungarische
Nordpol-Expedition,
Wien 1876

Marianne Klemun, Wien

Die Österreichisch-Ungarische Nordpolexpedition (1872-1874)

Euphorie der Gemeinsamkeit und die Positionierung der Akademie der Wissenschaften (Wien) im Vorfeld der Expeditionsplanung

Die Österreichisch-Ungarische Nordpol-Expedition[1] (1872-1874) war nicht nur wegen der Auffindung eines neuen Territoriums (des nach dem Herrscher programmatisch benannten Franz-Josef-Landes), sondern wegen des allgemeinen Beifalls und der ausdrücklichen Befürwortung durch viele maßgebende Institutionen, Einrichtungen, Vereine und Einzelpersonen eines der öffentlich erfolgreichsten unter vielen ähnlichen Projekten, die in der Habsburgermonarchie zustande gekommen waren. Für die außerordentliche

---

1 Ihr wichtigster Vertreter Carl Weyprecht begründete den eigentlich unkorrekten Namen der Expedition in einem Brief: *Ich habe die Expedition, die anfangs bloss ‚Österreichische' hiess, ‚Österreichisch-ungarische' genannt, erstens: weil ich unter der österreichisch-ungarischen Mercantilflagge segelte; zweitens weil Graf Edmund Zichy – ein edler Ungar – so viel Förderndes geleistet hat; drittens weil der Bordarzt Kepes ein Ungar war, und viertens, was ich eigentlich erstens nennen sollte, weil ich mir dachte, es werde die Ungarn freuen, dass die ungarische Seebehörde in Fiume die Anwerbung der Mannschaft so freundlich zugestanden hat, und weil sogar von einem ungarischen Nationalgeschenke die Rede war. Nun höre ich, dass sich dieses auf 120 fl, beschränkt,* […] Brief Weyprechts an Littrow (März 1875), abgedruckt in: F. Berger – B. Besser – R. A. Krause (Hrsg.), Carl Weyprecht (1838-1881), Wien 2008, 434.

Aktivität war der Kontinentalmacht erstmals unerwartet Landgewinn in arktischer Region beschert worden. Davon unabhängig zeigte sich jedoch bereits während der Planung der Hauptexpedition eine wie nie zuvor dagewesene Erwartungshaltung für das Projekt, was zwar mit der allgemeinen Popularität des arktischen Zieles[2] korrespondierte, aber besonders auch der symbolischen Bezugnahme auf hochrangige Wissenschaft zuzuschreiben ist.

Aufgabe der Expedition waren die Erreichung der nördlichen Barents-See bis in das eigentliche Eismeer und die Aufsuchung der Zentralarktis. Die Frage der nordöstlichen Passage nach Ostasien, die seit dem 16. Jahrhundert nahezu alle seefahrenden Mächte beschäftigt hatte,[3] schien nach einer Vorexpedition lösbar geworden zu sein. Für eine weitere Unternehmung engagierten sich Privatpersonen wie Graf Hans Wilzcek (1837-1922) sowie die unterschiedlichsten öffentlichen Einrichtungen und Vereinigungen aus vielen Territorien des Reiches. In einer nationalen gemeinschaftlichen Anstrengung, die sich gleich einem „Selbstkolonialisierungsakt"[4] gestaltete, wurden die Ressourcen für die Fahrt zu einem größeren Teil öffentlich und zu einem geringeren Teil staatlich zustande gebracht. Das machte die Basis der Popularität des Unternehmens aus. Diese Geschichte ist bestens bekannt und historisch bereits vielfach aufgearbeitet.[5] Dass

---

2 Vgl dazu: R. A. Potter, The Frozen North in Visual Culture, 1818-1875, Seattle/London 2007.

3 Vgl. J. Dörflinger, Die Nordwestpassage – Theorien von 1731-1823, ungedr. Phil. Diss, Wien 1969, 4 Bde.

4 Ruthner spricht bezüglich des in der schönen Literatur thematisierten Fremden im eigenen Staat von einem „Binnenkolonialismus". Vgl. C. Ruthner, „K. u. k. postcolonial"? Für eine neue Lesart der österreichischen (und benachbarten) Literatur/en, in: W. Müller-Funk – P. Plener – C. Ruthner (Hrsg.), Kakanien revisited, das Eigene und das Fremde (in) der österreichisch-ungarischen Monarchie, Tübingen/Basel 2002, 93-103, hier: 98.

5 Aus der reichen Literatur bes.: Österreichische Nationalbibliothek (Hrsg.), 100 Jahre Franz Josefs-Land. Zur Erinnerung an die Entdeckungsreise der Österreichisch-Ungarischen Nordpol-Expedition 1872-1874 unter Julius von Payer und Carl Weyprecht. Ausstellung im Prunksaal der Österreichischen Nationalbibliothek, Katalog (Biblos-Schriften 75), Wien 1973; Heeresgeschichtliches Museum Wien (Hrsg.), Die Schrecken des Eises und der Finsternis. Österreich und die Arktis. Katalog zur Sonderausstellung, Heeresgeschichtliches Museum, Wien 1996; E. Mazzoli, Dall'Adriatico ai ghiacci. Ufficiali dell'Austria-Ungheria con i loro marinai istriani, fiumani e dalmati

die Hauptexpedition jedoch nicht erst infolge ihrer Erfolge,[6] sondern bereits in ihrer Planungsphase als multifunktionale Hoffnungsträgerin fungierte, ist bisher kaum betont worden. In den habsburgischen Ländern erfüllten die Wissenschaften das, was die Politik bezüglich der nationalen Vielfalt zwar intendierte, aber nicht zustande gebracht hatte: eine einigende Sinngebung. Die Wissenschaften standen einerseits für den zivilisatorischen Aufstieg der Menschheit und definierten sich andererseits als ideale Projektionsfläche der Aufopferung für das eigene Land. Universalismus und Patriotismus konnten sich hier utopisch verbünden. Intellektuelle Vorherrschaft und allgemeine Veredelung des Geistes standen einander als hehre Begleiterscheinungen des Fortschritts nicht im Wege. So schrieb Wilhelm Haidinger Akademiemitglied und Direktor der geologischen Reichsanstalt an seinen Kollegen Franz von Hauer: *Den Fortschritt der Wissenschaften zu fördern. Die Lage, die Umgebung gab Anlaß zu gemeinschaftlichen Entschlüssen, deren Folgen immer wieder von einem Schritte zum anderen führten. Gleichzeitig mit dem was uns umgab sahen wir aber, eben auch in unserer Zeit, anderwärts in unserem Vaterlande den Antheil an wissenschaftlichen Fortschritt sich erweitern, wir sahen die Erfüllung einer hohen Pflicht für ein mächtiges Reich. Das ist das Große, das uns mit Freude, mit beruhigendem Bewußtseyn der Zukunft entgegenführt.*[7]

Viribus unitis,[8] die allseits in diesem Zeitraum von der Wissenschaft zunächst aus dem dynastischen Bezug entlehnte, beschworene

alla conquista dell'Artico, Triest 2003; S. Barr – H. Slupetzky (Hrsg.), Franz Josef Land, Oslo 1995.

6 Auf die Rolle der Medien und die Instrumentalisierung der Expedition für die Innenpolitik gehen besonders in ihrer originellen Analyse der Pressemeldungen ein: J. Schimanski – U. Spring, Polarwissenschaft und Kolonialismus in Österreich-Ungarn: Zur Rezeption der österreichisch-ungarischen Polarexpedition, in: M. Klemun (Hrsg.), Wiener Zeitschrift für Geschichte der Neuzeit 9 (2009), 53-71.

7 Brief Wilhelm von Haidingers an Hauer (3. Mai 1856), Bibliothek der Geologischen Bundesanstalt, Wissenschaftliches Archiv, Inv.-Nr. A 004-BM; Franz-Hauer-Nachlass; abgedruckt auch bei: K. Kadletz, Die Geologische Reichanstalt im Schicksalsjahr 1860. Genese und Ablauf des Konflikts um ihre Eingliederung in die Akademie der Wissenschaften, ungedr. Phil. Diss, Wien 2003, 126f., Brief 17.

8 Diesem Motto wurden zahlreiche bildliche und lyrische Darstellungen gewidmet. Der Grundgedanke ist bereits im bekannten Huldigungsgedicht Grillparzers an Radetzky enthalten. *Treue und Eintracht der österreichischen Völker, Viribus Unitis*, so betitelt es eine Lithographie von Franz Kollarz 1849

und instrumentalisierte Metapher der Habsburgermonarchie, repräsentierte nach ihrem ersten Auftreten zunächst das Zurücktreten liberalen Gedankengutes zugunsten unbedingter Loyalität gegenüber der Krone nach der gescheiterten Revolution 1848, nachdem sich das Wirtschafts- und Bildungsbürgertum Österreichs, das seine Chance für eine politische Umsetzung versäumt hatte, erneut mit der wiederhergestellten monarchischen Macht arrangieren musste. Nationale Gegensätze, wie sie 1848 erstmals radikal zutage gekommen waren, wurden zunächst harmonisiert, und die Antwort auf die evident gewordenen Probleme manifestierte sich nicht zuletzt auch in der staatlichen Zulassung und Förderung der von der bürgerlichen Elite gewollten und von ihr getragenen neuen wissenschaftlichen Einrichtungen, etwa der *k. k. Geologischen Reichsanstalt*, der *Akademie der Wissenschaften in Wien*, der *k. k. Meteorologischen Reichsanstalt* und der *Zoologisch-Botanischen Gesellschaft in Wien.* Alle standen über den Ländern der Monarchie. Es war kein Zufall, dass etwa die Geologische Reichsanstalt dieses Motto auf den Titelblättern ihrer Publikationen angebracht hatte, ging es doch um die geologische Kartierung des Gesamtstaates, die Fundierung einer politischen Einheit mittels Natur und des Einsatzes „vereinter Kräfte".[9] Die Aufnahme mündete in ein nach standardisierten Prinzipien gestaltetes Gesamtbild,[10] ein Kartenwerk, das wie die neue Verwaltung einheitlich den Raum von der Lombardei bis zur Bukowina, von Dalmatien bis zum Elbedurchbruch im Maßstab 1:144.000 einbezog.

Dass die Wissenschaften nach 1848 zur Aufrechterhaltung der multinationalen Habsburgermonarchie beitrugen und zunächst noch das Projekt einer quasi-nationalen Harmonisierung fundierten, beweisen beispielsweise die innerhalb des statistischen Büros vorge-

---

(nach einer Zeichnung von Josef A. Hellich). Es ist eine allegorische Darstellung der 18 in der Monarchie lebenden Völker. Männer in Volkstrachten sind um Kaiser Franz Josef geschart, darüber im Ornamentband sind die sieben Ministerien und unten die Wappen aller Länder symbolisiert. Vgl. S. Nasko (Hrsg.), Österreich unter Kaiser Franz Joseph I. (Historische Sonderaustellung im Schloß Pottenbrunn), Wien 1979, bes. 102.

9 Siehe dazu: M. Klemun, "National Agreement" as a Style of Thinking? The Geological Survey of the Habsburg Empire (1849-1967), in: M. Ash (Hrsg.), National(istic) Science, in Druck.

10 F. Hauer, Geologische Übersichtskarte der Österreichisch-Ungarischen Monarchie nach den Aufnahmen der k. k. geologischen Reichsanstalt 1: 576.000, Wien 1867-1874.

nommenen Arbeiten eines Karl Freiherr von Czoernig, der mit seiner *Ethnographische[n] Karte der Monarchie* (1857) deutlich machte, dass die einzelnen österreichischen Kronländer keine ethnisch-national homogenen Territorien bildeten. Dass Österreichs Bevölkerung ethnisch und national gemischt sei und keine Nation ein Gebiet alleine beanspruchen könne, korrespondierte mit einem höheren staatlichen Interesse. Die entscheidenden Begriffe, mit denen Czoernig operierte, waren jener der „Mischung" und jener der Harmonie,[11] und beide fungierten in der Folge als symbolisches Potential für die vielfältige rhetorische Anbindung der Naturwissenschaften. Der Präsident der Zoologisch-Botanischen Gesellschaft, Eduard Fenzl, formulierte es 1853 angesichts der Rückschau auf die erfolgreiche Gründung der bürgerlichen Einrichtung in diesem Sinne: *Meine Herren, unverbrüchlich an dem Einem festhalten: dem Interesse für das Kleinste, was Jeder von uns liefert,* […] *Unsere Gesammtsitzungen müssen der lebende Kitt dieser wechselseitigen Interessensverschmelzung bleiben.* […] *Zeigen Sie der Welt, dass Oesterreich's Männer der Wissenschaft trotz aller Nationalitäts- und Sprach-, Rang- und Standesverschiedenheit rascher im Erfassen höherer Zielpuncte, besonnener in der Wahl ihrer Mittel, einiger in der Verfolgung und zäher im Festhalten der ersteren sind als Andere, die sich eines Stammes zu sein mit Recht rühmen dürfen, einig geworden und geblieben zu sein, aber nie sich rühmen konnten.*[12]

Die Wissenschaften lebten allerdings von der Hierarchie innerhalb ihrer Reihen. Denn trotz aller Einigungsrhetorik herrschte Konkurrenz zwischen den Institutionen, und die 1847 gegründete Akademie der Wissenschaften hatte für sich die Vorrangstellung beansprucht. In seiner ersten Festrede hatte ihr Präsident bereits betont: Die Akademie ist *keine Schule, sondern ein Richterstuhl wissenschaftlicher Leistungen in letzter Behörde.*[13] In diesem Sinne wurde die Eingliederung der 1849 gegründeten Geologischen Reichsanstalt unter die Akademie durch viele Jahre, bis 1861, von ihr betrieben,

---

11 Vgl. dazu: B. Fuchs, „Rasse", „Volk", Geschlecht. Anthropologische Diskurse in Österreich 1850-1960, Frankfurt/ New York 2005, 154.

12 Verhandlungen der Zoologisch-botanischen Gesellschaft 2 (1853), 4f.

13 Hammer Purgstalls Festrede, zitiert nach: R. Meister, Geschichte der Akademie der Wissenschaften in Wien 1847-1947 (Denkschriften der Gesamtakademie 1), Wien 1947, 60.

aber von der Reichanstalt erfolgreich bekämpft.[14] In meiner Studie hier ist deshalb zu erläutern, wie sich besonders die Akademie der Wissenschaften im öffentlichen Diskursraum über das Nordpolarvorhaben positionierte. Damit verfolge ich die These, dass die durch die Expedition mobilisierten Anschlussmöglichkeiten von Sinndeutungen, in denen die Wissenschaften als Leitbild fungierten, auch von der Akademie der Wissenschaften in Wien genutzt wurden und sie sich sogar zur Teilhabe gezwungen sah. Ausgehend von der öffentlich gedruckten Befürwortung, einem Gutachten,[15] das von der Kaiserlichen Akademie der Wissenschaften für die Nordpolexpedition ausgestellt wurde, soll in meiner Studie die dynamische Verflechtung zwischen wissenschaftlichen und außerwissenschaftlich symbolischen Wertzuschreibungen erörtert werden. Dass sich die Akademie dieser einmaligen Chance bediente, um selbst als Institution ihrem Selbstbild gemäß auch öffentlich den ersten Rang unter allen Wissensträgern zu verteidigen, ist anhand der Diskussion über ihre finanzielle Beteiligung in Dokumenten archivalisch nachvollziehbar. Für wenige Momente verließ die Akademie zudem ihre bisherige elfenbeinturmartige elitäre Abstinenz von allen Popularisierungstendenzen, um sich innerhalb der national und kolonial dynamisierten Medienlandschaft zu positionieren. Da die Arktisfahrten stets als Forschungsfahrten und nicht als „Entdeckungsfahrten" im zeitgenössischen Verständnis gedeutet wurden, war es für die Akademie unverzichtbar, ein solches Projekt auch durch eine Befürwortung einzubegleiten. Die Diskussion um die finanzielle Beteiligung der Akademie wird auf den nächsten Seiten ausführlich dargestellt werden, um ein durchaus ambivalentes Verhältnis dieser Institution zur Öffentlichkeit nachzuweisen.

---

14 K. Kadletz, „Krisenjahre" zwischen 1849 und 1861, in: Geologische Bundesanstalt (Hrsg.), Die Geologische Bundesanstalt in Wien. 150 Jahre Geologie im Dienste Österreichs (1849-1999), Wien 1999, 78-92.

15 [Anonymus], „Die österreichische Nordpol-Expedition. Gutachten der Kaiserlichen Akademie der Wissenschaften über die Nützlichkeit der Fortsetzung der Polarforschung, sowie über die Zweckmäßigkeit des von den Herren Carl Weyprecht, k. k. Schiffs-Lieutenant, und Julius Payer, k. k. Oberlieutenant, für die österreichische Expedition vorgeschlagenen Weges (Wien, 25. Jänner 1872), in: Mittheilungen der kais. und königl. Geographischen Gesellschaft in Wien 15 (1873), 57-59.

Das anonym gezeichnete Gutachten der Akademie erschien in dem Publikationsorgan der 1856 gegründeten *k. k. Geographischen Gesellschaft in Wien,*[16] eines für die neue in Wien etablierte bürgerliche Institutionalisierung der Wissenschaften wichtigen Mediums, das zum Hauptträger geographisch-geopolitischer Interessensbekundungen avancierte und nicht zuletzt auch deshalb eine geeignete Plattform der Propagierung einer Teilhabe vom habsburgischen Zentrum aus an der geographischen *Lösung des größten noch übrig gebliebenen Problems der Erdkunde*[17] war, wie es in einem Beratungsprotokoll in der Akademiesitzung formuliert ist.

In der internationalen Landschaft der geographischen Zeitschriften, die sich alle bereits in der ersten Hälfte des 19. Jahrhunderts konstituiert hatten, nämlich dem *Bulletin de la Société de Géographie* (Paris, ab 1822) und dem *The Journal of the Royal Geographical Society* (London ab 1830), war die Wiener Zeitschrift zwar eine Epigonin, aber dennoch machte auch diese jene Entwicklung durch, die ihre Vorbilder vorgezeichnet hatten: Sie transformierte sich vom *Reise- und Expeditionsjournal zur Fachzeitschrift.*[18] War das Wissen der Geographie noch im 18. Jahrhundert Grenzen der Erkenntnis unterworfen

16 Es war kein Zufall, dass Erich Woldan sich darum bemühte, auch die Bände dieser Gesellschaftszeitschrift persönlich für seine (heute als *Sammlung Woldan* der ÖAW gehörende) Kollektion zu erwerben, obwohl er selbst als ehrenamtlicher Bibliothekar der *Geographischen Gesellschaft* fungierte und direkten Zugang zu dieser Lektüre hatte: Die *Mitteilungen der Österreichischen Geographischen Gesellschaft* waren und sind zentral für das Verständnis geographischer Forschung, sind zudem auch mit Karten ausgestattet und damit für eine derart ausgerichtete Sammlung wie jene Woldans unerlässlich.

17 Protokoll der gemischten Commission zur Begutachtung des von dem k. k. Oberlieutenant Herrn Julius Payer gestellten und vom Herrn Dr. A. Petermann unterstützten Ansuchens um eine Subvention (20. April 1871), Archiv der Österreichischen Akademie der Wissenschaften [AÖAW], Allgemeine Akten, Nr. 232 und 276/1871; Diese Formulierung geht auf den Brief August Petermanns an die Akademie der Wissenschaften (23. März 1871) zurück; vgl. AÖAW, Allgemeine Akten, Nr. 276/1871. Für die wertvolle Unterstützung bei meinen Recherchen im Archiv der Österreichischen Akademie der Wissenschaften möchte ich mich bei Dr. Stefan Sienell besonders bedanken.

18 A. Borsdorf, Motor des Erkenntnisfortschrittes und Dokument der Disziplingeschichte. Ein Rückblick auf 146 Jahrgänge der Zeitschrift der Gesellschaft für Erdkunde zu Berlin – und ein Ausblick auf die Zukunft, in: Die Erde, Sonderheft 1 (2003) 39-66, hier: 48.

gewesen, so hatte sich die Kenntnis über die Erde ab der Mitte des 19. Jahrhunderts in ihren Dimensionen nach innen (Geologie) und bezüglich der Umwelt erweitert. Gehörte die Karte lange zu ihrer ureigensten Erkenntnisform, so musste die Geographie diese nun mit anderen Wissenschaften wie Meteorologie etc. teilen.[19] Die Geographie war wohl die einzige Wissenschaft, die aus ihrer Verankerung in der Entdeckungsgeschichte ihren spezifischen Gewinn und ihre Nähe zur populären Wissenschaft gezogen hatte. Neben den Visualisierungen der Polarfahrten etablierten die in den geographischen Zeitschriften publizierten Berichte die Beliebtheit des Topos von der Arktis im hohen Norden. Nicht zuletzt durch das Scheitern der Franklin-Expedition 1845 und die erfolglos gebliebenen Suchexpeditionen der Folgezeit übten die damit verbundenen Gefahren und Schrecken des Eises einen Teil der Faszination aus.

Beginnen wir nun, die Geschichte der österreichischen Nordpolfahrten von einem historisch relevanten Punkt her zu verfolgen, bevor ich die Rolle der Akademie der Wissenschaften diskutiere! War Mitteleuropa an der Polarforschung lange nicht beteiligt gewesen, so änderte sich das ab den sechziger Jahren des 19. Jahrhunderts. Die treibende Kraft bildete bekanntlich der deutsche Kartograph und Geograph August Petermann (1822-1878),[20] der nicht nur die diesbezüglichen Aktivitäten in den norddeutschen Territorien initiierte, sondern auch auf die habsburgischen Anstrengungen einen entscheidenden Einfluss ausübte. Petermann setzte sich seit 1863 für die Beteiligung der Deutschen an der Polarforschung ein. Während des *Ersten deutschen Geographentages* in Frankfurt am Main des Jahres 1865[21] stellte er seine Theorie vom eisfreien Nordpolarmeer vor, wonach das Meer zwischen Spitzbergen und Nowaja Semlja infolge des Einflusses des warmen Golfstromes nicht zufrieren würde und eine mögliche Route bis zum Nordpol denkbar sei. Er beklagte die mangelnde Unterstützung durch die seefahrenden Mächte, aber auch der „Preußischen und Österreichischen Regierungen" und warb um

---

19 A. M. C. Godlewska, Geography unbound, Chicago/London 1999, bes. 313.

20 E. Weller – A. Petermann, Ein Beitrag zur Geschichte der geographischen Entdeckungen und der Kartographie im 19. Jahrhundert, Leipzig 1911.

21 Amtlicher Bericht über die erste Versammlung deutscher Meister und Freunde der Erdkunde in Frankfurt a/M im Heumonat 1865 (Freies deutsches Hochstift für Wissenschaften, Künste und allgemeine Bildung, hrsg. vom geschäftsleitenden Vorsitzenden), Frankfurt am Main 1865.

Spenden für die von ihm geplante Unternehmung. Folgen wir einem späteren Schreiben an die Wiener Akademie der Wissenschaften, in dem er die Entwicklung des polaren Expeditionswesens knapp memorierte: *Man wird sich erinnern, dass die gegenwärtige Polarforschung zunächst dadurch angeregt worden war, dass der Englische Kapitän Osborn vor beinahe 7 Jahren eine neue Englische Expedition vorgeschlagen hatte, die die Erforschung der arktischen Centralregion und die Erreichung des Nordpols bezweckte. Der Plan war, die Baffin-Bai hinauf und durch den Smith-Sund zu gehen, und wurde von den erfahrenen und wissenschaftlichen Kreisen und Autoritäten Englands aufs Wärmste gut geheißen und unterstützt, sodaß alle Aussicht vorhanden war, die Englische Regierung werde eine große wissenschaftliche Expedition ausrüsten und aussenden. Als ich aber den Englischen Autoritäten meine Ansichten und Projekt vorlegte, welche das Europäische Nordmeer zwischen Ostgrönland und Nowaja Semla zur Basis einer solchen Expedition empfahl, wurde der Osbornsche Plan nach den erschöpfenden Diskussionen der bedeutendsten Männer in vier Sitzungen der königlichen Geographischen Gesellschaft in London verworfen, und meine Ansichten die vollste Zustimmung und Anerkennung zu Theil. Während nun die Engländer 7 Jahre lang es bei bloßen Worten und Diskussionen bewenden ließen, ermöglichte die außerordentliche Bereitwilligkeit und Generosität, der hohe wissenschaftliche und nationale Sinn der deutschen Nation die Ausrüstung und Aussendung zweier deutschen wissenschaftlichen nordpolaren Expeditionen, die unter das Kommando des Kapitäns Koldwey gestellt wurden.*[22]

Fast wäre es auch zu einer Zusammenarbeit zwischen Preußen und Österreich gekommen. Der Krieg zwischen beiden vereitelte jedoch die Pläne einer gemeinsamen Nordpolfahrt im Jahre 1866. Zwei deutsche Expeditionen (1868 und 1869/70) folgten Petermanns Mobilisierung der Öffentlichkeit, die er mithilfe des von ihm geleiteten Verlages Perthes und von *Petermann's Mittheilungen*, der bedeutenden geographischen Zeitschrift, medial in Szene setzte. Diese Expeditionen erbrachten jedoch keine Bestätigung seiner Theorie, und deren Kapitän sprach sich öffentlich für die Route Osborns aus. In dieser Situation war es für Petermann eine Prestigefrage, weitere Ausführende für seine Überzeugung zu gewinnen.

---

22 August Petermann in dem handschriftlichen Memorandum *Die Entdeckung eines offenen Polarmeeres durch Payer und Weyprecht im September 1871* (Gotha, 9. Oktober 1871), AÖAW, Allgemeine Akten, Nr. 756/1871.

Für die habsburgischen Initiativen der Folge sind zwei zentrale Figuren zu nennen: Carl Weyprecht (1838-1881), der eigentlich in Darmstadt in Deutschland 1856 geboren wurde und der seit 1856 im Dienst der Österreichischen Kriegsmarine Erfahrungen als Seeoffizier in der Adria gesammelt hatte, zeigte sich bereits 1866 auf Petermanns Aufruf hin von sich aus interessiert, an einer Expedition führend mitzuwirken. Sein Interesse war – wie er es in einem Brief schildert[23] – durch Petermanns Darstellungen[24] angeregt worden. Jedenfalls wandte er sich persönlich bereits im November 1865 an Petermann und bot ihm seine Dienste für das Projekt an.[25] Weyprecht sah sich selbst infolge seiner harten Ausbildung bei der Marine als prädestiniert, an einer Nordpolunternehmung teilhaben zu können.[26] Sein Ziel eröffnete Weyprecht in einem Brief an seine Eltern: *Ich muß Euch als Weihnachtsangebinde einen Plan mittheilen, der wenn er in Erfüllung geht, meine kühnsten Wünsche erfüllt. Eingeleitet ist Alles, es fehlt zu seiner Ausführung nichts mehr, als das deutsche Volk. Lacht nicht, es ist wirklich so. Ich habe mich nämlich dem Dr. Petermann angeboten, bei der projektierten Deutschen Nordpolexpedition mitzuwirken und in einem äußerst freundlichen Schreiben schon vor drei Wochen von demselben seine volle Unterstützung zugesichert bekommen.* [...] *Ich habe jetzt sehr viel zu thun; außer Dienst besorge ich eine eingehende Korrespondenz mit allen Windrichtungen und benütze jeden freien Augenblick zu Privatstudien, um die* [...] *in gewissen Zweigen noch fehlenden Kenntnisse zu kompletiren. Es kommen Tage vor, an denen ich ununterbrochen 10-12 Stunden hinter den Büchern sitze.*[27] Weyprecht hatte die Tatsache, dass er an keiner Universität studiert

[23] Brief Carl Weyprechts an seine Eltern (Pola, Januar 1866): *Die Idee erfasste ich schon zu Hause, als ich den Bericht über die Versammlung zu Frankfurt las,* [...] *Hier wurde ich durch das Studium eines wissenschaftlichen Buches, in dem ich viele Lücken fand, die auf einer Expedition in den hohen Norden ausgefüllt werden können, wieder darauf gebracht.* Abgedruckt in: Berger, Weyprecht, 217.

[24] A. Petermann, Spitzbergen und die arktische Central-Region (Ergänzungsheft der Mittheilungen aus Justus Perthes' Geographischer Anstalt über wichtige neue Erforschungen auf dem Gesammtgebiete der Geographie), Gotha 1865.

[25] Handschriftlicher Brief Carl Weyprechts an Petermann (Pola, 2. November 1865), abgedruckt in: Berger, Weyprecht, 203-204.

[26] Brief Carl Weyprechts an seine Eltern (Triest, 16. Dezember 1859), abgedruckt in: Berger, Weyprecht, 125.

[27] Brief Carl Weyprechts an seine Eltern (Triest, 20. Dezember 1859), abgedruckt in: Berger, Weyprecht, 216.

hatte, zwar als Nachteil empfunden, sich aber erfolgreich selbst weiterbilden können. Die Grenze zwischen Laienforschern und Professionalisten war allerdings in dieser Zeit noch nicht merklich gezogen worden, und Selbstbildung bildete ein mögliches Aufstiegsprofil für Interessierte.

Julius Payer (1841-1915), ein gebürtiger Böhme, der sich als Kartograf und Alpinist in der Adamello- und Ortler-Gruppe hervorgetan hatte, kam mit diesem Engagement in Kontakt zu Petermann. So war Payer für die zweite deutsche Nordpolar-Expedition 1869/70 angeworben worden, und er hatte sich an der Ostküste Grönlands auf einer 600 km langen Schlittenexpedition besonders bewährt.

Beide Persönlichkeiten, Weyprecht und Payer, griffen 1871 Petermanns Ideen auf und wurden in Österreich initiativ. Petermann hatte ihnen 2000 Taler zugesagt, mit der Versicherung, dass alle Ansprüche auf Ergebnisse bei Österreich lägen. Zunächst musste jedoch um weitere Gelder geworben werden, um die geplante Expedition privat zu finanzieren. Bestärkt durch Petermann und begleitet von einem persönlichen von diesem an die Akademie der Wissenschaften in Wien gerichteten Schreiben,[28] erbat sich Payer[29] bei der Akademie eine Subvention für eine Vorexpedition, die ihn und Weyprecht in den Osten Spitzbergens führen sollte. Gleichzeitig wurde auch angesucht, dass die *Hohe Akademie beim k. k. Reichs-Kriegs-Ministerium (Marineabteilung)* sich dafür einsetzen möge, dass Instrumente und Chronometer ihnen leihweise überlassen würden. Petermann argumentierte im genannten Empfehlungsschreiben: *Durch die Initiativen unserer Nordpolar-Expeditionen ist allenthalben ein so reger Eifer für die Vollendung der Aufgabe wach geworden, daß sich verschiedene Nationen anschicken, uns* [er meinte die Deutschen] *den Rang auf dem rühmlich begonnenen Wege zur Lösung des größten und übrig gebliebenen Problems der Erdkunde streitig zu machen.* Die Akademie unterstützte das Unternehmen mit 1000 Gulden und verwies ausdrücklich in ihrem Protokoll, *daß sowohl Payer als Weyprecht geborene Österreicher sind.*[30] Die

[28] August Petermanns handschriftlicher Brief an die Akademie der Wissenschaften in Wien (Gotha, 23. März 1871), AÖAW, Allgemeine Akten, Nr. 276/1871.

[29] Handschriftlicher Brief Julius Payers an die Akademie der Wissenschaften (Wien 15. März 1871), AÖAW, Allgemeine Akten, Nr. 232/1871.

[30] Protokoll der gemischten Commission zur Begutachtung des von dem k. k. Oberlieutenant Herrn Julius Payer gestellten und vom Herrn Dr. A.

Tatsache, dass Weyprecht in österreichischen Diensten stand, machte ihn offensichtlich zum „Österreicher".

Diese Vorexpedition war sowohl in wissenschaftlicher wie auch finanzieller Hinsicht ein Probelauf: Sie zeigte, dass innerhalb kürzester Zeit genug Spenden zusammengebracht werden konnten und auch die Akademie sich wohlwollend gegenüber den Plänen zeigte. Noch im Jahre 1865 hatte Weyprecht persönlich die Möglichkeit einer größeren Resonanz des Polarthemas in der österreichischen Presse nicht für möglich gehalten.[31] Nun war das Eis gebrochen.

Die Vorexpedition ging von Tromsø aus, weil von dort Erkundigungen über Erfahrungen bei Praktikern eingeholt werden konnten. Weyprecht weilte bereits Wochen vor der Übernahme eines gecharterten Segelschiffes in Tromsø, wo er jene Kapitäne und Lotsen befragte, die mehr als 30 Sommer in den Gewässern von Spitzbergen oder Novaja Semlja zugebracht hatten. *Alle schildern das Eis in dieser Gegend als schwer und dicht aufgestaut und stellen dem Vordringen von Süden gegen Gilis-Land ein schlechtes Prognostikon. Hierzu muß ich jedoch bemerken, dass ich trotz aller Mühe, die ich mir gegeben habe, nicht im Stande war, irgend Jemand zu finden, der dieses Urtheil aus eigener Anschauung oder Untersuchung in der Nähe gefällt hätten. Es sind zwar schon oft Reisen von Spitzbergen nach Nowaja-Semlja oder umgekehrt gemacht worden, allein die Course derselben führen immer so weit Süd, dass das Eis entweder gar nicht oder nur aus großer Entfernung gesichtet wurde. Die Zustände oberhalb 76° in dieser Gegend sind absolut unbekannt; jede von uns gemachte Beobachtung wird etwas ganz Neues sein.* Weyprecht erläuterte sein ehrgeiziges Programm, *das Augenmerk nicht nur auf Oberflächentemperaturen,* [sondern] *hauptsächlich auf Tiefseetemperaturmessungen zu richten. Wir haben zu diesem Zwecke die Minimal- und Maximalthermometer von Casella, ganz vorzügliche Instrumente, bei uns.*

Petermann unterstützten Ansuchens um eine Subvention (20. April 1871), AÖAW, Allgemeine Akten, Nr. 232/1871.

31 *Stünde uns die Presse mehr offen, so ließe sich auf diese Art viel wirken, da es doch von allen Dingen darauf ankommt, das große Publikum so viel als möglich für die Sache zu interessieren: Leider verhält sie sich sehr indifferent; mit Ausnahme kurzer Notizen habe ich, außer zweier Artikel in der Triester Zeitung und des Berichtes der geogr. Gesellschaft in Wien in der Neuen Freien Presse, bis jetzt noch in keiner der bedeutsameren Zeitungen Österreich etwas entdecken können, was geeignet gewesen wäre, im großen Publikum Interesse für das Unternehmen zu erwecken.* […] *Für uns Offiziere sind, mit Ausnahme unserer Fachblätter, die Zeitungsspalten fast völlig verschlossen – Ich würde es kaum wagen, einem Wiener Redakteur einen Artikel zu versenden.* Brief Carl Weyprechts an Petermann (Pola, 13. Dezember 1865), abgedr. in: Berger, Weyprecht, 211.

Unseres Wissens wird dieß das erste Mal sein, daß im arktischen Meere solche Untersuchungen mit verläßlichen Instrumenten gemacht werden; alle früheren Resultate basieren auf Schöpfapparaten oder ähnlichen Instrumenten, die beim Heraufholen bedeutende Wärmeveränderung erleiden. Wir haben hier ein Loth konstruieren lassen, das uns ermöglicht, mit Steinen ähnliche Resultate zu erlangen; wie mit dem amerikanischen Lothungsapparat, der durchlöcherte sichere Kugeln bedarf, deren Transport hierher für unsere Mittel zu kostspielig gewesen wäre. Außer diesen Beobachtungen werden wir von 4 zu 4 Stunden alle meteorologischen Beobachtungen machen.[32] Carl Weyprecht kam also dem Anspruch nach, die Akademie als wissenschaftliche Einrichtung adäquat von seinen Tätigkeiten zu informieren.

Über die Vorexpedition, die fast vier Monate gedauert hatte, berichtete Weyprecht nach erfolgreicher Beendigung erneut persönlich an die Akademie: *Das Hauptresultat unserer Expedition ist die Entdeckung eines offenen Polarmeeres in einer Gegend, die wegen ihrer schauerlichen Eiszustände* [...] *förmlich verrufen war. Nördlich von 76 ½° war dieses Meer bisher ganz unbekannt. Dieß ist dadurch zu erklären, daß alle Versuche hier vorzudringen viel zu Früh im Jahre gemacht wurden, die günstigsten Eiszustände treten hier erst im Spätherbst ein.*[33]

Das Unternehmen bezog sich auf wissenschaftliche Vorüberlegungen, die nun durch Messungen der Tiefseetemperatur bestätigt wurden. Die Eisfreiheit des Polarmeeres auf 79° n. Br. erklärte Weyprecht mit dem Golfstrom, der im September die größte Stärke erreiche[34]. Petermann feierte das Ereignis dieser Probefahrt als *Die Entdeckung eines offenen Polarmeers durch Payer und Weyprecht im September 1871!*[35] Dieses Vorergebnis bewirkte den Optimismus für die nächste große Expedition.

Inzwischen war in Wien im Februar 1872 ein eigener Verein zur Förderung der großen nächsten österreichischen Nordpolexpedition gegründet worden, der die Finanzierung des eigentlich privat getra-

---

[32] Handschriftlicher Brief Carl Weyprechts an Akademie der Wissenschaften (Tromsø, 17. Juni 1871), AÖAW, Allgemeine Akten, Nr. 550/1871.

[33] Handschriftlicher Brief Carl Weyprechts an die Akademie der Wissenschaften (Tromsø, 5. Oktober 1871), AÖAW, Allgemeine Akten, Nr. 775/1871.

[34] Ebd.

[35] Carl Petermanns handschriftliches Schreiben an die Akademie der Wissenschaften (Gotha 9. Oktober 1871), AÖAW, Allgemeine Akten, Nr. 756/-1871. Das Schreiben wurde auch publiziert in: Hansa (1871), Nr. 20, 92.

genen Unternehmens ermöglichen sollte. Zum Präsidenten des Vereins wurde der ehemalige Leiter der Weltumsegelung der österreichischen Fregatte *Novara*, Minister Baron Bernhard vom Wüllerstorf Urbair, gewählt. Kriegsminister Kuhn, ein Förderer Payers, beteiligte sich als Vizepräsident, und Ferdinand Hochstetter als ehemaliger Teilnehmer der Novara-Fahrt und geologischer Erforscher Neuseelands setzte sich als einer der prominentesten Proponenten für die Expedition ein.[36] Als Präsident der *k. k. Geographischen Gesellschaft in Wien* und als Mitglied der Akademie der Wissenschaften stellte Hochstetter das Bindeglied zwischen den verschiedenen Wissensinstitutionen her, wobei die *k. k. Geographische Gesellschaft in Wien* die Frage der Polarforschung auch in ihr mediales Programm ambitioniert aufnahm.[37] Hochstetter hatte sich an dem von Petermann organisierten Geographentag in Frankfurt am Main 1865 bereits von Petermanns Überlegungen faszinieren lassen. Wie Payer und Weyprecht kannte auch er den großen Propagator des Arktis-Unternehmens persönlich.

Die Akademie öffnete nun sogar ihre Tore für einen Redner, der nicht aus ihren Reihen stammte: Carl Weyprecht erhielt die Möglichkeit, im „grünen Saal"[38] am 7. Dezember 1871[39] in der Wiener

36 Zu Hochstetter s. den Beitrag im 1. Band von G. Holzer, S. 197-232)

37 Auf bibliographische Hinweise muss hier aus Platzmangel verzichtet werden.

38 Bewilligung für die Überlassung des „grünen Saales": AÖAW, Allgemeine Akten, Nr. 63/1872.

39 [C. Weyprecht], Vortrag. Wissenschaftliche Resultate seiner letzten, in Gemeinschaft mit Herrn Oberlieutnenant Julius Payer unternommenen Nordpolarfahrt, in: Anzeiger der mathematisch-naturwissenschaftlichen Classe der Kaiserlichen Akademie der Wissenschaften in Wien 8 (1871), 230-234; mehrmals veröffentlicht: [C. Weyprecht], Bericht des k. k. Schiffs-Lieutenant Weyprecht über die Polarexpedition an die kais. Academie der Wissenschaften in Wien (Vorgetragen in der Academiesitzung am 7. December 1871), in: Mittheilungen der Kais. Königl. Geographischen Gesellschaft in Wien 14 (1871). 545-555; [C. Weyprecht], Bericht des k. k. Schiffs-Lieutenant Weyprecht über die Polarexpedition an die Kais. Akademie der Wissenschaften in Wien über seine und Payer's Expedition im Nowaja Semlja-Meere, Juni-September 1871 (vorgetragen in der Akademiesitzung am 7. December 1871), in: Petermanns geographische Mittheilungen 18 (1871). 69-74 ; [C. Weyprecht], Rapport du Lieutenant Weyprecht à l'Academie Imperiale des Sciences de Vienne, in: Le Globe. Bulletin de la Societé de Gérographie de Geève 11 (1872), 95-110; [C. Weyprecht], Weyprecht über die Eisverhältnisse im arktischen Norden (Vorgetragen in

Akademie ein Referat über die meteorologischen Ergebnisse der Vorexpedition zu halten. Er nützte die Gelegenheit, um persönlich und in einer patriotisch aufgeladenen Rede für das nächste Unternehmen innerhalb der Akademie zu werben: *Wir Österreicher sind dadurch erfolgreich in den Wettkampf eingetreten, der schon seit Jahrhunderten zwischen allen gebildeten Völkern um die Lösung des großen Problems gekämpft wird.*[40]

Als treibende Instanz für die Expedition agierte im Hintergrund Ferdinand Hochstetter, der als Präsident der *k. k. Geographischen Gesellschaft in Wien* schon am 18. November 1871 Payer und Weyprecht zu einer Monatsversammlung der Gesellschaft eingeladen hatte. Es wurde in dieser Sitzung eine Resolution verabschiedet, die eine finanzielle Zusage sogar als *eine moralische Verpflichtung* bewertete. Weyprecht schrieb über diese Entwicklung euphorisch an Petermann: *Lieber Freund! Seit acht Tagen befinden wir uns wieder in Wien, vor lauter Besuchen sind wir noch gar nicht zu Atem gekommen. Man interessiert sich sehr für die nächstjährige Reise, was umso mehr hervortreten wird, wenn wir unsere Vorträge gehalten haben werden.* [...] *Es steht hier gut und wir hoffen, die Mittel aufzubringen;* [...] *Payer hält am 28. In der geographischen Gesellschaft, ich am 7. December in der Academie der Wissenschaften einen Vortrag. Diese beiden Vorträge sollen den Ausgangspunkt der Agitation für das nächste Unternehmen bilden.*[41]

Im Februar 1872 erbat sich wegen der Verhinderung des Freiherrn von Wüllerstorff, des Präsidenten des *Vereines zur Förderung der österreichischen Nordpolexpedition*, Edmund Graf Zichy als dessen Vizepräsident bei der Akademie die finanzielle Unterstützung. In diesem Schreiben betonte er: *Das Beispiel der ersten wissenschaftlichen Körperschaft ‚Österreichs'* [!] *würde gewiss nicht verfehlen, dem Unternehmen im In- und Auslande den kräftigen Vorschub zu leisten.*[42] Die Akademie

der Sitzung der k. k. österreichischen Akademie am 7. Dec. 1871), in: Ausland 45 (1872), 25-31.

40 [C. Weyprecht], Bericht des k. k. Schiffs-Lieutenant Weyprecht über die Polarexpedition an die kais. Academie der Wissenschaften in Wien (Vorgetragen in der Academiesitzung am 7. December 1871), in: Mittheilungen der Kais. Königl. Geographischen Gesellschaft in Wien 14 (1871), 545-555, hier: 555.

41 Brief Carl Weyprechts an Petermann (Wien, 18. November 1871), abgedruckt in: Berger, Weyprecht, 338.

42 Handschriftliches Schreiben Edmund Graf Zichys an die Akademie (Wien, 21. Februar 1872) AÖAW, Allgemeine Akten, Nr. 169/1872.

könnte demnach bei der öffentlichen Werbung beispielgebend vorangehen, und umgekehrt profitierte die Akademie von der aufsehenerregenden Beteiligung Österreichs an der sogenannten Erschließung der Arktis.

Die Summe von 2000 Gulden, die von beiden Teilen der Akademie, den Philologen und Naturwissenschaftlern, bewilligt wurde, war in Anbetracht des Gesamtbetrages von 222.616 Gulden Kosten für die Expedition nicht groß, aber für die Institution bedeutete sie dennoch einen beträchtlichen Aufwand. Der Kassastand wurde vor der Entscheidung kritisch überprüft und Sparsamkeit eingefordert, aber dennoch die Bedeutung der Finanzierung nicht bezweifelt, sondern es wurden allfällig entstandene finanzielle Engpässe für die Zukunft infolge der Subventionszusage für die Expedition durch Sparmaßnahmen bei der Publikationstätigkeit einkalkuliert. So heißt es im Protokoll der Sitzung: *Es wird von Mehreren Mitgliedern der Commission geltend gemacht, dass das in Frage stehende Unternehmen ein wahrhaft patriotisches und von so eminent wissenschaftlicher Bedeutung sei, wie noch nie eines dagewesen, und dass daher die Akademie, selbst wenn sie ihre Publicationen beträchtlich vermindern sollte, um jeden Preis das Unternehmen mit einem namhafften Betrage unterstützen müsse. Die Ehre der Akademie sei dabei betheiligt und sie steige von ihrer Stellung herab, wenn sie eine solche Unterstützung ablehne, es würde dieß sicher im Publicum einen sehr üblen Eindruck machen, zumal man nicht wisse, dass sie vom Staate reichlich dotiert sei, um wissenschaftliche Unternehmungen zu unterstützen.*[43] Da die Gesamtdotation der Akademie zu diesem Zeitpunkt insgesamt 63.000 Gulden betrug, verstand man die bewilligte Summe durchaus nur als *moralische Unterstützung, oder höchstens, um ihren guten Willen zu zeigen.* Auch wurde bedacht, dass finanzielle Einschränkungen, welche sich auf die Publikationsarbeit der Akademie auswirken würden, abzulehnen seien, da Veröffentlichungen den eigentlichen Kern der Aktivitäten der Akademie ausmachten und diese Tätigkeit nicht eingeschränkt werden dürfe, da sie in den Statuten an erster Stelle stehe. Es meldeten sich auch negative Stimmen, die den guten Ruf

43 Bericht der in der Gesamtsitzung vom 25. Jänner 1872 ernannten gemischten Commission über den in dieser Sitzung gestellten Antrag: *Die Kaiserliche Akademie der Wissenschaften möge sich bewogen finden, aus ihrer Dotation für die österreichische Nordpol-Expedition einen namhaften Beitrag |:zu gleichen Theilen auf Rechnung beider Classen:| zu bewilligen, welcher Beitrag auf die Jahre 1872, 73 & 74 ertheilt werden könnte.* AÖAW, Allgemeine Akten, Nr. 986/1871.

der Akademie in einer breiteren Öffentlichkeit als irrelevant einschätzten: *Die Meinung des großen Publicums bedarf in dieser Beziehung einer Berichtigung, kann daher nicht als Bestimmungsgrund für die Akademie gelten*, wurde in die Auseinandersetzung eingeworfen.[44]

Die Diskussion zeigt die ambivalente Haltung der Akademiemitglieder, die sich zwischen einem Verharren im bewährten Raum der wissenschaftlichen Community und einem Gang darüber hinaus in eine breitere Öffentlichkeit bewegte. Der Historiker Julius von Ficker (1826-1902), der Mathematiker und Professor für Physik an der Universität Wien Josef Stefan (1835-1893) und der Geologe Ferdinand Hochstetter plädierten für eine Subvention von insgesamt 6000 Gulden, konnten sich aber nicht durchsetzen. Bemerkenswert ist, dass der Astronom Carl Ludwig von Littrow (1811-1877), Direktor der Wiener Universitätssternwarte, und der Meteorologe Carl Jelinek (1822-1876), Direktor der Zentralanstalt für Meteorologie und Erdmagnetismus in Wien, die beide dem Unternehmen vom Fach her nahe standen, sich aber dennoch nicht für eine großzügigere Förderung der geplanten Expedition gewinnen ließen.[45] In dieser Sitzung am 18. Februar konnte keine Einigung erzielt werden. Ferdinand Hochstetter kündigte sogar ein Separatvotum an.[46]

Nun wurde nochmals gesondert in den Klassen beraten und getrennt von jeder Klasse eine Summe von 1000 Gulden bewilligt. Hochstetter und Stefan waren mit ihren Vorgaben von 3000 Gulden unter den Naturwissenschaftlern erneut in der Minderheit und mussten sich dem Ergebnis beugen.[47]

Ansonsten waren an der Finanzierung Ministerien, die Armee, verschiedene Bankhäuser, Finanzinstitute, Handelskammern, Firmen, Stadtgemeinden und besonders maßgeblich spontan gegründete Bürgerkomitees, Schulsammlungen und eben auch wissenschaftliche Institutionen sowie auch private Mäzene und viele Bürger beteiligt. Diese Finanzierungsform verkörperte die Mobilisierung aller Länder, aller Einrichtungen, aller Kräfte, aller Stände und „Nationen" der Monarchie. Selbst 250 kg Erbswürste wurden von einem Koch für die Expedition versprochen.

---

44 Bericht, AÖAW, Allgemeine Akten, Nr. 986/1871.

45 Ebd.

46 Allerdings wurde diese Zeile durchgestrichen! Ebd.

47 Protokoll der Sitzung der math.-nat. Classe am 22. Februar 1872. AÖAW, Allgemeine Akten, Nr. 202/1872.

Der Finanzierungskampagne vorangegangen war das bereits oben erwähnte Gutachten der Akademie der Wissenschaften, das von der *k. k. Geographischen Gesellschaft in Wien* publiziert wurde. Payer hatte sich im Dezember 1871 erbeten, die Akademie solle [...] *ein geeignetes Gutachten über die Nützlichkeit der Fortsetzung der Polarforschung, und über die Zweckmäßigkeit des von uns betretenen Weges und aufgestellten Planes abgeben um dieses Gutachten zur öffentlichen Kenntniß gelangen zu lassen, darauf die von uns eingeleiteten Sammlungen* [gemeint sind Geldsammlungen] *beginnen würden.*[48] Die Seriosität der Expedition in der Öffentlichkeit zu belegen war dringlich, um danach Geld einwerben zu können. In der Tat bestand aber dieses Gutachten nicht aus einer Erörterung der wissenschaftlichen Ziele, sondern lediglich in der ausführlichen Darstellung der wichtigsten international getragenen polaren Unternehmungen der Jahre davor. Ferner ging es auch um die Einbeziehung bereits laufender oder geplanter Expeditionen aller Seemächte, welche von den Nordamerikanern, den Schweden, den Russen und den Franzosen initiiert wurden. Anstelle einer ausführlichen wissenschaftlichen Begründung stand die Rhetorik im Hinblick auf die Darstellung eines Wettkampfes zwischen den Mächten im Vordergrund. Die Akademie hatte sich mit diesem Gutachten auf die Ebene der nationalen Mobilisierung begeben und verzichtete auf eine inhaltlich differenziertere Erörterung von wissenschaftlichen Zielen. Es wurde gänzlich vermieden, detaillierte wissenschaftliche Fragen anzusprechen. Kurz zuvor hatte beispielsweise der Pflanzenphysiologe Julius Wiesner die von Payer übermittelten Treibhölzer im Auftrag der Akademie begutachtet und darüber einen Bericht für den Anzeiger der Akademie verfasst: Er wies nach, dass die Treibhölzer von der sibirischen Lärche stammten.[49] Solche Details waren nicht für eine breitere Öffentlichkeit vorgesehen, sondern vielmehr wollte dieses Akademie-Gutachten die nationale gemeinsame Anstrengung untermauern. So heißt es auch am Ende der Expertise, für deren Zustandekommen wohl Hochstetter verantwortlich ist:[50] *Jedenfalls*

48 Ansuchen Julius Payers, gerichtet an die Akademie der Wissenschaften (21. Dezember 1871), Bericht, AÖAW, Allgemeine Akten, Nr. 986/1871.

49 Untersuchung von Julius Wiesner: Bericht, AÖAW, Allgemeine Akten, Nr. 4/1872.

50 Es befinden sich bei diesen Akten auch handschriftliche Notizen, die Hochstetters Handschrift entsprechen und die als Vorlage für das Gutachten fungierten.

bietet das projectierte Unternehmen die schönste Aussicht, den Ruhm wichtiger geographischer Entdeckungen Oesterreichs zu sichern, und würde durch das Gelingen nur dazu beitragen, den Sinn für wissenschaftliche Forschungen und Unternehmungen bei uns zu heben und dadurch die Wissenschaft selbst zu immer größerer Ehre zu bringen. Die kaiserliche Akademie der Wissenschaften wird daher auch nicht ermangeln, das Unternehmen mit all ihr zu Gebote stehenden Mitteln zu unterstützen.[51]

Die Expedition bot der Akademie die Gelegenheit, in der breiten Öffentlichkeit zu unterstreichen, dass wissenschaftliches Tun per se von Bedeutung sei und nationale Relevanz in sich berge, aber nicht der Inhalt selbst zu beschreiben sei. Dass die erfolgreiche Expedition tatsächlich der Akademie eine Fülle von unterschiedlichsten Daten erbrachte, unzählige Publikationen in der Folge evozierte und ein eigener Band der Denkschriften der Veröffentlichung der Ergebnisse gewidmet wurde, ist eine andere Geschichte. Julius Payer selbst nannte nach seiner Rückkehr in seiner Monographie,[52] die mit seinen eigenen Zeichnungen ausgestattet worden war, an keiner Stelle mehr die Akademie, stattdessen erweckten die abenteuerliche Leistung der Mannschaft und ihr Überleben im Eis nun allgemeine Bewunderung. Die Arktis als Alpenland faszinierte nun die Österreicher.

Und Carl Weyprecht, der sich nicht nur der seriösen Publikation der gewonnen Daten widmete, sondern ebenfalls für die Popularisierung der Arktisforschung einsetzte, sorgte mit seiner Monographie, die den ansprechenden Titel *Metamorphosen des Polareises* trug, tatsächlich für die Vermittlung meteorologischen Wissens. Auf die Wichtigkeit der Bewilligung des Planes durch die Akademie hatte er, in seiner Einleitung explizit zu verweisen, nicht vergessen.[53] Dass er nicht mehr auf den Wettlauf zwischen den Nationen, sondern auf Kooperation setzte, belegt seine wissenschaftliche Weitsicht, die ihn zum Erfinder des internationalen Polarjahres werden ließ.

---

51 [Anonymus], Gutachten, 59.

52 Julius Payer, Die österreichisch-ungarische Nordpol-Expedition in den Jahren 1872-1874, Wien 1876.

53 Carl Weyprecht, Die Metamorphosen des Polareises, Wien 1879, VIII.

Quellen- und Literaturverzeichnis

Archiv der Österreichischen Akademie der Wissenschaften, AÖAW, Allg. Akten, Nr. 232/1871, 276/ 1871, 388/1871, 550/1871, 756/1871, 775/1871, 986/1871, 4/1872, 63/1872, 102/1872, 108/1872, 169/1872, 202/1872 und 299/1872.

Brief Carl Haidingers an Hauer (3. Mai 1856), Bibliothek der Geologischen Bundesanstalt, Wissenschaftliches Archiv, Inv.-Nr. A 004-BM.

Amtlicher Bericht über die erste Versammlung deutscher Meister und Freunde der Erdkunde in Frankfurt a/M im Heumonat 1865 (Freies deutsches Hochstift für Wissenschaften, Künste und allgemeine Bildung, hrsg. vom geschäftsleitenden Vorsitzenden), Frankfurt am Main 1865.

[Anonymus], Die österreichische Nordpol-Expedition. Gutachten der Kaiserlichen Akademie der Wissenschaften über die Nützlichkeit der Fortsetzung der Polarforschung, sowie über die Zweckmäßigkeit des von den Herren Carl Weyprecht, k. k. Schiffs-Lieutenant, und Julius Payer, k. k. Oberlieutenant, für die österreichische Expedition vorgeschlagenen Weges (Wien, 25. Jänner 1872), in: Mittheilungen der kais. und königl. Geographischen Gesellschaft in Wien 15 (1873), 57-59.

Barr S. – Slupetzky H. (Hrsg.), Franz Josef Land, Oslo 1995.

Berger F. – Besser B. B. – Krause R. A., Carl Weyprecht (1838-1881). Seeheld, Polarforscher, Geophysiker (Österreichische Akademie der Wissenschaften, math.-naturw. Kl., Veröffentlichungen der Kommission für Geschichte der Naturwissenschaften. Mathematik und Medizin 57), Wien 2008.

Borsdorf A., Motor des Erkenntnisfortschrittes und Dokument der Disziplingeschichte. Ein Rückblick auf 146 Jahrgänge der Zeitschrift der Gesellschaft für Erdkunde zu Berlin – und ein Ausblick auf die Zukunft, in: Die Erde, Sonderheft 1 (2003), 39-66.

Dörflinger J., Die Nordwestpassage – Theorien von 1731-1823, Bd. 1-4, ungedr. Phil. Diss, Wien 1969.

Fuchs B., „Rasse“, „Volk“, Geschlecht. Anthropologische Diskurse in Österreich 1850-1960, Frankfurt/ New York 2005.

Godlewska, A. M. C., Geography unbound, Chicago/London 1999.

Hauer F., Geologische Übersichtskarte der Österreichisch-Ungarischen Monarchie nach den Aufnahmen der k. k. geologischen Reichsanstalt 1: 576.000, Wien 1867-1874.

Heeresgeschichtliches Museum Wien (Hrsg.), Die Schrecken des Eises und der Finsternis. Österreich und die Arktis. Katalog zur Sonderausstellung, Heeresgeschichtliches Museum, Wien 1996.

Kadletz K., „Krisenjahre" zwischen 1849 und 1861, in: Die Geologische Bundesanstalt (Hrsg.): Die Geologische Bundesanstalt in Wien. 150 Jahre Geologie im Dienste Österreichs (1849-1999), Wien 1999, 78-92.

Kadletz K., Die Geologische Reichsanstalt im Schicksalsjahr 1860. Genese und Ablauf des Konflikts um ihre Eingliederung in die Akademie der Wissenschaften, ungedr. Phil. Diss, Wien 2003.

Klemun M., "National Agreement" as a Style of Thinking? The Geological Survey of the Habsburg Empire (1849-1967), in: Ash M. (Hrsg.), National(istic) Science, in Druck.

Mazzoli E., Dall'Adriatico ai ghiacci. Ufficiali dell'Austria-Ungheria cin i loro marinai istriani, fiumani e dalmati alla conquista dell'Artico, Trieste 2003.

Meister R., Geschichte der Akademie der Wissenschaften in Wien 1847-1947 (Denkschriften der Gesamtakademie 1), Wien 1947.

Nasko S. (Hrsg.), Österreich unter Kaiser Franz Joseph I. (Historische Sondersaustellung im Schloß Pottenbrunn), Wien 1979.

Österreichische Nationalbibliothek (Hrsg.), 100 Jahre Franz Josefs-Land. Zur Erinnerung an die Entdeckungsreise der Österreichisch-Ungarischen Nordpol-Expedition 1872-1874 unter Julius von Payer und Carl Weyprecht. Ausstellung im Prunksaal der Österreichischen Nationalbibliothek, Katalog (Biblos-Schriften 75), Wien 1973.

Payer J., Die österreichisch-ungarische Nordpol-Expedition in den Jahren 1872-1874, Wien 1876.

Petermann A., Spitzbergen und die arktische Central-Region (Ergänzungsheft der Mittheilungen aus Justus Perthes' Geographischer Anstalt über wichtige neue Erforschungen auf dem Gesammtgebiete der Geographie), Gotha 1865.

Potter R. A., The Frozen North in Visual Culture, 1818-1875, Seattle/London 2007.

Ruthner C., "K. u. k. postcolonial"? Für eine neue Lesart der österreichischen (und benachbarten) Literatur/en, in: W. Müller-Funk W. – Plener P. – Ruthner C. (Hrsg.), Kakanien revisited. Das Eigene und das Fremde (in) der österreichisch-ungarischen Monarchie, Tübingen/Basel 2002, 93-103.

Schimanski J. – Spring U., Polarwissenschaft und Kolonialismus in Österreich-Ungarn: Zur Rezeption der österreichisch-ungarischen Polarexpedition, in: Klemun M. (Hrsg.), Wiener Zeitschrift für Geschichte der Neuzeit 9 (Wien 2009), 53-71.

Verhandlungen der Zoologisch-botanischen Gesellschaft 2 (1853).

Weller E., August Petermann. Ein Beitrag zur Geschichte der geographischen Entdeckungen und der Kartographie im 19. Jahrhundert, Leipzig 1911.

[Weyprecht C.], Vortrag. Wissenschaftliche Resultate seiner letzten, in Gemeinschaft mit Herrn Oberlieutenant Julius Payer unternommenen Nordpolarfahrt, in: Anzeiger der mathematisch-naturwissenschaftlichen Classe der Kaiserlichen Akademie der Wissenschaften in Wien 8 (1871) 230-234; auch abgedruckt als: [C. Weyprecht], Bericht des k. k. Schiffs-Lieutenant Weyprecht über die Polarexpedition an die kais. Academie der Wissenschaften in Wien (Vorgetragen in der Academiesitzung am 7. December 1871), in: Mittheilungen der Kais. Königl. Geographischen Gesellschaft in Wien 14 (1871) 545-555; [C. Weyprecht], Bericht des k. k. Schiffs-Lieutenant Weyprecht über die Polarexpedition an die Kais. Akademie der Wissenschaften in Wien über seine und Payer's Expedition im Nowaja Semlja-Meere, Juni-September 1871 (vorgetragen in der Akademiesitzung am 7. December 1871), in: Petermanns geographische Mittheilungen 18 (1871) 69-74; [C. Weyprecht], Rapport du Lieutenant Weyprecht à l`Academie Imperiale des Sciences de Vienne, in: Le Globe. Bulletin de la Societé de Gérographie de Geève 11 (1872) 95-110; [C. Weyprecht], Weyprecht über die Eisverhältnisse im arktischen Norden (Vorgetragen in der Sitzung der k. k. österreichischen Akademie am 7. Dec. 1871), in: Ausland 45 (1872), 25-31.

Weyprecht C., Die Metamorphosen des Polareises, Wien 1879.

Zur Autorin

Klemun Marianne, Ao. Univ. Prof., dzt. Vizedekanin der Historisch-kulturwissenschaftliche Fakultät.

Forschungsgebiete: Kultur- und Wissenschaftsgeschichte des 18. und 19. Jahrhunderts.

Kontakt: Institut für Geschichte, Universität Wien, Dr. Karl Lueger-Ring 1, 1010 Wien, marianne.klemun@univie.ac.at

Hartmann Schedel,
Buch der Croniken und geschichten,
Nürnberg 1493

Walter Koch, München

Epigraphisches zur *Schedelschen Weltchronik*

Das Schriftgeschehen am Übergang vom Mittelalter zur Neuzeit ist von einem – über die Grundtatsachen hinaus noch viel zu wenig untersuchten – Beziehungsgeflecht von Paläographie, Epigraphik und Typographie geprägt. Es ist dies die Zeit einer epigraphischen Vielfalt[1]. Wir sehen neben der damals dominierenden Gotischen Minuskel ein Bündel vielfach individueller frühhumanistischer Schreibweisen bis hin zur bzw. neben der Frühhumanistischen Kapitalis[2], die Wiederaufnahme der alten antiken Kapitalis und manche Vorformen[3] am Wege dorthin, die Gotico-Antiqua auch im

---

1 Vgl. hierzu zum bayerisch-österreichischen Bereich W. Koch, Epigraphische Vielfalt am Übergang vom Mittelalter zur Neuzeit, in: Österreichische Zeitschrift für Kunst u. Denkmalpflege 54, 2/3 (2000), 367-376.

2 Siehe Round-table-Gespräch II: Epigrapische Schriften zwischen Mittelalter und Neuzeit (Teilnehmer: Renate Neumüllers-Klauser, Martin Steinmann, Rüdiger Fuchs und Walter Koch), in: W. Koch (Hrsg.), Epigraphik 1988. Fachtagung für mittelalterliche und neuzeitliche Epigraphik, Graz, 10.-14. Mai 1988, Referate und Round-table-Gespräche (Denkschriften der Österreichischen Akademie der Wissenschaften, Phil.-Klasse 213 = Veröffentlichungen der Kommission für die Herausgabe der Inschriften des Deutschen Mittelalters 2), Wien 1990, 313-346, weiters W. Koch, Das 15. Jahrhundert in der Epigraphik. Die Schriften „zwischen" Mittelalter und Neuzeit in Italien und nördlich der Alpen, in: F. Magistrale – C. Drago – P. Fioretti (Hrsg.), Libri, documenti, epigrafi medievali: Possibilità di studi comparativi (Atti del Convegno internazionale di studio dell'Associazione Italiana dei Paleografi e Diplomatisti), Bari (2-5 ottobre 2000), Spoleto 2002, 587-606.

3 Vgl. zu letzteren etwa F.-A. Bornschlegel, Die frühe Renaissance-Kapitalis in Augsburg, in: W. Koch (Hrsg.), Epigraphik 1988, 217-225.

epigraphischen Bereich[4], schließlich den Beginn der Fraktur mit gelegentlich noch unverfestigten schreibschriftlichen Bastarden im Vorfeld[5]. Die Gotische Majuskel war bekanntlich – sieht man von einem späten lokalen Auftreten im Wormser Bereich ab[6] – um etwa 1400 als Schrift gesamter Inskriptionen nördlich der Alpen aufgegeben worden, jedoch durch ihre Rolle als Auszeichnungsschrift und für Initialen (Lombarden) in Codices weiterhin bekannt und auch – wenn auch abnehmend – für Initialen in Inschriften gelegentlich noch in Verwendung. Diese Vielfalt ist uns aus älterer Zeit nicht bekannt, da wir es zumindest bis ins vorgerückte 14. Jahrhundert jeweils nur mit einem Schriftstrang zu tun haben. Damals trat die Gotische Minuskel neben die auslaufende Gotische Majuskel. Die Vielfalt zur Zeit des anbrechenden und sich schließlich durchsetzenden Humanismus, einer Periode des Überganges, ist die Voraussetzung für die Aufsplitterung des epigraphischen Geschehens in der Neuzeit, wo mit der Kapitalis, der humanistischen Minuskel (Minuskelantiqua) und der Fraktur – nach der auslaufenden Gotischen Minuskel in der zweiten Hälfte des 16. Jahrhunderts – zumindest drei organisch gewachsene Schriftstränge gegeben sind, bis einerseits der Historismus im 19. Jahrhundert ältere Schreibweisen (Gotische Majuskel und vor allem Gotische Minuskel) eine Zeit lang wieder aufgriff und andererseits nach dem Ende der Fraktur neben den bis heute lebenden Antiqua-Schriften (Kapitalis und Minuskel) zuneh-

---

4 Siehe R. Epp, Eine epigraphische Minuskel zwischen Mittelalter und Neuzeit. Die Gotico-Antiqua in den Inschriften, in: Archiv für Diplomatik, Schriftgeschichte, Siegel- und Wappenkunde 47/48 (2001/2002), 167-221, weiters Dies., Gotico-Antiqua oder „Gotico-Fraktur“, in: W. Koch – G. Mras – A. H. Zajic (Hrsg.), Epigraphische Schriften „zwischen“ Mittelalter und Neuzeit (in Druckvorbereitung). Dieser Sammelband wird im Rahmen der Österreichischen Akademie der Wissenschaften in der Reihe der Denkschriften der Phil.-hist. Klasse (Unterreihe: Forschungen zur Geschichte des Mittelalters) erscheinen.

5 Siehe Koch, Vielfalt, 374-376, und F.-A. Bornschlegel, Epigraphische Bastarden und/oder frühe Frakturen, in: W. Koch – G. Mras – A. H. Zajic (Hrsg.), Epigraphische Schriften „zwischen“ Mittelalter und Neuzeit (in Druckvorbereitung).

6 Siehe R. Fuchs, Wormser Inschriften. Zur Schriftgeschichte und Quellenkunde, in: K. Stackmann (Hrsg.), Deutsche Inschriften. Fachtagung für mittelalterliche und neuzeitliche Epigraphik, Lüneburg 1984. Vorträge und Berichte (Abhandlungen der Akademie der Wissenschaften in Göttingen, Philolog.-hist. Kl., Dritte Folge 151), Göttingen 1986, 94-99.

mend die zahllosen, bereits im 19. Jahrhundert einsetzenden und großem Wandel unterworfenen Designer-Schriften und schließlich auch elektronischen Schriften die Vielfalt des 20. Jahrhunderts und unserer Zeit ausmachen und auch ihren Einfluss im epigraphischen Bereich erkennen lassen.

Was nun das eingangs angesprochene Beziehungsgeflecht zwischen Paläographie, Epigraphik und Typographie betrifft, so sind die Bezüge zwischen den beiden ersteren – zwischen den Inschriften und Auszeichnungsschriften im Codex – immer wieder zu beobachten. Während Monumentalinschriften – von unterschiedlichem Aussehen bis hin zu höchster Qualität – etwa in der Karolingischen Renaissance für Zierschriften des höchsten Niveaus in den Codices zur Anwendung gekommen sind[7], so nahm der Gotisierungsprozess in der romanischen Zeit seinen Ausgang in den Büchern und wurde erst dann – zum Teil mit beträchtlicher Zeitverzögerung – in den epigraphischen Bereich aufgenommen, wobei gemalte Inschriften häufig die Brücke bildeten[8]. Massiv ist der Bezug in Spätmittelalter und früherer Neuzeit, als etwa mit der Textura oder dann mit der Humanistischen Minuskel Stile des *Schreibens* in die Epigraphik eindrangen. Aber auch erste Spuren frühhumanistischen Gestaltens sind wohl ebenfalls in Schriftversuchen – und zwar von Humanisten in Italien schon bald nach 1400 – anzunehmen. Ebenso hat der epigraphische Bastardbereich seine ersten Wurzeln in spätmittelalterlichen Schreibschriften. Zum vielbehandelten Thema der Lettern in den Frühdrucken und ihren Vorbildern in den diversen Schreibstilen ist vor allem die ausgezeichnete Übersicht durch Otto Mazal heran-

---

7 Siehe etwa W. Koch, Inschriftenpaläographie des abendländischen Mittelalters und der früheren Neuzeit. Früh- und Hochmittelalter, Wien-München 2007, 101-113. Zu den karolingischen Inschriften (Grabinschriften) generell siehe jetzt C. Treffort, Mémoires carolingiennes. L'épitaphe entre célébration mémorielle, genre littéraire et manifeste politique (milieu VIIIe-debut XIe siècle), Rennes 2007.

8 Siehe etwa W. Koch, Inschriftenpaläographie, 158-160, bzw. Ders., Auf dem Wege zur Gotischen Majuskel. Anmerkungen zur epigraphischen Schrift in romanischer Zeit, in: W. Koch – C. Steininger (Hrsg.), Inschrift und Material. Inschrift und Buchschrift. Fachtagung für mittelalterliche und neuzeitliche Epigraphik, Ingolstadt 1997 (Abhandlungen der Bayerischen Akademie der Wissenschaften, Phil.-hist. Klasse NF 117), München 1999, 237-240.

zuziehen[9]. Dem Themenbereich Typographie und Epigraphik widmete sich zuletzt Thomas Glöß[10].

An der Schnittstelle vom Mittelalter zur Neuzeit steht die berühmte und schon frühzeitig weit verbreitete Weltchronik (*Liber chronicarum*) des Nürnberger Arztes und Humanisten Hartmann Schedel (1440-1514). Der Druck dieses Werkes wurde in der Offizin des renommierten Nürnberger Druckers, Verlegers und Buchhändlers Anton Koberger (1440/1445-1512) hergestellt. Die lateinische Fassung erschien am 12. Juli 1493, die deutsche am 23. Dezember desselben Jahres[11]. Das Werk ist eine Kompilation in der traditionellen Einteilung nach den sechs Weltaltern und folgt im Charakter den mittelalterlichen Universalchroniken[12]. Schedel fügte – hierin noch vorhumanistischem, scholastischem Denken verhaftet – ein siebentes mit Endzeitvorstellungen hinzu. Neben dem zunächst biblischen Geschehen und hernach den historischen Ereignissen als rotem Faden – das sechste Weltalter gilt der Zeit von Christi Geburt bis in

---

9 O. Mazal, Paläographie und Paläotypie (Bibliothek des Buchwesens 8), Stuttgart 1984.

10 T. Glöß, Druckschrift und Inschrift. Formzusammenhänge und wechselseitige Einflüsse von frühen Druckschriften und epigraphischen Schriften der Mitte des 15. Jahrhunderts bis zum Anfang des 16. Jahrhunderts in Deutschland, Berlin-Leipzig 2006. Siehe zuvor schon F.-A. Bornschlegel, Druckschriften und epigraphische Schriften an der Schwelle zum Frühdruck am Fallbeispiel Augsburg, in: Inschrift und Material, 213-224.

11 Sehr zu danken habe ich Herrn Dr. Thomas Horst (München), der mir sein privates Exemplar der Faksimileausgabe von 2001 der deutschen Fassung zur Verfügung stellte, das meinen Ausführungen zugrunde liegt: H. Schedel, Weltchronik. Kolorierte Gesamtausgabe von 1493. Einleitung und Kommentar von Stephan Füssel, Ljubljana 2005. Diesem Werk sind auch die Abbildungen (Bildzitate) entnommen. Die Folioangaben im Rahmen des Beitrages folgen der deutschen Fassung. – Die Tatsache einer kolorierten Ausgabe bedeutet bei den Inschriften gelegentlich Rotstrichelungen, die zu abstrahieren sind: siehe bei GENVA (im Buchstaben G), DESTRUCCIO (im Buchstaben D), HERBIPOLIS (im Buchstaben H), AVGVSTA (beim ersten A), VENECIE (beim ersten E), COLONIA (bei CO), BRESSLA (im Buchstaben B), BASILEA (vor dem Buchstaben B), CONSTANCIA (beim ersten C) und CRACOVIA (beim ersten C).

12 Siehe zum Folgenden vielfach Füssels Einleitung in Schedel, Weltchronik, weiters P. Zahn, Hartmann Schedels Weltchronik, in: Bibliotheksforum Bayern 24 (1996), 230-248, sowie C. Ahlzweig, Kompilation und Komposition. Die Schedelsche Weltchronik: Ein Meilenstein ästhetischer Buchgestaltung, in: Buch & Bild 1 (1997), 58-64.

Schedels Gegenwart und ist doppelt so umfangreich wie die fünf vorangegangenen Zeitalter – bot dieser Typus vielerlei Exkursen Platz – von Naturereignissen und Naturkatastrophen bis hin zum Geschehen in anderen Ländern, von Aussagen über Kaiser und Päpste, Philosophen und Dichter bis hin zur Gründung und Geschichte von Städten mit ihrer Kultur und ihrer Wirtschaft – mit einem Wort eine Enzyklopädie des Wissenswerten, was die Menschen in einer im Geiste des sich durchsetzenden Humanismus inzwischen wesentlich offener gewordenen Welt interessieren konnte. Dieser *Liber chronicarum*[13], eine Kompilation aus mannigfachen geistes- und naturwissenschaftlichen Thematiken, trägt – zumindest im deutschen Bereich – seinen Namen nach Hartmann Schedel, der weniger Verfasser als vielmehr Redaktor des Textes war[14]. Der lateinische Text, gedruckt in einer Druck-Rotunda, setzt sich aus verschiedenen lateinischen Werken, vielfach italienischer Humanisten zusammen, wobei Werke des Jacobus Philippus Foresta da Bergamo, Bartolomeo Sacchi, Flavio Biondo, Petrarca, Boccaccio, aber auch Vinzenz von Beauvais u. a. die Vorlagen lieferten. Von besonderer Bedeutung war für Schedel Enea Silvio de Piccolomini, vor allem dessen Stadtbeschreibungen. Im Unterschied zur lateinischen Fassung stellt die deutsche Übersetzung durch Georg Alt (1440/1450-1510), gedruckt in Schwabacher, eine einheitlichere Arbeit dar. Wenn auch Schedel in seiner Weltchronik für den Aufbau seines Werkes den Typ der spätmittelalterlichen Weltchronik wählte, so sieht man seine humanistische Ausrichtung in der Mehrzahl der ausgewählten Quellen, vielfach zeitgenössischer oder zeitnaher italienischer Humanisten, und besonders an seinen sich weit erstreckenden geographischen Interessen.

Der besondere Rang dieses Werkes, eines Meisterwerkes des noch jungen Buchdrucks, beruht nicht zuletzt auf seinen Illustratio-

---

13 Die Chronik wird im Registertitel der lateinischen Ausgabe *Liber chronicarum cum figuris et ymaginibus ab inicio mundi* genannt. Im Registertitel der deutschen Ausgabe heißt sie *Buch der Chroniken und geschichten mit figuren und pildnussen von anbeginn der welt bis auf dise unsere Zeit.*

14 In der Nürnberger Stadtbibliothek befinden sich die handschriftlichen Druckvorlagen sowohl für die lateinische als auch die deutsche Fassung. Etwa 62% der lateinischen Fassung schrieb Schedel selbst. Die handschriftlichen Vorlagen enthalten bereits den Seitenumbruch, in den die Abbildungen in kursorischen Skizzen eingezeichnet sind.

nen. Es enthält über 1800 Abbildungen, wofür 652 Holzstöcke geschnitten wurden. Für die Anfertigung der Holzschnittillustrationen, die man dann auf die Holzstöcke übertrug, konnten die Werkstätten namhafter Nürnberger Kunsthandwerker, Michael Wohlgemut (1434/1437)[15] und Wilhelm Pleydenwurff (ca. 1450-1494), gewonnen werden, wobei zahlreiche Abbildungen – in unterschiedlichen Zusammenhängen – mehrfach Verwendung fanden. Dies gilt auch für zahlreiche Städtebilder, mit Kennzeichnung (außerhalb des Bildfeldes) durch jeweils andere Städte- bzw. gelegentlich auch Ländernamen. Von 76 Städtebildern liegen authentische Ansichten für 32 Städte vor: Augsburg, Bamberg, Basel, Breslau, Buda(pest), Eichstätt, Ensisheim, Erfurt, Florenz, Genua, Heraklion, Jerusalem (mit dem Tempel Salomons), Köln, Konstantinopel, Konstanz, Krakau, Lübeck, Magdeburg, München, Neisse, Nürnberg, Passau, Prag, Regensburg, Rhodos, Rom, Salzburg, Ulm, Venedig, Wien und Würzburg[16]. Sie wurden zum Teil nach Augenschein, zum überwiegenden Teil nach zeitgenössischen Ansichten hergestellt, was in einigen Fällen belegbar ist. Diese Stadtansichten sind fast zur Gänze – neben einer deutschen bzw. eingedeutschten Namensform in Textschrift oder vergrößert in Rotunda oberhalb des Bildfeldes wie bei den sonstigen Städtebildern (siehe oben) – zusätzlich durch eine lateinische oder latinisierte Namensform gekennzeichnet, die dekorativ im Rahmen des Holzschnittes ausgeführt ist[17]. Vgl. als Beispiel die fiktive Stadtansicht von Babylon (fol. XXIV^{v}):

---

15 In dessen Werkstatt arbeitete auch der junge Dürer. Siehe dazu vor allem P. Zahn, Albrecht Dürer und die Holzschnitte der Schedelschen Weltchronik, in: Gutenberg-Jahrbuch 77 (2002), 124-144.

16 Die Veduten dieser Städte erstrecken sich – mit Ausnahme von Jerusalem, Genua und Eichstätt – jeweils über eine Doppelseite im Unterschied zu den meisten sonstigen Städtebildern.

17 Diese Namensformen sind in der lateinischen und deutschen Fassung der Weltchronik identisch. Für beide Fassungen fanden die gleichen Holzschnitte Verwendung, allerdings offensichtlich mit einer Ausnahme: Auf fol. XX trägt die Stadtdarstellung in der deutschen Fassung im Unterschied zur lateinischen – bei gleicher bildlicher Darstellung – nicht die Beischrift NINIVE. – Von den oben aufgezählten Städten tragen keine Namensform im Rahmen des Holzschnittes: Ensisheim, Heraklion, Konstantinopel, Magdeburg, München und Rhodos. In diesen Fällen ist kein Name in das Feld des Holzschnittes eingearbeitet worden. Hingegen sind zusätzlich –

Als wesentlichstes Vorbild für die Aufnahme von Stadtveduten – für manches Bild selbst als unmittelbare Vorlage[18], aber auch für deren Integration in den Text – gilt Bernhard von Breydenbachs *Peregrinatio in terram sanctam*, deren drei Ausgaben in Mainz 1486 (lateinische und deutsche Ausgabe im Februar bzw. Juni) bzw. 1488 (niederländische Ausgabe) erschienen. Holzschnittillustrationen – Städtebilder nach der Wirklichkeit – waren etwas Neues damals in der Inkunabelzeit. Im Bild geboten wurden Venedig, von wo die von Ende April 1483 bis Februar 1484 dauernde Pilgerfahrt des Mainzer Domdekans zu Schiff ihren Ausgang nahm, weiters Parenzo (Poreč),

---

über die aufgezählten Städte hinaus – auch innerhalb des jeweiligen Bildfeldes benannt: Babylon, Straßburg, *Anglie provincia, destruccio Iherosolime*, zu Bamberg zusätzlich *Mons monachorum* (Michelsberg), zu Krakau zusätzlich *Casmirus*.

18 Vgl. etwa Breydenbachs Darstellung (siehe Anm. 19) von Venedig, die bei Schedel mit einem größeren Ausschnitt übernommen ist, oder auch die der Insel Kreta (Candia – siehe in Füssels Einleitung zur Weltchronik, Abb. 30), die als Vorlage für das Bild der Stadt Mainz herangezogen wurde. Auch Jerusalem geht zusammen mit der *Zerstörung der Stadt* auf die Zeichnung Reuwichs in Breydenbachs *Peregrinatio* zurück. Dies nur als Beispiele. – Die Holzschnitte von Rom und Genua wieder sind Foresti da Bergamo, *Supplementum chronicarum ab initio mundi usque ad annum 1482*, Venedig 1492, entnommen. Zu den allfälligen Bildvorlagen siehe generell Füssel in seinem Kommentar zur Weltchronik, passim (ab Seite 634).

Korfu, Modon, Kandia (Kreta), Rhodos und Jerusalem[19]. Der Reisegesellschaft gehörte auch der Maler Erhard Reuwich (1450-1500) an, der die wichtigsten Hafenstädte und die Heiligen Stätten in Zeichnungen festhalten sollte. In unserem Zusammenhang ist von Bedeutung, dass die Ortsnamen in die Holzschnitte integriert sind[20] und somit für die Schedelsche Weltchronik zweifellos als Vorbild dienen konnten, mag es auch zu diesem einige Jahre älteren Werk manche Unterschiede geben. Von Jerusalem abgesehen, werden die Namen auf Bändern, die zum Teil sehr stark verschlungen sind, oberhalb der jeweiligen Ansicht geboten. Dies gilt ganz besonders für Venedig[21], dem als Ausgangspunkt der Seereise überhaupt besondere Aufmerksamkeit gilt. Bei Schedel sind die Namen ohne solches Zierwerk geboten, was ihnen ein monumentaleres Aussehen verleiht. Vom Graphischen her sehen wir manch Uneinheitliches. So ist Parenzo in Gotischer Minuskel (Textura) ausgeführt. Ebenfalls in Minuskeln lesen wir Rodys, wobei in diesem Fall die Ausführung im Holzschnitt offenkundig missglückt ist. Es ist nicht eindeutig anzugeben, welcher Schrifttyp wohl geplant war. Majuskel-R am Wortbeginn weist eine senkrecht nach unten verlaufende geschwungene Cauda auf. O ist auf seiner linken Seite eingedrückt, d ist unzial, s am Wortende ist geschlossen, wobei sich die beiden Bögen um einen diagonal liegenden Schaft winden. In die in kapitalen Formen ausgeführten Namen MODON und CANDIA ist unziales, in die Oberlänge reichendes D aufgenommen. A weist einen gebrochenen, bis zur Grundlinie reichenden Mittelbalken auf, die Schäfte des M sind nach außen gespreizt, der Winkel des M reicht ebenfalls bis zur Grundlinie. Im kapital geschriebenen CORFVN ist das F mit einem beträchtlichen vertikalen Strich geschlossen, was an die Gotische

---

19 Siehe T. Glöß, *Peregrinatio in terram sanctam* – Bernhard von Breydenbachs Pilgerreise im Zusammenhang paläotypischer und epigraphischer Entwicklungen, in: Leipziger Jahrbuch zur Buchgeschichte 15 (2006), 12-15.

20 Zur graphischen Gestaltung innerhalb der sieben Panoramabilder vgl. Glöß, *Peregrinatio*, 26f. Die Kennzeichnung wichtiger Gebäude und Lokalitäten innerhalb der bildlichen Darstellungen in Textschrift begegnen in Schedels Weltchronik nur ausnahmsweise (siehe etwa bei Jerusalem [fol. XVII] oder der Stadt Nürnberg [fol. XCIX^{v}]). In unserem Interesse liegen nur die dekorativen Beischriften (Städtenamen).

21 Siehe Glöß, *Peregrinatio* 26, Abb. 8. Einen anderen Ausschnitt aus der Darstellung Venedigs bei Breydenbach siehe in Füssels Kommentar zur Weltchronik Schedels, 641.

Majuskel erinnert. Die Formen zeigen die Unsicherheit der Zeit in vielerlei Hinsicht. Am ehesten noch in die uns geläufige Ausprägung frühhumanistischen Schreibens sind CIVITAS VENECIARV(M) und CIVITAS IHERVSALEM einzuordnen. Basis sind sehr schmale kapitale Formen, in die Fremdelemente eingestreut sind. Bei *Venedig* sehen wir seitenverkehrtes N, schmales unziales E sowie I mit links in der Schaftmitte angesetztem Knoten. *Jerusalem* weist jeweils epsilonförmiges E auf. Der Bogen des R ist oben offen. In beiden Namen sind die Mittelschäfte des A entweder geknickt oder schräg gesetzt. Die S-Formen weisen verhältnismäßig lange, in die Ober- bzw. Unterlänge reichende Schlussstriche auf.

Unsere Aufmerksamkeit gilt im Folgenden den Inschriften, vornehmlich den Städtebezeichnungen[22], innerhalb der Holzschnitte in der Schedelschen Weltchronik[23]. Sie übertreffen die in Breydenbachs *Peregrinatio* an Qualität und Einheitlichkeit des Komplexes insgesamt – trotz mancher Unterschiede in Details (siehe unten) – bei weitem, repräsentieren frühhumanistisches Schreiben und weisen überwiegend die Merkmale der Frühhumanistischen Kapitalis auf[24].

Der frühhumanistische Bereich ist von einer beträchtlichen, vielfach individuellen Vielfalt geprägt, doch durch manche in der Zeit liegende Gemeinsamkeiten immer wieder miteinander verbunden.

---

22 Neben den geographischen Beischriften sind vereinzelt in der gleichen graphischen Gestaltung (Frühhumanistische Kapitalis) auch Personennamen wie die Söhne Noes Sem, Iaphet und Cham (fol. XII^{v} - XIII) sowie Noe mit seinen Söhnen (XV^{v}), Iob (fol. XXIX) und Salomon (fol. XLVII^{v}) ausgeführt, weiters Archa Noe (fol. XI). Siehe Anm. 29. Zu Casmirus (im Bild der Stadt Krakau, fol. CCLXIIII^{v}) siehe unten.

23 Zu dem vom Augsburger Drucker Johann Schönsperger 1496 bzw. 1497 veranstalteten verkleinerten Nachdruck der Weltchronik siehe in Füssels Einleitung zur Weltchronik, 34-36. Die Illustrationen in diesem Nachdruck besitzen keine Beischriften, vielmehr werden die Städtenamen jeweils oberhalb, d. h. außerhalb des Bildes, in vergrößerter Textura angegeben.

24 Siehe den kurzen Hinweis von R. Neumüllers-Klauser, Epigraphische Schriften zwischen Mittelalter und Neuzeit (Grundsatzreferat), in: Epigraphik 1988, 317, auf den Gebrauch dieser Schrift in Schedels Weltchronik. Siehe auch ihre Auflistung der Verwendung dieser Schriftformen – in der Tradition der Handschriften – für Kolumnentitel, Bildbeischriften, Namensbeischriften und Kolophone in anderen Werken des Frühdrucks. Vgl. auch F. G. Blay, *Suole l'usanza antiqua cavare la littera.* Auf der Suche nach einem graphischen Vorbild für die Inschriften, in: W. Koch – G. Mras – A. H. Zajic, Epigraphische Schriften (in Druckvorbereitung).

Wir sehen Schreibweisen, die als Übergangsschriften hinführen zur im Zeichen von Renaissance und Humanismus wieder aufgegriffenen alten Kapitalis, also Übergangsschriften, ebenso – und das sind die interessanteren Fälle – Schriften einer Übergangszeit, Specimina sui generis, die nicht in die Renaissancekapitalis einmünden, sondern – nach einer kurzen Zeit des Nebeneinanders – deren Übermacht weichen müssen. Dies gilt für die Schrift, für die sich die Bezeichnung „Frühhumanistische Kapitalis" eingebürgert hat, ein Schreibstil, der schließlich überregional Anwendung fand und trotz verfließender Ränder sich doch einigermaßen verfestigt hat. Die Vielfalt frühhumanistischer Ausprägungen liegt einerseits in den verschiedenen Ansätzen zu ihrer Entwicklung begründet, oftmals singulär von einem einzigen Meister getragen, auch in unterschiedlichen Regionen initiiert[25]. Diese Schreibweisen, geprägt von der Freude am Innovativen im Rahmen der neuen Geisteshaltung, tragen viel zur Buntheit des epigraphischen Bildes im 15. Jahrhundert bei, mag es sich auch nur um ausgesprochene „Minderheitenschriften" neben den dominierenden Schreibweisen – etwa nördlich der Alpen die Gotische Minuskel – handeln. Dies gilt auch für die Frühhumanistische Kapitalis. Wenn von polygenetischem Ursprung die Rede war, so ist für diese der Ansatz bei italienischen Frühhumanisten im ersten Viertel des 15. Jahrhunderts, die für ihre neue Minuskel in den Codices (Frühhumanistische Minuskel) nach geeigneten Majuskeln – vor allem für Titeleien, Incipits, Explicits und Kolophone – suchten, inzwischen wohl evident. Wenn sich in diesem Raum diese Schreibweise im Epigraphischen dann seltener findet, so dürfte dort das rasche Aufgreifen der Kapitalis, wenn auch noch geraume Zeit nicht der klassischen monumentalen Ausprägung, der Frühhumanistischen Kapitalis keine echte Möglichkeit, sich in größerem Maße zu etablieren, geboten haben. Ihr Hauptverbreitungsgebiet war, soweit wir bisher sehen, der deutsche Bereich, wo zunächst vornehmlich der Südwesten mit den Konzilsorten Konstanz und Basel, die eine internationale Begegnung von Personen und Ideen brachten, sowie der Südosten mit den Kaiserresidenzen Wiener Neustadt und Graz das

25 Zum polygenetischen Ursprung jener Schriften zwischen Mittelalter und Neuzeit im Majuskelbereich sowie zu ihrer Formensprache siehe die in Anm. 2 zitierte Literatur, weiters W. Koch, Variationsfreudige Majuskel, in: G. Ardenna – H. Houben (Hrsg.), Mediterraneo, Mezzogiorno, Europa. Studi in onore di Cosimo Damiano Fonseca, Bari 2004, 621-640.

neue Schreiben aufnahmen. Der Formenbestand setzt sich neben gotischen Relikten und schon manchen Elementen der wieder kommenden Kapitalis vornehmlich aus vorgotischen Formen des hohen Mittelalters, häufig auch Nebenformen, zusammen, in die Elemente, die wir aus der griechischen Welt kennen, aufgenommen wurden – alles insgesamt einem neuen gemeinsamen Stilgefühl unterworfen, wobei allerdings in der Ausführung Spezimina, die – aufgrund einer fetten und flächigen Gestaltung – im Gesamteindruck noch an das Gotische erinnern[26], anderen gegenüberstehen, die aufgrund einer linearen und dünnstrichigen Gestaltung „moderner“ anmuten[27]. Dies eröffnet eine breite Palette in der Gestaltung, nicht zuletzt durch die Formenvielfalt, wobei ein Großteil der Buchstaben durch mehrere, aus verschiedenen Bereichen herrührende Exemplare gegeben ist, die häufig nebeneinander – nach Laune des Meisters – herangezogen werden konnten. Das bedeutet ein hohes Maß an Dekorativem, das dieser Schrift innewohnt. Dies zeigt sich auch in ihrer Anwendung. Wir sehen sie vornehmlich etwa für Titeleien in Codices, als Beischriften auf Fresken und in der Tafelmalerei, insgesamt im kunstgewerblichen Bereich, etwa in Schnitzarbeiten, natürlich auch in der steinernen Epigraphik. Es sind – von eher wenigen Fällen abgesehen – überwiegend Schriftbeispiele von hoher Qualität, nicht selten auch für Auszeichnungspartien in Verbindung mit „kleineren“, als alltäglich empfundenen Schriften wie etwa der Gotischen Minuskel. Bei diesem Rückgriff auf ältere Formen geht es im Wesentlichen darum – es ist dies nicht anders als bei der Kreierung der Humanistischen Minuskel der Textschrift für die Codices nach dem Vorbild der karolingischen, als alt empfundenen Minuskel des Hochmittelalters -, die wenig geschätzte gotische Zeit möglichst zu überspringen. Dieses Streben *ad fontes* entspricht durchaus auch humanistischer Gesinnung

---

26 Vgl. etwa die Fastrada-Inschrift aus St. Alban, jetzt im Mainzer Dom. Siehe Die Deutschen Inschriften 2: Die Inschriften der Stadt Mainz von frühmittelalterlicher Zeit bis 1650, gesammelt und bearbeitet von Fritz Viktor Arens aufgrund der Vorarbeiten von Konrad F. Bauer, Stuttgart 1958, 1 Nr. 1.

27 Vgl. etwa die Inschrift auf der Grabplatte der Kaiserin Eleonore, der Gattin Friedrichs III., in der Neuklosterkirche zu Wiener Neustadt. Siehe Die Deutschen Inschriften 48: Die Inschriften des Bundeslandes Niederösterreich 2: Die Inschriften der Stadt Wiener Neustadt, gesammelt und bearbeitet von Renate Kohn, Wien 1998, 59-61, Nr. 98 (Abb. 39 und Frontispiz).

und das Geschaffene – in unserem Fall die Frühhumanistische Kapitalis – hat den Leuten damals durchaus als etwas Innovatives gegolten[28]. Ob dies auch noch in späterer Zeit, als sich die Kapitalis immer mehr durchsetzte, so gesehen wurde, muss mehr als bezweifelt werden. Neben dem herausragenden Wert als Dekorschrift könnte dann – wohl nach Region unterschiedlich früh – schließlich inzwischen auch etwas Konservatives dieser Schreibweise angehaftet haben.

Die Verwendung frühhumanistischen Schreibens und insbesondere der Frühhumanistischen Kapitalis – freilich kommen in den kurzen Texten nicht immer die herausragenden Leitbuchstaben vor – für die Beischriften in den Stadtveduten der Weltchronik Schedels hat sicherlich ihren Hauptgrund zunächst im Dekorativen dieses Schriftstils. Darüber hinaus passt die Frühhumanistische Kapitalis auch thematisch ausgezeichnet zum Charakter dieses Druckwerkes. Beide stehen an der Schnittstelle zwischen mittelalterlicher und humanistisch-neuzeitlicher Gesinnung. Die von der Anlage her noch spätmittelalterliche Weltchronik ist schon voll von humanistischem Geist bestimmt und auch die Schrift verbindet vielfach konservative Schreibformen mit dem Wunsch nach Innovativem. Eine nähere Analyse der einzelnen Tituli macht verschiedene Gruppierungen offenkundig, wohl wenigstens zum Teil verschiedene Hände, von denen die Druckvorlagen für den jeweiligen Text hergestellt wurden – ausführende Meister oder Gesellen, die frei und nach Belieben mit dem damaligen Reservoir an möglichen Schreibformen umgingen.

Zunächst ist eine größere Zahl von Texten zusammenzufassen, die einander von der Formensprache her ungemein nahestehen (Gruppe I). Dass alle von derselben Hand stammen, soll nicht grundsätzlich ausgeschlossen werden. Allerdings sind in der unmittelbaren Ausführung – im Gesamtcharakter – Differenzen festzustellen, die die Annahme einer zumindest zweiten Hand als möglich erscheinen

28 Nicht umsonst hatte der Münchener Arzt Sigmund Gotzkirchner auf einer Italienreise 1436 in eines seiner Notizbücher zwei Alphabete mit 250 bzw. 150 Formen – vielfach skurrilen Formen – eingezeichnet. Siehe die Abb. etwa bei Neumüllers-Klauser, in: Epigraphik 1988, Abb. 1 und 2 (nach Seite 328), oder Koch, Variationsreiche Majuskel, Abb. 5 und 6. Man hatte Interesse an diesem Spiel mit Formen, die viele individuelle Gestaltungen ermöglichten. Dieses differenzierte Sehen von Schriften und ihrer Herkunft wäre vor der Zeit des Humanismus nicht denkbar gewesen.

lassen. Eine Gruppe (Ia) weist ein gezierteres Aussehen auf und ist vor allem durch einen deutlichen Wechsel von Haar- und Schattenstrichen bestimmt, macht insgesamt also einen sorgfältigeren Eindruck: BAMBERGA (fol. CLXXIV^{v}), MONS MONACHORVM (Michelsberg) (fol. CLXXV), DESTRVCCIO IHEROSOLIME (fol. LXIII^{v}), RATISBONA (fol. XCVII^{v}), BABILONIA SEV BABILON (fol. XXIV^{v}), GENVA (fol. LVIII^{v}) und wohl auch FLORENCIA (fol. LXXXVI^{v})[29]. Ihnen stehen Ausprägungen gegenüber, die ein wesentlich fetteres, kräftigeres und insgesamt weniger gekonntes, eher klobiges Aussehen aufweisen. Es ist kaum ein Unterschied zwischen Haar- und Schattenstrichen gegeben (Gruppe Ib): NVREMBERGA (fol. XCIX^{v}), SALCZBVRGA (fol. CLII^{v}), AVGVSTA (fol. XCI^{v}), ROMA (fol. LVII^{v}), EISTETT (fol. CLXII) und merklich abweichend ERFORDIA (fol. CLV^{v}). Als dritte Gruppe (Ic) sollen zwei Beischriften angeführt werden, die ein charakteristisches Merkmal gemeinsam haben (siehe unten). Sie jedoch als eigenen Bereich zu benennen, steht freilich auf unsicheren Beinen, so dass eine Einordnung in die beiden Gruppen zuvor denkbar wäre.

Gruppe Ia: 

---

29 In dieses Umfeld sind auch die Namen der Söhne Noes CAM, IAPHET und SEM (fol. XII^{v}-XIII), weiters NOE, CHAM, IAPHET und SEM (fol. XV^{v}), ARCHA NOE (fol. XI), IOB (fol. XXIX) und REX SALOMON (fol. XLVII^{v}) einzuordnen (vgl. Anm. 22). Vgl. dazu auch die Beischrift TEMPLVM SALOMO(N)IS im Rahmen der Stadtansicht von Jerusalem (fol. XVII). In dem in verhältnismäßig geringer Größe geschriebenen ARCHA NOE sind C und das epsilonförmige E mit Schlussstrich – nach Art der Gotischen Majuskel – versehen.

Gruppe Ib:

Gruppe Ic:

Das Kriterium, das den gesamten Komplex vordergründig verbindet, ist fürs erste die Setzung jenes paragraphenartigen Zeichens, das ein Quadrangel in seiner Mitte hat[30]. Es rahmt die Inschriften und (oder) wird – bei mehr als einem Wort – als Trennzeichen verwendet (siehe bei BABILONIA SEU BABILON, MONS MONACHORUM bzw. DESTRVCCIO IHEROSOLIME). An Dekorelementen – dies gilt vornehmlich für die Gruppe Ia, aber nicht ausschließlich – sehen wir den Doppelansatz von Zierpunkten in der Mitte des I-Schaftes, einmal einen ausgesparten Kreis (siehe bei RATISBONA), weiters Ausbuchtungen am Diagonalschaft des N, vereinzelt Zierpünktchen an den Scheitelstellen des O (siehe bei IHEROSOLIME und MONACHORVM) und Haarlinien neben Schattenlinien (siehe in A bei BABILONIA SEV BABILON und BAMBERGA, bei letzterem auch beim Winkel des M). Der Buchstabe A ist so gut wie ausschließlich durch zwei Formen gegeben: Dreieckiges oder leicht trapezförmiges A mit gebrochenem Mittelbalken und mächtigem, nach beiden Seiten reichendem Kopfbalken[31] bzw. ein rechtslastiges dreieckiges A mit beträchtlichem Kopfbalken nach links. Für E findet sich die Epsilon-Form, ein markanter Buchstabe der Frühhumanistischen Kapitalis, aber auch im spielerischen Wechsel die kapitale Form. Einmal sehen wir offenes unziales D und unziales E (siehe bei DESTRVCCIO). Diese Form und ebenso C sind – wie häufig im frühhumanistischen Bereich – recht schmal. G ist verhältnismäßig breit, wobei die einstige gotische Rolle des Buchstabens zu einem mächtigen Winkel umgewandelt ist. Dies sehen wir immer wieder in frühhumanistischen Schriften und unterscheidet sich deutlich von der klassischen kapitalen Form. M ist hauptsächlich durch die Form mit den beiden senkrechten parallelen Außenhasten gegeben, wobei der Winkel nur geringfügig über die Mitte herunterreicht[32]. Allerdings findet sich auch das sogenannte griechische M – als erster Buchstabe von MONACHORVM[33]. Auch dies ist eine der

[30] Bei den Beispielen der Gruppen Ib und Ic ist die obere Schlinge des Zeichens nahezu stets ganz oder fast geschlossen.

[31] Das Fehlen dieses Kopfbalkens beim ersten A von RATISBONA ist wohl durch die Platzverhältnisse zu erklären.

[32] M mit nach außen gespreitzten Hasten ist im Wort MONACHORVM wohl durch Nexus litterarum VM bedingt. Siehe die Form – mit kleinem Winkel – aber auch in fol. XIII und XV^{v} (vgl. Anm. 22 und 29).

[33] Siehe die Form auch fol. XV^{v} (vgl. Anm. 22 und 29).

Leitformen der Frühhumanistischen Kapitalis, die sich jedoch mit den anderen M-Formen abwechselt. Die Cauda des R ist ein steifer, nach unten sich verbreiternder Stachel, gelegentlich leicht nach außen gekrümmt. Schäfte und Balken weisen beträchtliche ausgerundete Serifen auf, die bei der Gruppe Ib aufgrund der fetten und vielfach pfahlförmigen Schäfte und Balken meist weniger zur Geltung kommen[34]. Die beiden in Gruppe Ic genannten Beischriften werden durch zwei links und rechts übereinandergestellte paragraphenförmige Zeichen bestimmt. Ansonsten entsprechen sie im Wesentlichen den beiden vorigen Gruppen, wobei ARGENTINA eher zu Gruppe Ib tendiert, während PRAGA eher an die erste Gruppe erinnert.

Den im Wesentlichen gleichen Formenbestand weisen vier weitere Beischriften auf (Gruppe II): HERBIPOLIS (fol. CLIX^{v}), COLONIA (fol. XC^{v}), VENECIE (fol. XLIII^{v}) und BVDA (fol. CXXXVIII^{v}). Sie könnten von den Händen der vorigen Beispiele stammen. Den Unterschied macht freilich aus, dass sie nicht zwischen jene paragraphenartigen Zeichen gesetzt sind, in einem Fall (COLONIA) jedoch zumindest zwischen zwei bloße Quadrangeln. Die ersten beiden stehen der zarteren und verspielteren Ausführung von Gruppe Ia näher[35], die beiden anderen dem kräftigen fetten Stil von Gruppe Ib. Besondere Beachtung verdient die D-Form in BVDA. Oben offenes unziales oder kapitales D gehörten zum Standardbestand der frühhumanistischen Schriften und begegnet häufig[36]. Zwar eher selten, aber durchaus zu finden, ist jedoch jene

---

34 Aus dem Rahmen fällt merklich – wir deuteten es oben an – ERFORDIA. Insgesamt entspricht es dem etwas klobigen Charakter der Gruppe Ib. Die paragraphenartigen Zeichen muten jedoch missglückt an. Die Balken bei E und F weisen mächtige Schlussstriche auf. A hat den Kopfbalken nach rechts. Die Cauda des R trägt eine Schwellung wie in gotischer Zeit. I ist mit Punkt versehen.

35 In beiden Städtenamen sehen wir kleine eingeschriebene Buchstaben, vgl. dazu BABILON in Gruppe Ia. In HERBIPOLIS berühren die Bögen von B und R nicht den Schaft, vgl. dazu B in BAMBERGA in Gruppe Ia.

36 Zum offenen unzialen D vgl. im Wort DESTRVCCIO (siehe oben). Zum offenen kapitalen D – oben etwas verkürzter Schaft und meist überhängender Bogenansatz – vgl. etwa Neumüllers-Klauser, in: Epigraphik 1988 (wie in Anm. 2), Abb. 10 und 12 (nach Seite 328).

kapitale Form, bei der der Schaft zur Gänze fehlt[37]. Die Form im Worte BVDA ist davon abgeleitet, wobei bei einem rechteckig gestalteten Buchstabenkörper die gesamte linke Längsseite offen bleibt[38].

Gruppe II:

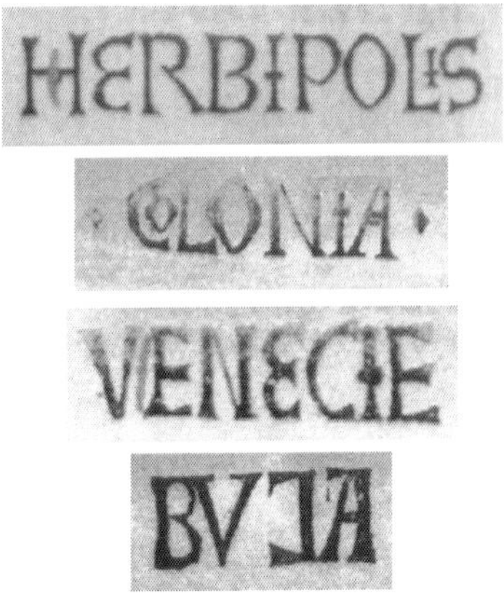

Die eigenwillige Gestaltung des A – fettes trapezförmiges A mit mächtigem Kopfbalken und gebrochenem Mittelbalken, dessen linker Teil mit Haarstrich bis zur Grundlinie verlängert ist – empfiehlt, vier Beischriften zusammenzustellen (Gruppe III), zunächst zwei im Duktus einander sehr nahestehende Schriftbilder: LVBECA (fol. CCLVI) und NISSA (fol. CCLVI^{v}), weiters BRESSLA (fol. CCXXXII^{v}) und sich abhebend VIENNA PANNONIE (fol. XCVIII^{v}). Wir sehen Zierpunkte im Wort NISSA, epsilonförmiges E in LVBECA und BRESSLA. Letzteres mutet stärker „gotisch“ an. Dies ergibt sich aus der mächtigen Schwellung an der Cauda des R

37 Vgl. etwa in der Fastrada-Inschrift (siehe Anm. 26, auch abgebildet bei Neumüllers-Klauser, in: Epigraphik 1988, Abb. 32 [nach Seite 328]).

38 Wie die Kontrolle im Original der lateinischen Fassung der Weltchronik erwartungsgemäß ergab, findet sich dort dieselbe etwas eigenwillige Ausführung des Buchstabens. In der Faksimile-Ausgabe der lateinischen Fassung (H. Schedel, Liber Chronicarum, hrsg. von der Bibliothek der Abtei Niederalteich, Puchheim 1967-1970) wurde die für den Leser ungewöhnliche Form jedoch durch das uns geläufige kapitale D ersetzt, und zwar durch die „Grotesk“-Form des Buchstabens, die erst im 19. Jahrhundert entwickelt wurde und bis zum heutigen Tag Verwendung findet. Es ist eine serifenlose Form, bei der alle Linien – ob Schaft oder Bogen – gleich stark gestaltet sind. Im Worte BVDA der lateinischen Faksimile-Ausgabe ist sie völlig unpassend.

sowie aus der Ausführung des L, dessen Balken spitz nach oben reicht und in einer gebogenen Haarlinie ausläuft. VIENNA PANNONIE weicht beträchtlich von den anderen ab. Es ist in fetten, langgestreckten und eng gesetzten ausschließlich kapitalen Formen geschrieben. Neben dem eingangs beschriebenen A weist diese Inschrift auch ein stark linkslastiges A mit Kopfbalken nur nach rechts auf. Das Schriftbild hebt sich im Gesamteindruck, vor allem in den Proportionen wie auch in der Verteilung von Haar- und Schattenlinien von den vorigen Inschriften, aber auch vom Bild der im Kommen begriffenen Kapitalis noch deutlich ab. Die Gestaltung des P, der einzigen breiten Form, mit dem mächtigen, nahezu bis zur Grundlinie reichenden Bogen reiht die Schrift letztlich doch in den frühhumanistischen Bereich ein. Die ersten beiden Namen stehen zwischen zwei Quadrangeln, die beiden anderen Texte verwenden ein Zeichen nach Art eines seitenverkehrten Blitzes.

Gruppe III:

Einen merklich anderen Charakter weisen eine Reihe anderer Texte auf (Gruppe IV). Sie sind in das Umfeld der wieder in Verwendung gekommenen Kapitalis zu setzen – freilich einer Kapitalis, die sich noch auf dem Wege zur klassischen Ausprägung befindet. Wir sehen da und dort noch Relikte aus früherer Zeit, wie etwa pfahlförmige Schäfte, manche Zierelemente (etwa in CONSTANCIA) oder frühhumanistische Anklänge: BASILEA (fol. CCXLIII^{v}), VLMA (fol. CXC^{v}), CONSTANCIA (fol. CCXLI), PATAVIA (fol.CXCIXv[39]), weiters ANGLIE PROVINCIA (fol.

39 Diese vier Veduten befinden sich im hinteren Teil des Werkes. Dies gilt auch für Krakau (siehe unten).

XXXXVI) und sich beträchtlich abhebend HIEROSOLIMA (fol. XVII). Zunächst springt ins Auge die ausschließliche Verwendung des dreieckigen kapitalen A, freilich mit pfahlförmigen Schäften. Was diesen Buchstaben betrifft wie auch das Wort insgesamt, so ist BASILEA schon am deutlichsten einer guten Kapitalis verpflichtet. In VLMA hingegen fällt der kleine Winkel des M, der aus leicht nach außen gekrümmten Haarlinien besteht, auf. Der Balken des L weist mit einem deutlichen Spitz nach oben. Typisch für die frühhumanistische Welt ist P von PATAVIA. Der übergroße Bogen reicht fast bis zur Grundlinie. Von diesen vier Inschriften ist abzusondern ANGLIE PROVINCIA. Wir sehen den Trend zur Setzung von dünnen Kopf- und Fußstrichen – auch beim kapitalen A. R hat eine Cauda mit übergroßer Schwellung, während G, was die Gestaltung der Cauda betrifft, der klassischen Form entspricht, während die schmale Proportion des Buchstabens an die frühhumanistischen Schriften erinnert. HIEROSOLIMA haben wir wegen der A-Form hier aufgenommen, es wäre ansonsten durchaus in Gruppe I einzuordnen[40].

Gruppe IV:

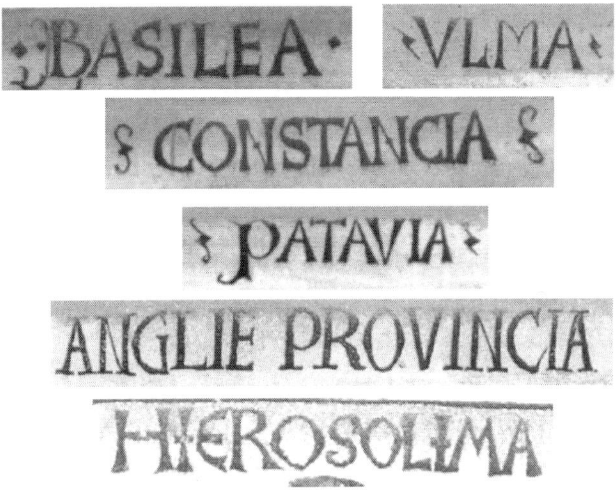

[40] Die beiden letzten Inschriften weisen keine Anfangs-, Schluss- bzw. Trennzeichen auf, die anderen vier wohl, doch lässt deren andere Schriftgestaltung nicht zu, sie früheren Gruppen zuzuordnen.

Es empfiehlt sich, die beiden Inschriften auf der Stadtdarstellung Krakaus eigens zusammenzufassen (Gruppe V): CRACOVIA (fol. CCLXV) und CASMIRVS (fol. CCLXIV^{v}). Nach den Proportionen des Gesamtbildes wie der Einzelformen, nach der Verteilung von Haar- und Schattenlinien (Linksschrägenverstärkung) haben wir die wieder gewonnene Kapitalis vor uns, mag sie auch in hochrangigen Inschriften in Stein wesentlich perfekter und monumentaler gestaltet sein. Was in den beiden Inschriften jedoch als abweichend auffällt, sind – neben den I-Punkten – die kleinen Kopf- und Fußstriche. Sie mögen als Element des „Zeichnerischen" bei der Herstellung der Druckvorlage zu werten sein[41].

Gruppe V: 

Diese kurze schriftkundliche Präsentation der Beischriften in den Holzschitten der Städtebilder in Schedels Weltchronik lässt erkennen, wie lohnend es ist, Auszeichnungspartien im frühen Buchdruck – wie im handschriftlichen Bereich – unter epigraphischen Gesichtspunkten zu untersuchen. Im vorliegenden Fall sehen wir das Formenrepertoire, das den ausführenden Meistern damals im vom Humanismus schon bestimmten ausgehenden 15. Jahrhundert in Deutschland zur Verfügung stand, wenn sie sich Neuem und zugleich Dekorativem zuwenden wollten. Sie hatten eine beträchtliche Auswahl, aus der sie mehr oder weniger frei schöpfen konnten. Diese reicht von den frühhumanistischen Schreibweisen und hierbei vor allem der Frühhumanistischen Kapitalis, den zeitgleichen frühen Ausprägungen der wieder aufgegriffenen Kapitalis bis zu deren weitgehender Ausgestaltung nach dem antiken Vorbild.

41 Die Schrift auf der Darstellung Krakaus entspricht in etwa der auf der Karte der *Germania magna* auf der letzten Doppelseite des Werkes. Zur Deutschlandkarte und ihrer Entstehung siehe im Kommentar Füssels zur Weltchronik, 666f.

Zum Autor

Walter Koch, geb. 1942 in Wien, Dr. phil., Habilitation 1978 an der Univ. Wien, 1982-2007 o. Prof. für Geschichtliche Hilfswissenschaften an der Univ. München, o. Mitgl. der Bayerischen Akademie der Wissenschaften (Vorsitzender der Kommission für die Herausgabe der Urkunden Kaiser Friedrichs II., Vorsitzender der Kommission für die Herausgabe der Deutschen Inschriften des Mittelalters und der frühen Neuzeit), korr. Mitgl. der Österreichischen Akademie der Wissenschaften im Ausland, o. Mitgl. der Zentraldirektion der Monumenta Germaniae Historica, Mitgl. der Commission Internationale de Diplomatique, Mitgl. des Comité International de Paléographie latine, Mitgl. der Interakademischen Kommission des deutschen Inschriftenwerks.

Kontakt: Universität, Historisches Seminar. Abt. Historische Grundwissenschaften und Historische Medienkunde, Geschwister-Scholl-Platz 1, 80539 München

Norbert Wend von Linck,
Mappa des March-Flus,
Wien 1719

Ivan Kupčík, München

Die Karte des March-Flusses mit der Einzeichnung des Oder-Donau-Kanals von ca. 1719

Ein wenig bekanntes Dokument zur kartographischen Darstellung der Markgrafschaft Mähren (seit 1182) und des früheren Kronlandes der Österreichischen Monarchie (seit 1526 unter habsburgischen Herrschaft) ist die in der Sammlung Woldan vorhandene *Mappa des March-Flus, wie solcher in die | Donau lauft, und wie die so grossen Nutzen | bringende Schiffart, unter Glorwürdig | ster Regirung Ihrer Römisch Kaiser zu His | panien, Hungarn und Böheim König, Mayst. | Caroli VI. durch den flus Beczwa, vermög eines Canals, oder Durchschnit in die | Oder gebracht wird, dadurch die Schiffart | und Handlschaft vom Teütschen bis in das Schwartze Meer eingerichtet werden.* Der Titel der Karte befindet sich in der Titelkartusche rechts oben.

Die geographisch-thematische Karte in zwei Blättern im Maßstab 1:158 000 und in montierter Form hat innerhalb des Kartenrahmens das Format 111 x 43 cm (Tafel S. 551). Das Kartenbild stellt den größten Teil des Landes mit reichem topographischen Inhalt und im noch größeren Maßstab als die Karte von Mähren des Johann Christoph Müller von 1716 im Maßstab 1:180 000 dar. Die Karte ist nach Südosten (oben) orientiert, mit Olmütz (Olomouc) in linker unterer Ecke und mit Preßburg (Bratislava) in oberer rechter Ecke. Es fehlt der obere Lauf der March mit dem Quellengebiet, wo die March in der Höhe von 1275 m am südlichem Hang von Králický Sněžník (Großer Schneeberg, 1422 m), des höchsten Berges des Glatzer-Schnee-Gebirges, entspringt.

Die Karte enthält keine Orientierungsangaben. Der Kartenrand besteht aus einer einfachen Randlinie ohne Graduierung der geographischen Koordinaten; die geographische Lage der dargestellten Orte lässt sich also nicht feststellen. Das Kartenbild erstreckt sich etwa von 48°00´ bis 49°50´ nördlicher Breite und von 16°50´ bis 18°00´ östlicher Länge. Auffällige Unstimmigkeiten zwischen Kartenbild und Realität waren eine Folge der mangelnden Überprüfung der geographischen Koordinaten und wurden erst später verbessert.

Links von der Titelkartusche ist eine komplette Windrose mit 32 Richtungen zu sehen: vier Hauptrichtungen sind mit Abkürzungen der lateinischen Termini *Meridies* (Süden), *Occidens* (Westen), *Septentrio* (Norden) und *Oriens* (Osten) bezeichnet.

In der zweiten Kartusche mit Ornamentrahmen in der Mitte der Karte ist im unteren Teil der folgende Text zu finden: *Norbert Wend von Linck, Obristwachtmaister bey der Kay. | Quarnison der granicz Vestung, und König. Stadt. Hradisch | hat mit der Kaÿ. hoch Löbl. Commission in August 1719 den March | Flus visitirt, und dise Mappam delinirt.* Die Karte wurde somit von Norbert Wend von Linck, dem Kommandanten der kaiserlichen Garnison und königlichen Stadt Hradist (Uherské Hradiště) wohl in Wien etwa 1719, hergestellt. Die Stadt Hradiště war nach dem kaiserlichen Reskript vom 15. September 1655 eine von sechs damaligen mährischen Festungen *Olmütz* (Olomouc), *Prerau* (Prerov), *Lipník* (Lipník nad Bečvou), *Tobischau* (Tovačov), *Kremsír* (Kroměříž), und *Hradisch* (Uherské Hradiště), die noch 1708/1712 als befestigte Städte galten. Im oberen Teil der Kartusche ist eine Maßstabsleiste: 3 Meilen [deutsche (geographische) Meile à 7 420 m] zu 14 mm angegeben.

Die Hydrographie bildet das Gerüst der Karte, denn sie stellt in erster Linie den Verlauf des mährischen Nationalflusses vom heutigen Naturschutzgebiet Litovelské Pomoraví mit fast unberührten Auwäldern ca. 15 km nördlich von Olmütz oberhalb des Dorfes *Cometau* (Chomoutov) bis zur Mündung in die Donau unter dem Grenzberg Děvín (Theben), 132 m über dem Meer, unweit der Donaustädte *Altenburg*, *Hainburg* und *Prespurg*, dar. Alle wichtigen Zuflüsse sind mit Flussnamen ausgestattet – *Beczwa Flus* (Bečva), *Ober Bezwa* (Horní Bečva), *Taya Flus* (Thaya, Dyje) und *Zeua flus* (Zaya) – genauso wie *Donau Strom* und die größten Teiche, die überwiegend in der Nähe der Flüsse March und Beczwa eingezeichnet sind (*Pisek Taich*, *Teich Nimersat*). Sonst benennt Linck keinen Nebenfluss oder

Bach, auch wenn viele andere Flüsse dargestellt sind. Die einzige Ausnahme ist der linke Zufluss *Rautenbach* aus der Slowakei. Der Fluss Beczwa ist eingezeichnet in voller Länge von beiden Zweigen des Oberlaufes bei *Rosenau* (Rožnov pod Radhoštěm) und bei *Rosinkof* (Starý Hrozenkov) bis zur Mündung in die March. Die Oder, die ein von der March unabhängiges Sammelgebiet bildet, ist von der Quelle bei *Kosel* (Kozlov) in der Nähe von Fidlův kopec, 680,1 m, bis zum Ort *Polanka* (Polanka nad Odrou) bei Ostrau wiedergegeben. Für die Oderquelle östlich von Olmütz benutzte Linck den speziellen Gewässernamen *Fontes Oder.*

Die Zuflüsse sind fast vollständig erfasst, nur findet man bei ihrer Darstellung sehr oft schematische, wellenförmige Abschnitte. Wir können nicht davon ausgehen, dass sie tatsächlich den realen Flussverlauf festhalten. Es fehlt die Einzeichnung des Flusses *Myjava*, des ersten slowakischen Zuflusses von links. Die Donau und die untere March sind horizontal schraffiert und ihre Breite ist überdimensional dargestellt. Die Brücken über die Flüsse sind nicht gekennzeichnet. Die einzige Ausnahme ist die Donaubrücke (eher eine Pontonbrücke) über den *Donau Strom* in Preßburg und westlich davon die Brücke über die Leitha. Die Darstellung der Straße zwischen den beiden Brücken ist die einzige Wegemarkierung auf der Karte. Wenn man die Flüsse der Karte mit der Müller-Karte von Mähren vergleicht, findet man nur z. T. passende Übereinstimmung, weil die ganze Karte um etwa 115° in Richtung Westen gedreht worden ist.

Neben der klassischen Hydrographie zeigt die Karte zum ersten Mal die Einzeichnung des geplanten Kanals zwischen Oder und March in Ostmähren durch den Fluss *Beczwa*. Die Idee, eine Wasserstraße zwischen Oder und Donau zu bauen und damit die Ostsee mit dem Schwarzen Meer zu verbinden, hatten die mährischen Stände bereits 1653 nach der Beendigung des Dreißigjährigen Krieges diskutiert. Der Weiterbau der Fortifikationen nach den Verwicklungen in Polen (1657-60) und nach den Kriegen mit der Türkei (1660-64), mit Frankreich (1673-79) und wieder mit der Türkei (1682-99) entleerten jedoch die kaiserliche Kasse zur Gänze. Aus gleichem Grund scheiterten 1700 auch die Pläne des Experten für Wasserbau, des Niederländers Vogemonte, eine Verbindung zwischen Elbe, Oder, Weichsel und Donau zu realisieren.[1] Kartographisch wurde das ehrgeizige Vorhaben zum ersten Mal 1719 durch Norbert Wend von Linck angegangen.

Der künstliche Kanal sollte in der Mährischen Pforte auf der Passhöhe von 310 m ü. M. im kürzesten Abschnitt zwischen den Orten *Lesna* (Lesná) und *Olmendorf* (Polouvsí) gebaut werden (Kartenausschnitt 1b). Zweck des Unternehmens wars, auf der March das kaiserliche Salz aus der rumänischen Region Marmarosch an der oberen Theiß in die Stadt *Nepagedel* (Napajedla, 200 m ü. M., oberhalb von *Hradisch* im zentralen Teil der Karte) zu befördern, wo man plante, ein großes Salzlager zu errichten. Die geographische Lage in der Nähe von Napajedla zwischen *Otrokowiz* (Otrokovice) und *Hradisch* war für den Bau des Kanals besonders günstig, denn die March steigt bei Napajedla in das schmale Napajedel-Tor ein, welches das Obermährische Tor vom Untermährischen Tor geomorphologisch trennt, und hatte schon damals ausreichend Wassertiefe.

Der Verkehr sollte an eine Privatperson vergeben werden, mit dem Auftrag auf eigene Kosten die Schifffahrtstraße zu errichten. Man plante die vorgeschlagene Wasserstraße mittels des Flusses Bečva bei *Chorin* (Choryně) zu beginnen und nach etwa einer deutschen Meile nach dem Dorf Vysoká bei Olmendorf (Polouvsí) in den kleinen Fluss Luka münden zu lassen, der die weitere Verbindung zur Oder herstellen sollte. Zu diesem Zweck wurde bereits in August 1719 eine lokale Untersuchung durch eine Sonderkommission durchgeführt, der Norbert Wend von Linck angehörte. Da Lincks eigenes Fachgebiet der Schanzenbau war, also eine rein technische Angelegenheit, konnte er gleich auch die Aufsicht über den Bau eines schiffbaren Flusses übernehmen. Lincks Teilnahme sollte sicherstellen, dass die Flussbereinigung keineswegs die Abwehraufgabe der Stadt Hradiště gefährden sollte. Von der Messung dieser Sonderkommission am Fluss March im Sommer 1719 hat sich auch eine handgezeichnete Vedute der Stadt *Nepagedel* (Napajedla) in Farbe

1 Der letzte Versuch den Bau des Kanals zu beleben stammt aus den Jahren 1938-1940. Übrig geblieben ist nur eine unterbrochene, etwa neun Kilometer lange und schmale Ringe in Wien-Lobau, die gleich an der Donau am Hubertusdamm, südöstlich des Napoleons Hauptquartier von 1809, beginnt und bei Thavonhof kurz vor der Bahndamm bei Glizendorf an der Strecke von Wien nach Marchegg endet.

erhalten (Farbtafel S. 552 a). Die Ansicht, datiert vom 11. August 1719, dokumentiert den Stand der Visitation am Ort mit der Einzeichnung des geplanten Salzlagers (B) unter dem langen ebenerdigen wirtschaftlichen Gebäude und der steinernen Mühle (C). In der Mitte der Stadtansicht sind ein altes Schloss mit drei Flügeln und oberhalb die barocke Kirche St. Bartholomäus dargestellt. Bereits im 17. Jahrhundert war die March ab *Nepagedel* ein schiffbarer Fluss, im 18. Jahrhundert kam aus Napajedla nach *Göding* (Hodonín) bereits geflößtes Holz.

Der Schwerpunkt des Projektes eines Kanals wird durch die Umrandung des Textes der Hauptkartusche betont. In dieser Kartusche reichen zwei personifizieren Flussgötter, der Danuvius und der Viadrus, einander die Hände. Zu ihren Füßen vermischen sich die Wasser der Donau, March und Beczwa mit denen der Oder von rechts. Ein kleiner Abriss des Kanals an der Wasserscheide mit zwei Lastschiffen und dem Text *Canal oder durchschnit in die Oder* stellt die Verbindung her. Die Embleme am Rande des Parergons bestehen aus einem Adler mit der kaiserlichen Krone, zwei Segelschiffen, einem Anker sowie aus Kaufmannswaren und geben dem Ganzen damit eine reichliche barocke Kontur.

Die Ortsnamen sind statt mit altdeutschen Lettern schon durch lateinische Schriftzeichen ausgedrückt. Die Namenschreibung ist einsprachig und der deutschen Schreibart angepasst. Auf keinen Fall Linck bemühte sich keineswegs tschechisierten Namen zu benützen bzw. diese als Dublette den deutschen Namen beizufügen. Auf der gedruckten Karte findet man eine beträchtlich größere Zahl an Ortsnamen und hydrographischen Namen sowie eine reiche Reliefdarstellung.

Zu Anfang des 18. Jahrhunderts fehlte der kartographischen Darstellung des Gebietes noch immer die nötige Verlässlichkeit, vor allem im Bereich des Terrains. Bei der Reliefdarstellung hält auch diese Karte an der Hügelmanier fest. Einen speziellen Geländeformennamen trägt der Vlara-Pass (*Vlars Pass*), der in die Slowakei führt. Auffallend ist das Fehlen der unterschiedlichen Darstellung des Laub- und Nadelwaldes durch damals bereits bekannte neue graphische Methoden, sowie die nur symbolische Darstellung von Feldern (Ländereien) und im südlichen Teil der Karte von Weinbergen.

Über den mährischen Raum ist die Beschriftung gleichmäßig verteilt, sodass das Relief hierdurch nahezu verschwindet und das größere, aber relativ wenig beschriftete Niederösterreich fast ganz in den Hintergrund tritt. Bei den Siedlungen wird mit Signatur und Schriftart zwischen Städten und Dörfern unterschieden. Die Siedlungen sind mit drei Typen von Ortssignaturen dargestellt:

Der erste Typ – für die größeren Städte – ist durch Signaturen von in Form und Größe verschiedenen Grundrissen von Befestigungsmauern gekennzeichnet (*Olmütz*, *Tobischau*, *Kogetain*, *Kremsir*, *Hradisch*, *Hungris Brod*, *Stramberg*, *Prerau*, *Lípník*, *Elfenstein*, *Weiskirch*, *Oder*, *Meseriz*, *Fulnek*, *Neu Tischein*, *Wisovitz*, *Marchek*, *Hainburg* und *Altenburg*). Preßburg ist sogar durch eine Kombination der Seitenansicht der Burg und der mit einem Grundriss angedeuteten Fortifikation eingezeichnet.

Der zweite Typ zeigt kleinere Städte oder Marktflecken wie zum Beispiel *Holeschan* (Holešov), *Stadt liebe* (Město Libavá), *Sternberg* (Šternberk), *Bistriz* (Bystriče pod Hostýnem) und *Strasnitz* (Strážnice) in meist realitätsnaher Seitenansicht.

Die Mehrheit der Signaturen weist aber nur eine symbolische Darstellung – meist in Form einer Kirche – auf.

An anderen Signaturen findet man viereckige Symbole für *hin und wieder zerstreute Bauernhöfe* (Wehrhöfe) im Gebirge Nízký Jeseník (Niederes Gesenke), wie zum Beispiel *Hoff* und *Goldensaifer Hoff*. Ansonsten ließ Linck kleine Ortschaften wie auch sonstige Kleinsiedlungen ohne Bennenung; die einzige Ausnahme davon bildet Schlosshof südlich von Marchegg in Niederösterreich. Neben den bereits aufgezählten Signaturen ist die durch eine Kirche individualisierte Darstellung des Wallfahrtsortes Svatý Kopeček (*Heilig berg*) bei Olmütz beachtenswert. Diese Wallfahrtskirche wurde im Mai 1995 zur Basilika minor ernannt. Ansonsten führt der Autor aber die individuellen Illustrationen nur in seltenen Fällen, wie zum Beispiel *Prespurg*, *Holeschau* und *Hradisch,* bis ins Detail aus.

Hinsichtlich der Orthographie verursacht der unerfahrene Stecher eine eigentümliche Verschmelzung von Buchstaben, wie z. B. *Sibenhofen* (Siebenhofen, Sedm Dvorů) am linken Kartenrand. Einige Ortsnamen verstümmelte der offenbar mit der Ausfertigung der Schrift nicht vertraute Kupferstecher bis zur Unkenntlichkeit, wie zum Beispiel *Hostoczalkoff* (Hošťálková) bei *Meseriz* (Valašské Meziříčí).

Die Karte ist eine seltene Symbiose kartographischer Gestaltung der Landschaft, der Darstellung des bautechnischen Vorhabens und dekorativer Ausstattung zu Beginn des 18. Jahrhunderts und es ist durchaus offen, inwieweit Linck als Kartograph selbstständig arbeitete. Das üppige Parergon seiner March-Karte diente als Kompositionsvorlage für die Homannsche Karte des Prerauer Kreises *Marchionatus Moraviae circulus Preroviensis* [...] *exhibet J. C. Müller S.C.M. capit. et ingen. editore Joh. Bapt. Homanno Norimbergae*, im Maßstab ca. 1 : 240 000. Es handelt sich um eine von sechs Spezialkarten der mährischen Kreise, die von der Nürnberger Firma nach der Müllerschen Karte von Mähren (1716) ca. im Jahre 1720 gestochen und herausgegeben wurde (Farbtafel S. 552 b). Die Übereinstimmung der beiden dekorativen Beiwerke (Parerga) ist auffällig, selbst wenn die allegorische Auffassung bei jedem der beiden Parerga eine andere ist. Auf der Homannschen Karte ist das Bild ungewöhnlich ähnlich in eine Projektionsfläche in Form einer Höhle komponiert, die in senkrechter Richtung durch Personifizierung der Quellen der Flüsse March und Oder abgegrenzt ist. Beide Quellen liegen innerhalb entlegener Berglandschaften und kommen so durch kaskadenartige Gefälle in Bewegung. Die Vertreter des Wassergeflügels ergänzen den unteren Rand des Bildes. Die Oder wurde in dieser Allegorie keineswegs ohne Grund dargestellt, denn die Quellen der Oder befinden sich im nördlichen Teil des ehemaligen Prerauer Kreises und wurden auch durch Müller in seiner Karte von Mähren mit der Bemerkung *Fontes Odra* aufgeführt.

Weitere Exemplare der gedruckten Karte von 1719 sind in der Sammlung Moll der Universitätsbibliothek in Brno (Brünn, Sign. 90900-G.T. XXII., 52) und in der Kartensammlung *Laudon* in Bystřice pod Hostýnem (*Bistriz unter Hostain*) aufbewahrt (Sign. 1164).

Von der March-Karte sind zwei handgezeichnete Unikate erhalten geblieben, beide datiert mit 1719. Auf die erste Handzeichnung der March-Karte von Linck hat der aus Gablonz an der Neiße stammende böhmische Militärkartograph, Kartenhistoriker, Kartenbibliograf und langjährige Leiter der Kartenabteilung des Kriegarchivs in Wien, Josef Paldus (1863-1937), aufmerksam gemacht. Die farbige Karte liegt bis heute im Österreichischen Staatsarchiv, Abteilung Kriegsarchiv, in Wien. Ein anderes Unikat der handgezeichneten March-Karte desselben Datums gehörte noch im Jahre 1954 zur Kartensammlung des bekannten österreichischen Feldmarschalls Gideon Ernst Loudon (Laudon, 1717-1790). Seine Sammlung wurde nach dem Zweiten Weltkrieg im damals staatlichen Schloss in

Bystřice pod Hostýnem und nachher im staatlichen Schloss Vranov nad Dyjí in Südmähren aufbewahrt. Beide Schlossbibliotheken stehen heute unter der Verwaltung des Nationalmuseums in Prag, die Manuskriptkarte aus der Sammlung Loudon gilt heute allerdings als verschollen.

Wesentlich mehr über damalige Überlegungen bei der Vorbereitung des Kanals verrät der Text in der Manuskriptkarte von Linck aus der Kartensammlung des Wiener Kriegsarchivs. Die Karte hat den Titel *Mappa Des March fluss von der Statt Olmitz bis in die donau in welcher ist zu ersehen So wol der Lauff dieses Flus sambt der umligenten Situation als auch der Orth, wo der Fluss Betczwa mit der Oder Coniungieret und dadurch die Schiff-fahrt von dem Teutschen bis an das Schwartze Meer brauchbar gemacht werden kann* (Farbtafel S. 553).

Der dazugehörige Text lautet:

Der Fluß Oder fliesset in der eken durch die felder mit viel fältigen umbschweift, jedoch langsamb und ver= | blaibt jederzeit in seinem alten lauf und stromb in diesem orth ist der fluß von 5. bis 6. klafter weit mit genung= | samben tiefe, der halben kan man ihme leichtlich schiefreich machen dagegen ist der fluß Beczwa reißend und ganz welches | bald zu bald abnimbt, fangt an seinem lauf zu Meseritz, in einem waiten ebenen feld daß wegen der waithe des orths | auch unstand häftigkeit des flußes selbigen, schiefreich zu machen unmöglich, zwischen Hustopetz und Jasnik ist ein berg, wo baide flüß | von ainnander heist nicht sehr hoch sondern lang, der flus Beczwa ligt höher den der flus Oder so man nun die Schiffahrt der march | in die Oder richten wil, vermeinen ich das solches gar nicht könne volzogen werden, wo nemblich der Fluss Beczwa, so in die March seinem | lauf hat, bis auf Prerau schifreich gemacht werden, von der nötig ein Kanal bis auf Waiskirchen und von dannen bis auf Hustopetz sintermalen oberhalb Brerov der Fluß nicht tauglich ist schifreich gemacht zu werden | von nun die Schiffahrt bis auf Hustopetz zugerichtet worden ist, manglt allein der Schif über den berg Porop bis auf Jasnik zu gehen, welches nicht geschehen kann, und zwar auf diese weiß, kann | man, den fluß Beczwa hin ein halbe meil ober Meseritz wo der Buchstaben **A** *ligt lannt und aldorten ein Kanal anfanget, welcher das Wasser in die Höhe hebt, und bis auf Porop wo der Buchstaben* **B** *ist traget | von* **B** *bis* **C** *muß der berg aus gehauen werden und dieses wird in der großen Höhe des bergs so hernach beydersaits abnimbt, nicht über 25. klafter ein das orth zwischen* **B** *und* **C** *wird jederzeit mit waser so immer der Sub Assignirte Canal Sub ministriret | über schwembt und erfüllet sain, solches waser kann man in dem fluß Oder oder in dem fluß*

*Beczwa wo hernachen die noch die herauf oder hinab zu bringen mann es ervordert wirt laufen lassen von dem Buchstaben **E** bis **B**. müssen die hiermit so viel clausen als nötig | ab und auf staigen welches ebener machen gegen der Oder von dem Buchstaben **C** und **D**, geschehen wirt, und auf diesr manier wird ohne alle abgang auf der Oder und March ein ewige Schiffahrt sein.*

Auch die eigentliche Umgebung des geplanten Kanals auf der Karte gibt dem Kartenleser Gelegenheit zum Studium der Abweichungen zwischen Lincks Manuskript und dessen Druck. Interessant ist, dass der Inhalt der Manuskriptkarte sich vorwiegend auf die Umgebung der Flüsse Oder, March, Bečva und Donau und ihre Zuflüsse beschränkt, und die Darstellung der Situation, des Terrains und der Hydrographie in den gedruckten Exemplaren wesentlich reichhaltiger ist. Ansonsten ist der Inhalt einer handgezeichneten Karte bedeutend opulenter.

Die Manuskriptkarte im Kriegsarchiv wurden durch eine später hergestellte bräunliche Skizze der weiteren Umgebung des Kanals im Maßstab 1 : 200 000 (Tafel S. 554) ergänzt. Die Skizze ist begrenzt durch die Städte *Olmitiz* (Olmütz), Stadt *Oder*, *meseriz* (Valašské Meziříčí) und *Otrokowitz* (Otrokovice). Das hier ausgewählte Gebiet enthält eine Namenschreibung, die die Rekonstruktion des zeitgenössischen Textes erleichtern soll.

Norbert Wenck von Linck hat 1739 auch einen Grundriss und einen Prospekt der Prager Zitadelle Wyschehrad angefertigt (Kartensammlung der Karlsuniversität in Prag, Signatur 93eA94 und 93eA95).

Quellen- und Literaturverzeichnis

Sammlung Woldan, Cartographica K-V B1 OE Boh 94, 1-2.

Moravský zemský archiv, Palachovo náměstí 1, 626 00 Brno. Fond D 22, Signatur 96.

Bayerische Staatsbibliothek München, Abt. für Musik, Karten und Bilder, Mapp. IX, 82ab.

Österreichisches Staatsarchiv Wien, Abteilung Kriegsarchiv, B IXb169.

Roubík F., Loudonova sbírka map, in: Kartografický přehled 8 (1954), 164-165.

Honl I., Linckova mapa Odersko-dunajského pruplavu z r. 1719, in: Kartografický přehled 10, (1956), 34-36.

Zum Autor

Ivan Kupčík, geb. 1943 in Pardubice/Ostböhmen, Dr. rer. nat., Geograph und Kartenhistoriker. Absolvent der Naturwissenschaftlichen Fakultät der Karlsuniversität Prag, Schüler von Univ. Prof. Dr. Karel Kuchař. 1972-78 in der Geographischen Anstalt und 1979 in der Historischen Anstalt der Tschechoslowakischen Akademie der Wissenschaften tätig. Seit 1980 in München, Bayerisches Hauptstaatsarchiv, Archiv des Deutschen Museums, 1993-2006 Lehrbeauftragter für Historische Kartographie am Historischen Seminar der Ludwig-Maximilians-Universität.

Wichtigste Arbeiten: *Alte Landkarten*, (Hanau, 7. Ausgabe 1992), *Cartes géographiques anciennes* (Paris, 4. Ausgabe 1989), *Mappae Bavariae. Thematische Karten von Bayern bis zum Jahr 1900* (Weissenhorn, 1995), *Münchner Portolankarten/Munich Portolan Charts* (München, 2000), Alte Landkarten. Handbuch zur Geschichte der Kartographie (Stuttgart, 2010, im Druck).

Itinerarium Europae
provincias sontinens,
[Antwerpen] ca. 1590

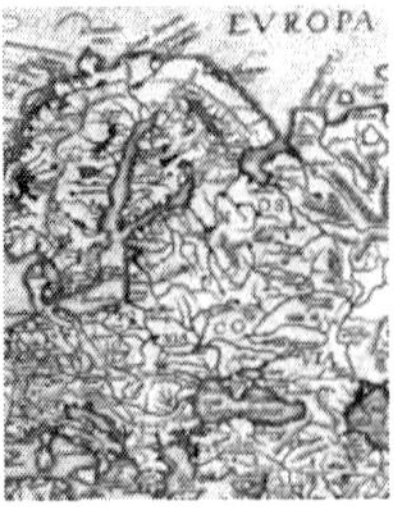

Peter H. Meurer, Heinsberg

Das Woldan-Exemplar des Kölner *Itnerarium Europae*

Der aus Mechelen stammende Kupferstecher Frans Hogenberg (um 1538-1590) war seit etwa 1565 zeitweilig, erst seit etwa 1585 dauerhaft in Köln ansässig. Kontakte zur führenden niederländischen Kartographieszene der Zeit hatte er bereits durch sein Elternhaus. Seit etwa 1568 stach er die Karten für die ersten Auflagen des epochalen Atlas *Theatrum Orbis Terrarum* (Antwerpen 1570 ff.) von Abraham Ortelius (1527-1598). In Köln versammelte sich um Hogenberg eine hochaktive Gruppe von Kupferstechern, Gelehrten und Publizisten. Im hier vorliegenden Zusammenhang sind drei Leute hervorzuheben:

– Persönlicher Patron Hogenbergs und Redakteur des wichtigsten Gemeinschaftsprojektes, des Städteatlas *Civitates Orbis Terrarum* (Köln 1572–1617), war der aus Köln gebürtige Humanist und katholische Theologe Georg Braun (1541-1622).

– Eine Art humanistisch-gelehrte Oberinstanz war der aus Burgund gebürtige, seit 1563 als freier Autor und Wissenschaftsagent in Köln lebende liberale Jurist Jean Matal (Johannes Metellus, um 1517-1597).

– Der Österreicher Michael von Aitzing (um 1535-1598) war als vermutlich politischer Flüchtling aus den Niederlanden seit 1581 in Köln ansässig. Als Nachrichtenhändler und freier Autor publizierte er mit Hogenberg die ersten periodischen Relationen und illustrierte Werke über die Niederlande und deren Kampf gegen die spanische Krone.

Zusammen mit weiteren einheimischen und zugewanderten, katholischen und protestantischen Autoren und Kupferstechern bildet dieser Kreis die *Kölner Schule* der Kartographie. Dadurch wurde Köln im letzten Viertel des 16. Jahrhunderts zu einem Zentrum der Kartenherstellung in Europa. Insgesamt erschienen in diesem Umfeld bis um 1610 an die 1000 verschiedene Kupferstiche mit kartographischem und topographischem Inhalt.

Ein zentrales Produkt dieser Kölner Schule ist das *Itinerarium Orbis Christiani.* Dieser Zyklus besteht aus 84 bzw. später 88 Karten europäischer Länder und Regionen im Format um 14 x 19 cm. Die kartographische Eigenständigkeit ist recht gering. Die Karten sind sämtlich kopiert nach größerformatigen Vorlagen in den Antwerpener Folioatlanten von Abraham Ortelius (1570 ff.) und Gerard de Jode (1578). 54 Karten zeigen Einträge des Verlaufes der wichtigsten Fernstraßen. Aus diesem Grunde ist das Werk als erster europäischer Straßenatlas in Teile der Literatur eingegangen.

Die Publikationsgeschichte dieses Kartenzyklus, die sich über ein Vierteljahrhundert erstreckt, ist recht komplex. Die Grundzüge habe ich 1988 in meiner Bibliographie der Kölner Atlanten und atlasähnlichen Drucke erstmals dargelegt. Von den späteren Neufunden, deren Anzahl insgesamt gering ist, befindet sich ein Exemplar einer seltenen Auflagenvariante in der Sammlung Woldan. Dies gibt Anlass, das Thema nach zwanzig Jahren nochmals in konzentrierter Form zusammenzufassen.

Gleichzeitig sei damit – en passant und ohne mehr – an ein wissenschaftliches Kind erinnert, dem auch Johannes Dörflinger viel Liebe und Vorarbeit geschenkt hat: das leider mit großer Kunst zum Waisen gemachte Projekt *Atlantes Germanici.*

Die Erstausgabe als *Itinerarium Orbis Christiani* (1579 ff.)

Zu Autor, Verlagsort und –jahr des *Itinerarium Orbis Christiani* (nachfolgend: IOC) gibt es in der älteren Literatur viele Spekulationen. Zuschreibungen an Michael von Eitzing und Jean Matal basieren auf der späteren Verwendung der Kupferplatten in deren Werken. Beide lebten zu der Zeit mit Hogenberg in Köln. Man kann nicht ausschließen, dass sie an der Idee einen Anteil hatten. Die kartographische Konzeption und die Realisierung sind aber allein das Werk Frans Hogenbergs bzw. seiner Werkstatt. Als Erscheinungsort ist in jedem Falle Köln (und nicht Antwerpen!) gesichert.

Anlass für die Publikation war wahrscheinlich ein konkretes Ereignis, das einen entsprechenden potentiellen Käuferkreis bot. Von Mai bis November 1579 fand in Köln ein Friedenskongress statt. Hier versuchten Teilnehmer aus dem Heiligen Römischen Reich, Spanien, Frankreich und den Niederlanden vergeblich, einen Weg zur Beendigung des Aufstandes der niederländischen Provinzen gegen die spanische Krone zu finden.

Die Originalausgabe mit 84 Karten erschien als textloser Atlas im Format 4°-quer. Sie hat vier Teiltitelblätter mit Daten 1579 und 1580.

1. Haupttitel in Kupferstich: *Itinerarium Orbis Christiani | Itinerario di tutti in Paesi Christiani | Wegweiser des gantzen Christenthumbs | La Guide des chemins de tous les Pais de la Chrestienté.*
2. Kupferstichrahmen mit leerem Mittelfeld, in das ein Zettel mit typengedrucktem Titel zu Teil II eingeklebt wurde. Hier sind bekannt Versionen in Latein (*Tomus II. Itinerarii Orbis Christiani* [...]), Deutsch (*Das ander Theil des Reyßbüchleins* [...]), Niederländisch (*Dander Deel des Reysboexkens* [...]), Französisch (*Le second volume de l'itineraire* [...]) und Spanisch (*Tomo segundo del Itinerario* [...]), datiert 1579 oder 1580.
3. Titel zu Teil III in Kupferstich: *Tomus tertius Itinerarii orbis christiani | Terzo tomo del Itinerario de la christianita | Das dritte theil des reisbuchlins des christenthums | Le troisieme tome de l'itineraire de la chretienté.*
4. Titel zu Teil IV in Kupferstich (mit Datum 1580): *Tomus quartus itinerarii orbis christiani Quarto tomo del itinerario de la christianita | Das Vierdte theil des reisbuchlins des christenthum | Le quatrieme tome de l'itineraire de la chretienté.*

In dieser Form ist der Atlas bis in die 1580er Jahre hinein gedruckt worden. Zu Beginn sind die Karten nicht nummeriert. Später sind auf den Platten unten rechts eine oder zwei Nummerierungen hinzugefügt worden. Sie ergeben unvollständige Zahlenfolgen von 1 bis 36 und 1 bis 84. Für die Kollation sind diese Nummerierungen aber unrelevant, denn es ist kein Exemplar bekannt, in dem die Abfolge der Karten einer dieser Zählungen entspricht.

Zwei Regionalatlanten der Niederlande (1587)

Eine Auswahl aus dem IOC-Zyklus hat Frans Hogenberg 1587 zur Publikation von zwei niederländischen Regionalatlanten verwendet. Eine lateinische Ausgabe erschien anonym, als Herausgeber ist durch Sekundärquellen Georg Braun zu erschließen.

Itinerarium Belgicum.

M. D. LXXXVII Agrippinae Coloniae.

4°, 22 Karten.

Der Band enthält 20 bzw. 21 Karten des IOC-Zyklus. Einige Exemplare enthalten statt der IOC-Übersichtskarte der Niederlande eine neue Karte, kopiert nach einer Vorlage im Atlas Gerhard Mercators von 1585. Neu ist weiters eine von dem Kölner Kartographen Gerhard Stempel (1546–1619) aufgenommene Karte der brabantischen Exklaven Kerpen und Lommersum westlich von Köln. Alle Karten haben verso einen typengedruckten Text in Latein mit einer Zählung von 1 bis 22.

Ebenfalls 1587 erschien ein Atlas in deutscher Sprache, den Michael von Eitzing als Zusammenfassung seines bei Hogenberg erschienenen Kompendiums zur niederländischen Zeitgeschichte *De Leone Belgico* (1583 ff.) konzipiert hat:

Belgici Leonis Chorographia. *Das ist ein Newe Landt beschreibung des Belgischen Lewen, in XVII Provincien des gantzen Niderlandts außgetheylet. Durch Michael Eyzinger Austriacum. 1587.*

Köln: Frans Hogenberg, 1587.

4°, 17 Karten und 7 Porträtstiche.

Der Band enthält 17 Karten des IOC-Zyklus. Sie haben verso in Typendruck einen deutschen Text und Zählungen I bis XVII und C bis T. Eine Übersichtskarte der Niederlande fehlt. Stattdessen ist in den meisten Exemplaren außerhalb der Zählung die berühmte Eitzing-Karte der Niederlande in Gestalt eines Löwen eingeklebt.

Aus Anlass der Verwendung zum Druck dieser beiden Atlanten ist auf den entsprechenden Karten am Rand ein Findegitter hinzugestochen worden.

Der Eitzing-Atlas von Europa (1588)

1588 erschien der IOC-Zyklus erneut als Atlas in Quartformat ohne Rückentexte, aber mit einer Vorrede Michael von Eitzings. Das zweiteilige Werk hat seine Begründung in den unterschiedlichen mythologisch-allegorischen Deutungen der Europa.

– Typengedruckter Haupttitel: *Ad Hispaniae et Hungariae reges termaximos, De Europae Virginis, tauro insidentis, topographica atque historica descriptione, liber. Quatuor orbis partibus distinctus; insuper et octaginta quinque, cum tribus, chorographicis tabulis, a Francisco Hogenbergio insigni artifice ornati. Michaele Aitsingero | Austriaco authore.*
Coloniae Agrippinae, Apud Godefridum Kempensem, Anno M.D.Lxxxviii.

– Zwischentitel zu Teil II in Kupferstich: ITINERARIVM GERMANIAE PROVINCIAS CONTINENS, IIS QVI ITER ALIQVO VEL LEGATIONIS, VEL MERCATVRAE VEL ALIA EX CAVSA SVSCIPIVNT LONGE NECESSARIVM. *Provincae, in qua facturus es iter, Chartam inspice, in qua, locum ubis, et eum quo tendis, inquire. Videbis, quae loca tibi pertranseunda sint, quinam superandi montes, quae transmittendae silvae; unde pericula. in quae aliqui inciderunt, vites. Itinera praeclariora geminatorum punctorum linea* [Doppellinie] *duximus. Arces et pagos, circuli nota* [Kreis], *Urbes et Oppida, duplicati circuli signo* [Doppelkreis] *distinximus. Chartae autem margini; vicinarum Civitatum et Emporiorum nomina; ut oculis magis exponentur, non ut inde aliqua locorum distantia Geographica colligatur, adscripsimus. Provinciae illa, quibus superiori margine nulla caeli plaga additur: eae omnes, ad Septemtriones versae sunt.*
4°, [4] Bl. + 1 Zwischentitel + 88 Karten.

Der Atlas enthält im ersten Teil aus dem IOC-Zyklus 52 Karten zu den Britischen Inseln, Südeuropa, Italien und den Niederlanden, einschließlich der Karte Stempels von Kerpen-Lommersum. Der spezielle Teil II enthält 36 Karten zum Heiligen Römischen Reich (ohne die Niederlande und die rheinischen Territorien) einschließlich Dänemarks, Polen und Livlands. Darunter stammen 33 aus dem IOC-Zyklus. Hinzugefügt sind drei neu gestochene Karten von Oberelsass, Unterelsass und Pommern-Brandenburg.

Für die Verwendung in diesem Eitzing-Atlas ist auf allen Karten unten links eine weitere Nummerierung von 1 bis 88 hinzugestochen worden. Ihr entspricht die Abfolge der Karten in allen bekannten Exemplaren des heute überaus seltenen Werkes.

Die Ausgabe als *Itinerarium Europae* (um 1590)

Mit dem Eitzing-Atlas hat der IOC-Zyklus durch die Nummerierung eine feste Abfolge erhalten. Auch ist die Zahl der Karten nun auf 88 angewachsen. Die Druckgeschichte des Werkes in den anschließenden Jahren ist jedoch sehr unübersichtlich. Grund waren vermutlich der Tod Frans Hogenbergs im Jahre 1590 und nachfolgende familiäre Wirren um die Nachfolge im Unternehmen.

In dieser Zeit wurden Exemplare mit einigem Belieben auf der Basis vorhandener Abzüge zusammengestellt. In zwei Beispielen:

– Ein Exemplar der Bibliothèque Nationale de France Paris (Sign: Ge.FF.720) als Variante des Eitzing-Atlas hat als alleiniges Haupttitelblatt den Zwischentitel ITINERARVM GERMANIAE, gefolgt von den 88 Karten, in Abfolge der Nummerierung.

– Ein Exemplar der Hessischen Landesbibliothek Wiesbaden (Ra 832) beginnt mit dem Haupttitel *Itinerarium Orbis Christiani* der Erstausgabe, hat als Zwischentitel dann ebenfalls das Titelblatt ITINERARIVM GERMANIAE. Die Abfolge der Karten weicht von der Nummerierung ab.

In einem nächsten Schritt wurde auf der Kupferplatte des Zwischentitels des Eitzing-Atlas das Wort GERMANIAE durch EVROPAE ersetzt. Mit diesem neuen Titel ITINERARIVM EVROPAE PROVINCIAS CONTINENS kamen wohl kurz nach 1590 weitere Exemplare auf den Markt.

Allerdings waren die Nachfolger im Hause Hogenberg auch damit zunächst noch weit von einer editorischen Regelmäßigkeit entfernt. Beleg ist ein Exemplar der Privatsammlung Fritz Hellwig in Bonn. Es hat zwar den neuen Titel ITINERARIVM EVROPAE, dann folgen aber nur die 84 Karten des originalen IOC-Zyklus durchweg im ersten Plattenzustand ohne jegliche Nummerierung; nur acht Karten zeigen eine gestochene Zählung aus dem Eitzing Atlas. Die 82 Karten des unvollständigen Exemplars der Universitäts- und Landesbibliothek Düsseldorf (Gu St.2591) liegen sogar komplett im ersten Plattenzustand vor.

Von der Ausgabe ITINERARIVM EVROPAE mit 88 Karten und in einer der Nummerierung folgenden Anordnung der Karten sind derzeit vier mehr oder minder vollständige Exemplare bekannt:

– Universiteitsbibliotheek Amsterdam (O 60-658);
– Royal Geographical Society, London (MR 264.E.8);
– Sammlung Klaus Stopp (†), Mainz;
– Österreichische Akademie der Wissenschaften, Wien (Sammlung Woldan: K-II: EU 49).

Abb.1: Das Titelblatt Itinerarium Europae,
ÖAW-Sammlung Woldan)

Von diesen gehört das Exemplar der Sammlung Woldan, dessen ausführliche Kollation nachstehend gegeben wird, zu den besten. Die Karten liegen in guten Abdrucken vor. Das Exemplar ist mit 88 Karten vollständig und zeigt eine korrekte Abfolge entsprechend der Nummerierung unten links wie im Eitzing-Atlas. Allerdings weist auch dieses Exemplar jene leichten Unregelmäßigkeiten auf, die für diese Spätausgabe typisch sind:

– Von den 88 Karten liegen nur 75 im letzten Plattenzustand mit gestochener Nummerierung unten links vor.

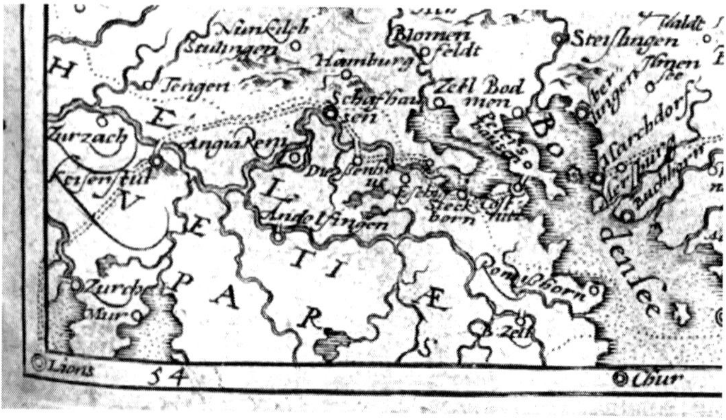

Abb. 2: Die linke untere Ecke der Schwaben-Karte (Nr. 54) mit gestochener Nummerierung., ÖAW-Sammlung Woldan

– 12 Kartenblätter zeigen den allerersten Plattenzustand im *Itinerarium Orbis Christiani*, ohne jegliche Nummerierung.

– Eine Karte (Nr. 58 Franken) zeigt den Plattenzustand 2 (Zählung unten rechts) aus den späteren Ausgaben mit dem Titel *Itinerarium Orbis Christiani.*

Die unten links fehlenden Nummerierungen der 13 Karten sind handschriftlich nachgetragen worden.

Abb. 3: Die linke untere Ecke der Germania-Karte (Nr. 53) mit handschriftlicher Nummerierung, ÖAW-Sammlung Woldan

Dies bestätigt den Befund, der sich bereits aus der Betrachtung des Exemplars der Sammlung Hellwig ergeben hat. Anscheinend war im Hause Hogenberg kurz nach 1590 immer noch ein Vorrat an Abzügen des IOC-Zyklus aus der etwa 15 Jahre zurückliegenden Originalausgabe vorhanden. In der Kompilation von Exemplaren des ITINERARIVM EVROPAE bot sich dann eine Möglichkeit zur Restverwertung dieses Bestandes. Sehr viel Mühe hat man sich damit aber nicht gemacht.

Wohl wegen der familiären Streitigkeiten haben die Erben Hogenbergs die 88 Druckplatten des IOC-Zyklus bald danach verkauft. Zusammen mit neuen Kartenstichen wurde der Großteil von ihnen im Rahmen eines von Jean Matal konzipierten Atlaswerkes verwendet, das ab 1594 in Lieferungen bei verschiedenen Verlagen in Köln und später Oberusel erschienen ist.

Kollation des Exemplars der Sammlung Woldan

ITINERARIVM | EVROPAE PROVINCIAS CONTINENS, IIS QVI ITER ALIQVO | VEL LEGATIONIS, VEL MERCATVRAE VEL ALIA EX CAVSA | SVSCIPIVNT LONGE NECESSARIVM. | *Provinciae, in qua facturus es iter, Chartam inspice, in qua, locum ubis, et eum quo tendis, inquire. Videbis, quae* | *loca tibi pertranseunda sint, quinam superandi montes, quae transmittendae silvae; unde pericula. in quae aliqui incide-* | *runt, vites. Itinera praeclariora geminatorum punctorum linea* [Doppellinie] *duximus. Arces et pagos, circuli nota* [Kreis], *Urbes et* | *Oppida, duplicati circuli signo* [Doppelkreis] *distinximus. Chartae autem margini; vicinarum Civitatum et* | *Emporiorum nomina; ut oculis magis exponentur, non ut inde aliqua* | *locorum distantia Geographica colligatur, adscripsimus. Pro-* | *vinciae illa, quibus superiori margine nulla caeli plaga* | *additur: eae omnes, ad Septemtriones* | *versae sunt.*

Kartenfolge mit Angabe der gestochenen und [handschriftlichen] Zählung

1 Europa
2 Anglia
3 Cambria
4 Hibernia
5 Scotia

6	Suedia
7	Cyprus
8	Grecia
9	Illyricum
10	Zara et Sebenici descriptio
11	Italia
12	Puglia – Terra di Otranto
13	Marchia Anconitana olim Picenum
14	Territorium Patavinum
15	Mediolanensis Ducatus
16	Pedemontanus Principatus
17	Veronae Urbis Territorium
18	Agri Cremonensis typus
19	Parmensis Placentinique Ducatus descriptio
20	Thuscia
21	Urbis Romane Territorium
22	Senensis Ditio
23	Regnum Neapolitanum
[24]	Hispania
25	Gallia
26	Pictavia – Poictou
27	Andegavensis Ducatus
28	Vermanduorum Comitatus
29	Picardia
30	Ponthieu Comitatus
31	Burgundia
[32]	Artesia
[33]	Nobilis Comitatus Hannoniae
[34]	Co. Namurcum
35	Lutzenburgensis Ducatus
36	Episcopatus Leodiensis
[37]	Belgium
[38]	Brabantia
[39]	Limburgensis Ducatus
40	Territorii et Dominii Kerpensis et Lommerschum
41	Iulia
42	Clivia
43	Geldriae Ducatus
44	Zutphaniae Comitatus

45 Transisulana Ditio
46 Frisia Orientalis
47 Groningense Territorium
[48] Frisia Occidentalis
49 Hollandiae Comitatus
50 Ultraiectensis Episcopatus
[51] Zelandia
52 Flandria
[53] Germania
54 Sweviae Circulus
55 Alsatia Superior
56 Alsatia Inferior
57 Du. Wirtenberg
[58] Franconia
59 Episcopatus Salczburgensis
60 Tyrol
61 Palatinatus Superior Bavariae
62 Bavaria
63 Stiria
64 Carinthia
65 Histria
66 Forum Iuly Friuli
67 Helvetia
68 Basiliense Territorium
[69] Trier
70 Comitatus Waldeckensis
71 Hassia
[72] Westfalia
73 Thuringia
74 Misnia
75 Lusatiae Marchionatus
76 Mansfeldia
77 Silesia
78 Moravia
79 Austriae Archiducatus
80 Bohemia
81 Hungaria
82 Tractus Danuby pars
83 Polonia

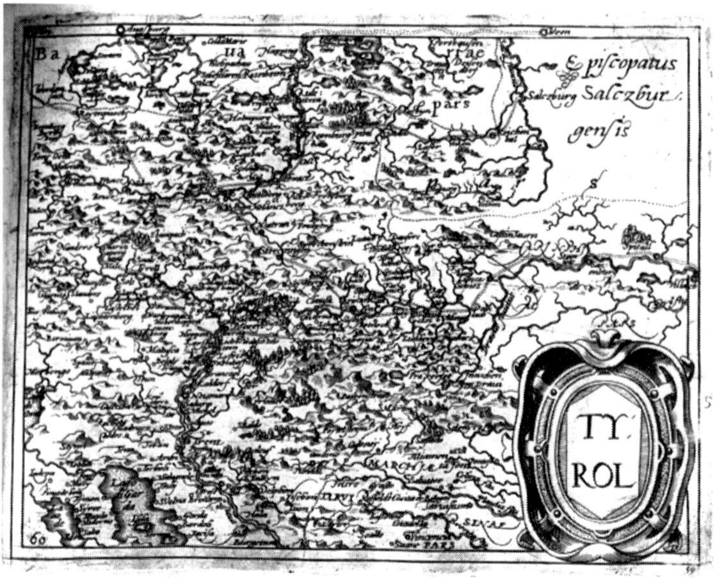

Abb. 4: Die Tirol-Karte im Itinerarium Europae (Karte Nr. 60), ÖAW-Sammlung Woldan

Abb. 5: Die Österreich-Karte im Itinerarium Europae (Karte Nr. 79), ÖAW-Sammlung Woldan

84 Livonia
85 Prussia
86 Pomerania et Brandenburg
87 Holsatia
88 Dania

Literaturverzeichnis

Heuser P. A., Jean Matal. Humanistischer Jurist und europäischer Friedensdenker *(um 1517–1597)*, Köln, Weimar und Wien 2003.

Meurer P. H., Atlantes Colonienses. Die Kölner Schule der Atlaskartographie 1570–1610, Bad Neustadt an der Saale 1988.

Mielke U., Remigius and Frans Hogenberg (in: The New Hollstein. Dutch and Flemish Etchings, Engravings and Woodcuts 1450–1700), Ouderkerk aan den IJssel 2009.

Das Exemplar der Universitäts- und Landesbibliothek Düsseldorf ist als Digitalisat zugänglich unter der Adresse: http://digital.ub.uni-duesseldorf.de/content/titleinfo/298378

Zum Autor

Peter H. Meurer, geb. 1951, Dr. rer. nat., Kartenhistoriker, Tätigkeiten im Antiquariats- und Sammlerbereich sowie als Projektleiter an den Universitäten Trier, Duisburg und Köln. Unter den etwa 150 Fachveröffentlichungen sind *Atlantes Colonienses* (1988), *Fontes Cartographici Orteliani* (1991) und *Corpus der älteren Germania-Karten* (2001) besonders hervorzuheben.

Kontakt: Rochusstr. 41, D-52525 Heinsberg, petermeurer@online.de

John Smith,
Virginia,
London 1612

Jan Mokre, Wien

Grundbesitz für eine Karte

Die Geschichte der Virginia- und Maryland-Karte des Augustin Herrman von 1673

Einleitung

Während meines Studiums an der Universität Wien interessierte mich besonders die Geschichte der Kartographie, das Spezialgebiet von Univ.Prof. Dr. Johannes Dörflinger, und ich beendete meine Studien mit einer bei ihm eingereichten und von ihm approbierten Diplomarbeit. Danach empfahl mich Johannes Dörflinger an die Kartensammlung der Österreichischen Nationalbibliothek. Deren damaliger Direktor, Hofrat Dr. Franz Wawrik, ein Studienkollege Johannes Dörflingers, suchte einen Mitarbeiter für die Vorbereitung einer von ihm kuratierten Ausstellung, welche unter dem Titel „Die Neue Welt. Österreich und die Erforschung Amerikas" vom 15. Mai bis zum 26. Oktober 1992 im Prunksaal der Österreichischen Nationalbibliothek präsentiert wurde.[1]

Im Zuge dieser Ausstellungsvorbereitung beschäftigte ich mich unter anderem mit dem aus Böhmen nach Nordamerika ausgewanderten Kaufmann, Kolonisten, Landvermesser und Kartenzeichner Augustin Herrman. Johannes Dörflinger und Franz Wawrik ermu-

1 F. Wawrik – E. Zeilinger – H. Hühnel – J. Mokre (Hrsg.), Die Neue Welt. Österreich und die Erforschung Amerikas. Ausstellung im Prunksaal der Österreichischen Nationalbibliothek, 15. Mai - 26. Oktober 1992, Wien 1992.

tigten mich, die Ergebnisse meiner Recherchen auf dem von der Stiftung Preußischer Kulturbesitz am 20. und 21. Februar 1992 in Berlin veranstalteten Symposium „Deutsche und mitteleuropäische Beiträge zur Kartographie beider Amerikas“ vorzutragen. Sie nahmen mich mit, führten mich ein und unterstützten mich.

Der auf der Grundlage des Vortrags für einen vom Veranstalter geplanten Tagungsband verfasste Artikel wurde jedoch, wie auch die anderen Beiträge des Symposiums, nicht publiziert. Nun, an dieser Stelle veröffentlicht, soll er vor allem der Erinnerung und Würdigung dieser durch fachliche Toleranz und Förderung sowie Ermutigung und großzügige Unterstützung geprägten Zeit dienen. Den letzten Stand der Forschung bietet die im Jahr 2007 veröffentlichte Arbeit über Augustin Herrmans kartographisches Wirken von Dr. Ivan Kupčik.[2]

Biographisches zu Augustin Herrman

Die Karte von Virginia und Maryland des aus Böhmen gebürtigen Augustin Herrman stellt ein in der Kartographiegeschichte des frühen kolonialen Nordamerikas außerordentlich bedeutsames Werk dar. Es muss daher etwas verwundern, dass über diese Karte und ihren Autor in der deutschsprachigen Literatur bis in die 1990er Jahre so gut wie nichts erschienen ist, obwohl sich namhafte Kartographiehistoriker aus den USA bereits intensiv mit ihr beschäftigt haben und eine umfangreiche amerikanische Sekundärliteratur zum Thema vorliegt.[3]

Augustin Herrman ist eine der interessantesten Personen im Nordamerika des 17. Jahrhunderts. Er war ein bedeutender Kaufmann in Neu-Amsterdam, dem heutigen New York und engagierte

---

2 I. Kupčik: Die kartographische Tätigkeit von Augustin Herman (ca. 1621-1686) — Realität und Fiktion. In: Mappae Antiquae. Liber Amicorum Günter Schilder. Essays on the occasion of his 65th birthday. Edited by Paula van Gestel-van het Schip and Peter van der Krogt with the collaboration of Marco van Egmond, Peter H. Meurer, Paul van den Brink and Edward H. Dahl (= Utrechtse Historisch-Cartografische Studies, Utrecht Studies in the History of Cartography, 6, ’t Goy-Houten 2007, 203-220.

3 Für die freundliche Unterstützung bei der Beschaffung der zum großen Teil aus den USA stammenden Sekundärliteratur bin ich Herrn Dr. Franz Wawrik zu großem Dank verpflichtet.

sich in Landspekulationen wie auch in der Landwirtschaft. Er spielte eine wesentliche Rolle in der Politik der Neu-Niederlande während der Verwaltung durch Gouverneur Peter Stuyvesant, zeitweilig als dessen Gegenspieler und wirkte in den häufigen Konflikten mit den im Norden und Süden der niederländischen Kolonie befindlichen englischen Ansiedlungen als recht erfolgreicher Diplomat. In einem Schreiben an Stuyvesant bezeichnete er sich selbst einmal als den Begründer des Virginia-Tabakhandels.

Über Herrman wurden im letzten Jahrhundert einige Artikel verfasst und in unserem erschienen nicht nur zwei Monographien über sein Leben, sondern auch eine belletristische Verarbeitung. Dass er eine bekannte Persönlichkeit war, geht auch aus der Tatsache hervor, dass er in den größeren amerikanischen biographischen Nachschlagewerken aufscheint. Nicht alle Arbeiten über ihn werden indes seiner Rolle als bedeutender Kartograph gerecht, was vielleicht daran liegt, dass die von ihm angefertigte Karte außerordentlich selten ist und daher lange nicht in ihrer Bedeutung erkannt wurde.

Über Herrmans Leben vor seiner Ankunft in Nordamerika existieren kaum gesicherte Informationen, was dazu geführt hat, dass in den Publikationen über ihn eine Anzahl von unbelegten und unbelegbaren Vermutungen und Phantasien aufscheinen. Ziemlich sicher scheint nur sein Geburtsort Prag zu sein.[4] Bei der Angabe des Geburtsdatums kursieren vier verschiedene Versionen, 1605, 1608, 1621 und 1623[5] – von denen nur eine auf eine nachvollziehbare Quelle zurückgeht, und zwar ist das die Angabe 1621.

Diese stammt von seinem 1664 aufgesetzten Testament, dort unterschrieb er mit dem Zusatz *Aetatis 63*. Angesichts einer anderen Quelle erscheint es jedoch als recht unwahrscheinlich, dass wirklich 1621 als Geburtsdatum anzusehen ist.[6]

Herrmans Biographen gehen in der Regel von der Annahme aus, dass er als Kind wohlhabender protestantischer Eltern in Prag aufwuchs und gemeinsam mit diesen im Zuge der Vertreibungen der Protestanten aus Böhmen in den 30er Jahren des 17. Jahrhunderts

---

4 Diese Angabe stammt aus dem, seine Einbürgerung in die Kolonie Maryland betreffenden Akt – Vgl. T. Capek, Augustine Herrman of Bohemia Manor, Praha 1930, 23.

5 Rattermann – Wilson: 1605; Cumming: 1623; Mathews erwähnt: 1605, 1608 und 1621.

6 Siehe unten die Passage zum Schuylkill -Vertrag.

nach Amsterdam emigrieren musste. In Amsterdam gelangte er irgendwie zur Westindischen Kompanie, in deren Dienst er Seefahrten nach Amerika unternahm.[7]

Herrmans erster belegter Aufenthalt in Nordamerika datiert von 1633; in diesem Jahr scheint er als Zeuge eines Vertrages auf, der einen Landkauf von den Indianern im Gebiet der niederländischen Kolonie betrifft (Schuylkill-River-Vertrag).[8]

Wäre nun – wie aus seinem Testament ableitbar – 1621 das Jahr seiner Geburt, müsste er zu diesem Zeitpunkt 12 Jahre alt gewesen sein. Es erscheint unwahrscheinlich, dass er in diesem Alter als Zeuge einen Vertrag unterzeichnete.

Herrman siedelte sich in Neu-Amsterdam an und lebte vom Handel mit den Indianern (Pelzhandel), mit den englischen Besitzungen, mit Westindien und mit Europa. Vor seinem Aufenthalt in Neu-Amsterdam hat Herrman wahrscheinlich in Virginia gelebt[9] – belegt ist diese Vermutung allerdings nicht. In den 1640er Jahren war er als Agent des bekannten Amsterdamer Handelshauses Gabry u. Comp. tätig.[10] Ab 1647 spielte er eine bedeutende Rolle im politischen Leben Neu-Amsterdams – und zwar als ein Mitglied des *Rates der Neun*, der Vertretung der Bürgerschaft, die als Gegengewicht zur autokratischen Regierung Gouverneur Stuyvesants eingerichtet worden war. 1651 heiratete er die in Neu-Amsterdam lebende Niederländerin Jannettje Varlet[11], mit der er bis zu ihrem Tod um 1662 fünf gemeinsame Kinder hatte.

In den 1650er Jahren war er dann zu einem der führenden Kaufleute und Bürger Neu-Amsterdams geworden und stand in enger, wenn auch nicht immer konfliktfreier Beziehung zu Gouverneur Stuyvesant. Aufgrund einer vom *Rat der Neun* in Amsterdam vorgebrachten Beschwerde über die autokratische Regierungsform

---

7 Für Herrmans Aufenthalt in Amsterdam und eine eventuelle Tätigkeit für die WIC gibt es keine Belege. 1648 ist jedoch eine Beatrix Heermans (seine Mutter?) in Amsterdam bezeugt – Vgl. H. A. Rattermann, Augustin Herrman. Eine Karakterfigur aus der Begründungsgeschichte von New York und Maryland, in: Deutsch-Amerikanisches Magazin. Vierteljahresschrift für Geschichte, Literatur, Wissenschaft, Kunst, Schule und Volksleben in Amerika 1 (1887), 202–225 und 524–538.

8 Vgl. Ebd., 206.

9 Vgl. Ebd., 203; T. Capek, Herrman, 13.

10 Vgl. Ebd., 211.

11 Vgl. H. A. Rattermann, Herrman, 529; T. Capek, Herrman, 14.

Stuyvesants kamen sie einmal in derart heftige Gegensätze, dass der Gouverneur versuchte, Herrman wirtschaftlich zu ruinieren und ihn sogar in Schuldhaft nahm.[12]

Herrmans Verhältnis zu Stuyvesant scheint sich dann wieder beruhigt zu haben. Er wurde nicht nur aus der Haft entlassen, sondern wirkte in den 50er Jahren mehrmals als diplomatischer Vertreter der niederländischen Kolonie in Konflikten mit den Engländern.

Eine dieser diplomatischen Missionen war auch der Anlass für die Entstehung seiner später so berühmt gewordenen Karte. 1659 eskalierte der Konflikt zwischen Neu-Niederlande und der englischen Kolonie Maryland um die ehemalige schwedische Kolonie (Neu-Schweden) am Delaware River. Diese ehemalige niederländische Besitzung war nach ihrem Intermezzo als schwedische Siedlung 1635-1655 wieder an Neu-Niederlande angegliedert worden. Inzwischen erhob aber auch Maryland Besitzansprüche auf das Gebiet.

Die Niederländer, die einer militärischen Auseinandersetzung nicht gewachsen waren, versuchten, den Konflikt auf diplomatischem Wege zu klären. Zu diesem Zweck wurde Augustin Herrman mit einem zweiten Kommissär nach Maryland gesandt. Im Verlaufe seiner Verhandlungen entwickelte er die Idee, dass zur Unterstützung der niederländischen Ansprüche die Anfertigung einer genauen Karte des umstrittenen Gebietes von Nutzen wäre, da ihm die von den Engländern verwendeten Karten zu ungenau erschienen. Er schlug daher Stuyvesant die Herstellung einer solchen genauen und zeitgemäßen Karte vor.

Ob nun Stuyvesants ablehnende Haltung gegenüber diesem Projekt oder Herrmans Einsicht, dass die Tage der niederländischen Herrschaft in Nordamerika gezählt seien, für die folgende Entwicklung ausschlaggebend waren, lässt sich nicht klären – auf alle Fälle bot Herrman die Anfertigung einer modernen Karte auch Lord Baltimore, dem *Lord Proprietor* der Kolonie Maryland, an.

Dieser erkannte durchaus den Wert des Vorschlages für seine Interessen und Besitzansprüche und zeigte sich sofort interessiert. Baltimore versprach Herrman im Gegenzug für die Herstellung der Karte das Bürgerrecht in seiner Kolonie, die Verleihung eines fruchtbaren Landgutes auf der nördlichen Landenge zwischen der

[12] Vgl. H. A. Rattermann, Herrman, 220; T. Capek, Herrman, 14.

Chesapeake und der Delaware Bay (Cecil County) sowie die erblichen Privilegien eines Grundherrn.

Herrman übersiedelte 1661/62 mit seiner Frau und seinen 5 Kindern aus Neu-Amsterdam nach Maryland und gründete dort auf dem ihm von Lord Baltimore zugewiesenen Landstrich das Gut *Bohemia Manor*.

Neben der Kultivierung des Landes, der Errichtung der Gutsgebäude und aufrecht erhaltener Handelstätigkeit arbeitete Herrman insgesamt 10 Jahre lang an der versprochenen Karte. Aufgrund der weitgehenden Unberührtheit des Landes gestalteten sich die Aufnahmearbeiten entsprechend aufwendig. 1670 vollendete Herrman sein Werk zur vollen Zufriedenheit Lord Baltimores. Es gibt übrigens keine Anhaltspunkte, wo Herrman die Kenntnisse zur Verfertigung einer solchen großmaßstäbigen Karte erworben hatte, wohl aber eine Reihe unbelegter Vermutungen von Seiten seiner Biographen.

Die fertige Manuskriptkarte sandte Herrman dann nach London, wo sie von dem für seine akkuraten Kupfersticharbeiten bekannten William Faithorne (1616–1691)[13] in Kupfer gestochen wurde. 1673 lag die Karte dann fertig vor. Sie erhielt 1674 ein königliches Privileg gegen Nachdruck[14] und wurde durch den königlichen Hydrographen und bekannten Instrumentenbauer John Seller († 1697)[15] vertrieben, der auch in Form von Zeitungsannoncen in der *London Gazette*[16] auf sie aufmerksam machte.

Herrman vergrößerte bis zu seinem Lebensende das Gut *Bohemia Manor* und ließ es von Kolonisten besiedeln.

Obwohl die Karte von Virginia und Maryland das einzige nachweisbare kartographische Produkt Herrmans darstellt, wird sie oft mit der frühesten gedruckten Ansicht Neu-Amsterdams auf der bekannten Visscher-Karte von Neu-Niederlande in Zusammenhang

---

13 Vgl. U. Thieme – F. Becker, Allgemeines Lexikon der bildenden Künstler von der Antike bis zur Gegenwart 11, Leipzig 1915, 204.

14 Vgl. T. Capek, Herrman, 25; P. L. Phillips, Virginia Cartography. A Bibliographical Description (Smithonian Miscellaneous Collection 37/1039), Washington 1896, 9.

15 Vgl. S. Tyacke, London Map Sellers 1660–1720. A collection of advertisements for maps placed in the London Gazette 1668–1719 with biographical notes on the map-sellers, Tring / Hertfordshire 1978, 139f.

16 Ebd., 11.

gebracht;[17] von manchen seiner Biographen wird sogar eine Mitarbeit an der Visscher-Karte in Erwägung gezogen.[18]

Augustin Herrman starb 1685 oder 1686. An ihn und sein Landgut erinnert heute noch die Bezeichnung *Bohemia River*, an dem früher sein Besitz gelegen war, und auf manchen US-amerikanischen Karten wird auf dem Gebiet des früheren *Bohemia Manor* der Besuch der alten böhmischen Kirche empfohlen. 1956 wurde ein Teil eines Highways in Maryland nach Augustin Herrman, dem prominenten ersten Eingebürgerten der ehemaligen Kolonie, benannt.[19]

Die Karte von Virginia und Maryland

Die in Kupferstich gedruckte Karte von Augustin Herrman stellt das Gebiet der Chesapeake Bay in einer Ausdehnung von etwa 400 x 350 km dar und ist nach Westen ausgerichtet. Diese Region umfasst die heutigen Staaten Virginia, Maryland und Delaware sowie Teile von North Carolina, New Jersey und Pennsylvania. In ihr befinden sich so bekannte Städte wie Philadelphia, Baltimore, Washington, Richmond und Norfolk.

Die Karte ist auf vier Platten[20] graviert und ist zusammengesetzt 94 x 79 cm groß. Ihr Maßstab nach den beiden Maßstableisten[21] beträgt etwa 1:480.000. Der Kupferstich ist sauber ausgeführt, die Beschriftungen sind gut zu lesen.

Der Titel *Virginia and Maryland As it is Planted and Inhabited this present Year 1670 Surveyed and Exactly Drawne by the Only Labour & Endeavour of Augustin Herrman Bohemiensis* befindet sich auf einer Platte, die von einem Sockel getragen wird, auf dem eine Buchstaben- und Zeichenlegende angebracht ist.

---

17 N. J. Visscher, *Novi Belgii Novaeque Angliae Nec Non Partis Virginiae Tabula multis in locis emendata a Nicolao Joannis Visschero* [Amsterdam 1655] 55,5 x 46,5 cm; *Nova Belgica sive Nieuw Nederlandt* in der zweiten Auflage von: Adriaen van der Donck, *Beschryvinge van Nieuw-Nederlant* (Amsterdam 1656).

18 Vgl. H. A. Rattermann, Herrman, 213.

19 Vgl. The Library of Congress, Geography and Map Division, The Bibliography of cartography 3 (1973), 25.

20 Die oberen beiden Platten sind 39 x 47,3 cm und die unteren 39,7 x 47,3 cm groß.

21 Die eine Maßstableiste gibt das Verhältnis von 8 Seemeilen zu 3¼ Zoll und die andere das Verhältnis von 24 englischen Meilen zu 3¼ Zoll an.

Auf der oberen Kante der von einem Indianer und einer Indianerin flankierten Titelplatte befindet sich ein Schild mit Neptunkopf und Blätter- und Früchtegirlanden. Die Karte zeigt weiters auf einem Sockel ein Portrait Herrmans in einem Oval, das von der Aufschrift *Avgustine Herrman Bohemian* umrahmt ist.

Oben in der Mitte der Karte ist das königlich englische Wappen und im oberen rechten Viertel auch das Wappen der Baltimore's abgebildet. Eine Windrose und ein Zirkel, zwischen dessen beiden Spitzen die Maßstableisten untergebracht sind, sowie Abbildungen von europäischen Schiffen und eines indianischen Kanus vervollständigen den ornamentalen Schmuck der Karte.

Es fallen drei verschiedene Schreibweisen des Autorennamens auf, die wohl auf den Stecher Faithorne zurückgehen dürften.[22] Am unteren Rand der Karte finden sich die Vermerke *Published by Authority of his Ma.ties Royall Licence and particular Priviledge to Aug. Herman and Thomas Withinbrook his Assignee for fourteen years from the year of our lord 1673* und *W: Faithorne Sculp.* – das königliche Privileg und den Stecher betreffend. Auf einem Exemplar der Karte, das in der John Carter Brown Library in Providence/Rhode Island aufbewahrt wird, befindet sich auf dem Sockel des Herrman-Portraits ein Hinweis auf den Herausgeber John Seller: *Sold by John Seller. Hydrographer to the King at his Shop in Exchange ally in Cornhill. London.*[23]

Hunderte Symbole zeigen die Lage von Siedlungen und Farmen der Kolonisten entlang der Küsten und der Flüsse an; nur wenige davon sind mit Namen bezeichnet. Die dargestellten Indianerdörfer sind häufig mit den Stammesbezeichnungen versehen.

Auf der Karte sind mehrere Textblöcke untergebracht, die Bemerkungen zur Kolonialgeschichte, zu Grenzfestlegungen, zur europäischen Namensgebung, zum Klima und zu Bodenverhältnissen enthalten. Ein Text bezieht sich auf den Susquehanna-River und einer auf die Kanus der Indianer.

Die Darstellung der Küstengewässer ist recht aufwendig gestaltet und mit Angaben zur Meerestiefe und zur Lage von Sandbänken versehen. Die geographischen Bezeichnungen sind z. T. mehrsprachig – englisch, niederländisch und indianisch.[24]

---

22 Avgustine Herrman; Augustin Herrman und Aug. Herman.

23 Vgl. J . C. Brown Library, 14.

24 Z. B.: *Wigho alias Potomac River*; *Schuylkill in dutch; Nanayunck in Indian.*

Herrman scheint die westlichen Gebiete der Kolonien nicht besucht zu haben; diese enthalten kaum topographische Angaben sondern sind mit beschreibenden Texten und der Darstellung des großen englischen Wappens ausgefüllt.

Wie wenig Herrman vom Westen der englischen Kolonien bekannt war, lässt sich an einem der Kartentexte ablesen, in dem er schreibt: *Certain it is that as the Spaniard is possessed with great Store of Minneralls at the other side of these Mountaines* [...]. Gemeint sind die im Nordwesten der Karte symbolisch angedeuteten Berge, die das Appalachengebirge darstellen sollen. Dass die Siedlungen der Spanier an der Westküste Nordamerikas mehrere Tausend Meilen entfernt lagen, scheint ihm nicht bewusst gewesen zu sein.[25]

Zur Anfertigung der Karte führte Herrman sicher eigene Vermessungsarbeiten durch; er verarbeitete aber auch frühere kartographische Erzeugnisse, wie die berühmte John Smith-Karte[26] und vermutlich auch Erfahrungen und Aufzeichnungen von lokalen Lotsen und Kapitänen.

Die Karte von Virginia und Maryland von Augustin Herrman ist eine der drei bedeutsamsten Karten der Region vor der offiziellen Landesvermessung 1833. Nachdem die Virginia-Karte von John Smith von 1612 fast 60 Jahre lang die Grundlage aller Karten dieses Gebietes gebildet hatte, übernahm die Herrman-Karte diese Rolle als Prototyp bis zur Mitte des 18. Jahrhunderts. Dann wurde sie von der dritten bedeutsamen Karte der kolonialen Periode, der Fry-Jefferson-Karte[27] abgelöst.

Das Original, das als Vorlage für den Stich Faithornes gedient hatte, ist bisher nicht aufgetaucht. Bis zum Beginn unseres Jahrhunderts war überhaupt nur ein Exemplar des Kupferstiches, das der Kartenabteilung des Britischen Museums, bekannt. Auch heute noch ist die Karte außerordentlich selten; bisher sind nur fünf Exemplare nachgewiesen. Neben dem Exemplar in London befinden sich zwei

---

25 Eine der Vorläufer von Herrmans Karte, John Farrer, *A mapp of Virginia discouered to the Falls* 1651, in: E. William, *Virgo Triumphans: or Virginia Richly and Truly Valued*, London 1651, zeigt eine kuriose Darstellung des Kontinents – sie stellt eine durch den Hudson gebildete Durchfahrt zum Stillen Ozean dar und gibt die Entfernung zum Pazifik mit einigen wenigen Tagreisen an.

26 John Smith, *Virginia*, 1612.

27 Joshua Fry – Peter Jefferson, *Map of the Inhabited part of Virginia containing the Whole Province of Maryland Drawn* [...] *in 1751*.

in der Bibliothèque Nationale in Paris und je eines in der John Carter Brown Library in Providence/Rhode Island und in der Library of Congress in Washington D.C. Es ist nicht eruierbar, wie viele Exemplare überhaupt gedruckt wurden. Eine sehr kleine Auflage, wie man aufgrund der wenigen erhaltenen ableiten könnte, würde eigentlich im Gegensatz zu den Annoncen in der *London Gazette* stehen.

Aufgrund der Bedeutung der Herrman-Karte wurden seit dem Ende des letzten Jahrhunderts mehrere Reproduktionen angefertigt.[28] Es existieren vier zeitgenössische Kopien, die wahrscheinlich um 1677 von Angehörigen der *Thames-Shool of chart-makers* in London für den offiziellen Gebrauch in Amerika hergestellt wurden – vermutlich erschien der Kupferstich aufgrund seiner Größe als zu unhandlich. Diese Manuskriptkarten stammen nicht vom Kupferstich ab, sondern wurden nach einer früheren Version (vielleicht dem Originalmanuskript) kopiert. Die gedruckte Version enthält Fehler und Zusätze, die die handgezeichneten und kolorierten Exemplare nicht aufweisen.[29]

Trotz des königlichen Privilegiums erschienen bald nach der Veröffentlichung der Karte Verarbeitungen derselben. Die erste stammt von John Speed[30] – eine Karte, die topographisch noch der Smith-Karte von 1612 folgt, aber bereits die Toponyme Herrmans verarbeitet. Sie stellt einen interessanten Übergang von einem Prototyp zum folgenden dar. 1698 folgte dann die Darstellung von Thornton und Fisher[31] und zwischen 1690 und 1696 erschien die erste französische Version, auf Herrman und auf Thornton-Fisher

---

28 In: Maps to accompany the report of the commissioners on the boundary line between Virgina and Maryland, Richmond 1873; in: B. F. Stevens, Stevens's Facsimiles, London 1896; in: P. L. Phillip, The Rare Map of Virginia and Maryland by Augustine Herrman, First Lord of Bohemia Manor Maryland, Washington 1911; und ein Faksimile der John Carter Brown Library, Providence 1948.

29 Vgl. W. P. Cumming, Early Maps of Chesapeake Bay Area; their relation to settlement and society, in: Early Maryland in a Wilder World. Ed.: David B. Quinn, Detroit 1982, 293.

30 John Speed, *A Map of Virginia and Maryland*, in: John Speed, The Theatre of the Empire ... , London 1676.

31 John Thornton – William Fisher, *Virginia. Maryland. Pennsylvania, East & West New Jersey,* 1689 in: The English Pilot: The Fouth Book.

basierend.[32] Weitere Verarbeitungen stammten von Herman Moll, John Senex und, mit einigen Veränderungen versehen, bildete die Herrman-Karte auch die Basis für die Abbildung im *Atlas Geographicus Maior* von Homann.[33]

Literaturverzeichnis

Beck H. H., Augustine Herrman, Lancaster Country's First Map Maker, in: Historical Papers and Addresses of the Lancaster Country Historical Society 35 (1931), 261–266.

Capek T., Augustine Herrman of Bohemia Manor, Praha 1930.

Cope G., Copy of the Will of Augustine Herrman, of Bohemia Manor, in: The Pennsylvania Magazine of History and Biography 13 (1889),. 321–326.

Cronau R., Drei Jahrhunderte Deutschen Lebens in Amerika. Eine Geschichte der Deutschen in den Vereinigten Staaten, Berlin 1909.

Cumming W. P., British Maps of Colonial America, Chicago/London 1974.

Cumming W. P., Early Maps of Chesapeake Bay Area; their relation to settlement and society, in: Early Maryland in a Wilder World. Ed.: David B. Quinn, Detroit 1982, 267–310.

Heck L. W., Augustine Herrman. Beginner of the Virginia Tobacco Trade, Merchant of New Amsterdam and First Lord of Bohemia Manor in Maryland, Englewood (Ohio) 1941.

Herrman, Augustin. In: The National Cyclopaedia of American Biography being the History of the United States 4, New York 1897, 509.

Herrman, Augustine. In: Dictionary of American Biography. Ed.: Dumas Malone, 8, London/New York 1932, 592/593.

Koudelka J., Pán na Ceské Rece (The Lord of Bohemian River), Praha 1946.

Kupčík I., Die kartographische Tätigkeit von Augustin Herman (ca. 1621-1686) — Realität und Fiktion, in: Mappae Antiquae. Liber Amicorum Günter Schilder. Essays on the occasion of his 65th birthday. Edited by Paula van Gestel-van het Schip and Peter van der Krogt with the collaboration of Marco van Egmond, Peter H. Meurer, Paul van den Brink and Edward H. Dahl (= Utrechtse Historisch-Cartografische Studies, Utrecht Studies in the History of Cartography 6), 't Goy-Houten 2007, 203-220.

Library of Congress, Geography and Map Division, The Bibliography of Cartography 3, Boston 1973.

---

32 *Carte Particulière de Virginie, Maryland, Pennsilvanie, La Nouvelle Iarsey* [...] *chez P. Mortier* – diese Karte stammt entweder von Nicolas Sanson oder von A. H. Jaillot und wurde von Peter Mortier herausgegeben.

33 Zu den Verarbeitungen vgl. E. B. Mathews, Maps and Map-Makers of Maryland, in: Maryland Geographical Survey 2 (1898), 220.

Mathews E. B., Maps and Map-Makers of Maryland, in: Maryland Geographical Survey 2, Baltimore 1898, 337–488.

New York Colonial Manuscripts. Journal of the Dutch Embassy to Maryland, in: O'Callaghan E. B. (Hrsg.), Documents relative to the Colonial History of the State of New York; procured in Holland, England and France by John Romeyn Brodhead, Esq. 2 (Albany 1858), 88–98.

Papenfuse E. C. – Coale J. M., The Hammond-Harwood Atlas of Historical Maps of Maryland, 1608 – 1908, Baltimore/London 1982.

Phillips P. L., Virginia Cartography. A Bibliographical Description (= Smithonian Miscellaneous Collection 37/1039), Washington 1896.

Phillips P. L., The Rare Map of Virginia and Maryland by Augustine Herrman, First Lord of Bohemia Manor Maryland. A Bibliographical Account with Facsimile Reproduction from the Copy in the British Museum, Washington 1911.

Rattermann H. A., Augustin Herrman. Eine Karakterfigur aus der Begründungsgeschichte von New York und Maryland, in: Deutsch-Amerikanisches Magazin. Vierteljahresschrift für Geschichte, Literatur, Wissenschaft, Kunst, Schule und Volksleben in Amerika 1 (1887), 202–225 und 524–538.

Ristow W. W., Augustine Herrman's Map of Virginias and Maryland, Acquired by the Library of Congress, in: Surveying and Mapping 21/1961, 57–62.

Ristow W. W., Augustine Herrman's Map of Virginia and Maryland, in: A la Carte. Selected papers on maps and atlasses. Ed.: Walter W. Ristow (Reprinted from the August 1960 issue of the Library of Congress Quarterly Journal of Current Acquisitions), Washington 1972, 96–101.

Sanchez-Saavedra E. M., A Description of the Country. Virginia's Cartographers and Their Maps 1607–1881, Richmond 1975.

Schwartz S. I. – Ehrenberg R. E., The Mapping of America, New York 1980.

Svejda M. N., Augustine Herman of Bohemia Manor, in: The Czechoslovak Contribution to World Culture. Ed.: Miloslav Rechcigl, The Hague/London/Paris 1964, 500–504.

Thieme U. – Becker F., Allgemeines Lexikon der bildenden Künstler von der Antike bis zur Gegenwart 11, Leipzig 1915.

Tyacke S., London Map Sellers 1660–1720. A collection of advertisements for maps placed in the London Gazette 1668–1719 with biographical notes on the map-sellers, Tring (Hertfordshire) 1978.

Vallandingham E. N., Lord of Bohemia Manor. Augustine Herrman, an early hero of New York and Maryland. Story of a seventeenth Century Merchant Who became a Great Landowner and Baron on the Delaware Peninsula, in: The New York Sun, 23. 10. 1892. Abgedruckt in: P. Lee Phillips, Virginia Cartography. A Bibliographical Description (Smithonian Miscellaneous Collection, 37/1039), Washington 1896, 37–41.

Verner C., The First Maps of Virginia 1590–1673, in: The Virginia Magazine of History and Biography 58 (1950), 3–15.

Wawrik F. – Zeilinger E. – Hühnel H. – Mokre J. (Hrsg.), Die Neue Welt. Österreich und die Erforschung Amerikas. Ausstellung im Prunksaal der Österreichischen Nationalbibliothek, 15. Mai - 26. Oktober 1992, Wien 1992.

Wroth L., John Carter Brown Library Providence, Rhode Island, in: Report to the Corporation of Brown University July 1, 1930, Providence 1930, 10–15.

Wilson J. G., A Maryland Manor (Maryland Fund Publications 30), Baltimore 1890.

Zum Autor

Jan Mokre, geb. 1961 in Berlin, Studium der Geschichte an der Universität Wien, Spezialisierung auf Kartographiegeschichte, Entdeckungsgeschichte und Geschichte außereuropäischer Völker. Direktor der Kartensammlung und des Globenmuseums der Österreichischen Nationalbibliothek, Generalsekretär der Internationalen Coronelli-Gesellschaft für Globenkunde.

Kontakt: Österreichische Nationalbibliothek, Kartensammlung und Globenmuseum, Josefsplatz 1, A-1015 Wien; jan.mokre@onb.ac.at

Jean Lattré,
Carte des États-Unis de l'Amérique Suivant
le Traité de Paix
Paris 1784

Mary Sponberg Pedley, Ann Arbor

A Map for Benjamin Franklin

Paris in 1784 buzzed with the news of the Peace Treaty negotiated between the victorious Americans and the defeated British, as the world watched a powerful nation lose her colonies. One of the chief negotiators of the Treaty was Benjamin Franklin, printer, essayist, scientist, and diplomat, who had arrived in Paris in late 1776 to act as minister plenipotentiary for the fledgling United States, shortly after the official break with Britain had been made via the Declaration of Independence. Franklin was a popular figure in Paris, lionized by salon society, the darling of the urban chatelaines; he also pursued his scientific interests while in the city, using his time to attend meetings of the *Académie des Sciences* and to continue to conduct scientific experiments. One particular interest was his research into the Gulf Stream, that warm ocean current running close to the eastern shore of North America across the Atlantic to northern Europe. Franklin's map of the Gulf Stream was first produced in Paris, engraved and printed by Georges-Louis LeRouge.[1]

It is not surprising, therefore, to find another Paris mapmaker paying his respects to Franklin and the young United States of America. In 1784, Jean Lattré, a well known and prolific engraver, printer, and mapseller, published the *Carte des États-Unis / de l'Amérique | Suivant le Traité de Paix de 1783./ Dediée et Presentée | A S.*

---

1 E. R. Cohn, Benjamin Franklin, Georges-Louis Le Rouge and the Franklin/Folger Chart of the Gulf Stream, in: Imago Mundi *52* (2000), 124-142.

Excellence Mr. Benjamin Franklin / Ministre Plénipotentiaire des Etats-Unis de l'Amérique près la Court de France,/ anc[ien]. Présid[ent]. de la conventio[n] de Pensilvanie et de la Société Philosophique de Philadelphie. &c &c./ Par son très humble et très obeissant / Serviteur Lattré. | 1784. | avec privilege du Roi.

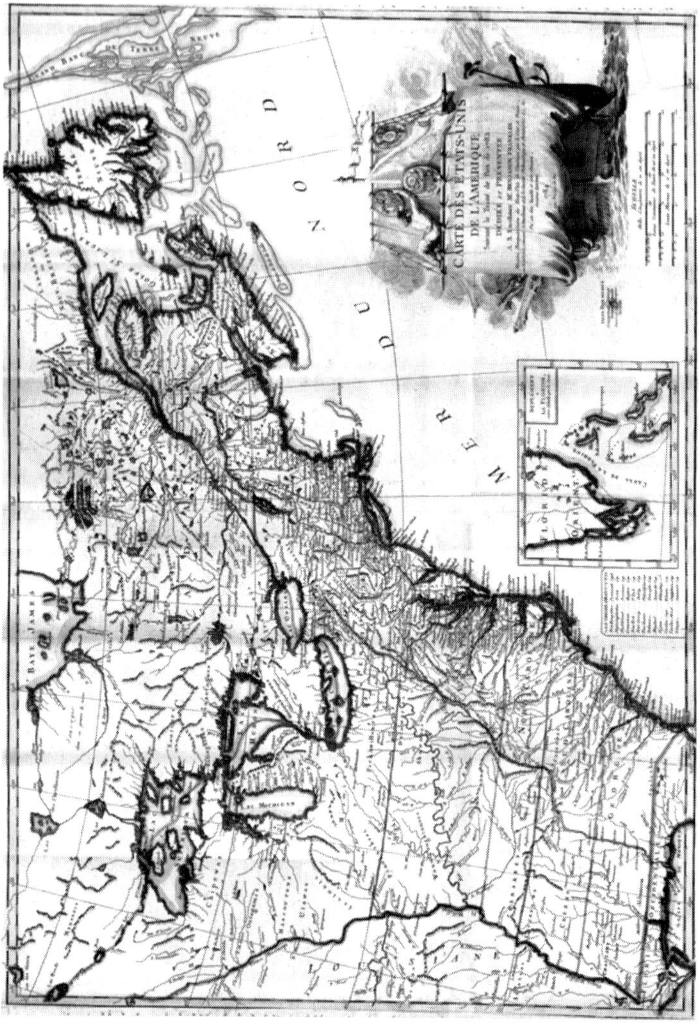

Fig.1: Image of complete map from Woldan collection, K-V(B1) : AM/Nor 171

Lattré's map has been described as *the first French map of the United States of America.* As with so many claims of *first*, this statement misses a more essential purpose of the map, which was to celebrate the connections between France and the United States and to pay homage to Franklin for his services both to his native country and to his hosts in France.[2]

Geographically, the map covers that part of North America from which the young United States were carved: the eastern seaboard from Newfoundland (*Terre Neuve*) in British Canada to the northern edge of Florida (Ft Moos), westward to the boundary of the Mississippi River, north beyond the Great Lakes to the southern end of Canada's Hudson Bay and south to the Gulf of Mexico. The remainder of *La Floride* is shown in the inset *Supplément.* The large map measures 55.7 cm by 77.5 cm, on a single plate, 58 x 78.5 cm. Its scale is c.1:4 500 000. Three local scales are shown: *Milles d'Angleterre 70 au degré*; *Lieues communes de France 25 au degré*; and *Lieues marines de France 20 au degré.* Some versions of the map are accompanied by two engraved texts, appended to each side of the map, listing the *Principaux Evenemens Militaires entre les Americains et Les Anglois.*

The map's cartouche displays a view of the bow of a ship under full sail. Two seals hang from the sails: on the far right, the seal of the Society of the Cincinnati; in the middle, the seal of the United States of America. On the left, a sailor balances precariously atop the sail's spar in order to hang the seal of Benjamin Franklin onto the sail itself.

The coloring of the map depended entirely on the owner: most copies highlight only the boundaries between the fledgling United States and Canada and between the individual states in watercolor; other copies use a water color wash to distinguish the states from each other and to shade the water.

---

2 A facsimile of this map appeared in 1973 published by R. Donnelly & Co., accompanied by a pamphlet by Lester Cappon: *The First French Map of the United States of America.* Cappon's title is mis-leading, since there were several maps of the *United States* which preceded Lattré's map, including one by Lattré himself, dated 1783. See note 18 below. As this article will show, the significance of this map is not in being a *first* but in its dedication to Benjamin Franklin, and its early use of the Great Seal of the United States in conjunction with the seals of Franklin and the Society of the Cincinnati.

The author of the map is unknown, though it is likely to have been a cartographer associated with workshop of the publisher and engraver, Jean Lattré. Likely candidates might be Jean Janvier or Rigobert Bonne, both of whom produced maps for Lattré, particularly for his *Atlas Moderne* (1762). Lattré's other cartographic colleague, Giovanni Antonio Rizzi-Zannoni had by this time, 1783, left Paris and re-settled in his native Italy.

Jean Lattré

Jean Lattré (c. 1722? – 1788) was a well-known engraver in Paris during the second and third quarter of the eighteenth century. The quality of his work garnered him titles as *graveur ordinaire du roi*; *graveur ordinaire de monseigneur le Dauphin* (1767) and *graveur ordinaire de monseigneur le duc d'Orléans et de la Ville* [*de Paris*] (1771).[3] From early in his career, he specialized in engraving and publishing maps and city plans; he was also an expert engraver of calligraphy.[4] His fine work is particularly evident in the *Plan de la ville de Nantes* […] surveyed by François Cacaut, engraved by Lattré, and published in 1759.[5] On this map, Lattré has carefully differentiated a variety of types of land and land use, using gradations of line and shading in his engraving, without the application of color.

In his early years, Lattré worked for Gilles Robert de Vaugondy (*Terre de Chanaan,* 1743), for A.F. Gervais de Palmeus (*Malta*, 1741) and for the abbé Jean de la Grive, engraving the script for his well known *Plan de Paris* (1744). Later in his career, he worked occasionally for Jacques Nicolas Bellin in the Dépôt de la Marine, engraving some of the charts found in the *Hydrographe Françoise*.[6] He seems to

3 Bibliothèque nationale de France (Estampes), Inventaire du fonds français; graveurs du dix-huitième siècle 12, Paris 1973, 430-463.

4 F. Glachant, *Nouveau traité de l'écriture*, Paris 1754. The plates are engraved by Aubin, J. Lattré and L. Baillieu. Chicago, Newberry Library, Wing oversize ZW 739.G452.

5 Paris, Bibliothèque nationale de France: Ge DD 2987 (1183) B, collection d'Anville. The exceptionally rich cartouches, abundant with putti, were designed by J. A. Volaire and executed by J. N. Tardieu and J. J. Flipart: http://visualiseur.bnf.fr/ark:/12148/btv1b7711344c/f2 [24.7.2009].

6 *Mer de l'Inde, l'Archipel au nord de l'Isle de France, Découvertes faites au Nord-Est de l'Isle de France*, Volume I, sheet 95.

have become an independent publisher sometime in the 1750s; he published the *Plan géometral de la ville de Bordeaux*, surveyed by the geographers Santin and Mirail, in 1755.[7] Lattré engraved the map and employed Paris' foremost engraver and ornament designer, Charles Nicolas Cochin fils, for the cartouche. Lattré's attention to design and fine detail of the aesthetic aspect of his maps would become a defining feature of his work.[8]

The success of Lattré's maps from the 1750s must have encouraged him to launch into the far more complex and financially precarious business of atlas production. Enlisting the aid of well known geographers, Rigobert Bonne and Giovanni Battista Rizzi-Zannoni, his first productions focused on the coasts of France[9] and Germany, where the recently battles of the Seven Years' War engaged market attention.[10] These efforts may have inspired him to produce the *Atlas moderne ou collection de cartes sur toutes les parties du globe terrestre,* in 1762.[11] This atlas comprised maps by Giovanni Antonio Rizzi-Zannoni, Rigobert Bonne, and Jean Janvier. Lattré joined with another bookseller, Jean Thomas Hérissant, to share the costs of production; for a later edition (1771) he shared the publication rights with Nicolas

---

7 Washington D.C., Library of Congress: G5834.B6 1755.

8 Lattré's wife, Marie Françoise Vérard, was an equally gifted engraver, adept in calligraphy and map engraving. Contemporary critics remarked upon her map of the solar eclipse, a fine example of two-color printing, as it showed the map of Europe in sepia or red, with the visibility of the passage of the solar eclipse overprinted in black. The data for the eclipse were based on the astronomical work of Nicole-Reine Lépaute Etable de la Briere, wife of Jean André Lepaute, the royal *horloger*. The ornamental cartouche was engraved by Mme (Elisabeth-Claire) Tardieu, wife of Nicolas Tardieu of the illustrious engraving family. *Passage de l'Ombre de la Lune au travers de l'Europe dans l'Eclipse de Soleil centrale & annulaire qui doit arriver le premier avril 1764.* See *Mémoires de Trévoux*, juin 1762, 1530-1534, for favorable comment on this map, as an example of what women could produce, if only they were educated.

9 *Atlas Maritime, ou cartes réduites de toutes les côtes de France,* 1762, with maps by Rigobert Bonne.

10 *Atlas géographique et militaire, ou Théâtre de la guerre présente en Allemagne*, 1761, maps by G.A. Rizzi-Zannoni. This Atlas was published along with the book dealer Ballard, minimizing some of the risk for Lattré.

11 While it might seem that this atlas was produced in competition with Rizzi-Zannoni's *Atlas géographique, contenant la mappemonde et les quatre parties, avec les différents états d'Europe, dressés sous les yeux de M. Rizzi-Zannoni*, the latter atlas was published in 12°, while Lattré's atlas was between quarto and folio sized.

Augustin Delalain, bookseller.[12] The following year, 1763, Lattré published the *Petit Neptune Anglois ou Carte Marine des Côtes d'Angleterre, d'Ecosse et d'Irlande* with maps by Rigobert Bonne. Twenty years later, his marketing instincts encouraged him to publish a school atlas, the *Petit atlas moderne ou collection de cartes élémentaires dédié à la jeunesse* (1783).

Like other engravers during the latter half of the eighteenth century, Lattré saw economic survival in the diversity of his stock. While his inventory seemed to concentrate on urban plans (e.g., Nantes, Rouen), atlases of general maps, the occasional specialty map such as the eclipse map of 1764, and globes, he also expanded into a more general market with the publication of the *Almanach Iconologique*, published in fascicules from 1765 to 1782. It was completed in 17 small (in 16°) volumes, the images designed and engraved by Charles Nicolas Cochin fils and Hubert Gravelot, brother of Jean Baptiste Bourguignon d'Anville. It also was printed in a larger version (in 8°) in four volumes.[13] In all these productions one discerns Lattré's refined sense of design and carefully tuned engraving hand.

During the period 1774 to 1781, Lattré spotted a particularly eager market in Paris: maps of the war of American Independence being fought between the British and their revolutionary colonists in North America. Lattré advertised English maps, imported *from London, relative to the present war in North America.*[14] The map of the newly minted United States was published very much in this vein: as an *aide-mémoire* for up-to-date journalism.

Advertising for the map

The *Mercure de France* advertised the Carte *des Etats-Unis* in 1784, in the July-August issue (page 91) and again in the November – December issue (page 43), giving the price as two *livres* ten *sous*, three *livres lavée* (lightly colored).

---

12 For a fuller discussion regarding the need for booksellers to join with printsellers in order to publish atlases, see M. Sponberg Pedley, The Commerce of Cartography, Chicago 2005, 90-93.

13 See announcements in *Mercure de France,* December 1763, 98.

14 *Journal de Paris* jeudi 15 mars 1777 (no. 135): *Lattré* […] *avertit le Public qui'il vient de recevoir un assortiment complet de Cartes originales angloises sur tout ce qui est relatif à la guerre présente dans l'Amérique septentrionale.*

Une Carte des Etats-Unis de l'Amérique, avec la nouvelle division suivant le Traité de Paix de 1783, dédiée & présentée à M. Franklin Ministre des Etats-Unis de l'Amérique près la Cour de France. Cette Carte est sur une feuille & demie d'Aigle, & est accompagnée d'un Abrégé Historique des Expéditions Militaires. Prix. 2 liv. 10 sols, & tout ce qu'on peut desirer en Géographie; des Ecrans de tous genres proprement faits, & plusieurs objets d'étrennes curieux & Utiles. A similar advertisement was carried in the *Journal de Paris* of 9 July with a similar price.

This price was very much in line with French maps in general published in Paris: one *livre* per sheet was an average price and the addition ten *sous* for simple outline color was an average additional cost. Thus the map fell within the price range for a buyer interested in the outcome of the War for American Independence. Furthermore, the map also did not present any new geographical information about the fledgling United States: it followed the outlines of other English and French maps in its delineation of the North American continent, the Great Lakes, the eastern seaboard, and the Florida peninsula. What gave the map a new dimension was the addition of the advertised *Abrégé historique des Expéditions Militaires*». This *Abrégé* materialized as *Principaux Evenemens militaires entre les Americains et les Anglois*, a chronological list of battles and events of the war printed from engraved plates and glued as long strips to either side of the map, forming a marginal frame for the map itself. Indeed, almost all of the sites of battles and conflicts, however small and obscure, are shown on the map, as are the routes of the American army under Washington, and the French forces who came to his aid.

Sources for the Map

Like most geographical engravers, Lattré probably did not construct this map himself, relying instead on the work of more capable geographers. His earlier maps, beautifully engraved by his workshop, were prepared by skilled geographical compilers, such as Jean Janvier or Rigobert Bonne. Lattré had relied on both these geographers for the compilation of maps in his *Atlas Moderne* of 1762, along with the specialized skills of Giovanni Antonio Rizzi-Zannoni. Yet it is unusual that Lattré did not credit either the map's creator or the designer of the sumptuous cartouche. In his *Atlas moderne*, most of the cartouches are signed by P.P. Choffard, Arrivet, Morillier and Berthaut. Lattré and his workshop may well have engraved the geographical section of the map, as was done for the map of the Holy Land in the *Atlas moderne.*[15]

The general geography of Lattré's map of North America seems to be based upon the landmark map of 1755 by John Mitchell, printed in London by Thomas Kitchen.[16] Mitchell's map served as the archetype for many, if not most, later maps of eastern North America, supplemented by the results of other British surveys conducted during the period of the Seven Years' War (1756-1763), which provided large scale information to the European community, through the distribution network of London cartographers. The use of the Mitchell map during the boundary negotiations in the 1780s between the British, the Americans, the French and the Spanish is well known.[17] Continental mapmakers sometimes relied upon the 1756 copy of the Mitchell map published in Paris by Georges Le Rouge.

Lattré himself was in touch with London mapmakers; his letters to William Faden during the War of American Independence testify to his interest in up to date information about North America and about the ongoing war. In 1777, he ordered the Mitchell map, maps of the Theatre of war, of Canada, of the province of New York, of the area around Philadelphia, of specific military operations.[18] He

---

15 *Terrae Sanctae Tabula* […], based on the work of Guillaume and Joseph Nicolas Delisle. The map is signed *gravé par Lattré* while the cartouche was signed *Morillier inv*[enit] and *Berthaut sculp*[sit].

16 M. Edney, John Mitchell's map of North America (1755). A study of the use and publication of official maps in eighteenth century Britain, in: Imago Mundi 60 (2008), 63-85.

17 J. Dunbabin, Red lines on maps: the impact of cartographical errors on the border between the United States and British North America, 1782-1842, in: Imago Mundi 50 (1998), 105-125. – The map used in the Course of our Negotiations was Mitchells; cf. The American Commissioners (i.e., John Adams, Benjamin Franklin, John Jay, Henry Laurens) to Robert R. Livingston, Paris, December 14, 1782, in: E. Cohn (Ed.), The Papers of Benjamin Franklin 38, New Haven 2006.

18 M. Pedley, The Map trade in the late eighteenth century: Letters to the London map sellers Jefferys and Faden, SVEC, 2000:06, Voltaire Foundation, Oxford, Letter I: 24, 84-85. He sold these maps directly in the Paris market, as advertised in the *Journal de Paris* (jeudi 15 mars 1777, no.

himself had published a *Theatre de la guerre en Amérique* compiled by Rigobert Bonne in 1779.[19]

Upon the signing of the preliminary draft of the Treaty of Paris in late 1782[20] which determined the boundaries of the fledgling United States, mapmakers in London quickly moved into high gear. By April 3, 1783, John Wallis at the Map Warehouse had published his *United States of America* [...] *Agreeable to the Peace of 1783.* Thomas Kitchin soon followed on May 2nd with *Map of the United States in North America* [...] *according to the Treat of 1783.* August 6th saw the appearance of John Cary's *Accurate map of the United States of America* [...] *agreeable to the Treaty of Peace of 1783.* William Faden himself produced *The United Stated of America* [...] *according to the Treaty,* which is dated 1783 in the cartouche, but, unusually for Faden, lacks the usual copyright statement and date *As the Act directs* engraved on the map.

Paris mapmakers were not slow to emulate their cross Channel comrades, creating new publications such as Mondhare's edition of the J.B. Eliot *Carte generale des États Unis de l'Amérique Septentrionale, avec les limites de chacun des dits Etats, convenus par le traité provisional du mois de novembre 1782.* Or they revised old maps with new boundaries: Dezauche/Buache/Delisle, *Carte du Canada qui comprend la partie septentrionale des Etats Unis d'Amérique.* [...] *Revue et augmentée en 1783, a Paris, Chez Dezauche, successeur des Srs. Del'Isle et Buache.* One of the most elegant was published by Brion de la Tour: *Carte des États-Unis d'Amérique, et du cours du Mississipi*[21]; Even Lattré took his old map of

135): *Lattré* [...] *avertit le Public qui'il vient de recevoir un assortiment complet de Cartes originales angloises sur tout ce qui est relatif à la guerre présente dans l'Amérique septentrionale.*

19 Dated 1779 on LC G3300 1779. B6 Vault. A copy in the Boston Public Library is dated 1781. The same map may be found with a new title, *Partie de l'Amérique septentrionale qui comprend le Canada, la Louisiane le Labrador le Groenland, les Etats-Unis et la Floride.* [...] *par Mr Bonne. Chez Lattré* [...] *revue et corrigée 1783*, cf. http://www.davidrumsey.com [24.7.2009].

20 The preliminary draft was agreed November 30th, 1782; formally ratified September 3, 1783; ratification documents exchanged in Paris May 12, 1784.

21 [...] *rédigée d'après différentes cartes et relations anglaises, et les opérations de la dernière guerre: avec les nouvelles limites générales fixées par les articles préliminaries de paix, signés tant à Paris qu'à Versailles, le 30 9bre., 1782 et le 20 janr. 1783, et confirmées par le traité définitif du 3 7bre. 1783. Cette carte composée par le Sr. Brion de la Tour, ingr. géographe du roi.* [Washington, D.C., Library of Congress: G3700, 1787, B7

North America from the *Atlas Moderne* of 1762 and re-titled it *Theatre de la Guerre* in 1779, and a further new title in 1783: *Partie de l'Amérique septentrionale qui comprend* [...] *les Etats-Unis.*

The Peace Treaty, finally ratified formally in May of 1784, was available to the French public via the limited number of copies printed at Passy from Franklin's own press, printed by printer, Philippe-Denis Pierres. The boundaries established by the treaty could also be seen in the maps of William Faden and Brion de la Tour, on which the lines of demarcation followed the wording of the pertinent articles on boundaries in the treaty.[22] It is surprising that Lattré's *Carte des États-Unis,* dedicated to Franklin, does not follow the boundary lines according to the terms of Article II of the Peace Treaty, which describe the boundary in detail. The northern boundary on the *Carte des États-Unis* follows the treaty line accurately westwards to the Lake of the Woods in Canada whence it disappears off the left edge of the map. Thus the western boundary, running south from the Lake of the Woods along the course of the Mississippi River is initially invisible, only appearing at about the 45th parallel to follow the great river southwards. Similarly, the southern boundary, which according to Article II turns east from the Mississippi at 31 degrees north latitude travels along this parallel to the Apalachicola River. By the terms of the Treaty, the line should turn north to the junction of the Apalochiola with the Flint River, thence turning east to meet the headwaters of the St. Mary's River, and thence down along the middle of St. Mary's River to the Atlantic Ocean. On Lattré's *Carte des États-Unis*, the St Mary's River is not to be found, and the boundary does follow the prescribed jog in the line, but rather extends due east to the Atlantic.

Clearly Lattré did not intend the *Carte des États-Unis* to be used to examine closely the boundaries established by the Peace Treaty of 1783.

Vault, a unique copy purportedly with the author's annotations made in 1787 on the 1783 map].

22 See M. Edney, The Mitchell Map, 1755-1782: An Irony of Empire. Part 4: The Mitchell Map in Paris, available as a web document, http://www.usm.maine.edu/maps/web-document/2/home/the-mitchell-map-1755-1782-an-irony-of-empire [22.12.2009] for the exact wording of the treaty.

Fig 2: Detail of the southern boundary

What then was his purpose? The clue is provided by the text pasted on either side the map: *Principaux evenemens militaire entre les Americaines et les Anglois.* This chronological account of the "principal military events" lists the actions of the war grouped geographically, beginning with Massachusetts (1775-1776), Virginia (1775), Canada (1775-1776), New York (1776-1777), Virginia again (1777 though all the place names are in Pennsylvania), York County [Virginia] (1777, but place names that follow are all in New York), Pennsylvania (1778), Georgia (1778), New York (1779), Connecticut, Georgia and South Carolina (1779), South Carolina (1780) Massachusetts (1780), and finally Virginia (1781), ending with the surrender of Cornwallis on the 18th [*sic!*, 19th] of October. Almost all the place names mentioned in the description of the action, with only a few exceptions, may be found on the map, despite its small scale. While relatively obscure scenes of battles, such as Linch's Creek and Tugley in South Carolina are shown, other place names, more iconic in American history, such as *Walley* [i.e., Valley] *Forge*, are not labeled, though they appear in the *Principaux Evenemens.* The source of Lattré's *Principaux Evenemens* is not evident; because of his correspondence with William Faden, one might surmise that he had taken advantage of the printed text accompanying many of Faden's battle maps published soon after conflicts. The London map publishers Robert Sayer and John Bennett also provided textual summaries of the war to accompany their maps. Nonetheless, Lattré's succinct chronological and geographical account may be a unique summary of the war's events.

The choice of several of the "principal events" highlights the role of the French as they assisted the Americans in their insurgency against the British. The marquis de Lafayette plays a large part here: first described as "beating an English detachment larger than his own. (Virgina, October 25, 1777)". Lafayette takes up considerable space in the description of his efforts with Washington in Philadelphia, including his close escape from capture by the British General Howe, who had boasted to the ladies of Philadelphia that he would bring this young warrior back from battle with him to be their guest at dinner (Pennsylvanie, 1778, juin):

La Fayette prend post à Baren hill à 4 ms. De Philadelphie. L'Armée Angloise marche pour enveloper ce détachemt. Howe avoit promis aux Dames de la Ville de les faire souper avec ce jeune Guerrier, la Fayette se retire sans perdre un homme quoiqu'au milieu de l'armée Angloise. Le 18 les Anglois évacuent Philadelphie.

Only Lafayette is given this singular treatment in the list of "Principal Events". The arrival of the French army with General Rochambeau is mentioned in 1780 only as *Juillet le 15. arrivé des Troupes Françoises à Rode-Island.* Similarly, French naval action along the Virginia coast is remarked, but the only commanders mentioned are Ternay (*1781 Mars: L'escadre de Ternay qui portoit des Troupes de Terre à la Chesapeak rencontre la flotte Angloise. Combat opiniatre*) and d'Estaing (*1779, Septemb.: Le 9 D'Estaing jette l'ancre à l'Isle Tybee, le 16 les troupes son débarquées et les Americs. Se joignent aux François*). De Grasse, who kept the British navy at bay and away from Chesapeake Bay, is nowhere to be found.

The French action and support described in the *Principaux Evenments* provided the necessary support for the Americans to win their independence from Britain. This alliance and bond between the two countries is celebrated in the cartouche, with its rhythmic array of seals.

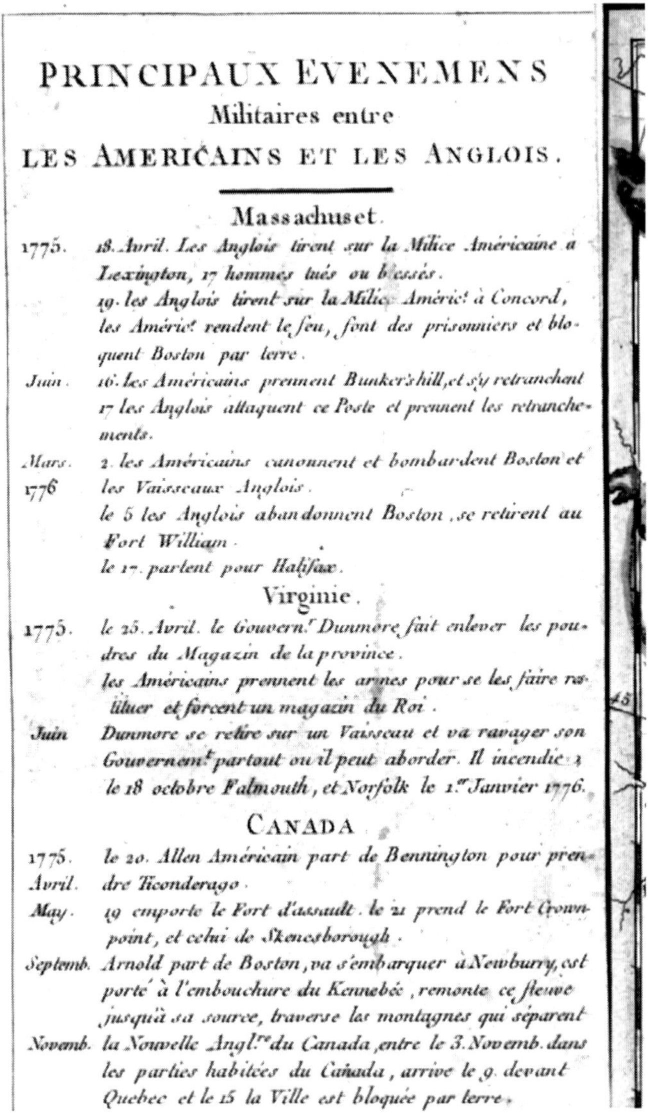

PRINCIPAUX EVENEMENS
Militaires entre
LES AMERICAINS ET LES ANGLOIS.

Massachuset.

1775. 18. Avril. Les Anglois tirent sur la Milice Américaine a Lexington, 17 hommes tués ou blessés.
19. les Anglois tirent sur la Milic. Améric. à Concord, les Améric. rendent le feu, font des prisonniers et bloquent Boston par terre.
Juin. 16. les Américains prennent Bunker's hill, et s'y retranchent 17 les Anglois attaquent ce Poste et prennent les retranchements.
Mars. 1776 2. les Américains canonnent et bombardent Boston et les Vaisseaux Anglois.
le 5 les Anglois abandonnent Boston, se retirent au Fort William.
le 17. partent pour Halifax.

Virginie.

1775. le 25. Avril. le Gouvern.r Dunmore fait enlever les poudres du Magazin de la province.
les Américains prennent les armes pour se les faire restituer et forcent un magazin du Roi.
Juin Dunmore se retire sur un Vaisseau et va ravager son Gouvernem.t partout ou il peut aborder. Il incendie le 18 octobre Falmouth, et Norfolk le 1.er Janvier 1776.

CANADA

1775. Avril. le 20. Allen Américain part de Bennington pour prendre Ticonderago.
May. 19 emporte le Fort d'assault. le 21 prend le Fort Crownpoint, et celui de Skenesborough.
Septemb. Arnold part de Boston, va s'embarquer à Newbury, est porté à l'embouchure du Kennebéc, remonte ce fleuve jusqu'à sa source, traverse les montagnes qui séparent
Novemb. la Nouvelle Angl.re du Canada, entre le 3. Novemb. dans les parties habitées du Cañada, arrive le 9. devant Quebec et le 15 la Ville est bloquée par terre.

Fig. 3: Detail of the *Principaux evenemens*

Cartouche

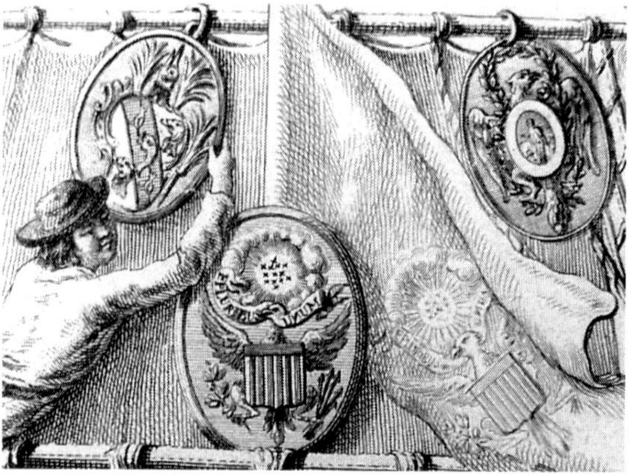

Fig. 4: Detail of the cartouche

The cartouche comprising ship under full sail demonstrates Lattré's economical use of decorative ideas. He had employed a smaller version of the same device on his map of the Northern Netherlands in the *Atlas Moderne* of 1762. He has doubled the size of the ship and sail to fit the much larger *Carte des Etats-Unis* and added a sailor who is busy hanging up the three seals onto the sails.

These three seals connect the subject matter of the map to its dedicatee, Benjamin Franklin, and to the French government. In the center, the seal of the United States commemorates the legitimization of the young country by the terms of the Treaty of Paris, recently signed. The history of the seal's design took root during the deliberations over the signing the Declaration of Independence in 1776. Franklin, John Adams, and Thomas Jefferson had worked on a design of an earlier seal, which was not accepted by the Continental Congress. A later design was proposed by the Secretary of the Congress, Charles Thomson assisted by William Barton of Philadelphia. It consists of the eagle grasping the arrows of war and the olive branch of peace in its talons; he holds the Latin motto of the United States in his beak: *e pluribus unum* (out of many, one). Above the eagle, clouds surround the thirteen stars, representing the thirteen original British colonies, now independent states. The design was adopted in 1782 and a die cut from brass in 1782.[23] Franklin had ordered another block to be cut in Paris for use on the published version of the Treaty of Paris. Franklin's printer, Philippe-Denis Pierres, used the seal on the title pages of both the *Constitutions des Treize Etats-unis de l'Amérique*, and the *Definitive treaty between Great Britain and the United States,* both published in 1783 at Passy.[24] Franklin also used the seal for other official documents issued by him as Minister Plenipotentiary. On the cartouche ship sail, the seal is repeated in faint outline imprinted on the flag billowing in the wind.

Fig. 5: Further detail of cartouche with three seals

23 *The Great Seal of the United States*, United States Department of State, Bureau of Public Affairs, September 1996.

24 E. Cohn, The Printer at Passy, in: Page Talbott (Ed.), Benjamin Franklin: In search of a better world, New Haven and London 2005, 235-272.

Until he had his own printing block of the official seal of the United States, Franklin had used his own seal for identifying his papers and books and as his official seal of office. This is the seal which the young sailor is hanging onto the sail. Franklin had devised this seal when he was a printer's apprentice in Boston, designing his *blazon* in imitation of a coat of arms he had found associated with an English Franklin family: the arms display a pike's head facing upward, two lion heads facing to their right, and a diagonal cross bar with two martins either side a dolphin. Such crests were usually associated with landed gentry and nobility, not men of the lower social orders, such as Franklin; but he justified the creation of his seal: *The first thing requisite in an almanac-writer is, that he should be descended of a great family and bear a coat of arms, this gives luster and authority to what a man writes, and makes the common people believe, that certainly this is a great man.*[25]

The third seal on *Carte des Etats-Unis,* to the right, is that of the Society of the Cincinnati, a group which honors the participation of the French army and navy in the War of American Independence. The Society was founded in May 1783, with George Washington as its first president. Named in honor of the ancient Roman farmer and patriot, Cincinnatus, who twice abandoned his plough to take up arms in defense of his country, the society was open to the officers of the Continental (American) army and their French counterparts, in order *to perpetuate the memory of the revolution and their mutual devotion to it by aiding the officers according to the three principles of financial, fraternal, and patriotic support.*[26] The society's seal was designed by Pierre Charles l'Enfant (a member of the original society and architect who laid out the city of Washington D.C.); it consists of an eagle with wings slightly spread and a roundel on its chest, in which Cincinnatus is

---

25 From *The Pennsylvania Gazette*, October 20, 1737: *The first thing requisite in an almanac-writer is, that he should be descended of a great family and bear a coat of arms, this gives luster and authority to what a man writes, and makes the common people believe, that certainly this is a great man. I have known almanac writers so curious and exact in this particular, that they have been at the expense and charge of a wooden cut in the frontispiece, with their arms emblazoned, and surrounded with a label, expressing the name of the family. This, sir, made a great impression, I confess, upon myself and others, and made those works to go off well.*

26 *L'ordre de Cincinnatus*, in: Benjamin Franklin: un Américain à Paris, [exposition] Paris, Musée Carnavalet, 2008, 226-228.

depicted with his plough and the name of the society written in Latin and its foundation date around the edge of the circle. The seal, also known as an "eagle", was first manufactured sometime between December 1783 and April 1784 by the Paris goldsmiths Duval and Francastel.[27] Lattré's depiction of the seal on this map must be a very early representation of it in print.

Franklin himself felt certain ambivalence about the Society of Cincinnati, as he was concerned, rightly, that it would become a hereditary society, rather than merely honoring the courage and patriotism of the participants in the war. Critics of the Society accused it of creating an aristocracy in a country which valued liberty and equality. Franklin supported this view privately and in his support of the pamphlet concerning the Society written by the comte de Mirabeau.[28]

The dedication to Benjamin Franklin

The viewer's attention is drawn particularly to the sailor attaching the personal seal of Benjamin Franklin onto the spar, reminding the audience that the map was dedicated to Franklin. Benjamin Franklin (1709-1790) was an immensely popular ambassador for the fledgling United States. His presence in Paris elicited the same delirium that certain modern celebrities (sports figures, film and music stars, political or religious leaders) enjoy. He had visited Paris on earlier occasions, but arrived as a minister plenipotentiary in 1777 to do the hard work of encouraging French participation in the war as allies of the young republic and of acting as the foreign face of the new United States. In these roles he succeeded admirably: the French provided integral support both by land and sea to the military operations of the Continental Army and financial support to a county whose trade was cut off from its colonial mother. While Franklin worked tirelessly for the cause of the United States, he also found time and energy to pursue his scientific interests, making presentations to the Académie des Sciences, and doing his own bit of

27 F. Bucher, American Footprints in Paris, New York 1921, 97.

28 *L'ordre de Cincinnatus* (note 26), 228: Mirabeau, H.G. de Riqueti, (comte de), *Considérations sur l'ordre de Cincinnatus, ou Imitation d'un pamphlet anglo-américain* […], Londres, J. Johnson, 1784.

map mapmaking. He refined his ideas about the Gulf Stream and employed the Paris geographer George Le Rouge to visualize the Gulf Stream in a map.[29] As a printer himself, Franklin was eager to set up his own press at Passy, where he stayed as the guest of Jacques-Donatien Le Ray de Chaumont, a wealthy slave trader and merchant who personally contributed to the American cause by providing the uniforms for the Army of Virginia.

Nothing in the voluminous Benjamin Franklin papers connects him with Jean Lattré, the publisher of the map. Although the title of the map says "dedicated and presented" to Franklin, no record of Lattré's presentation has been recovered. Nonetheless, Franklin's acquaintance with the circle of printers in Paris, his interest in mapmaking and acquaintance with Le Rouge could easily have included an introduction to Jean Lattré who, like Le Rouge, had an interest in the current events of the War of American Independence and in placing these events on maps. The map's publication in 1784, the year before Franklin's departure (12 July 1785) from France to return to the United States and in the year of the formal ratification of the Peace Treaty between the United States and England, serves as an *homage* to the one man who successfully represented the alliance between France and the young nation. With his dedication to Franklin and the innovative display of the three seals, Lattré summarized and represented the extraordinary victories of the War of American Independence.

Epilogue

I first met Johannes Dörflinger at the International Conference for the History of Cartography in 1983 in Dublin where we shared our mutual interest in eighteenth century cartography and the creation of atlases. We discovered that we both enjoyed the experience of archival research and its potential to reveal the story behind every map. We both have been fascinated by the social and economic background of mapmakers and have understood the importance of the history of the events surrounding a map's appearance. My correspondence with Johannes Dörflinger has always

---

29 E. R. Cohn, Benjamin Franklin, in: Imago Mundi 52 (2000), 124-142.

been valuable and fruitful and deepens my admiration for his work. His assiduous attention to the historical context of maps continues to be revealed in the contribution of his students to the history of cartography.

About the author

Mary Sponberg Pedley is adjunct assistant curator of maps at the William L. Clements Library at the University of Michigan, a Latin instructor in the Ann Arbor Public School system, and the associate editor of *Imago Mundi: The International Journal for the History of Cartography*. With Matthew Edney, she is co-editor of Volume Four of The History of Cartography: Cartography in the European Enlightenment (University of Chicago Press).

Contact: William L. Clements Library, The University of Michigan, Ann Arbor, Michigan USA; mpedley@umich.edu

Josef Sucher,
Grafische Zeit-Darstellung
zur Weltgeschichte,
Wien 1878

Franz Reitinger, Salzburg

Strom der Zeiten

Zur Kartierung von historischen Entwicklungen

Der von seinen Zeitgenossen mehr als Pädagoge denn als Romancier und Komödiendichter geschätzte Christian Weise (1642-1708) war während seiner Rektoratszeit am Zittauer Gymnasium dazu übergegangen, Landkarten im Unterricht einzusetzen. Sein Schüler Johannes Hübner (1668-1731) nahm die Anregung auf und entwickelte den didaktischen Ansatz systematisch weiter. Als erstes Ergebnis seiner Studien legte der junge Schulmann ein geographisches Lehrbuch vor, das er seinem Lehrer widmete. Dieses ahmte die äußere Form eines Katechismus nach, indem es den Stoff in kurze Fragen und Antworten gliederte. Trotz der unmittelbar daran anschließenden Kontroversen wurden Hübners Fragen aus der *Neuen und Alten Geographie* zu einem durchschlagenden Erfolg.[1] Allein bis

1 M. Grünwald, Richtige Beschreibung der beyden Marggraffthümer Ober- und Nieder Lausitz durch allerhand geographische, historische u. politische Discourse von den häuffigen Irrungen, welche Auctor Anonymus in den Kurtzen Fragen aus der neuen und alten Geographie begangen hat, möglichsten Fleißes gesaubert, Görlitz 1696; Waremundo [M. Grünwald], Vertrautes Schreiben an den demasquirten Autorem der geographischen Fragen über die letzten fünff Paragraphos seiner neuen Vorrede zu der Anno 1696 verfertigt sechsten Edition, [Görlitz] 1696; J. Hübner, Gründliche Vertheidigung der Kurtzen Fragen aus der alten und neuen Geographie, Leipzig 1696; [M. Grünwald], Rechtmäßige Verantwortung der Lausitzschen Beschreibung, wider M. Johann Hübners, Rect. in Merseburg, grundlose Vertheidigung der geographischen Fragen, [Görlitz] 1696; J. Hübner,

1735 erlebten sie siebenunddreißig Auflagen und wurden in die meisten europäischen Sprachen übersetzt.[2] Aktiven Anteil an dieser nachhaltigen Wirkung des Buches hatte Johann Friedrich Gleditsch, der dank seines Erfolgsautors seine marktführende Stellung unter den deutschen Wissenschaftsverlagen ausbaute und den Grund für Leipzigs Ansehen auf dem Gebiet der Lexikographie und der Enzyklopädik legte.[3]

Wesentlicher als der publizistische Erfolg war die Nachhaltigkeit der von Hübner initiierten schulischen Reform.[4] Tatsächlich gelang es dem sächsischen Gymnasialdirektor mit seinem Buch, die Geographie als Lehrfach im Rahmen des Geschichtsunterrichts an den deutschen Schulen zu etablieren. Die verstärkte Nachfrage in Kartenwerken rief erste Landkartenverlage auf den Plan und setzte am Beginn des 18. Jahrhunderts einen allgemeinen Aufschwung der Kartographie in Gang, von dem zunächst die protestantischen Kunstverleger der Reichsstadt Nürnberg zu profitieren wussten. Innerhalb dreier Jahrzehnte erschienen zwei planmäßig auf Hübners *Kurtze Fragen* aufbauende Kartenwerke, für die der Autor mit Johann Baptist Homann einen herausragenden Kartographen gewinnen konnte.[5] Das eigentliche Lehrbuch enthielt hingegen nur einen einzigen Kartenstich, dessen Platte sich mit jeder Auflage abnutzte und durch detailgetreue Reproduktionsstiche ersetzt werden musste (Abb. 1). Hübners *Nova et Accurata Pantotopiae Descriptio* verlieh dem trocke-

---

Abgenöthigte Vertheidigung seiner herausgegebenen Schriften wieder die unbilliche Censur, welche sich Johann Burckhard Mai in seiner Deutschen Stats-Kunst über dieselben angemasset hat, Leipzig 1710; s. weiters H. J. Kämmel, Martin Grünwald. Ein Beitrag zur Geschichte der Pädagogik, Zittau 1859.

2 Ausgaben in holl., Amsterdam 1707 ([5]1736); engl., hrsg. v. J. Cowley, London: T. Cox, 1738 ([7]1777); franz., hrsg. v. Frantz Bertaud, Magdeburg 1709, Amsterdam 1735; italienisch, schwedisch und russisch.

3 Zur Person des Verlegers s. A. Brauer, Nachkommen des Leipziger Verlagsbuchhändlers Johann Friedrich Gleditsch. Vorfahren, Verwandtschaftskreis und soziale Struktur, in: Archiv für Geschichte des Buchwesens 3 (1961), 77-96.

4 F. Brachmann, Johann Hübner. Johannei Rector 1711-1731. Ein Beitrag zur Geschichte der deutschen Litteratur, Hamburg 1899.

5 J. B. Homannn, Atlas scholasticus, Nürnberg 1710; Ders., Atlas methodicus, Nürnberg 1719.

nen Lehrstoff des Buches und der an Pedanterie grenzenden Herangehensweise des Autors geradezu den Anstrich des Genialischen.[6]

Abb.1: Nova et Accurata Pantotopiae Descriptio, in: Johann Hübner: Kurtze Fragen aus der Neuen und Alten Geographie, Leipzig 1693;
Archiv für Bildforschung, Salzburg

Die ersten Versuche, die alte Welt im Rahmen einer umfassenden *Sacra Geographia* zu erfassen, reichten in die Frühzeit des Protestantismus zurück. An sie konnte Abraham Ortelius anschließen, der in

6 Nova et Accurata Pantotopiae Descriptio, Kupferstich, 14,3 x 16,1 cm, in: J. Hübner, Kurtze Fragen aus der Neuen und Alten Geographie, Leipzig: Johann Friedrich Gleditsch, 1693 (1764), Regensburg und Wien 1735 (1767), (doppelseitiges) Titelfaltblatt, vgl. C. Castellani, Catalogo ragionato delle piu rare o piu importante opere geografiche a stampa che si conservano nella Biblioteca del Collegio romano, Rom 1876, 95; W. Grieß – S. Luber (Hrsg.), Reiseliteratur und Geographica in der Eutiner Landesbibliothek, Heide 1990, 325; T. Brüggemann – O. Brunken, Handbuch zur Kinder- und Jugendliteratur. Von 1570 bis 1750, Stuttgart 1991, 1425-1426; Verk.-Kat.:, J. Potter, Choice Items from Stock 25, London 2005, Nr. 322.

seinen letzten Lebensjahren mit der Zusammenstellung eines Geschichtsatlasses befasst war und dabei die historische Kartographie praktisch als auch theoretisch begründete. Bibel- und welthistorische Kartographie hielten sich im Wesentlichen an den realgeographischen Rahmen, der ihnen durch den frühneuzeitlichen Zugewinn an Welterfahrung vorgegeben war.[7] Erfolgte durch sie eine Kartierung von geschichtlichen Räumen, so wurde dem 18. Jahrhundert durch die allmählich einsetzende Dynamisierung des neuzeitlichen Geschichtsbildes die Aufzeichnung der historischen Zeit zum eigentlichen Problem. Erste Ansätze zur Darstellung makrohistorischer Prozesse im Kartenbild kündigten sich in den großangelegten Universalgeschichten des Barock an. In dem *Mappamondo istorico* (Parma 1690-1693) des Jesuiten Antonio Foresti etwa, dessen bildhafter Titel von Georg Schlüter mit *Historische Welt-Cart* (Augsburg 1716-1748) ins Deutsche übersetzt wurde, gab die Landkarte den konzeptuellen Erzählrahmen für eine Epochenreise vor.

Eine umfassende Visualisierung der historiographischen Kartenmetaphorik strebte der mailändische Historiker Girolamo Andrea Martignoni an. In seiner *Imago Romani imperii* von 1717 verschmolz Martignoni Geographie und Chronologie zu einer Gesamtdarstellung, auf der das Mittelmeerreich des alten Roms die Drehscheibe für diverse zeitparallele Geschichtsverläufe abgab. Um dieses Reich liefen zweimal sechs historische Zeitströme konzentrisch auseinander. Die linke Hälfte des Kreisschemas erlaubte, die Entstehung erster Nationalstaaten aus der Verschmelzung einzelner Stämme und Völkerschaften von der Gründung Roms im Jahr 753 in Hunderterschritten nachzuvollziehen. Die rechte Hälfte war der jüngeren Zeit nach Christi Geburt mit ihren feinen dynastischen Verflechtungen gewidmet.[8] Die Übertragung der Karte in Kupfer besorgten der

---

7 Das sogenannte *Parergon* ging aus den Supplementen zu Ortelius' Hauptwerk *Theatrum Orbis Terrarum* hervor und lag seit 1624 als selbständiger Geschichtsatlas im Buchhandel auf, vgl. J. Black, Maps and History. Constructing Images of the Past, New Haven/Conn. 1997; W. A. Goffart, Historical Atlases. The First Three Hundred Years, 1570-1870, Chicago/Ill. 2003; F. Reitinger, Geschichtsdeutung auf alten Karten. Archäologie und Geschichte, hrsg. v. Dagmar Unverhau, Wiesbaden 2003, in: Medievistik 19 (2006), 505-507.

8 G. A. Martignoni, Saggio di un'opera di nuova invenzione intitolata l'immagine dell'Impero romano dedicata alla santita di n.s. papa Clemente undecimo, Roma 1717; Ders., Imago Romani imperii. SS.P., et D.N.

Turiner Stecher Bartolomeo Giuseppe Tasniere (1675-1752) und sein Römischer Kollege Giovanni Petroschi (1715-1766).[9] Diese erschien zusammen mit der dazugehörigen Erklärung im Verlag von Antonio de'Rossi (1695-1755) in der Strada del Seminario Romano nahe des Pantheons in Rom. In zwei Nachfolgewerken arbeitete Martignoni die Zeitströme von *Germania Pars*, *Italia*, *Gallia* und *Insula Britannica* aus dem ursprünglichen Übersichtsplan bis ins Detail aus.[10] Die beiden letzten Sektoren des *Imperium Orientalis* und der *Hispania* sollten nicht mehr zur Ausführung gelangen.

Wissend, dass jede Karte immer nur ein Abriss der Wirklichkeit sein kann, bringt Hübner die monumentale Geschichtsmetapher italienischer Historiker in dem verkürzten Maßstab seiner *Pantotopiae Descriptio* zur Ausführung. Schon der Titel der Karte evoziert die Allgegenwart der Geschichte, die weniger als Agent des Vergangenen denn als eine universale, über die Jetztzeit hinaus wirkende Kraft begriffen wird. Es ist dieses substantielle Verständnis von Geschichte, das Hübners Karte zu den synchronen Querschnitten der Geographie als auch zu den diachronen Längsschnitten der

---

Clementi XI. P.O.M. Rom. Imp. imagin., Barth. Ioseph Tasniere sculp. Taurini, Roma 1718.

9 E. Esposito, Annali di Antonio De Rossi stampatore in Roma (1695-1755) (Biblioteca di bibliografia italiana 67), Milano 1972, Nr. 318 u. 319.

10 G. A. Martignoni, Spiegazione della carta istorica dell'Italia, e di una parte della Germania. Dalla nascita di gesù cristo | fino all'anno MDCC. La quale dimostra in compendio gli eventi principali di queste due Regioni finchè durò l'Impero Romano; e dappoi l'origine, l'accrescimento, la diminuzione, il fine, e la grandezza de' Regni, e delle Signorie più riguardevoli, che in esse derivarono; i titoli de' Dominj, Le Dignità, la Serie de' Re, le Arme, la Geografia, e la Cronologia. Con un triplicato metodo di rinvenire le storie, cioè da successo in successo da secolo in secolo, e da signoria in signoria. Dedicata alla santità di n.s. Papa Innocenzo XIII da Girolamo Andrea Martignoni Milanese, Roma 1721; Ders., Explication de la carte historique de la France et de l'Angleterre depuis la naissance de jesus-christ jusqu'à l'an MDCC. Qui contient en abrégé les evenements principaux de ces deux royaumes pendant l'Empire Romain; l'origine, l'acroissement, la diminution, la fin, la grandeur d'eux-memes, et des autres, royaumes, et principales seigneuries; les rois, les titres de chaque domaine, les dignitez, les blasons la géographie, et la chronologie: Avec un triple moyen d'apprendre les histoires; sçavoir, de succes en succes de siecle en siecle, & de seigneurie en seigneurie. Dediee a notre tres-saint pere le Pape Innocent XIII par Jerome Andre Martignoni Milanois, Roma 1721

Chronologie in produktive Spannung treten lässt.[11] Der Rekurs auf die Rhetorik des Allegorischen ergibt sich dabei aus dem Umstand, dass die Zeit in den Repräsentationsvorgang einfließt, ohne von außen, sei es als neutraler Richtwert, sei es als objektive Maßeinheit, an historische Tatsachen angelegt zu werden.

Hübner gliedert das Studium der Universalgeschichte nach ihren Teildisziplinen in *Genealogia*, *Historia* und *Geographia* und macht deren Verlaufsform anhand eines Flusses und des Einzugsgebiets seiner Nebenflüsse anschaulich. Reichs-, Herrschafts- und Kulturgeschichte werden von ihm anhand einer Abfolge von historischen Großreichen (*Assyria*, *Persia*, *Gracia*, *Roma*), Staatsformen (*Eugenia*, *Archontina*, *Basilea*, *Caesarea*) und Weltgegenden (*Asia*, *Africa*, *Europa*, *America*) anschaulich gemacht. In dieser Dreiteilung des Geschichtsfaches kündigt sich die Trilogie von Hübners eigenen didaktischen Schriften an, der den *Kurtzen Fragen aus der neuen und alten Geographie* zu Lebzeiten weitere *Fragen aus der politischen Historia* (1698/99) und der *Genealogie* (1708) folgen ließ.

Hübners *Pantotopiae Descriptio* leistet einer Historisierung der Geschichte als Fach Vorschub, das sich mit seinen Wissensfeldern sukzessive, das heißt in aufeinanderfolgenden Etappen konstituiert: Die Geschichte mit den antiken Weltreichen in alter Zeit, die Genealogie mit den aristotelischen Staatsformen in mittlerer und die Geographie mit den vier Kontinenten in neuerer Zeit. Schließlich vereinigen sich alle drei Fachbereiche zu dem auf eine „Weltstadt des Schicksals" (*Tychopolis*) hinstrebenden „Strom der Politik" (*Politica*), in dem Hübner ein Bild des Geschichtsverlaufs als solchen entwirft.[12] Jüdischer Ursprungsmythos und christlicher Providenzglaube büßen in Hübners säkularisiertem Geschichtsbild ihren Platz ein. Statt ihrer wird die Politik der europäischen Herrscherhäuser zur geschichtsmächtigen Kraft und damit zum eigentlichen Träger der historischen Entwicklung aufgewertet. In gewisser Weise lässt sich die von Hübner propagierte Geschichtsideologie als Parabel auf eine

---

11 Zum christlichen Gebrauch des griechischen „Pantotopia" vgl. J. Lavater, Pantotopias atopon, sive dissertatio philosophica. Qua demonstratur, omnipraesentiam corporis Christi esse figmentum, Zürich 1671.

12 Hübner verwendet *Tychopolis* zwar als lateinischen Namen auch für die dänische Gründung Glücksstadt. Angesichts der eudämonistischen Karriere des Glücksbegriffs in der Moderne scheint es uns aber angebracht, Tyche hier in Anlehnung an einen älteren Glücksbegriff mit Schicksal zu übersetzen.

zwischen Traditionalisten und Reformern ausgetragene *Querelle* um die epochalen Errungenschaften von antiker und moderner Welt lesen.

Noch in ihrer Modalform gewinnt die Karte indes an Aussagekraft. Indem Hübner *cives* und *urbs* zusammendenkt, trägt er der Tatsache Rechnung, dass die ersten Zivilisationen aus einzelnen individuellen Stadtgründungen hervorgingen. Mit dem Fortschreiten der historischen Zeit kommt es zu einer Verdichtung der verschiedenen urbanen Kräfte, an deren Ende die moderne Metropole steht. Der „Strom der Politik“ zeichnet diesen Prozeß fortschreitender Integration in seinem mächtig anschwellenden Lauf nach.

Gegenüber den planparallelen Flußläufen auf den Karten des sächsischen Pietismus legen die Schlingen des Stroms einen ausgeprägt dialektischen Geschichtsverlauf nahe. Sollten die Flüsse im Pietismus die Funktion von Landesgrenzen oder Wasserwegen erhalten, so bezeichnen sie hier territoriale Systeme.[13] Anstatt durch Verzweigung und Ausdifferenzierung besetzen sie ihr Territorium, sich vorwärtswindend, in einer ausgreifenden Eigenbewegung. Diese geht in eine von innen her sich beschleunigende Rotation über, welche die ganze Karte erfasst und die Dynamik des Geschichtsprozesses in Gang setzt.[14] Im Zentrum dieser schraubenförmig rotierenden Bewegung befindet sich das antike Rom, das sich als ‚ewige Stadt’ gewissermaßen um die eigene Achse dreht. Indes drohen die nach außen wirkenden, exzentrischen Kräfte einer neuen Zeit ihrerseits auf die Stadt am Zusammenfluss von klassisch-antiker *Historia* und biblischer *Genealogia* zurückzuwirken und letztere aus ihrer idealen Mitte zu heben.

Hübner ging in seinen *Fragen aus der politischen Historia* auch auf das Alter der Welt ein und legte sich für 1702 auf fünftausendsechshunderteinundfünfzig Jahre fest. Die Koordinaten der Karte folgen dieser Annahme, indem sie einen Zeitraum von rund viertausend Jahren aufschlagen, die nach Hübners Daumenrechnung seit der *Sündflut* verstrichen sind.[15] Wollte man die Linien geographisch im

---

13 Vgl. hierzu meinen Vortrag über Pietismus und Kartographie auf dem 12. Kartographiehistorischen Colloquium, Frankfurt/M., 30. Sept. - 2. Okt. 2004.

14 Zur Symbolform von Swastika und Triskeles vgl. I. Schwarz-Winklhofer – H. Biedermann, Das Buch der Zeichen und Symbole, Graz 1975, 24-27.

15 J. Hübner, Kurtze Fragen aus der politischen Historia 1, Leipzig 1702, Einleitung I u. 4.

Sinne von Meridianen deuten, dann käme das Land *Pantotopia* etwa auf Höhe des Mittelmeeres zu liegen. Die zeitliche Ausdehnung würde in diesem Fall nach Westen zu erfolgen.

Hübners Karte wurde zum Ausgangspunkt für großangelegte Kartierungsversuche der Weltgeschichte in jüngerer Zeit. Die *Charte der Weltgeschichten* des württembergischen Pfarrers Friedrich C. Fulda etwa, die 1782/3 in Augsburg erschien, war als ein in hundert und vier Spalten unterteiltes Wandrollbild konzipiert, das sich mittels einer Zugvorrichtung aus- und wieder einrollen ließ: *Daß die Tabelle die völlige Einrichtung einer Landcharte habe*, ergab sich dem Autor zufolge allein schon aus ihrer *Einteilung in geographische Länge und Breite.*[16] Als unbefriedigend fand der preußische Geschichtsprofessor Johann Gottlieb Friedrich Strass (1766-1845), dass die chronographischen Werke Fuldas und anderer sich der Karte im Titel als einer wirkungsvollen Metapher bedienten, obgleich es sich im Grunde dabei lediglich um die „Columnen" und „Fächer" von herkömmlichen Tabellen handelte. Strass entwickelte in Antwort darauf eine eigene Flusskarte, die in zahlreiche Sprachen übersetzt wurde und von allen einschlägigen Versuchen die größte Bekanntheit erlangte (Farbtafel S. 555).[17] Strass knüpft darin an Herders Metapher vom „Strom der Zeiten" an, die der Hegelschen Geschichtsphilosophie Pate stand.[18] Seine aus drei Bögen sich zusammensetzende Tafel umfasste eine Zeitspanne vom Beginn der eurasischen Hochkulturen

16 F. C. Fulda, Überblik [*sic!*] der Weltgeschichten, zur Erläuterung der Geschichtcharte [*sic!*], von dem Verfasser derselben, Augsburg 1783, 31.

17 F. Strass, Der Strom der Zeiten oder bildliche Darstellung der Weltgeschichte von den ältesten Zeiten bis zum Ende des 18. Jahrhunderts, Radierung, Berlin 1803/1804 (Leipzig ²1819, ³1828), dreiteilige Wandkarte; Ders., Neuester Strom der Zeiten oder bildliche Darstellung der Weltgeschichte, hrsg. v. E. T. Hohler, Wien 1817; Ders., Allgemeine Übersicht der Weltgeschichte von den ältesten Zeiten bis auf gegenwärtige in Strömen und Flüssen, Augsburg 1818; Ders., Erläuterungen zur allgemeinen Weltgeschichte, in einem bildlichen Strome der Zeiten dargestellt, Augsburg 1818; s. weiters H. Harms, Themen alter Karten, Oldenburg 1979, 238-239; Ausst.-Kat., Idee Europa. Entwürfe zum „Ewigen Frieden", hrsg. v. M.-L. v. Plessen, Deutsches Historisches Museum, Berlin 2003, 171.

18 A. Demandt, Metaphern für Geschichte, Sprachbilder und Gleichnisse im historisch-politischen Denken, München 1978, 166-167, 176, 182-184, 186-187.

bis zur französischen Revolution und bot eine figurative Alternative zur horizontalen Zeilenordnung abstrakter *Timelines.*[19]

Der *Strom der Zeiten* erweist sich als ein kompliziertes Flußystem aus zu- und abfließenden Wasserläufen. Ausgehend von neun Quellflüssen macht Strass den Ursprung der Weltgeschichte in *Kleinasien* aus. Zu dessen Seiten verteilen sich westliche (*Italier, Griechen*) und östliche Kulturen (*Assyrer, Syrer, Phoenicier, Juden, Aegypter, Chineser*). Von Westen her treten sukzessive neue Völkerschaften in die Geschichte ein. Die Großreiche des mazedonischen Feldherrn Alexander und des römischen Kaisers Augustus muten wie synthetische Gebilde an. Diese bilden massive Querriegel, die sich über die ganze Breite des Blattes ziehen, sich dem Zeitfluss querlegen und ihn zum Stauen bringen. Ein progressiver Zeitraster gewährt den jüngeren Ereignissen größeren Raum und erlaubt so die Konstruktion eines perspektivischen Blicks auf die Vergangenheit. Den Gradleisten der Kartographie analog gestatten verschiedene Jahrestabellen an den Rändern eine Konkordanz der unterschiedlichen historischen Zeitrechnungen. Zu diesen kommt nun auch noch eine neue Weltzeit hinzu, die von einem hypothetischen Anfang im Jahr Null ausgehend die kulturelle Entwicklung mit dem Jahr 1300 einsetzen lässt.

Alle historisch bekannte Völker erscheinen hier unter dem Bilde von Strömen, die sich bald vereinigen, bald trennen, bald in mehrere Arme vertheilen, je nachdem die Völker selbst ähnliche Schicksale hatten, befand ein zeitgenössischer Rezensent. *Mit ihrer Selbständigkeit verlieren die Völker zugleich ihre Columne, oder ihren eigenen Fluß, und dadurch wird für die Hauptvölker, in deren Strom sich nach und nach so viele andere stürzen, oder mit fortgerissen werden, desto mehr Platz gewonnen. So überschauet man den Ursprung, die Erweiterung und Abnahme, die Fortdauer und das Versinken der Nationen sehr bequem. Ein einzeln fortlaufender, aber doch auch mit der Chronologie in Harmonie gebrachter Strom enthält die Erfindungen, Entdeckungen und Fortschritte der Kultur.*[20]

Dem Rezensenten gelingt es hier in überzeugender Weise, das elastische Prinzip der neuen Bildfindung, durch das sich der *Strom der*

19 Beispielhaft seien hier die sechs Tafeln von J. C. Gatterer, Synopsis historiae universalis, Göttingen 1766, hervorgehoben. Zur Geschichte einer ausschließlich typographischen Tabellarik s. weiters B. Steiner, Die Ordnung der Geschichte. Historische Tabellenwerke in der Frühen Neuzeit (1500-1800), Weimar und Wien 2008.

20 Neue allgemeine deutsche Bibliothek 93 (1804), 2, 99-101.

Zeiten von den starren Tabellengittern älterer Geschichtsdiagramme auszeichnete, in Worte zu fassen. Seine Darstellung macht das der hydrographischen Geschichtskarte innewohnende dramaturgische Moment explizit, das dem faktographischen Ansatz chronographischer Zettelwirtschaften etwas wie einen Sinn abzuringen vermochte und gesichtslosen Zahlen und Daten ein Antlitz verlieh. Unbeantwortet blieb bei alldem die Frage, wie man sich das Verhältnis des *Versinkens der Nationen* und der *Fortschritte der Kultur* vorzustellen habe, ja ob die beiden Phänomene überhaupt miteinander vereinbar seien. Der Grundton konnte so heroisch oder elegisch ausfallen je nachdem, ob man den Zeitenfluss von seinem Anfang oder seinem Ende her besah. *Der unaufhaltsame Strom der Zeit reißt alles mit sich fort ins dunkle Meer der Vergessenheit* lautete nicht umsonst ein zeitgenössischer Stammbuchvers.[21] Auch Strass' Darstellung war von seiner Anlage her ein Deszendenzmodell, das den sonnengleichen Aufgang des Menschengeschlechts, dessen Entfaltung und Auseinanderentwicklung in verschiedenartige Weltkulturen als erhabene Geschichte des Verfalls wiedergab.[22]

Der *Strom der Zeiten* wurde 1805 unter Zuhilfenahme der Begriffe *Représentation figurée*, *Tableau* und *Carte chronologique* ins Französische[23]

---

21 J. K. Hechtel, Denkmale der Freundschaft, Nürnberg 1809; W. T. Krug, Allgemeines Handwörterbuch der philosophischen Wissenschaften 4, Leipzig ²1934, 583.

22 *Der Strom der Zeiten von Frdr. Straß* […] *ist eine anschauliche Darstellung der Weltgeschichte in Gestalt eines großen Flusses, der sich allmählich in mehrere Nebenflüsse theilt, weil auch das Menschengeschlecht im Laufe der Zeit sich nach und nach in mehrere Stämme, Völker und Staaten zertheilt hat*, vgl. Krug, Allgemeines Handwörterbuch 4, 583.

23 F. Strass, Le Fleuve du temps ou représentation figurée de l'histoire du monde jusqu'à la fin du XVIIIe siècle, ed. v. Vernies, Paris 1805; Le Cours des tems ou tableau de l'histoire depuis l'antiquité la plus reculée jusqu'à nos jours, d'après la carte chronologique de F. Strass, L. Aubert, père (sc.), Paris 1809 (Paris 1818; London: ²1812, ³1831; Cornelis van Baarsel (sc.), Den Haag 1819); Explication du Tableau de l'histoire universelle, Paris: ²1816, ³1825); Nouveau tableau de l'histoire universelle, d'après celui de Strass, E. Collin (sc.), Paris 1822; Carte chronologique de l'histoire universelle formant e tableau de tous les regnes connus, d'après la carte de F. Strass, Amsterdam 1830; s. weiters Le Télégraphe littéraire (Paris), 8 (1809), 1, 5. April, 19; M. A. J., Atlas général de l'histoire de France, in: Revue encyclopédique (Paris), 3 (1821), 10, 91-96, insbes. 95.

und vom Französischen weiter ins Englische[24] übersetzt. Englische Pädagogen gaben dem *Stream of Time* des Karten- und Vedutenstechers William Jefferys vielfach den Vorzug vor den älteren Geschichtsdiagrammen eines William Priestley und anderer.[25] Kirchliche Autoren glaubten dem Stromverlauf gar bis ans Ende aller Zeiten flussabwärts folgen zu müssen.[26] Passend für den Unterricht wurde der Zeitenstrom in Form eines gedruckten Rollbildes angeboten.[27]

Strass bildete auch die Grundlage für eine niederländische,[28] eine schwedische[29] und eine ganze Reihe von amerikanischen Fassungen mit modern anmutenden Titeln wie *History Made Visible,* die lange vor dem Entstehen einer Avantgarde-Kunst in Europa auch unkünstlerischen Bildmedien die Fähigkeit der Sichtbarmachung zuschrieben,[30]

---

24 W. Bell, Stream of Time. A Figurative Representation of Universal History, London ⁵1818; Descriptive Guide to the Stream of Time [...] continued to the Year 1818 by Rev. B. Godwin, London 1818; Charles Claude Hamilton: An Epitome of Universal Chronology, History and Biography, Forming a Companion to Irving's Stream of History, From the German of F. Strass, London 1826; Stream of Time, or Chart of Universal History, from the German of Strass, London 1833.

25 [...] *far superior to Priestley's,* M. Edgeworth – R. L. Edgeworth, Essays on Practical Education, Bd. 2, London ³1811, 49, vgl. D. Rosenberg, Joseph Priestley and the Graphic Invention of Modern Time, in: Studies in Eighteenth Century Culture 36 (2007), 55-103. – William Jeffreys war wie der Kartograph Thomas Jeffreys für kirchliche Kreise tätig und dürfte mit dem Kupferstichverleger *W. Jeffryes* in Blackfriars, London, identisch sein, vgl. I. Maxted, The London Book Trades 1775-1800. A Preliminary Checklist of Members, Old Woking/Surrey 1977, 121; T. Clayton, The English Print 1688-1802, New Haven/Conn. 1997, 242.

26 E. Denny, A Prophetical Stream of Time, or, an Outline of God's Dealings with Man, from the Creation to the End of all Things. Companion to a Chart [...] Intitled A Prophetical Stream of Time, London 1849.

27 F. M. C. Raimondi, Stream of Time. A Paper Roll Marked in Ink with Notable Dates from 4000 BC to 1851 (Great Exhibition) Intended for Young Persons to be Fitted during a Course of Historical Reading, London 1844.

28 P. J. Prinsen, Wegwijzer tot het spoedig aanleeren der algemeene geschiedenis, overeenkomst de geschiedkundige kaart, Haarlem 1820.

29 Förklaring hörande till en historisk karta eller figurlig uppställning af werldshistorien. Öfversättning, med rättelser och tillägg, Stockholm 1826.

30 F. Reitinger, Substance. The Mapping of an Early Ism-Movement, in: How to Write Art History – National, Regional or Global? International Conference of the History of Art, Budapest, 21.-25. November 2007 (=Acta Historiae Artium 49), Budapest 2008, 125-137.

und dem Hinweis *Essentially Modified, Continued Down to the Present Year*, aus dem das Bemühen sprach, den Fluss der Universalgeschichte auf aktuellem Stand zu halten.[31] Zu Sass, Strauss oder Stork verballhornt erlebte Strass' hydrographische Geschichtskarte bis in die 1880er Jahre zahlreiche Ausgaben.[32] Besonders reizvoll ist ein universalhistorisches Schaubild, auf dem das in dunkles Tintenblau getauchte obere Ende der Graphik die mediale Vorstellung von einer fernen Bilderwelt vermittelt, während der davon abgehende abstrakte Zeitfluss ein buntes Durcheinander aufeinanderfolgender Epochen ergibt (Tafel S. 556).[33] Die sich über die ganze Breite des Blattes ziehende Titellandschaft macht den voneinander unabhängigen Ursprung der Kulturen in Europa, Kleinasien und Afrika sinnfällig. Die weitere historische Entwicklung nimmt sich als ein Prozess der Diversifizierung aus, im Zuge dessen sich die Kulturblöcke in einunddreißig Flussläufe, davon mehrheitlich europäische Nationalstaaten, ausdifferenzieren. Die Art, wie das geschieht, bringt freilich erst die ganze Problematik von Strass' bildstatistischer Methode an den Tag.

Politisch blieb die Karte hinter den in sie gesteckten Erwartungen zurück, indem es ihr nicht gelang, das weltumspannende Großreich Kaiser Karls V. adäquat wiederzugeben. Geographisch wiederum musste unzureichend erscheinen, wenn England am gegenüberliegenden Ende von Amerika zwischen Staaten wie Portugal und Grie-

---

31 I. Hitchcock, History Made Visible on the Basis of Professor Strass's Stream of History which has been Essentially Modified, Continued Down to the Present Year, Philadelphia 1839; S. G. Goodrich (Hrsg.), Universal History Illustrated or the Stream of Time made visible, from the Original of Strass, New York 1841; D. Haskel (Hrsg.), Stream of Time, or Chart of Universal History, from the Original German of Strass, Lithographie, New York 1845 (^{2}1853); A Chronological View of the World or Chart of Universal History [...] Intended to [...] illustrate Strass's Stream of Time or chart of Universal History [...] adapted to English Literature by Dr. Irving, New York 1847 (New York 1866).

32 R. S. Fischer – H. A. Ford, Stream of Time. A Panorama of Universal History, Detroit/Mich. 1887; Dies., Manual of the World's History to Accompany "The Stream of Time", a Chart of Universal History, in Colors, Detroit/Mich. 1888. – Noch Goffart, 559, zitiert den *River of Time* eines gewissen *Sass*.

33 [V. Clérot], Cosmorama universel des peuples depuis la création du monde et leurs marches jusqu'à nos jours. Par une société de bibliophiles et historiens, Kupferstich, Longuet, Suce.'r de Charles Simonneau, graveur-editeur de cartes geographiques, 6 Rue de la Paix, Paris 1841, s. weiters Bibliographie de la France, ou Journal général de l'Imprimerie (Paris), 50 (1841), 11. Dez., 612.

chenland zu liegen kam. Die verschiedenen Teilstränge blockierten einander regelrecht und waren sich gegenseitig im Weg. Auch entsprach die Strombreite in keiner Weise der realen Ausdehnung der verschiedenen Staaten. Die Ströme von Holland und Belgien übertrafen nicht nur Russland und Preußen zusammen an Macht, sondern auch England, das sich als vereinigtes Königreich von Großbritannien im 18. Jahrhundert längst über die ganze britische Insel und Irland erstreckte. Anders als Persien, Afghanistan und China hatten Japan, Indien und Afrika in der Originalausgabe noch keinerlei Berücksichtigung gefunden. In späteren Ausgaben wurde der Flusslauf der *Ägypter* bis in die Vorzeit hochgezogen und nach oben hin verlängert, sodass Indien und Äthiopien als Vorläuferkulturen in Strass' universalgeschichtlicher Übersicht Platz finden konnten (Tafeln S. 557 a und b).

Die Koordinierungsfragen, die Strass' *Strom der Zeiten* aufwarf, mochten mit ein Grund sein, warum Edouard-Auguste-Patrice Hocquart (1789-1870) sich zunächst auf die Kartierung separater Teilgeschichten beschränkte. Von 1819 an gab Hocquart eine ganze Reihe von Flussdiagrammen zur christlichen, römischen und französischen Geschichte in Druck, später dann auch zur Vor- und Universalgeschichte, die 1873 der sogenannten *Encyclopédie Bouasse Lebel*, einer Folge von *synoptischen Tafeln zum Unterricht in den Künsten, den Wissenschaften und den verschiedenen Berufszweigen* einverleibt wurden. Auf Hocquarts Tafeln wird die Kontinuität eines Zentralstaats – sei es Rom oder Frankreich – betont, der gewissermaßen das Rückgrat für die Gesamtentwicklung bildet. Die neu hinzukommenden oder davon abzweigenden Territorien erscheinen als vergleichsweise schmale Kanäle, die kaum je in ihrer wahren territorialen Größe wiedergegeben und damit weitgehend marginalisiert werden.[34]

---

[34] Tableau de l'histoire Romaine depuis les tems les plus reculés jusqu'à la division de l'Empire. Rédigé d'apres l'art de vérifier les dates, les tables chronologiques de John Blair, Lenglet-Dufresnoy, Kupfertstich, Paris o. J.; Tableau de l'histoire sainte depuis la création du monde jusqu'à la dispersion des juifs. Rédigé d'après Prideaux, Usserius, John Blair, Lenglet-Dufresnoy, Kupferstich, Paris o. J.; Tableau de l'histoire de France depuis les Gaulois jusqu'à nos jours, Kupferstich, Paris 1819 (³1867); Panorama de l'histoire universelle ou tableau de l'histoire de toutes les nations jusqu'à nos jours, Paris 1820 (Paris ³1854); Explication du Panorama de l'histoire universelle, Paris 1833; Tableau historique de la mythologie, rédigé après Chompré et les meilleurs mythologues, Paris 1826 (Bouasse-Lebel, ³1867).

Mit über einem Meter Höhe und an die siebzig Zentimeter Breite gehört Ferdinand Ritter von Dresdnerns *Historischer-Zeitstrom des heutigen Kaiserthums Oesterreich* von 1825 zu den größten und eindrucksvollsten hydrographischen Geschichtstafeln (Tafel S. 558).[35] Das im neugotischen Stil gehaltene Schaubild des Aktuars der k. k. Zollgefällen-Administration aus Linz ist ein Beispiel für die fortschreitende Vernationalstaatlichung der universalhistorischen Metapher des Zeitenflusses. Dreiundzwanzig Ströme symbolisieren von *Salzburg* bis *Schwarz-Kumanien*, dem heutigen Moldawien, die habsburgischen Territorien. Diese durchfließen einen tausendjährigen Zeitraum zwischen Kaiser Karl dem Großen und dem Abschluss des Münchener Staatsvertrags im Jahr 1816, um sich über die Jahrhunderte zu dem mächtigen Strom *Kaiserthum Oesterreich* zu vereinigen. Der *Zeitstrom* wird von einem neu errichteten Brückenbogen überfangen, von dem der historisch Versierte den Stromverlauf in seiner Gesamtheit überblickt.[36]

Geschichte erscheint auf Dresdnerns Schautafel nicht so sehr als kosmoramatischer Prozess der Herausbildung und Ausdifferenzierung moderner Nationalstaaten, sondern als Einverleibungs- und Fusionierungsvorgang, an dessen Ende die hegemoniale Großmacht steht. Den Fluss als Wachstumsmetapher zu nutzen, musste sicherlich nahe liegen. Die Schautafel macht aber sinnfällig, wie rasch und überstürzt diese Entwicklung erfolgte, und auch, dass die Diversität des Anfangs als Hypothek weiter auf dem amorphen Staatsgebilde lastete.

Aktuellen Zeitereignissen verdankte sich eine Tafel zur Geschichte der Revolutionen zwischen 1787 und 1830, in der Hocquart Jahr für Jahr die territoriale Entwicklung Frankreichs nachzeichnete. Von der Annektierung Korsikas an erlebt die *France* einen kontinuierlichen Zustrom von Nebenflüssen, um 1814 schließlich

---

35 Mit Wohnadresse in der neuen Bethlehemgasse Nr. 392, vgl. Kaiserlich-königlicher Schematismus des Erzherzogthums Oesterreich ob der Enns 1, Linz 1825, 191.

36 Ritter von Dresdnern, Historischer-Zeitstrom des heutigen Kaiserthums Oesterreich. Von den Regierungs-Jahren Kaiser Karls des Grossen im achten Jahrhunderte angefangen, bis zum Abschlusse des Münchner Staatsvertrages im Jahre 1816, Lithographie, Ernst Karl Frühwirth (sc.), Ferdinand Cosandier (sc.), Wien 1826, dreiteilige Wandkarte; vgl. Allgemeines Repertorium der neuesten in- und ausländischen Literatur bis 1825, 3, Leipzig 1825, 226.

wieder auf ihre ursprüngliche Größe zusammenzuschrumpfen. Indem Hocquart die Ära Napoleons als glorreiche Epoche der jüngeren Geschichte in seinem *Tableau des révolutions* sinnfällig machte, bereitete er den Boden für das Zweite Kaiserreich unter Napoleon III.[37]

Hocquarts Flow Chart bot nicht das erste Beispiel einer Kartierung von Kurzzeitverläufen. Schon dem ersten Band zu Thomas Clarksons *History of the Rise, Progress, and Accomplishment of the Abolition of the African Slave Trade* von 1808 war ein Faltblatt in Form eines chronologischen Flussdiagramms beigelegen, das als einmaliger Versuch zur Visualisierung einer politischen Bewegung zu werten ist (Tafel S. 559 a). Clarksons hydrographische Karte rekonstruiert die Antisklaverei-Bewegung und ihre Frühgeschichte von 1650 bis ins Jahr 1787, in dem William Wilberforce (1759-1833) die Sklavenfrage erstmals auf die Tagesordnung des englischen Unterhauses setzte. Clarksons Karte bildete nicht wie die meisten übrigen Geschichtskarten einen linearen Zeitverlauf ab. Ausschlaggebend war das morphologische Gesamtbild eines lockeren Gespinsts an Flüssen, Armen und Kanälen, das eine umfassende Darstellung der Bewegung und der daran beteiligten Gruppierungen erlaubte, denen gleich den Wassermassen eines anschwellenden Stromes nichts zu widerstehen vermochte. Die Verlaufsform des Flusses bot dabei ein dynamisches Modell von Kooperationen und Zusammenschlüssen.[38]

Dem irischen Schulreformer Thomas Wyse (1791-1862) schwebte ein regelrechtes System an Karten vor, das nach dem Prinzip der chronologischen Geschichtskarte den vermeintlichen Fortschritt der menschlichen Kultur dokumentieren sollte: *Karten und dergleichen sollen nicht nur benutzt, sondern auch entworfen werden. Lassen Sie den Schüler beispielshalber mit England beginnen: Zunächst bringt er eine allgemeine Übersicht mit den Schlüsseldaten der Geschichte aufs Papier. Nach und nach kann er die Zwischenräume mit neu hinzukommenden Persönlichkeiten, Ereignissen und weiteren Einzelheiten auffüllen, wodurch schließlich eine Karte oder eine Tafel zur 'Allgemeinen Geschichte Englands' entsteht. Dann soll*

---

37 E. Hocquart, Tableau des révolutions de la France de 1787 jusqu'à nos jours. Pour servir de suite au tableau de l'histoire de France, Kupferstich, Paris 1832.

38 M. Wood (Hrsg.), The Poetry of Slavery. An Anglo-American Anthology, 1764-1865, Oxford 2003, XVI; C. L. Brown, Moral Capital. Foundations of British Abolitionism, Chapel Hill/N.C. 2006, 5; F. Reitinger, Kleiner Atlas amerikanischer Überempfindlichkeiten, Klagenfurt 2008, 276-282.

er mit der nächsten Karte, sagen wir beispielsweise zur 'Rechtsprechung', einer dritten, etwa zu den 'Moralischen und intellektuellen Errungenschaften', und so weiter fortfahren, die allesamt im gleichen Maßstab auszuführen wären. Diese würden den allgemeinen, physischen und politischen Karten der Geographie weitgehend entsprechen und wohl auch nicht weniger genau sein. Die Karte der Rechtsprechung, zum Beispiel, könnte in Kirchen- und Zivilrecht und dieses wiederum in eigentliches Zivil- und Kriminalrecht mit ihren jeweils eigenen Kolumnen untergliedert werden. In gleicher Weise könnte die Karte der moralischen und intellektuellen Errungenschaften eine Spalte für Religion enthalten (den wichtigsten Glaubensgemeinschaften entsprechend mitunter in kleinere Spalten aufgeteilt); eine andere für Ethik (ihrerseits untergliedert in Theorie und Praxis); eine dritte für Geist und Kultur unter den verschiedenen Stichworten der Wissenschaft und der Kunst, und so fort.[39]

Eine entsprechende Spezialkarte der Kunst lässt sich für das 20. Jahrhundert ausmachen. Für den Katalog der ersten großen Ausstellung deutscher Nachkriegskunst in Paris stellte der Berliner Galerist und Fluxusaktivist René Block mit Hilfe von Wolfgang Siano eine fünfzig Seiten umfassende, mehrspaltige Chronologie der Zeitereignisse zusammen.[40] Seinen anschließenden Textbeitrag leitete Block mit einem Chronogramm ein, das auf einer Seite die großen Namen synoptisch zusammenfasst (Tafel S. 559 b). Die Dokumentation der zeitgenössischen Kunst und ihrer jüngsten Entwicklungen in Karten und Diagrammen war Fluxus-Künstlern wie George Maciunas und

---

39 *But maps, etc. must not only be used, but constructed. Let the pupil begin for instance, with England: first, he forms a general outline, – the great headlines only of history: by degrees the intervals are filled up with new personages, new events, new details. He thus completes a map or table of 'General English history'. He then proceeds to a second, or a map of 'Legislation': to a third, or a map of 'Moral and Intellectual Progress'; and so on, all on the same scale. These correspond to the general, physical, and political maps in geography, and may be quite as minute. For instance, that of Legislation may be divided into Ecclesiastical and Civil – and Civil, again, into Civil properly so called, and Criminal, etc., each arranged in parallel columns. In like manner, that of Moral and Intellectual Progress may have one column for Religion (subdivided, if necessary, into smaller ones for the chief sects); another for Morals (also subdivided into Science and Practice); a third for Intellectual Culture, under the several heads of science, art, etc.*, T. Wyse, Education Reform, or, the Necessity of a National System of Education 1, London 1836, 151-152.

40 R. Block, Le Fleuve caché de l'art allemand, Offset, 30,5 x 22,8 cm, in: Ausst.-Kat.: Art Allemagne aujourd'hui. Différents aspects de l'art actuel en république fédérale d'Allemagne, hrsg. v. Suzanne Pagé, Musée d'Art Moderne de la Ville, Paris 1981, 78 ; s. weiters A. Schmidt-Burkhardt, Stammbäume der Kunst. Zur Genealogie der Avantgarde, Berlin 2004, 346.

Nam June Paik ein besonderes Anliegen gewesen. Wenngleich als Galerist nicht frei von persönlichen Interessen, zeigte sich Block gegenüber künstlerischen Gruppenidentitäten unbefangen. Im Zuge seiner Aktivitäten als Ausstellungsmacher bot sich ihm wiederholt die Gelegenheit, die Geschichtsfähigkeit der Kunst seiner Generation durch die visuelle Aufbereitung ihres vermeintlichen Verlaufes zu demonstrieren.

Von Berlin ausgehend zeichnete Block die Entwicklung der deutschen Kunst nach 1945 in Form eines Flusses. Das Berlin Hitlerdeutschlands legt sich gleich einem Riegel den großen Strömen der klassischen Moderne wie *Expressionismus*, *Dadaismus* und *Bauhaus* quer. Einige unterirdische Kanäle führen von dort nach 1945 zu einem Neuanfang. Der Beginn der Nachkriegskunst wird mit dem im französischen Exil lebenden Künstler *Wols* bis in die dreißiger Jahre zurückverlegt. Zu den Künstlern der ersten Stunde zählt Block außerdem die heute vielfach vergessenen Namen *K. O. Götz*, *Hofer*, *Baumeister*, *H. Uhlmann* und *E. Mataré*. Einen eigenen Zweig begründet *J. Beuys*, der erst Mitte der sechziger Jahre in den Mainstream einmündet. Als brückenschlagende Ereignisse wertet Block insgesamt drei der seit 1954 periodisch in Kassel stattfindenden Leistungsschauen zeitgenössischer Kunst: Die *Dokumenta 1* stand unter dem Zeichen der Pariser Schule der abstrakten Malerei. *Dokumenta 4* machte die New Yorker Schule der Pop- und Op-Art in Deutschland bekannt. *Dokumenta 6* scheint dem Ausstellungsmacher weniger wegen der Medienkunst und der Fotografie als wegen des neuen Realismus in der deutschen Malerei bedeutsam gewesen zu sein. Am Ende dreht der *Verborgene Fluss der deutschen Kunst* in eine andere Richtung, ohne dass sich für die achtziger Jahre ein spezifischer Trend abzeichnen würde. Fest steht nur: das Interesse an der deutschen Kunst der Nachkriegsgeneration wird – so naturhaft wie der Lauf eines Flusses – auch fernerhin anwachsen.

Die Verfolgung von Nebenflüssen dieser Art mochte nicht unwesentlich zur Atomisierung des Historischen beigetragen haben, die Jakob Burckhardts Traum von einer „riesigen Geisteslandkarte“, auf der sich anhand des Gradnetzes der Chronologie jedem historischen Phänomen ein spezifischer Ort zuweisen lasse, zunichte werden ließ.[41] Hingegen belegt das noch vor dem Ende des Zweiten Welt-

41 A. Demandt, Metaphern für Geschichte, München 1978, 220.

kriegs in der Schweiz publizierte Hauptwerk des belgischen Historikers Jacques Pirenne, *Les Grands courants de l'histoire universelle*, dass die sprachbildliche Kraft der Metapher zur Erfassung der Hauptzüge historischer Entwicklung auch für eine analytische Geschichtsschreibung moderner Prägung nichts an Attraktivität eingebüßt hat.[42]

Zum Autor

Franz Reitinger, Kunsthistoriker in Salzburg. Von seinen zahlreichen Publikationen sei hier nur sein *Kleiner Atlas der österreichischen Gemütlichkeit. 1800-1918* (Klagenfurt 2003) sowie sein *Kleiner Atlas amerikanischer Überempfindlichkeiten* (Klagenfurt 2008) hervorgehoben.

Forschungsschwerpunkte: Bildressourcen und ihre sachliche Erschließung, Randformen des Bildes.

Kontakt: theRevaluatedImage@sbg.at

42 J. Pirenne, Les Grands courants de l'histoire universelle, Neuchâtel 1944-1956 (dt., Bern 1944-56).

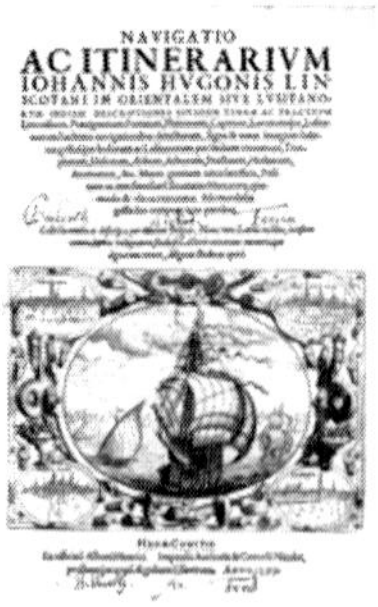

Jan Hygen Van Linschoten,
Navigatio ac itinerarium,
's Gravenhage 1599

Günter Schilder, Utrecht

Neue Horizonte

Die erste niederländische Reise nach Ost-Indien (1595-1597)

In der Sammlung Woldan befinden sich zwei gedruckte Karten welche einen Einblick in die niederländische Reise von Cornelis de Houtman nach Ost-Indien (1595-1597) geben und eine Zusammenfassung der erreichten Resultate bieten. Auf der Übersichtskarte (Sammlung Woldan K-V(Bl):AS 19), die der Frankfurter Verleger Théodore de Bry nach dem Vorbild eines niederländischen Originals im Jahre 1599 publizierte, ist der gesamte Verlauf der Reise mittels der eingezeichneten Route zu verfolgen. Die niederländische Originalkarte der Java-See (Sammlung Woldan K-V (Bl): AS/Sos 15 [1-2]) von Willem Lodewijcksz., welcher als Kaufmann die Reise mitmachte, vermittelt zudem eine exzellente Sicht auf die erzielten Ergebnisse dieser Reise.

Um diese beiden Karten in einen historischen Kontext zu stellen, wird zunächst eine Zusammenfassung von De Houtmans Reise geboten. Hiernach folgt eine Beschreibung der originalen Übersichtskarte, die 1598 vom Amsterdamer Verleger Cornelis Claesz. Herausgegeben wurde, und der deutschen Kopie durch De Bry in Frankfurt. Schließlich wird die Amsterdamer Originalkarte von Willem Lodewijcksz. näher behandelt.

Die Reise des Cornelis de Houtman (1595-1597)

Die Vorbereitung

Die Niederländer wollten gerne einen direkten Zugang zum gewinnbringenden Handel in Asien haben. Am Ende des 16. Jahrhunderts wurden aus Holland und Zeeland – auf dem Handelsgebiet die wichtigsten Provinzen – entsprechende Reisen ausgerüstet, welche verschiedene Segelrouten nach Asien ausprobieren mussten, um dieses Ziel zu erreichen. Im Prinzip waren hier vier Routen möglich: über Nordosten (entlang dem nördlichen Asien), über Nordwesten (entlang der Nordküste von Amerika), via Südosten (die Route um das Kap der Guten Hoffnung) und über Südwesten (durch die Magellan Strasse). Der Startschuss hierfür wurde 1594 gegeben, als die erste Polreise auf der Suche nach einer nordöstlichen Durchfahrt unternommen wurde.[1] Obwohl diese Reise große Hoffnungen weckte (dass die zweite Reise von 1595 misslang und die dritte schließlich in einer harten Überwinterung enden sollte, musste sich noch erweisen), wurden zum gleichen Zeitpunkt in Amsterdam bereits Pläne gemacht, um über das Kap der Guten Hoffnung – die von den Portugiesen bewährte Route – Asien zu erreichen. Die Initiative dieser Reise erfolgte von einer Gruppe von neun Amsterdamer Kaufleuten, welche sich später zur *Compagnie van Verre* zusammenschlossen.

Man war sich durchaus bewusst, dass die Kenntnis der portugiesischen Kaufleute und Seeleute für eine erfolgreiche Unternehmung unentbehrlich war, denn es waren die Portugiesen, die bereits ein Jahrhundert zuvor die Kaproute befuhren und Handel mit Asien betrieben. Die Niederländer mussten nun versuchen, genauere Informationen über die Produkte zu erlangen, die im Osten zu bekommen waren und herausfinden, welche europäische Waren in Asien besonders verlangt wurden. Außerdem mussten die

1 Die beste Quellenübersicht bietet immer noch S. P. L'Honoré Naber, Reizen van Willem Barents, Jacob van Heemskerck, Jan Cornelisz. Rijp en anderen naar het noorden (1594-1597) (Werken uitgegeven door de Linschoten Vereeniging XIV und XV), 2 Bde., 's-Gravenhage 1917. – In der Sammlung Woldan (R-IV: PO / Ark 1) befindet sich ein Exemplar der französischen Ausgabe von Gerrit de Veers Bericht der drei Polreisen, welcher durch Cornelis Claesz. in 1598 in Amsterdam publiziert wurde.

niederländischen Steuermänner – abgesehen von der notwendigen Kenntis der theoretischen und praktischen Navigation – vor allem mit der besten nautischen Ausrüstung (Segelbeschreibungen und Karten) ausgestattet sein, welche zu diesem Zeitpunkt erhältlich war.

Die *Compagnie van Verre* bekam mit Hilfe von nach Lissabon eingeschleusten, getarnten Informanten Material zugespielt, aber auch andere Niederländer lieferten Informationen und Unterstützung bei der Vorbereitung. So kümmerte sich besonders Petrus Plancius (1562-1622) um die Ausrüstung der Schiffe mit Karten und Segelanweisungen.[2] Bereits 1592 hatte er in Zusammenarbeit mit dem Amsterdamer Verleger Cornelis Claesz. (ca. 1551-1609) eine Wandkarte der Welt[3] und eine Serie von Seekarten (1592-94) publiziert, welche vollständig auf iberischen kartographischen Materialien basierten.[4] Wir wissen auch, dass Plancius sich persönlich mit der Ausbildung der Steuermänner befasste und die Kapitäne und Steuermänner unterrichtete und examinierte. Eine andere umfangreiche Quelle, aus welcher man schöpfen konnte, war das Werk von Jan Huygen van Linschoten (1562-1611).[5] Vor allem seine *Reys-Gheschrift* (1595) und sein *Itinerario* (1596) waren für die praktische Seefahrt von großer Bedeutung.

Die *Compagnie van Verre* brachte das benötigte Kapital von 290.000 Gulden zusammen, um vier Schiffe auf Kiel zu legen und auszurüsten.[6] Im Sommer 1594 war die Flotte segelklar und bestand

---

2 Eine Übersicht über Plancius' Werk bietet der Ausstellungskatalog: Plancius (1552-1622). Maritiem Museum 'Prins Hendrik', Rotterdam (23.12.1972-30.04.1973).

3 R. W. Shirley, The Mapping of the World: Early printed world maps 1472-1700 (The Holland Press Cartographica Series 9), London 1983, Entry 183.

4 G. Schilder, Monumenta Cartographica Neerlandica 7, Alphen aan den Rijn, 2003, Cap.V.

5 Die beste Übersicht betreffend Van Linschotens *Itinerario* findet sich bei H. Terpstra, Itinerario. Voyagie ofte Schipvaert van Jan Huygen van Linschoten naer Oost ofte Portugaels Indien 1579-1592 (Werken uitgegeven door de Linschoten-Vereeniging LVII, LVIII nd LX), 3 Bde.,'s-Gravenhage 1955-1957. – In der Sammlung Woldan (R-III: AS 18) befindet sich auch eine lateinische Ausgabe von Van Linschotens *Itinerario* von 1599.

6 Siehe die Quellenpublikation über die Reise von Cornelis de Houtman von G. P. Rouffaer – J. W. Ĳzerman, De eerste schipvaart der Nederlanders naar Oost-Indië onder Cornelis de Houtman, 1595-1597 (Werken uitgegeven door de Linschoten Vereeniging VII, XXV und XXXII), 3 Bde., 's-Gravenhage 1915-1929.

aus der *Mauritius* und der *Hollandia* (beide 230 last)[7], der *Amsterdam* (130 last) und der *Pinas Duyfken* (25 last). Die Tonnage aller Schiffe zusammen entsprach nur ungefähr dem Inhalt einer einzigen portugiesischen *Nau*. Auf den beiden großen Schiffen musterten 85 Mann an, auf der *Amsterdam* 59 und der Pinasse 20, insgesamt also 249 Mann. Als die Flotte im August 1597 an die Reede von Texel zurückkehrte, war die Bemannung auf 89 Personen geschmolzen. Cornelis de Houtman (1565-1599) wurde zum *oppercommies* für Handelsangelegenheiten ernannt. Die oberste Entscheidung der Flotte lag jedoch beim Schiffsrat, welcher sich aus den Kapitänen und Kaufleuten der vier Schiffe zuammensetzte. Dies hatte Streitigkeien während der ganzen Reise zur Folge. De Houtman hatte wenig Zugriff auf die Situation und konnte sich bei der Mannschaft nicht gut durchsetzen. Auch sein Bruder, Frederick de Houtman (ca. 1557-1627), machte die Reise mit und verrichtete unter anderem für Plancius Beobachtungen am südlichen Sternenhimmel. Diesbezüglich muss auch der Obersteuermann Pieter Dircksz. Keyser genannt werden, der eine Anzahl Sternbilder in der südlichen Hemisphäre entdeckte, die Jodocus Hondius später auf seinem Himmelglobus (1600) aufnahm. Keyser starb aber bereits am 12. September 1596 vor Bantam. Ein anderer, sehr wichtiger Teilnehmer der Expedition war der Unterkaufmann Willem Lodewijckz. (†1604) auf der *Amsterdam*, welcher bereits auf früheren Seereisen viel Erfahrung gesammelt hatte. Wiederholt leitete er während De Houtmans Reise mehrere Erkundungsfahrten, fertigte Küstenansichten an und notierte nautische Daten. Von seiner Hand stammt der ausführlichste und vielseitigste Reisebericht, welcher 1598 von Cornelis Claesz. in verschiedenen Sprachen publiziert wurde. Sein Name ist auch mit der später zu nennenden Spezialkarte der Java-See verbunden.

Der Verlauf der Reise

Die Schiffe stachen am 2. April 1595 von Texel aus in See, passierten am 16. dieses Monats Madeira, erreichten drei Tage später die Kanarischen Inseln und von wo aus man in der Ferne den 3700 m hohen Teide auf Teneriffa sah. Danach wurde der Kurs in Richtung der Kapverdischen Inseln gesetzt, wo man am 26. April auf der Ilha de

---

7 2 Lasten sind 1 Tonne.

Maio Wasser aufnahm. Bereits am nächsten Tag wurden die Anker gelichtet und bis zum 4. Mai ein südöstlicher Kurs genommen (parallel zur afrikanischen Küste). Nun segelte man weiter zum Nordost-Kap Brasiliens und nach der Querung des Äquators wurde mit einem weiten Bogen Anfang August das Kap der Guten Hoffung erreicht. Etwas östlicher landete man zum ersten Mal in der Bucht von S. Bras, später Mosselbaai (Mossel Bay) genannt, wo man sich vom 4. bis 11. August erholte. Auf den Schiffen war durch den langen Aufenthalt auf See ein Mangel an Vitamin C entstanden und Skorbut ausgebrochen. Man kam hier in Kontakt mit den Hottentotten, von welchen man im Tauschhandel den Fleischvorrat auffüllen konnte. Auch konnte man genügend Frischwasser bekommen, jedoch kein Obst und Gemüse zur Bekämpfung des Skorbuts. Man hoffte deshalb auf Madagaskar eine bessere Situation vorzufinden und die Antongil-Bucht an der Ostküste der Insel anzulaufen. Die Anker wurden am 11. August gelichtet und Kurs nach Madagaskar genommen, welches Anfang September erreicht wurde. Wegen des kräftigen Ostwindes gelang es jedoch nicht, das Kap Andavaka zu umrunden. Aufgrund des desolaten Zustandes der Mannschaft wurde durch den Schiffsrat beschlossen, Westkurs zu nehmen und an der Südküste Erholung zu suchen. Man verbrachte hier an verschiedenen Stellen fast drei Monate. Auf einer kleinen Insel in der weiten Ampalaza-Bucht wurden so viele Mannschaftsmitglieder begraben, dass sie den Namen *Hollantsche kerckhof* (Holländischer Friedhof) bekam.

Am 13. Dezember wurde schließlich die Reise fortgesetzt und das Südkap von Madagaskar mit dem Ziel umrundet, den Kurs zur Sunda-Strasse zu nehmen. Der Südostmonsun trieb die Schiffe jedoch nach Norden ab und Skorbut begann erneut auszubrechen. Der Schiffsrat musste deshalb abermals beschliessen, auf der Insel Madagaskar zu landen. Vom 23. Jänner bis 12. Februar 1596 ankerte die kleine Flotte bei der Insel S. Maria und der etwas nördlicher gelegenen Antongilbucht an Madagaskars Nordost-Küste.

Am Abend des 12. Februar 1596 gingen schließlich die Schiffe erneut unter Segel und begannen die Überquerung des Indischen Ozeans Richtung Sunda-Strasse. Das erste Land, das man sah, war am 5. Juni die Insel Engano im Südwesten von Sumatra. Nach der Durchseglung der Sunda-Strasse, wobei von Inseln und Teilen von Sumatra Küstenprofile in Lodewijcksz.' Bericht aufgenommen wrden, erreichte man am 26. Juni die wichtige Handels- und

Hafenstadt Bantam an Javas Nordwest-Spitze. Nach einer Reise von fast 15 Monaten hatte man das eigentliche Reiseziel erreicht. Bantam zeigte sich als ein sehr lebendiger Handels- und Stapelplatz, ein Schmelztiegel von Kaufleuten aus allen Welteilen. Die Stadt war ein Knotenpunkt des lokalen und internationalen Handels. Die Niederländer kauften hier Gewürze, insbesonders Pfeffer, und schlossen mit dem Gouverneur (trotz des Widerstandes der in der Stadt wohnenden Portugiesen) einen Handelsvertrag. Durch De Houtmans ungeschicktes und grobes Auftreten war das Verhältnis mit den Javanesen jedoch nicht immer freundschaftlich. So wurde er bei einem Landbesuch gefangen genommen und erst nach einiger Zeit gegen ein hohes Lösegeld wieder freigelassen.

Nach einem Aufenthalt von fast fünf Monaten verließen die Schiffe am 6. November Bantam mit östlichem Kurs. Die Fahrt verlief längs den Nordküsten von Java und Madura. Die Flotte ankerte vor Jakarta, Tuban, Sedaju (wo ein Überfall der feindlichen Bevölkerung zurückgeschlagen wurde) und Arosbaja. Hier entstand ein Konflikt über den weiteren Kurs: Während De Houtman und die Kaufleute zu den Molukken weitersegeln wollten, um noch mehr Gewürze kaufen zu können, hatten die Kapitäne und Steuermänner kein Vertrauen mehr in den Erfolg der Mission. Wegen des desolaten Zustandes der Schiffe und der stark ausgedünnten Bemannung war es ihrer Meinung nach nicht zu verantworten, die Reise fortzusetzen. Schließlich wurde die Heimfahrt beschlossen. Da die Schiffe nicht mehr ausreichend bemannt waren – es waren zusammen nur noch 94 Überlebende – wurde bestimmt, die Ladung und Bemannung der *Amsterdam* auf die anderen Schiffe zu verteilen und das bereits sehr lecke Schiff zu verbrennen. Am 17. Jänner 1597 erreichte man die Bali-Straße und die Schiffe ankerten noch einen Monat an verschiedenen Stellen vor der Südküste Balis, um sich mit den notwendigen Vorräten für die Heimreise zu versorgen. Der König von Bali war sehr hilfsbereit und gewährte den Niederländern alle benötigte Hilfe. Er war auch sehr neugierig und bat ihm zu zeigen, wo Holland lag. Daraufhin wurde dem König als eines der Geschenke eine Weltkarte angeboten, ihm die darauf eingezeichneten Länder und Städte erklärt und die von den Niederländern verfolgte Reiseroute gezeigt.

Am 25. Februar lichtete man schließlich die Anker und begann mit der Heimfahrt. Man bemerkte, dass das Südost-Kap von Java viel nördlicher liegen musste als es auf den portugiesischen Karten verzeichnet war und dass Java eine verhältnimässig schmale Insel sein musste. Die Überquerung des Indischen Ozeans ging sehr rasch; bereits nach zwei Monaten sichtete man Natal an der afrikanischen Ostküste. Im Verlauf der weiteren Reise verlor die *Hollandia* die beiden anderen Schiffe aus dem Auge, man traf sich jedoch wieder vor S. Helena. Eine Proviantierung konnte hier jedoch nicht stattfinden, weil an der Reede vier portugiesische Kraaken lagen. Notgedrungen musste man ohne Zwischenstopp die Reise fortsetzen. Ohne weitere Zwischenfälle wurde dann Mitte August die Heimat erreicht.

Die kommerziellen Resultate erfüllten nicht die Erwartungen, die sich die *Compagnie van Verre* von De Houtmans Reise gemacht hatte, doch seine Reise gab große Impulse für zukünftige niederländische Handelsunternehmungen, jedoch auf maritimem Gebiet war das wichtigste Ziel erreicht: Das Kap der Guten Hoffnung wurde zum ersten Mal von niederländischen Schiffen umrundet, der Indische Ozean überquert und die ersten Handelskontakte auf Java geknüpft. Van Linschotens Voraussage, dass hier wenig portugiesischer Widerstand zu erwarten war, hatte sich somit als richtig erwiesen.

PREMIER LIVRE
DE LHISTOIRE DE LA
NAVIGATION AVX INDES
ORIENTALES PAR LES HOLLANDOIS
ET DES CHOSES A EVX ADVENVES ENSEMBLE
LES CONDITIONS, LES MEVRS, ET MANIERES DE VIVRE DES NA-
tions, par eux abordees. Plus les Monnoyes, Espices, Drogues, & marchandises, & le pris d'icelles. Davantage les descouvrements & apparences, situations, & costes maritimes des contrees, avec le vray pourtraict au vif des habitans: Le tout par plusieurs figures illustré: tresrecreatif a lire a tous navigans & amateurs des navigations lointaines, et terres estrangeres. Par G. M. A. W. L.

Abb. 1: Titelseite der französischen Ausgabe von Willem Lodewijcksz.'s Bericht, Amsterdam: Cornelis Claesz., 1598. ÖAW- Sammlung Woldan

Die Rückkehr von De Houtmans Schiffen aus Ost-Indien weckte natürlich große Neugier und so erschienen bereits ab 1597 verschiedene Reiseberichte in Druck.[8] Der ausführlichste wurde durch Cornelis Claesz. publiziert. Er war in den Besitz des Reiseberichts von Willem Lodewijcksz. gekommen, welcher in Zusammenstellung, Umfang und Anzahl von Illustrationen alle anderen Berichte in den Schatten stellte. Die Ausgabe von *D'Eerste Boeck. Historie van Indien, waerinne verhaelt is de avontueren die de Hollandtsche Schepen bejeghent zijn* (Amsterdam, 1598) fand nicht nur im eigenen Land großes Interesse, sondern die noch im selben Jahr publizierte lateinische und französische Übersetzung (Abb. 1) fand auch im Ausland raschen Absatz.[9]Cornelis Claesz. begnügte sich nicht mit der Ausgabe von Lodewijcksz.' Journal in drei verschiedenen Sprachen, sondern publizierte außerdem im selben Jahr 1598 eine Übersichtskarte mit der darauf eingezeichneten Route von De Houtmans Reise und Lodewijcksz.' Spezialkarte der Java-See.

Die Übersichtskarten mit De Houtmans Route (1595-1597)

Die niederländische Originalkarte (1598) (Abb. 2)

Als Pieter Burger im Jahre 1915 eine Beschreibung aller ihm bekannt gewordener Karten anfertigte, die Cornelis Claesz. separat zwischen 1591 und 1598 publiziert hatte, vermutete er bereits, dass dieser Verleger auch eine große Karte von der ersten niederländischen Reise nach Osten herausgegeben haben musste.[10] Diese Vermutung basierte auf zwei Fakten: Einerseits auf einer kleinen, aber sehr instruktiven Übersichtskarte auf der Titelseite von Lodewijcksz.' gedrucktem Bericht,[11] andererseits stützte Burger seine

---

8 P. A. Tiele, Mémoire bibliographique sur les journaux des navigateurs Néerlandais, Amsterdam 1867. Reprint: Amsterdam 1960, 116-128.

9 Ebd., 124-128. In der Sammlung Woldan (R-IV: AS 16) befindet sich ein Exemplar der französischen Ausgabe von Lodewijcksz.' Bericht, welcher von Cornelis Claesz. 1598 in Amsterdam publiziert wurde.

10 E. W. Moes – C. P. Burger, De Amsterdamsche boekdrukkers en uitgevers in de zestiende eeuw, 4 Bde., Amsterdam 1900-1915 (Reprint: Utrecht 1988, IV, 443).

11 Der Kartentitel lautet: *TYPUS expeditionis nauticæ Batavorum in Iavam, absolutæ.*

Vermutung auf die Tatsache, dass auch Théodore de Bry im dritten Band seiner *Petits Voyages* eine Karte aufnahm, auf der die gesamte Reiseroute von Cornelis de Houtman eingezeichnet war. Es musste also ein in Amsterdam publiziertes Original gegeben haben, wonach De Bry kopiert hatte. Burgers Annahme hatte sich bestätigt, denn in den vergangenen vierzig Jahren wurden drei Exemplare dieser Karte aufgefunden.[12]

Abb. 2 Übersichtskarte mit der Route des Cornelis de Houtman 1595-1597, Amsterdam 1598, Maritiem Museum, Rotterdam

Titel:

DESCRIPTIO HYDROGRAPHICA accom= | modata ad Battavorum navigationem in Javam insulam Indiæ Orien= | talis, factam; ad quam postridie Calendas Aprilis ann. 1595 ex Hollan= | dia solverunt et ex qua domum redierunt 3. Idus Augusti an. 1597. | Horum exitus reditusq; via his notis ooo demonstratur. De hac navi | gatione extat descriptio lectu perquam admiranda

12 Nederlands Scheepvaartmuseum, Amsterdam (S. 3258 [Z I-20]); Maritiem Museum, Rotterdam (K 270) und Verhoeff Sammlung (NL) (ex Sotheby's Sale, London, 14 December, 2000, lot 424; beigebunden in Willem Barentsz's *Description de la Mer Mediterrannee*, 1607). Reproduktion in tatsächlicher Größe in: G. Schilder, Monumenta, Faksimile 25.

Caerte gemaeckt op de Navigatie der Hollanders gedaen op Java, | gelegen in Oost-Indien, t"seyl gegaen uit Holland den tweeden van | April 1595 ende wederom anno 1597 den 14den Augusti te huys | gecomen. Haeren wegh is afgeteekent met dese ooo rondeekens int heen | ende wederom keeren. Hier van is een seker beschryvinge seer vremt | om lesen.
- Maßstab: *Milliaria Germanica* [150= 37mm]
- [Amsterdam: Cornelis Claesz, 1598]
- Kupfergravur und Radierung, 2 Blätter, 40,5 x 67,5 cm.

In der rechten unteren Ecke befindet sich in einer Rollwerkkartusche dieser Kartentitel in lateinischer und niederländischer Sprache. Hierin wird auf die eingezeichnete Route und die ausführliche Beschreibung der ersten niederländischen Reise nach Ost-Indien verwiesen. In einer einfachen Kartusche entlang des unteren Randes ist zudem eine Maßstabsleiste in deutschen Meilen angegeben. Als dekorative Elemente sind Abbildungen im Kartenbild aufgenommen worden, welche dem Reisebericht von Willem Lodewijcksz. entlehnt sind. In der linken unteren Ecke ist ein Kärtchen der Mossel Bay in Süd-Afrika und von der Antongil Bucht auf Madagaskar zu sehen (beide von einem lateinischen Text begleitet). In der rechten oberen Ecke ist eine Tafel mit drei Abbildungen der Einwohner von Bantam und dessen Umgebung abgebildet. An den Seitenrändern befindet sich eine Einteilung für Breitengrade (55° N-54° S), in der Mitte der Kartenränder sind die Himmelsrichtungen auf Latein angegeben. Von einer zentralen Kompassrose ziehen sich Kompasslinien zu acht sekundären Kompassrosen, wovon vier vollständig eingezeichnet sind. Der Stecher der Karte wird leider nicht genannt, jedoch zeigt der Stil eine große Übereinstimmung mit anderen Karten, die von dem für Cornelis Claesz. arbeitenden englischen Stecher Benjamin Wright signiert sind.

Die Karte zeigt das Gebiet zwischen dem Ostpunkt von Brasilien im Westen und der Westspitze von Neu Guinea im Osten. Im Norden reicht das Kartenbild von Emden in Europa und Korea in Asien bis zum Ausläufer einer grossen *TERRA AUSTRALIS INCOGNITA* im Süden. Als Quellen für die Darstellung wurde kontemporanes Material benutzt, vor allem Karten, die durch Cornelis Claesz. publiziert worden waren. So basiert die Zeichnung der Küsten, Inseln und Untiefen im Indischen Ozean auf den Karten in Van Linschotens *Itinerario* (1596), während der größte Teil des Indonesischen Archipels und der Philippinen in Zeichnung und Toponymen mit Plancius' Molukken-Karte (1592) übereinstimmt. Für die Abbildung von Korea und Japan, welche auf Plancius' Karte nicht mehr vorkommen, stand Luis Teixeiras Karte von diesem Gebiet in Ortelius' *Theatrum Orbis Terrarum* (1595) Modell. Plancius' hypothetisch gezeichnete Küste von Süd-Java wurde weggelassen. Auffallend ist, dass im gesamten Kartenbild keine niederländischen Namen vorkommen, also auch keine Namen, die direkt von De Houtmans Reise stammen.

Hauptziel dieser Übersichtskarte war die Wiedergabe der mit Erfolg absolvierten ersten niederländischen Handelsfahrt nach Ost-Indien (1595-97). Die gesamte, von Cornelis de Houtman gewählte Route ist in der Karte eingezeichnet: Entlang der Hinreise (*Exitus*) sind seine vier Schiffe zweimal abgebildet, bei der Rückreise-Route (*Reditus*) die drei übrig gebliebenen Schiffe ebenfalls zweimal dargestellt. In Cornelis Claesz.' *Const ende Caert-Register* (1609) wird in der Sektion 'De Caerten van alle Navigatien' diese Karte (zusammen mit 50 Abbildungen aus dem *Journal*) für einen Preis von 26 stuivers angeboten.

Die deutsche Kopie dieser Karte, 1599

(Sammlung Woldan K-V(Bl):AS 19)

Cornelis Claesz.'s Prototyp-Karte von 1598 stand Modell für eine Kopie, die Théodore de Bry im dritten Band seiner *Petits Voyages* (Frankfurt 1599) aufnahm (Abb. 3).[13] Auf De Brys Karte verzichtete man jedoch auf das Kopieren der fünf Illustrationen aus dem von Lodewijcksz. publizierten Bericht. Der Kartentitel, der sich im Original in der rechten unteren Kartenecke befand, steht nun in einer Kartusche, wo auf der Karte von Cornelis Claesz. ursprünglich drei Abbildungen der einheimischen Bevölkerung standen:

13 J. G. Garrat, The maps in De Bry, in: The Map Collector 9 (1979), 2-11, hier: 10.

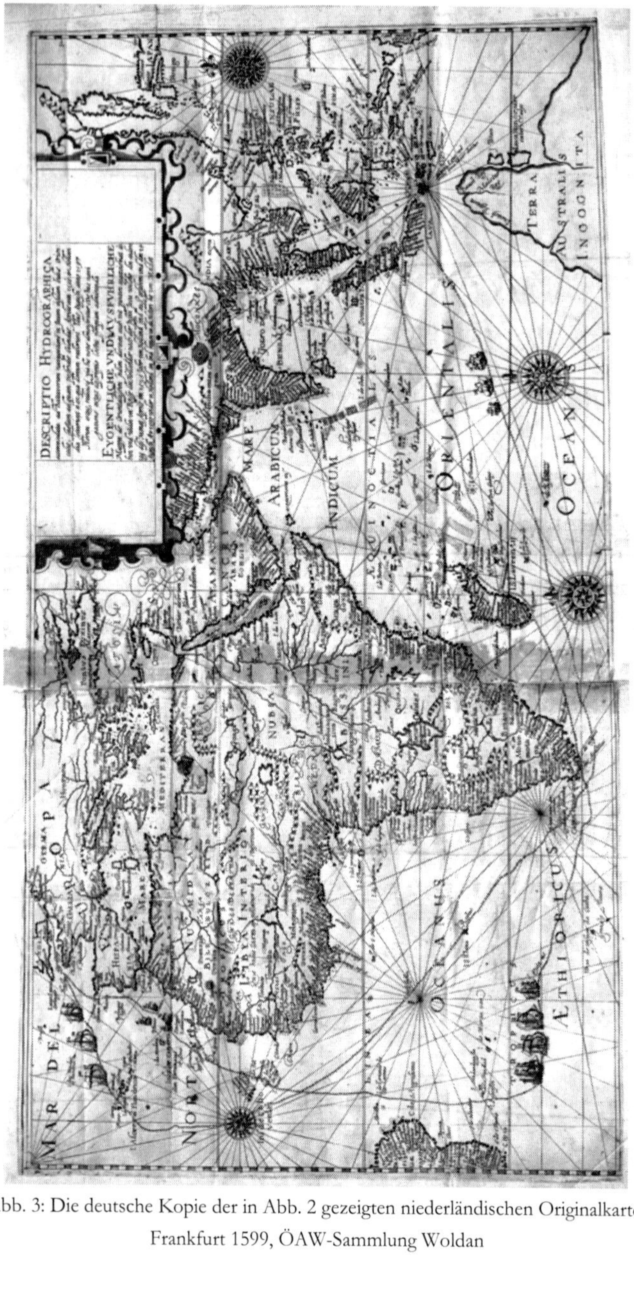

Abb. 3: Die deutsche Kopie der in Abb. 2 gezeigten niederländischen Originalkarte. Frankfurt 1599, ÖAW-Sammlung Woldan

DESCRIPTIO HYDROGRAPHICA | accommodata ad Battavorum navigationem in Javam insulam Indiæ Orien= | talis, factam ad quam postridie Calendas Aprilis ann. 1595 ex Hollan= | dia solverunt et ex qua domum redierunt Idus Augusti anno 1597 | Horum ex[i]tus reditusq' via his notis demonstratur. De hac navi | gatione extat descriptio lectu perquam admiranda / EYGENTLICHE VND AVSFVHRLICHE | Mappa de Orientallischen Indien darmen auch mit puntien angezeichnet di | hin vnd wider um Reisse der Hollander nach der Inseln Java welche den andern | tag dess Monat Aprils in 1595 Jahr von Holland sind aussgefahren und den 14. | Augusti A° 1597 wider in Holland an sind komen wi di Historie hie van meldet.
- [Frankfurt: Théodore de Bry, 1599]
- Kupfergravur, 2 Blätter, 35 x 66 cm.

Der lateinische Titel wurde also wörtlich vom Original übernommen. Der niederländische Text ist in einer sinngemässen deutschen Übersetzung wiedergegeben.

Die originale Karte der Java-See von Willem Lodewijcksz., 1598

(Sammlung Woldan K-V (Bl): AS/Sos 15 (1-2) (Abb. 4)

Außer der oben beschriebenen Übersichtskarte mit der Wiedergabe der durch Cornelis de Houtman befahreren Route, erinnert kartographisch an diese Reise auch eine in Folio-Format von der Hand von Willem Lodewijcksz. gefertigte Karte der Java-See:[14]

Nova tabula, | Insularum Iavæ, Su= | matræ, Borneonis et ali= | arum Malaccam usque, deli= | neata in insula Iava, ubi ad vi= | vum designantur vada et brevia | scopulique interjacentes descrip= | ta a G.M.A.L.
Nieuwe caerte op Java ghe= | teeckent, van de eylanden van | Java, Sumatra, Borneo tot Ma= | lacca toe, daer in dat alle de bij | leggende eylanden naer t'leven | gestelt, ende op haer plaetse | geleyt zyn door G.M.A.L.
Hispanicæ leucæ 17½, uni gradui competentia [30=36 mm]; Miliaria Germanica, quorum15.uni gradui respondent. [30=42 mm]
[Amsterdam:] Cornelius Ni= | colai excud. [1598] / Baptista à Doe= | techum sculp.
Kupfergravur und Radierung, 38,5 x 54,5 cm.

14 G. Schilder, Monumenta, Facsimile 26.

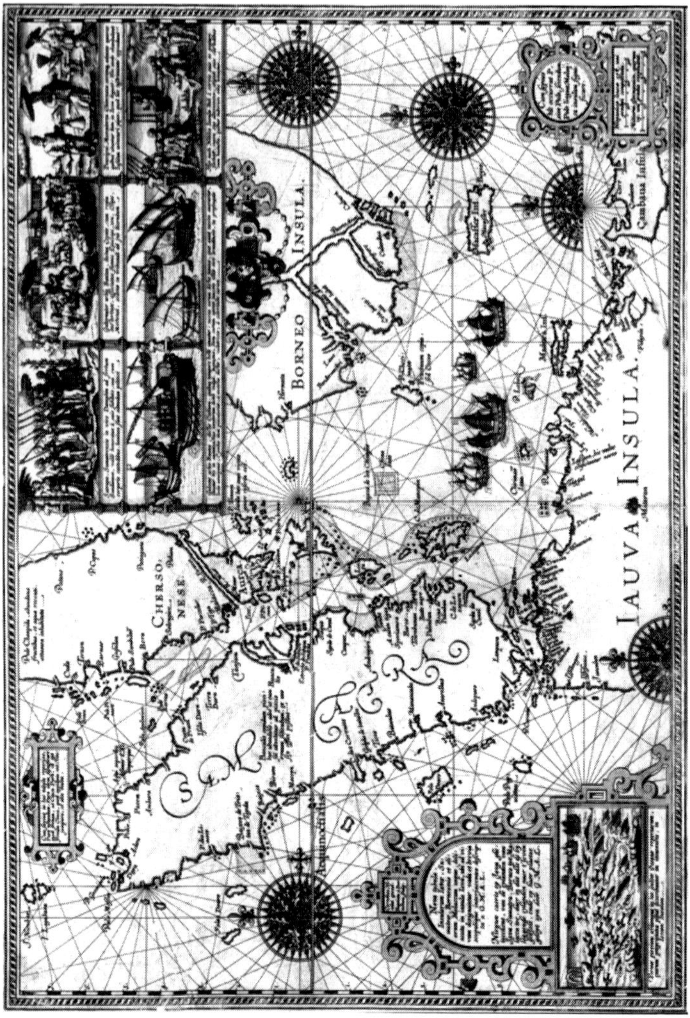

Abb. 4: Willem Lodewijcksz., Karte der Java-See. Amsterdam 1598,
ÖAW-Sammlung Woldan

In der linken unteren Ecke befindet sich hier eine dreiteilige Kartusche, wovon das Mittelstück den lateinischen und niederländischen Kartentitel enthält wird. Im obersten Teil dieser Kartusche werden die Verleger Cornelis Claesz. und Graveur Baptista van Doetecum genannt. Die rechteckige Basis der Kartusche zeigt einen verkleinerte Abbildung von Fischen und Vögeln aus Lodewijcksz.' Reisebericht (Tafel 6) mit einem lateinischen Begleittext. Eine kleine Kartusche in der rechten unteren Kartenecke enthält eine Erklärung des Wortes *Pulo* (malayisch für Insel) und die Maßstableisten in spanischen und deutschen Meilen. Oben links wird das Wort *Puro* erklärt (malayisch für Straße oder Stadt). Der dekorative Blickfänger ist aber die große Kartusche rechts oben mit sechs verkleinerten Abbildungen aus dem *Journal* von Lodewijcksz. im Spiegelbild: *Satrapa Sumatrensis* (Tafel 8); *Gubernator urbis Bantan* (Tafel 13); *Præcipui Mercatores Chinenses* (Tafel 19); *Galeotæ urbis Bantan* (Kombination Tafel 27 und 28); und *Hoc modo Nobiles Insulæ Bali* (Tafel 40).

Entlang der Seitenränder befindet sich die Einteilung der Breitengrade (7°N-9°S). Um eine zentrale Kompassrose in der Kartenmitte ist ein Kranz von elegant ausgeführten sekundären Rosen graviert. Das Netz der Kompasslinien beschränkt sich auf die Wasserflächen. In der Java-See werden die vier Schiffe von De Houtman abgebildet.

Das Kartenbild reicht von den südlichen Nikobaren (*I. Nicubar*) im Nordwesten bis zum westlichen Teil von Sumbawa (*Çambaua Insula*) im Südosten. Die Karte zeigt den südlichen Teil von Malakka (*CHERSONESE*), die Strasse Malakka, ganz Sumatra (*SAMATRA*), den Riau und Linga Archipel, Bangka (*Banca*), den südlichen Teil von Kalimantan (*BORNEO INSULA*), die Nordküste von Java (*IAUVA INSULA*), Madura (*Madura Ins.*), Bali (*Bali Ins.*) und den westlichen Teil von Sumbawa (auf Van Linschotens Karte noch *Java Menor* genannt).

Im Kartentitel wird zudem darauf hingewiesen, dass diese neue Java-Karte von Willem Lodewijcksz. gezeichnet wurde und dass die Länder und Inseln durch ihn *naer t'leven gestelt, ende op haer plaetse geleyt zyn* (nach der Wirklichkeit und auf der richtigen Position gezeichnet sind). Die Karte dokumentiert somit die lokale Kenntnis, die man im Osten während der ersten Reise gesammelt hatte – eine Kenntnis die im Grunde genommen noch recht bescheiden war.

De Houtmans Reise hat nur direkte geographische Erkenntnisse gesammelt: so in der Umgebung der Sunda-Straße, entlang Javas Nordküste und von der Insel Bali. Lodewijcksz.' Bericht enthält zudem Küstenprofile der Sunda-Straße, Pläne vom Hafen und der

Stadt Bantam, Küstenprofile von Java und Bali, sowie eine Übersichtskarte von letzterer. Im Vergleich mit den Karten von Plancius und Van Linschoten ist auf Lodewijcksz.' Karte – die in einem größeren Maßstab gezeichnet ist – die West- und Nordküste von Java viel detaillierter eingezeichnet und weist eine größere Zahl von Ortsnamen auf. Ausser *Bantan* (Bantam, wo De Houtman die erste Ladung Pfeffer kaufte), sieht man die Toponyme *Iacatra* (Jakarta), wo das spätere Zentrum der VOC entstehen sollte, *Taggal* (Tegal) und *Surabaya* (Surubaja). Im Landesinneren von Java wird mit einer Vignette die Stadt Mataram gekennzeichnet, eine Stadt, die sich zu einem wichtigen Zentrum des islamischen Handels entwickeln sollte. Vor dem Westeingang der Sunda-Strasse liegen zahlreiche Inseln.

Bantam liegt auf unserer Karte auf demselben Längengrad wie die Stadt Malakka, obwohl die Portugiesen wussten, dass dieser Ort vier Grade östlicher lag. Wie sind nun die Fehler auf den portugiesischen Karten zu erklären? Plancius gab hierfür in seinem 1598 aufgestellten Memoire eine interessante Erklärung. *Die Portugiesen*, bemerkte er, *zeichnen die Distanz von verschiedenen südlich und östlich von Malakka gelegenen Orten kleiner ein als dies in Wirklichkeit ist.* Und er zog die Schlussfolgerung: *Es ist zu vermuten, dass die Portugiesen die genannten Orte deshalb so einzeichnen, um die Gebiete von Banda, den Molukken, Gilolo und Japan in die Hemisphäre einzubeziehen, welche ihnen durch den Römischen Papst zugeteilt worden waren.* Man wollte also unbedingt, dass die Molukken im portugiesischen Einflussgebiet zu liegen kamen. Der portugiesische Pilot Pedro de Tayde, der mit den Holländern zusammenarbeitete (und darum durch seine portugiesische Landsleute am 16. August 1596 in Bantam ermordet wurde), wusste jedoch über die richtigen Entfernungen genau Bescheid, denn auf einer von ihm gezeichneten kleinen Karte war Bantam 60 deutsche Meilen östlicher als Malakka eingezeichnet.

Die Insel Bali wird durch Lodewijcksz. in einer durchaus erkennbaren Form gezeigt. Nach Osten werden Lombok und Sumbawa noch als zusammenhängende Insel gezeichnet. Merkwürdig ist die Position von Macassar. Auf Plancius' Karte wird diese südliche Region von Celebes durch eine schmale Landbrücke vom Rest abgeschnitten. Lodewijcksz. geht noch einen Schritt weiter und trennt diesen Teil von der Hauptinsel ab und zeichnet ihn als kleine Insel (*Macasser Ins.*) mitten in das Meer zwischen Borneo und

Sumbawa. Ein anderes Phänomen ist in Malaysien zu sehen, wo der südliche Teil von der Malakka-Halbinsel mit *C. de Cincapura* (Singapore) durch einen Fluss vom Rest abgeschnitten wird.

Bemerkenswert ist zudem, dass die Lodewijcksz.-Karte bereits einen Teil des Archipels zeigt, der sich zum Mittelpunkt der niederländischen Macht in Ost-Indien zu entwickeln sollte. Ursprünglich hatte Cornelis Claesz. die Absicht, eine verkleinerte Version dieser Karte im *Journal* von Lodewijcksz. Aufzunehmen, denn am Ende des 18. Kapitels seines Berichts steht: *Hier stelt het Caertgien van Iava ende Sumatra* (hier steht das Kärtchen von Java und Sumatra), aber keine einzigen Ausgabe des *Journals* enthält eine derartige kleine Karte.[15] Anscheinend sah sich der Verleger zunächst von den Reedern, welche die Reise von Cornelis de Houtman ausgerüstet hatten, genötigt, hiervon abzusehen.

Die direkte Anleitung zum Handel muss wahrscheinlich in Exemplaren der Lodewijcksz.-Karte in Folio-Format gesehen werden, welche auf der Rückseite einen gedruckten Text aufweisen (Abb. 5).[16] Dieser Text ist in derselben gotischen Lettertype gedruckt wie die von Cornelis Claesz. Für die Ausgabe des Berichts von Lodewijcksz. benutzte. Dass der Text auf der Rückseite der Karte ebenfalls von Lodewijcksz. stammt, ergibt sich vor allem aus der Tatsache, dass er zweimal auf sein eigenes *Journal* verweist. Aus diesen Verweisen kann man schließen, dass Cornelis Claesz. diese Karte mit Text verkleinert im Bericht von Lodewijksz. aufnehmen wollte.

---

15 In der französischen Ausgabe von 1598 in der Sammlung Woldan: *Icy doibt estre mis la Carte de Iava & Sumatra.*

16 Die folgenden Exemplare sind bekannt: Bibliothèque de l'Arsenal, Paris (Fol. H 181 [73]); Bibliothèque nationale de France (Ge D 15088); Bibliothèque de Sorbonne, Paris (A 507.I [113]); Maritiem Museum, Rotterdam (W.A. Engelbrecht Sammlung 687); Österreichische Akademie der Wissenschaften, Wien (Sammlung Woldan K-V (Bl): AS/Sos 15 (1-2); und Umanski Sammlung (USA). Beim Woldan-Exemplar, welches einmal Teil eines Sammelatlas gewesen sein muss, war der Text auf der Rückseite aus unbekannten Gründen abgedeckt. Der Versuch die Karte zu *spalten* misslang, sodass diese nun aus zwei Teilen besteht: das Kartenbild ist nun auf einem dünnen Papierträger, während der Text von der Kartenrückseite nach wie vor abgedeckt ist (jedoch deutlich erkennbar, wenn man das Blatt gegen das Licht hält)

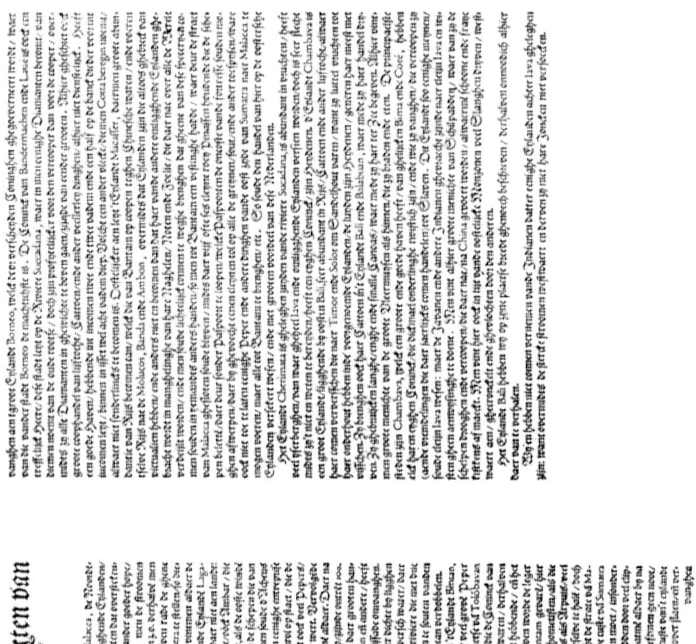

Beschryvinge vande Straten ofte engten van Malacca ende Sunda/ met haer omliggende Eylanden/ Bancken/ Ondiepten ende Sanden.

Abb. 5: Text auf der Rückseite der Willem Lodewijcksz.-Karte der Java-See, Amsterdam 1598, Maritiem Museum, Rotterdam

Der Text enthält allerdings eine Anzahl von Mitteilungen, die die Amsterdamer Reeder nicht publiziert haben wollten, sodass die

Folio-Karte vorläufig nicht distribuiert wurde.[17] Auf diese Weise signaliserte Lodewijcksz. unter anderem, dass man im Falle einer Errichtung eines Forts in Bantam ohne viel Mühe die Fahrt zwischen den Molukken und Malakka beherrschen könnte: So könnte man von hier aus den Handel mit den östlichen Inseln mit grossem Vorteil für die Niederlande sichern. Solche wichtigen Informationen wollten die beteiligten Amsterdamer Kaufleute bestimmt nicht an eventuelle Konkurrenten preisgeben. Letztendlich konnten sie jedoch die Entwicklung nicht aufhalten, denn noch im selben Jahr 1598 nahmen verschiedene Flotten aus Holland und Zeeland Kurs nach Osten. Deshalb konnte Cornelis Claesz. schliesslich die Lodewijcksz.-Karte – mit oder ohne Text in verso – auf den Markt bringen.

Epilog

Einer der Berührungspunkte, die das Studium der Kartographiegeschichte mit anderen Disziplinen hat, ist die Geschichte, vor allem die Entdeckungsgeschichte. Gerade kartographische Quellen haben für das Zeitalter der niederländischen Entdeckungen mit ihren Beschreibungen und Darstellungen unschätzbaren Wert und dürfen bestimmt nicht nur als Füllmaterial betrachtet werden; sie können manches Mal wegen des Fehlens von anderen Quellen die einzigen Zeugen einer Entdeckungsreise darstellen, sodass allein mit Hilfe von Karten der Verlauf einer Entdeckung rekonstruiert werden kann. Die niederländische Expansion nach allen Windrichtungen ab dem Ende des 16. Jahrhunderts hat dazu geführt, dass auch die Niederländer einen Anteil an der Entdeckung der Welt für sich beanspruchen können, wobei die zeitgenössischen Reisebeschreibungen und Karten besonderes wichtige Quellen sind.

---

17 Der Text beginnt mit: *Beschryvinge vande Straten ofte engten van Malacca ende Sunda, met haer omligghende Eylanden, Bancken, Ondiepten ende Sanden.* Transkription und Erklärung in: G. P. Rouffaer – J. W. Îjzerman, De eerste schipvaart der Nederlanders naar Oost-Indië 2, 206-221.

Schlussbetrachtung

Bereits während meines Studiums in Wien bekam ich die Gelegenheit, den berühmten Atlas Blaeu – Van der Hem in der Österreichischen Nationalbibliothek zu studieren, welcher einen außergewöhnlicher Kulturspiegel der Entwicklung des niederländischen Unternehmungsgeistes im 17. Jahrhundert darstellt. Dies war auch mein erster Kontakt zur niederländischen Kartographie, die einen dermaßen grossen Einfluss auf mich ausübte, dass das intensive Studium dieses Fachgebietes meine Berufswahl bestimmte und ich meine Lebensinteressen in die Niederlande verlegte. Mehr als vierzig Jahre später, bei der Vorbereitung zu dieses Artikels, befasste ich mich näher mit der Sammlung Woldan und war sehr erstaunt über den Reichtum ihrer geographischen Bestände und deren Fülle an Material zur Entdeckungsgeschichte. Dass darunter auch der niederländische Anteil relativ groß ist und einige Kleinode umfasst, erfüllt mich mit besonderer Freude. Wir sind meinem Studienfreund Prof. Dr. Johannes Dörflinger deshalb sehr viel Dank schuldig, dass es ihm, zusammen mit unserem gemeinsamen Promotor Prof. Hamann gelang, diese wertvolle Sammlung beisammen zu halten, in der Österreichischen Akademie der Wissenschaften unterzubringen und der Forschung zugänglich zu machen.

Literaturverzeichnis

Garrat J. G., The maps in De Bry, in: The Map Collector 9 (1979), 2-11.

Moes E. W. – Burger C. P., De Amsterdamsche boekdrukkers en uitgevers in de zestiende eeuw, 4 Bde., Amsterdam 1900-1915. Reprint: Utrecht 1988.

Naber P. S. L'Honoré, Reizen van Willem Barents, Jacob van Heemskerck, Jan Cornelisz. Rijp en anderen naar het noorden (1594-1597) (Werken uitgegeven door de Linschoten Vereeniging XIV und XV), dl. 1-2, 's-Gravenhage 1917.

Rouffaer G. P. – Îjzerman J. W., De eerste schipvaart der Nederlanders naar Oost-Indië onder Cornelis de Houtman, 1595-1597 (Werken uitgegeven door de Linschoten Vereeniging VII, XXV und XXXII), dl. 1-3., 's-Gravenhage 1915-1929.

Schilder G., Monumenta Cartographica Neerlandica, VII, Cornelis Claesz. (c. 1551-1609): Stimulator and Driving Force of Dutch Cartography, Alphen aan den Rijn, 2003.

Shirley R. W., The Mapping of the World: Early printed world maps 1472-1700 (The Holland Press Cartographica Seriesl. 9), London 1983.

Terpstra H., Itinerario. Voyagie ofte Schipvaert van Jan Huygen van Linschoten naer Oost ofte Portugaels Indien 1579-1592 (Werken uitgegeven door de Linschoten-Vereeniging LVII, LVIII und LX), dl. 1-3,'s-Gravenhage 1955-57.

Tiele P. A., Mémoire bibliographique sur les journaux des navigateurs Néerlandais, Amsterdam 1867. Reprint: Amsterdam, Nico Israel, 1960, 116-128.

Zum Autor

Günter Schilder, geb. 1942 in Wien, studierte von 1960 bis 1966 Geographie und Geschichte an der Universität Wien. Mit Hilfe eines Stipendiums forschte er zwei Jahre in den Niederlanden für seine Dissertation über den Anteil der Niederländer an der Entdeckung Australiens. Er promovierte 1970 über dieses Thema in Wien bei Prof. Hamann. Anschließend kehrte er nach Holland zurück, wo er zunächst in einem kartographischen Verlag in Amsterdam arbeitete. Von 1971 bis 2007 war er an der Universität Utrecht im Fachbereich Geschichte der Kartographie angestellt, seit 1981 als Professor. Er schrieb bedeutende Publikationen auf dem Gebiet der Geschichte der niederländischen Kartographie und Entdeckungen.

Kontakt: g.schilder@geo.uu.nl

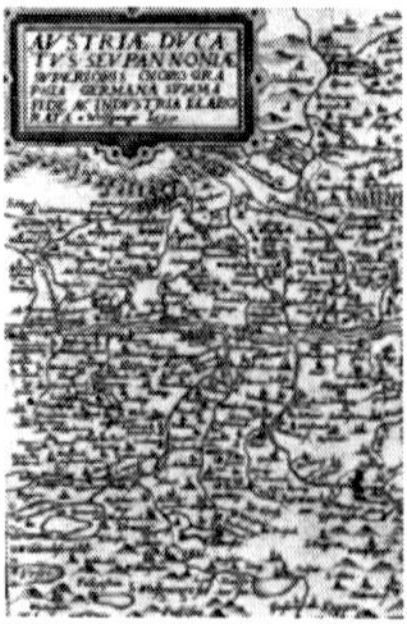

Wolfgang Lazius,
Austriae Chorographia,
Straßburg 1620

Petra Svatek, Wien

Die *Austriae Chorographia* des Wolfgang Lazius

Einführung

Wolfgang Lazius, einer der bedeutendsten österreichischen Kartographen des 16. Jahrhunderts, wurde am 14. Oktober 1514 als Sohn des aus Stuttgart stammenden Mediziners Simon Lazius und der Wienerin Ottilie (ehemals Schallauczer) in Wien geboren. Er studierte ab 14. April 1528 an der Wiener Artistenfakultät und setzte seine Studien nach Erlangung des Magisteriums (1532) an den Medizinischen Fakultäten der Universitäten Wien und Ingolstadt fort. 1542 wurde Lazius zum Universitätsprofessor für Medizin an der Universität Wien berufen. Zweimal bekleidete er das Amt des Rektors der Universität (1546 und 1560). Zudem war er Leibarzt Ferdinands I., Hofhistoriograph und Leiter der kaiserlichen Münzen- und Antiquitätensammlung. Lazius' Vorliebe galt der Geschichte und der Kartographie. In diesen Bereichen hat er ohne Zweifel ein großartiges Lebenswerk hinterlassen, wie beispielsweise die erste publizierte Geschichte der Stadt Wien (*Vienna Austria*, 1546), umfangreiche Abhandlungen zu den Habsburgern (*Commentariorum in Genealogiam Austriacam libri duo*, 1564; *Arbor genealogiae Austriacae*, 1564/65), Monographien zur alten Geschichte (*Commentariorum Reipublicae Romanae illius in exteris provinciis bello acquisitis constitutae libri XII*, 1551; *Commentariorum vetustorum numismatum maximi scilicet operis et quatuor sectionibus multarum rerum publicarum per Asiam, Aphricam et Europam antiquitatis historiam nodosque Gordianis difficiliores comprehendentis* 1558; *Exempla aliquot S. vetustatis Rom.*, 1560 usw.) und den ersten

Atlas der österreichischen Lande (*Typi chorographici provinciarum Austriae*, 1561). Von ihm selbst verfasste größere medizinische Abhandlungen sind hingegen nicht überliefert.[1]

Dieser Artikel beschäftigt sich im Speziellen mit der Karte *Austriae Chorographia*, die Lazius 1563 angefertigt hatte, heute aber nur mehr als ein Nachdruck aus dem Jahr 1620 erhalten ist. Dieser Nachdruck wird in der Sammlung Woldan unter der Signatur K-V(Bl): OE 20 verwahrt. Bis heute ist diese Karte lediglich bei Franz Wawrik in seinem 2004 publizierten Artikel über die österreichische Kartographie der frühen Neuzeit[2] etwas eingehender besprochen worden. Die Autorin beschäftigt sich seit 2000 mit den kartographischen Arbeiten des Wolfgang Lazius. In diesem Jahr erhielt sie von Univ. Prof. Dr. Johannes Dörflinger die Möglichkeit, im Rahmen eines am Institut für Geschichte der Universität Wien abgehaltenen Seminars einen kurzen Vortrag und eine Seminararbeit über Lazius zu verfassen.

Wolfgang Lazius als Kartograph

Von wem und wo Lazius eine Einführung in die Kunst des Kartenmachens erhalten hatte, kann heute nicht mehr ganz nachvollzogen werden. Nach Ernst Bernleithner soll Lazius während des Studiums in Ingolstadt von Peter Apian (1495-1552) unterrichtet worden sein.[3] Doch ist das nur sehr schwer vorstellbar, da Apian im Gegensatz zu Lazius sehr exakte Karten unter Einbeziehung bestimmter Projektionen entwarf. Auch Gemma Frisius (1508-1555), den Lazius wahrscheinlich Mitte der 1530er Jahre während einer

---

1 Zur Biographie des Lazius siehe vor allem: P. Svatek, Wolfgang Lazius. Leben und Werke eines Wiener Gelehrten des 16. Jahrhunderts, in: Wiener Geschichtsblätter 61/1 (2006), 1-22; P. Svatek, Wolfgang Lazius und seine kartographischen Werke. Kartenanfertigung – künstlerische Kartenelemente – wissenschaftlicher Stellenwert der Karten im internationalen Vergleich, in: Mensch – Wissenschaft – Magie. Mitteilungen Österreichische Gesellschaft für Wissenschaftsgeschichte 25 (2008), 5-8.

2 F. Wawrik, Von den Anfängen der österreichischen Kartographie bis zur Zweiten Türkenbelagerung Wiens (1683). in: I. Kretschmer – J. Dörflinger – F. Wawrik, Österreichische Kartographie. Wiener Schriften zu Geographie und Kartographie 15, Wien 2004, 38f.

3 E. Bernleithner, Wolfgang Lazius. Austria, Amsterdam 1972, VI.

Reise in die Niederlande an der Universität Loewen kennen gelernt hatte, kommt als Lehrer eher nicht in Frage. Als wahrscheinlich gilt die These, dass er in Wien von Georg Tannstetter (1482-1535) in das Wesen der Kartographie eingeführt worden ist. Tannstetter bearbeitete zusammen mit Johannes Cuspinian in den 1520er Jahren die Ungarnkarte des Lazarus und gab wahrscheinlich auch eine Karte über Österreich heraus, die heute allerdings verschollen ist.[4] Lazius kannte Tannstetter seit seiner Kindheit. Sein Vater Simon war nämlich Mitglied in der *Sodalitas Collimitiana*, einer Vereinigung Wiener Humanisten unter der Leitung Tannstetters.[5] Es gilt als wahrscheinlich, dass Tannstetter Lazius in diversen Fächern Unterricht erteilt hatte. Aus diesem Grund könnte er von ihm auch Informationen zur Kartenherstellung erhalten haben.

Bis 1563 fertigte Lazius nach unserem heutigen Wissensstand insgesamt 24 Karten und einige Kartenskizzen an, worauf Gebiete zwischen Brandenburg im Norden und Kreta im Süden sowie dem Elsass im Westen und Siebenbürgen im Osten dargestellt sind:[6]

Manuskriptkarten:

- *Charta Chorographia* (1545): aufgelöst in den beiden Blättern *Tabula prima superioris Austriae explicatio* und *Tabula secunda Austriae inferioris explicatio* (Karten über Ober- und Niederösterreich, heute verschollen)
- Zwei Karten über den Schmalkaldischen Krieg (wahrscheinlich Ende 1540er oder Anfang 1550er Jahre entstanden)

Holzschnittkarten:

- *Regni Hungariae descriptio* (große Ungarnkarte 1552/56)

4 F. Graf-Stuhlhofer, Humanismus zwischen Hof und Universität (Schriftenreihe des Universitätsarchivs 8), Wien 1996, 153ff.

5 H. Grössing, Humanistische Naturwissenschaft. Zur Geschichte der Wiener mathematischen Schule des 15. und 16. Jahrhunderts, Baden-Baden 1983, 291; M. Vancsa, Quellen und Geschichtsschreibung. in: Geschichte der Stadt Wien 4, Wien 1911, 5f.

6 Zu den Karten des Lazius siehe: P. Svatek, Wolfgang Lazius und seine kartographischen Werke, 8-31; F. Wawrik, Historische und kulturhistorische Informationen in den Werken österreichischer Kartographen des 16. Jahrhunderts, mit besonderer Berücksichtigung des Wolfgang Lazius, in: D. Unverhau, Geschichtsdeutung auf alten Karten (Wolfenbütteler Forschungen 101), Wiesbaden 2003, 193-212.

- *Rei contra Turcas gestae anno MDLVI brevis descriptio* (Westungarn-Karte, 1556)

Kupferdrucke:

- Marchfeldkarte (undatiert)
- *Austriae Chorographia* (1545, nur noch drei Fragmente erhalten)
- *Chorographia Helladis* (1558, Nordgriechenlandkarte)
- *Peloponnesus Peninsula* (1558, Karte über Südgriechenland)
- *Peloponnesus ex Pausania et Strabone descriptus* (undatierte Peloponneskarte)
- *Austria infra Anisum* (undatierte Niederösterreichkarte)
- Elf Karten der *Typi chorographici provinciarum Austriae* (1561):
    - R. *Austrasia ad Rhenum cum Edelsassia et ducat. Alemaniae* (Elsass)
    - *Regni Francorum orientalis sive Austriae ad Danubium alteri descriptio* (Ober- und Niederösterreich)
    - *Marcha orientalis* (östliches Niederösterreich)
    - *Austria supra Anisum* (Oberösterreich)
    - *Boiorum Regni una cum comitatibus suis descriptio* (Bayern)
    - *Regni veteris Suevorum una cum pagis descriptio* (Schwaben)
    - *Rhetiae alpestris in qua Tirolis com. Description* (Tirol)
    - *Carinthiae ducatus cum palatinatu Goricia* (Kärnten und Görz)
    - *Ducatus Stirae marchiae* (Steiermark)
    - *Ducatus Carniolae et Histriae una cum Marcha Windorum* (Krain)
    - *Principat. Goricens. cum Karstio et Chaczeola descriptio* (Görz)

Austriae Chorographia (1563, Karte über Nieder- und Oberösterreich)

Allgemeine Informationen zur Karte *Austriae Chorographia* und ihre Entstehungsgeschichte

Die Karte *Austriae Chorographia autore Wolfg. Lazio Viennensi Austrio. M. D. consiliar. et Historico Regio*, ein Kupferstich, wurde 1620 von dem in Straßburg lebenden Matthias Bernegger herausgegeben. Das Gesamtformat dieser Karte beträgt ungefähr 102,5 x 47 cm. Das Kartenbild erstreckt sich vom Salzkammergut im Westen bis zur Raab im Osten sowie vom südlichen Böhmen und Mähren im Norden bis zu den Niederen Tauern im Süden.

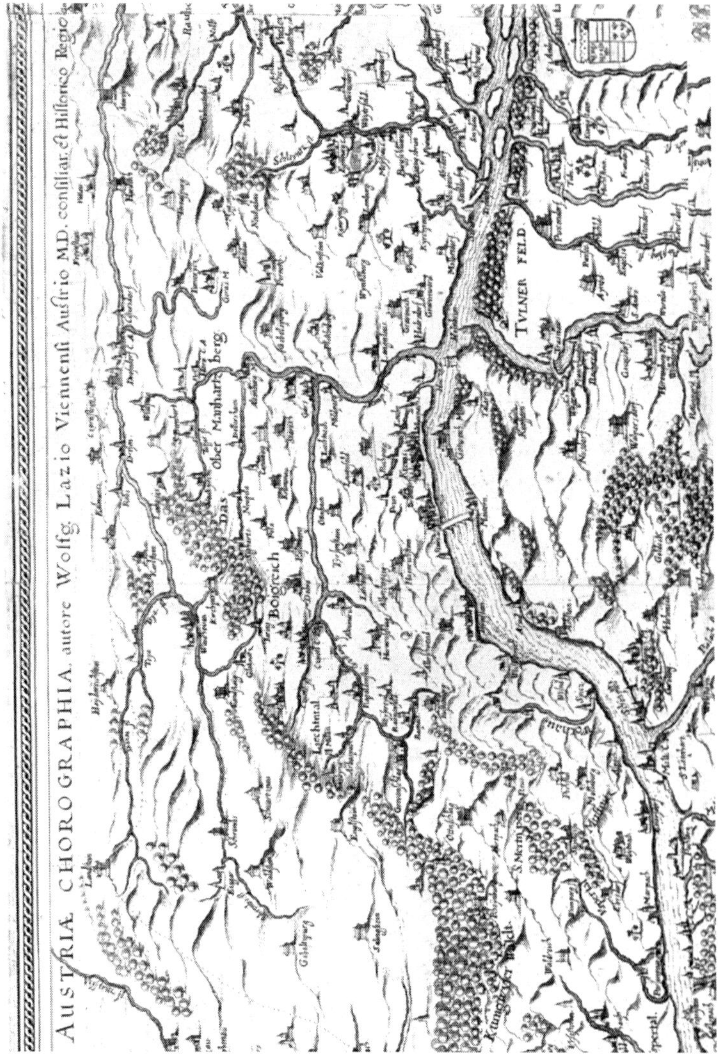

Abb. 1: Ausschnitt nördliches Niederösterreich, südliches Tschechien.
Aus: M. Bernegger – W. Lazius, *Austriae Chorographia*, Straßburg 1620,
ÖAW Sammlung Woldan K-V(Bl): OE 20

Über die Genauigkeit der *Austriae Chorographia* verlor Franz Wawrik im Jahre 2004 bereits einige Worte. Er stellte fest, dass die Donau sehr stark in die Länge gezogen wurde und dadurch die West-Ost-Ausdehnung der Karte gegenüber jener von Nord nach Süd um ungefähr ein Drittel zu groß geraten ist. Eine Maßstabsberechung entlang des Breitenkreises von Ried nach Preßburg ergab einen Wert von ca. 1:300.000, entlang des Meridians von Karlstein bis Bruck an der Mur ungefähr 1:415.000. Extrem kleine Maßstäbe fände man an den Kartenrändern.[7] Auch in den anderen Laziuskarten variiert der Maßstab zum Teil beträchtlich, wodurch viele Lageungenauigkeiten von Städten, Flüssen und Seen zustande gekommen waren (siehe auch unten das Kapitel „Karteninhalt der *Austriae Chorographia* im Kontext anderer Karten des Lazius“).

Lazius fertigte ein Manuskript der *Austriae Chorographia* im Jahre 1563 an und übermittelte es dem Wiener Magistrat. Lazius' Ableben 1565 verhinderte allerdings vorerst den Druck der Karte, bis der protestantische Theologe und Historiker Oseas Schad (1586-1626) aus Straßburg die Handzeichnung wiederentdeckte. Dem Humanisten Matthias Bernegger (1582-1640) haben wir schließlich den Druck der Karte zu verdanken, der von Jakob van der Heyden (1573-1645), einem flämischen Maler und Kupferstecher, besorgt wurde.[8] Matthias Bernegger kam 1582 in Hallstatt zur Welt. Nach seinem Geschichte-, Mathematik- und Sprachenstudium erhielt er in Straßburg einen Lehrstuhl für Geschichte. Freundschaftliche Kontakte pflegte er unter anderem zu Johannes Kepler und Wilhelm Schickard.[9] Die Originalzeichnung des Lazius gilt heute als verschollen.

Lazius standen für die Anfertigung der *Austriae Chorographia* und seiner anderen Karten einige Hilfsmittel zur Verfügung. Erstens benützte er nachweislich Namensverzeichnisse, wie zum Beispiel das große handschriftliche Ortsverzeichnis der österreichischen Lande aus dem beginnenden 16. Jahrhundert. Fritz Eheim konnte nachweisen, dass Lazius aus diesem Verzeichnis Namen in seine Karten integriert hatte, die zu seiner Zeit bereits ungebräuchlich waren, in

---

7 F. Wawrik, Kartographie, 39.

8 Ebd., 39.

9 F. X. Wegele, Mathias Bernegger, in: Allgemeine Deutsche Biographie 2, Leipzig 1875, 412f.

der Topographie aber noch zu lesen sind.[10] Zweitens stützte sich Lazius auf bereits vorhandene Karten, wie beispielsweise auf die verloren gegangene Manuskriptkarte des Augustin Hirschvogel. Aber auch die Werke antiker Autoren (Plinius, Strabon usw.) verwendete Lazius für die Anfertigung seiner Karten als Quelle. Aus diesen entnahm er eine ganze Reihe an lateinischen Orts-, Berg- und Gewässerbezeichnungen. Viertens prägte sich Lazius auf seinen vielen Reisen Landschaften und die Lage der Siedlungen genau ein und fertigte Skizzen an. So könnten vielleicht jene über Carnuntum und Umgebung, den Neusiedlersee und das nieder- und oberösterreichischen Alpenvorland bei der Anfertigung der *Austriae Chorographia* als Hilfsmittel gedient haben.[11]

Kartusche der *Austriae Chorographia*

Der Text der Kartusche südlich des Gesäuses gibt uns über die Entstehung der Karte Auskunft und enthält Namen von Persönlichkeiten, denen die Karte gewidmet wurde. In dieser Namensliste kommen unter anderem Vertreter der Familien Breuner (altes Herrschaftsgeschlecht deutscher Abstammung, das im 14. Jahrhundert nach Österreich eingewandert war)[12], Herberstein (steirisches Geschlecht, welches 1290 die namengebende Burg Herberstein erwarb),[13] Kainach (steirisches Adelsgeschlecht aus dem Raum Voitsberg),[14] Saurau (im 12. Jahrhundert erstmals urkundlich erwähntes Adelsgeschlecht aus dem Raum Unzmarkt)[15] und

---

10 F. Eheim, Die älteste Topographie von Österreich, in: Jahrbuch für Landeskunde von Niederösterreich 33 (1957), 77, 11-13, 17.

11 Lazius' Skizze über Carnuntum ist heute nicht mehr vorhanden. Die Skizze über das nieder- und oberösterreichische Alpenvorland befindet sich in der Handschriftensammlung der Österreichischen Nationalbibliothek Codex 7960 fol. 17v-18r. Die Skizze über den Neusiedlersee wurde bei Bernleithner (1972, s. Anm. 3) abgedruckt (S. XX).

12 J. Siebmacher, Die Wappen des Adels in Salzburg, Steiermark und Tirol, Neustadt 1979, 296f.

13 J. Siebmacher, Die Wappen des Adels in Niederösterreich 1, Neustadt 1983, 181.

14 Ebd., 214f.

15 J. Siebmacher, Die Wappen des Adels in Niederösterreich 2, Neustadt 1983, 22f.

Starhemberg (oberösterreichisches Herrschergeschlecht, welches 1150 zum ersten Mal genannt wurde)[16] vor. Von allen aufgezählten Persönlichkeiten ist heute Siegmund Herberstein (1486-1566) am bekanntesten. Er war Staatsmann, Historiker, Geograph und erster Russlandforscher.[17]

Im vorletzten Absatz der Kartusche erwähnt Bernegger, dass er gegenüber dem originalen Kartenbild nicht viel verändert hätte. Diese Aussage ist allerdings nur schwer vorstellbar, da gegenüber den anderen Karten des Lazius zum Teil sehr wohl große Unterschiede feststellbar sind (siehe dazu das nächste Kapitel „Karteninhalt der *Austriae Chorographia* im Kontext anderer Karten des Lazius"). Auch die Liste der Adeligen dürfte nicht ausschließlich Lazius zuzuschreiben sein, da ein Georg Ulrich Kainach meinen Recherchen zufolge erst im 17. Jahrhundert geboren wurde.[18] Im folgenden Kapitel soll nun die *Austriae Chorographia* mit anderen Karten des Lazius verglichen werden, wobei vor allem Lazius' Karten über Nieder- und Oberösterreich heranzuziehen sind. Es sind dies die 1545 entstandene *Austriae Chorographia*, von der nur mehr drei Fragmente in der Bibliothèque Nationale in Paris erhalten sind, sowie die Karten der *Typi* aus dem Jahre 1561 *Regni Francorum orientalis sive Austriae ad Danubium alteri descriptio* (42,6 x 86,4 cm), *Marcha orientalis* (27,3 x 36, 9 cm) und *Austria supra Anisum.*[19.]

Graduierungsangaben sind auf der *Austriae Chorographia* nicht vorhanden, ebenso wenig eine Maßstabsleiste. Aber auch die anderen Laziuskarten wurden mit Ausnahme der drei zuvor genannten Karten mit solchen Informationen nicht ausgestattet. Angaben zur geographischen Ausrichtung enthalten lediglich die *Austriae*

---

16 J. Siebmacher, Die Wappen des Adels in Oberösterreich, Neustadt 1984, 390f.

17 H. Egghardt, Österreicher entdecken die Welt, Wien 2002, 22-25.

18 J. Siebmacher, Niederösterreich 1, 215.

19 Ausgaben der *Typi chorographici provinciarum Austriae* findet man in Wien heute unter anderem in der Universitätsbibliothek (UB III-259182ES) und in der Kartensammlung der Österreichischen Nationalbibliothek (393221-D-E. Kar). Eine Faksimileausgabe, die von Eugen Oberhummer und Franz von Wieser angefertigt wurde (Wolfgang Lazius' Karten von Österreich und Ungarn aus dem Jahre 1545-1563, Innsbruck 1906) verwahrt die Sammlung Woldan der Österreichischen Akademie der Wissenschaften mit Signatur K-V(L): OE 2322a.

Chorographia von 1545, die *Marcha orientalis* und die *Rei contra Turcas gestae anno MDLVI brevis descriptio.*

Karteninhalt der *Austriae Chorographia* im Kontext anderer Karten des Lazius

Dekorative Ausgestaltung

Die künstlerische Ausgestaltung der *Austriae Chorographia* fiel gegenüber allen anderen Lazius-Karten etwas dürftig aus. Neben der eher schmucklosen Randbordüre und einigen wenigen Ornamenten der Kartusche wurden innerhalb des Kartenbildes die Wappen von Bayern, Oberösterreich, Böhmen, Mähren, Ungarn, Steiermark und Österreich bzw. Niederösterreich abgebildet. Lazius hätte wahrscheinlich den Kartenrand und vor allem auch die Kartusche etwas schöner gestaltet: Bis auf die kleine Marchfeldkarte (15 x 3,5 cm) können in allen anderen Laziuskarten mehr dekorative Elemente gefunden werden. So fallen zum Beispiel auf der *Austriae Chorographia* von 1545 auf allen drei Fragmenten neben einigen Wappen Kartuschen auf, in denen jeweils Engel die Wappen der dargestellten Region tragen. Besonders viele Ausschmückungen weist die Ungarnkarte *Regni Hungariae descriptio* auf. Der Kartenbetrachter kann dort sieben Kartuschen finden, deren Rahmen Lazius besonders kunstvoll ausgestaltet hatte. In diesen findet man Ornamente, Tier- und Pflanzendarstellungen und Heiligenfiguren sowie einige Wappen. In die drei Geschichtskarten über das antike Griechenland integrierte Lazius alte griechische Münzen, Meeresungeheuer, karavellenähnliche Schiffe, Tempelanlagen und Szenen aus griechischen Sagen (die zwölf Taten des Herakles, Theseus und Minotaurus, Perseus usw.). Besonders kunstvoll gestaltete Lazius die *Typi*-Karten. Das Kartenbild wird dort jeweils vom österreichischen Doppeladler gehalten, auf dessen Schwingen die Wappen der benachbarten Gebiete veranschaulicht werden. Im unteren Bereich bildete Lazius die Wappen jener Adelsfamilien ab, denen er die Karten widmete.[20]

[20] Zu den dekorativen Elementen der Laziuskarten samt deren Symbolik siehe auch: P. Svatek, Wolfgang Lazius und seine kartographischen Werke, 13-17, 25-28.

Geschichtlicher Karteninhalt

Sehr auffällig ist in der *Austriae Chorographia* das Fehlen von geschichtlichem Karteninhalt. Bis auf die bereits zuvor erwähnte Marchfeldkarte enthalten alle Laziuskarten geschichtliche Eintragungen oder sind sogar als Geschichtskarten (Karten über den Schmalkaldischen Krieg, Griechenlandkarten, *Rei contra Turcas gestae anno MDLVI brevis descriptio*, *Regni Francorum orientalis sive Austriae ad Danubium alteri descriptio*, *Marcha orientalis*) konzipiert worden. Besondere Aufmerksamkeit schenkte Lazius der Eintragung alter Völker- und antiker Gebietsbezeichnungen sowie mittelalterlicher Schlachten. Auf der *Regni Francorum*-Karte, die dem Titel nach jene von Karl dem Großen um 800 begründete Mark an der Donau darstellen soll, können zum Beispiel insgesamt 18 Hinweise auf Schlachten gefunden werden. Die *Marcha orientalis*, in der Lazius jene nach der Schlacht am Lechfeld (955) unter Otto I. errichtete Mark abbilden wollte,[21] beinhaltet römische Stadtnamen (*Comagenum*, *Carnuntum*), Hinweise auf den heiligen Severin und auf ein römisches Munizipium bei Zeiselmauer sowie römische Gebietsbezeichnungen (*Pannoniae superioris pars, Noricum ripense*). Nahe der Grenze zu Böhmen scheint auf der *Austria supra Anisum* zum Beispiel der Name des Volkes der *Mugilones* auf, das bereits bei Strabon genannt wurde. Auf den drei Fragmenten der *Austriae Chorographia* von 1545 kann der Kartenbetrachter einen Vermerk zur Schlacht bei Dürnkrut, bei der König Rudolf I. im Jahre 1277 König Ottokar II. von Böhmen besiegen konnte, und einen etwas längeren Text über die Auseinandersetzungen zwischen dem römischen Imperator Tiberius (42 v. Chr. – 37 n. Chr.) und dem König der Sueben und Markomannen (Marbod) sehen.[22]

Geländewiedergabe

Die Geländewiedergabe sieht auf der *Austriae Chorographia* von 1620 etwas anders aus als auf den authentischen Lazius-Karten. Die einzelnen Hügel erscheinen auf der Karte nicht so mächtig wie in

21 F. Wawrik, Historische und kulturhistorische Informationen, 204.

22 Nähere Informationen zu den geschichtlichen Karteneintragungen und den Geschichtskarten siehe: P. Svatek, Wolfgang Lazius und seine kartographischen Werke, 21 ff., bes. 25-31; P. Svatek, Die Geschichtskarten des Wolfgang Lazius – Die Anfänge der thematischen Kartographie in Österreich, in: Cartographica Helvetica 37 (2008), 37-42.

jenen der *Typi*, was zum Vorteil hat, dass die *Austriae Chorographia* nicht so überladen wirkt. Allerdings gingen dabei einige Feinheiten von Lazius verloren, wie etwa die Unterscheidung von Flysch- und Kalkwienerwald mittels sanften und schroffen Bergen. Die Flyschalpen wurden von Bernegger bei Wien bis auf den Kahlenberg gar nicht abgebildet. Auch die Bergwelt des Salzkammergutes erscheint gegenüber anderen Gebirgslandschaften als nicht besonders schroff. Eine typische Lazius-Darstellung ist das Leithagebirge (*Leyterberg*), das auf der *Austriae Chorographia* wie auf der *Regni Francorum*-Karte fälschlicherweise als kleiner Hügel zwischen Schwadorf (*Schadorf*) und Rohrau (*Roraw*) erscheint. In Wirklichkeit erstreckt sich dieses Gebirge vom Raum Bruck an der Leitha (*Pruck*) bis in die Umgebung von Müllendorf und Hornstein nahe Ebenfurth (*Ebnfurt*). Insgesamt sind in der *Austriae Chorographia* 26 Berge mit Namen versehen worden.[23] Auch viele Täler haben Namen eingezeichnet.[24]

Abb. 2: *Austriae Chorographia* (1620) – Ausschnitt Wiener Becken, ÖAW-Sammlung Woldan

23 *Artzberg, Der Peilstain, Der Pijen, Der Semering, Der Strenberg, Der Yrach, Drasmperg, Draunstaÿn, Eisnperg, Golichperg, Hardperg, Haÿnburgerperg, Kauneck, Kaunperg, Khÿrnperg, Kulnperg, Leÿsperg, Leÿterberg, Massamperg, Puechperg, S. Mertnperg, Schneperg, Tanberg, Teÿfl steÿg, Vetzerperg, Wienherberg.*

24 *Celltal, Christental, Enstal, Gastental, Hakntal, Kucheltal, Liechtntal, Muertal, Muertztal, Petzental, Pielachtal, Pottal, S. Bernhartsthal, Saltztal, Siedingtal, Tragösstal, Trattigisttal, Turnautal, Uspertal* (2x), *Vaÿtztal, Zÿmichertal.*

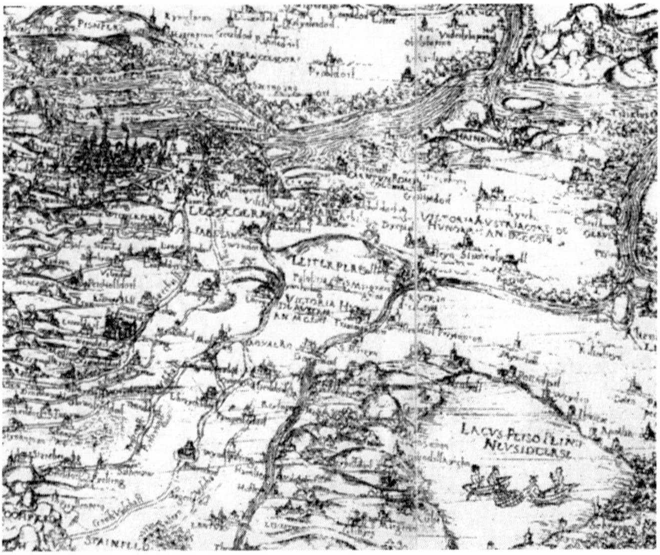

Abb. 3: *Regni Francorum orientalis sive Austriae ad Danubium* (1561), Ausschnitt Wiener Becken[25], ÖAW-Sammlung Woldan

Flussdarstellung

Das Hauptelement für die Orientierung bilden auf den Laziuskarten die Flüsse, die sehr zahlreich eingetragen worden sind. Gemäß der Zeit sind sie allerdings viel zu breit geraten, was man in der Karte sehr gut bei Donau, March, Enns und Traun erkennen kann. Die Laufrichtungen der Flüsse sind auf der *Austriae Chorographia* meist etwas anders eingetragen worden als in den *Typi*-Karten. Doch entsprechen auch hier viele nicht den realen Gegebenheiten, wie beispielsweise die Traun (*Draun fl.*), deren Unterlauf eher von Süd nach Nord gerichtet ist anstatt von Südwest nach Nordost, oder die Pielach, die auf der *Austriae Chorographia* statt von Südwest nach Nordost von Südost nach Nordwest fließt. Einige Fehler konnten gegenüber den *Typi* -Karten nicht korrigiert werden. So mündet die Laming auch hier in die Mur (*Muer fl.*) statt in die Mürz und die Ischl in den Traunsee. Besonders viele Fehler können im

25 E. Oberhummer – F. v. Wieser, Wolfgang Lazius' Karten von Österreich und Ungarn aus dem Jahre 1545-1563, Innsbruck 1906, ÖAW Sammlung Woldan K-V(L): OE 2322a.

Wiener Becken und im Wienerwald gefunden werden. Der Mauerbach entspringt auf der *Austriae Chorographia* bei Mauerbach und mündet fälschlicherweise im Bereich von Greifenstein in die Donau. Betrachtet man eine moderne Karte, so ist die Mauerbachquelle beim Tulbingerkogel zu sehen. Die Mündung erfolgt bei Weidlingau in den Wienfluss. Weiters mündet die Piesting (*Piesting fl.*) in die Triesting (*Driesting fl.*) anstatt bei Gramatneusiedl in die Fischa. Die Triesting mündet statt in die Schwechat in die Fischa usw.

Auf der *Austriae Chorographia* ergeben sich nun zusätzlich neue Fehler, wie zum Beispiel der Verbindungsfluss zwischen Traun- und Attersee. Die Mündung der Ager nahe Lambach konnte hingegen verbessert werden. Doch scheint nun Lazius Ager und Vöckla (*Vekla*) als einen einzigen Fluss angesehen zu haben. Insgesamt findet man 59 verschiedene Flussnamen in der *Austriae Chorographia* gefunden werden.[26]

Seendarstellung

Die Seen haben sich an Aussehen und Gestalt gegenüber jenen in den *Typi*-Karten kaum verändert. Der Traunsee (*Gmundersee*) ist noch immer der größte Salzkammergutsee, auch wenn er nicht mehr so lang gezeichnet wurde wie etwa auf der R*egni Francorum-Karte*. Der Attersee wurde in der *Austriae Chorographia* etwas breiter dargestellt, der Mondsee etwas länger. Die Umrisse des Neusiedlersees (*Neusydlersee*) ähneln vor allem der Abbildung in der *Regni Hungariae descriptio*. Allerdings sind die beiden kleinen Landzungen bei Rust und Illmitz (*Ilmicz*), die eine Verengung des Sees zur Folge haben, auf keiner anderen Lazius-Karte zu finden. Seenbezeichnungen kommen in der Karte insgesamt acht vor.[27]

---

26 *Alm fl., Antzespach fl., Blanitz fl., Daxen fl., Dintnpach, Drasam fl., Draun fl., Driesting fl., Dyming fl., Ens fl., Erlauh fl., Gentnpach fl., Gostnbach fl., Groskamp fl., Gyhs fl., Hirspach fl., Inczespach fl., Ips fl., Kaltngang fl., Kamp fl., Krembs fl., Lampach fl., Lauffnitz fl., Launitz fl., Lauspach fl., Markh fl., Mays fl., Melck fl., Muer fl., Muhel fl., Myda fl., Narn fl.* (2x), *Palthen fl., Pielach fl., Piesting fl. Pulka fl., Pynka fl., Raab fl., Rabnicz fl., Ramÿngpach fl., Rechnitz fl., Rorbach fl., Rottl fl., Schleyntz fl., Schrembs fl., Schwechat fl., Steyr fl., Sultz fl., Teya fl., Tuln fl., Tyerna fl., Varnpach, Vekla fl., Velt aÿst, Veystritz fl., Vischa fl., Waldeÿst, Wien fl., Zeya fl.*

27 *Atter See, Aussee, Gesterm See, Gmundersee, Helwisersee, Neusydlersee, S. Wolfg. See, Seewise.*

Abb. 4: *Regni Francorum orientalis sive Austriae ad Danubium* (1561), Ausschnitt Salzkammergut, ÖAW-Sammlung Woldan

Straßen- und Grenzeintragungen

An Liniensignaturen sind in der *Austriae Chorographia* Straßen und Grenzen abgebildet worden. Zwei Straßenzüge kann der Kartenbetrachter im Nordwesten des Kartenblattes finden. Sie sind mit den Namen *Gulden Steg* und *der Neuwe weg* versehen worden. Der „Goldene Steig" war jahrhundertelang ein bedeutender Handelsweg für den Salztransport von der Donau zur Moldau[28]. Zusätzlich wird in der Karte auf Straßen verwiesen, welche zu Städten außerhalb des Kartenblattes führen. So liest man zum Beispiel nahe Körmend *Weg auff Scharwar* (Weg nach Sárvár) und auf der Schütt-Insel *Weg auf Komorn* (Weg nach Komárno). Straßensignaturen integrierte Lazius auch in die *Marcha orientalis* und die *Austriae Chorographia* von 1545 (vor allem im Fragment Nordwest). Diese geben allerdings in allen drei Karten keine Auskünfte über Entfernungen.

Als Neuerung erscheinen auf der *Austriae Chorographia* Grenzen in Form punktierter Linien, die sonst auf keiner anderen Lazius-Karte eingetragen wurden. Sehr deutlich sind diese beispielsweise zwischen

---

28 F. Wawrik, Kartographie, 39.

Böhmen und Oberösterreich erkennbar. Auf den anderen Laziuskarten wird lediglich auf Grenzen hingewiesen, wie beispielsweise am Ausschnitt Nordost der 1545er Karte *Hungariae et Moraviae fines* sowie am Fragment Nordwest im Tal der Kleinen Mühl *Limites Austriae et Pataviae.*

Siedlungsdarstellung

Die Siedlungen erscheinen auf der *Austriae Chorographia* durch Vignetten mit Mauern, Türmen und Häusern markiert. Im linken unteren Eck findet der Kartenbetrachter nahe dem Attersee einen Zeichenschlüssel (Abb. 5), der Auskunft über unterschiedliche Siedlungsgrößen – unterschieden werden *Statt, Marck, Closter, Schloß* und *Dorff* – und das Aussehen der Signatur für den Weinbau (32mal in der Karte vorhanden) gibt. Von den anderen Laziuskarten enthalten lediglich die *Austriae Chorographia* von 1545, die *Regni Hungariae descriptio* und die Karte über den Süddeutschen Feldzug des Schmalkaldischen Krieges einen Zeichenschlüssel.

Abb. 5: *Austriae Chorographia* (1620) – Ausschnitt Salzkammergut, ÖAW-Sammlung Woldan

Betrachtet man die einzelnen Signaturen in der Karte, so fallen vor allem bei den Städten Variationen auf. Vor allem die Großstädte haben zum Teil sehr individuelle Charakteristika. Eine besonders schöne Ausführung erhielt die Stadtvignette von Wien, in der man sogar einzelne Sehenswürdigkeiten der Stadt ausfindig machen kann. So sticht vor allem der Stephansdom ins Auge. Aber auch die Hofburg kann der Kartenbetrachter im südlichen Teil der kleinen Stadtansicht erkennen. Auch andere Städte haben eine individuelle Eintragung erfahren, wie zum Beispiel Linz: Hier ist etwa der Pöstlingberg durch seine Kirche angedeutet worden. In Bratislava (*Preßburg*) ist die Burg gut erkennbar.

Die Siedlungsvignetten sind auf der *Austriae Chorographia* sorgfältiger ausgeführt worden als auf allen anderen Laziuskarten. Das hängt mit der guten Qualität des Kupferdruckes zusammen. Die anderen Karten, die in der Regel Radierungen waren, stellte Lazius selbst mit der Begründung her, dass es zu seinen Lebzeiten in Wien keine geeigneten Kupferstecher gab.[29] Seine Ausführung des Kupferdrucks kann aber zum Großteil als äußerst mangelhaft bezeichnet werden. Dadurch wird das Lesen seiner Karten erheblich erschwert. In den Abbildungen 1 bis 4 ist sehr schön der Unterschied zwischen dem Druck des Lazius und jenem Jakob van der Heydens zu erkennen.

Andere Eintragungen (Weinbau, Brücken, Wald)

Dass Lazius als ein Weinliebhaber bezeichnet werden kann, ist durch diverse Briefe belegbar. So bat er zum Beispiel Bischof Friedrich Nausea mehrmals um die Übersendung von *zwei bis drei Krügen heurigen Weines.*[30] In fast allen seinen Karten trug er die Verbreitung des Weinbaus durch eine Signatur in Form eines kleinen Weinstockes mit Blättern und Trauben ein. Auch in der *Austriae Chorographia* sind solche Signaturen erkennbar, wie zum Beispiel im südlichen Wiener Becken nahe dem Wienfluss oder südlich des Ortes Himberg (*Hymperg*). Sonstige wirtschaftsgeographische Hinweise sind auf der *Austriae Chorographia* nicht vorhanden. In manchen anderen Karten hat Lazius noch auf Bergbauaktivitäten hingewiesen.

---

29 W. Lazius, Commentariorum rerum Graecarum libri II, Viennae 1558, unpaginiert.

30 M. Mayr, Wolfgang Lazius als Geschichtsschreiber Österreichs. Ein Beitrag zur Historiographie des 16. Jahrhunderts, Innsbruck 1894, 76.

Abb. 6: *Austriae Chorographia* (1620) – Ausschnitt Donautal bei Linz, ÖAW-Sammlung Woldan

Brückensignaturen kann der Kartenbetrachter in der *Austriae Chorographia* 33 finden. Lazius bildete sie über die Flüsse Enns, Traun, Ager, Moldau, Luschnitz, Donau, March, Thaya, Mur, Traisen, Tulln, Wien und Leitha ab. Auch alle anderen Laziuskarten enthalten Brückensignaturen. Manche von ihnen sind sogar mit Namen versehen worden, wie zum Beispiel in der *Marcha orientalis* die Brücken bei Wien mit *Langnpruck* und *Wolfspruck*.

Als wichtiges Element der Naturlandschaft enthält diese Karte, wie auch alle anderen Karten des Lazius den Wald. Dabei handelt es sich in der Regel nicht um bloße Flächenfüllungen, sondern um die Wiedergabe realer Waldgebiete, die zum Teil auch mit Namen versehen worden sind.[31] Zwischen Laub- und Nadelwald gibt es keine Unterscheidung. Die Schreibweise von Namen scheint Bernegger seiner Zeit angepasst zu haben. Lazius schrieb beispielsweise das Wort *Wald* nie mit *-dt*, was auf der *Austriae Chorographia* aber einige Male zu lesen ist.

31 Czerwald, Im lange wald, Künigswyser Waldt, Passawer Wald, Wiener Walt.

Schlussbetrachtung

Die *Austriae Chorographia*, welche Lazius 1563 als Manuskript angefertigt hatte, uns heute aber nur in einem Nachdruck aus dem Jahre 1620 erhalten ist, enthält viele charakteristische Elemente, die auch in anderen Laziuskarten zu finden sind. Als Beispiel sei hier auf die Fluss- und Seendarstellung, auf die Eintragung des Leithagebirges oder etwa auf die Ausweisung bedeutender Weinanbaugebiete mittels einer Weintraube verwiesen. Allerdings kann angenommen werden, dass Bernegger den Karteninhalt etwas generalisierte. So ist zum Beispiel das Nichtvorhandensein von geschichtlichen Eintragungen für Laziuskarten sehr untypisch. Auch die dekorative Ausgestaltung und die Geländewiedergabe lassen auf eine Generalisierung schließen. Dafür wirkt die *Austriae Chorographia* von 1620 im Gegensatz zu den meisten anderen Laziuskarten nicht so überladen, wodurch sie für den Kartenbetrachter leichter lesbar ist. Der große Vorteil der Karte ist auch die gute Qualität des Druckes.

Lazius' bzw. Berneggers *Austriae Chorographia* wurde im 17. Jahrhundert von niederländischen Kartenproduzenten als Vorlage für deren Karten über Nieder- und Oberösterreich verwendet. So erschien sie ab 1630 in den Atlanten von Willem Janszoon Blaeu[32] und später seines Sohnes Joan, unter anderem auch im *Atlas Maior*, der Grundlage des weltberühmten *Atlas Blaeu – Van der Hem* wurde. Diese erweiterten das Blatt nach Norden und Süden und gaben der Karte eine Randgraduierung. Auffallend ist auch eine dem barocken Empfinden entsprechende dekorative Ausgestaltung rund um die Kartuschen. Der Inhalt der Karte divergiert gegenüber der 1620er Ausgabe hingegen kaum.

Literaturverzeichnis

Bernleithner E., Wolfgang Lazius. Austria, Amsterdam 1972.

Blaeu W. J.., Appendix Theatri A. Ortelii et Atlantis G. Mercatoris continens Tabulas Geographicas diversarum Orbis regionum nunc primum editas cum descriptionibus, Amsterdam 1631.

32 W. J. Blaeu, Appendix Theatri A. Ortelii et Atlantis G. Mercatoris continens Tabulas Geographicas diversarum Orbis regionum nunc primum editas cum descriptionibus, Amsterdam 1631, Bl. 81.

Egghardt H., Österreicher entdecken die Welt, Wien 2002.

Eheim F., Die älteste Topographie von Österreich, in: Jahrbuch für Landeskunde von Niederösterreich 33 (1957), 7-25.

Graf-Stuhlhofer F., Humanismus zwischen Hof und Universität (Schriftenreihe des Universitätsarchivs 8), Wien 1996.

Grössing H., Humanistische Naturwissenschaft. Zur Geschichte der Wiener mathematischen Schule des 15. und 16. Jahrhunderts, Baden-Baden 1983.

Lazius W., Commentariorum rerum Graecarum, Viennae 1558.

Mayr M., Wolfgang Lazius als Geschichtsschreiber Österreichs. Ein Beitrag zur Historiographie des 16. Jahrhunderts, Innsbruck 1894.

Siebmacher J., Die Wappen des Adels in Salzburg, Steiermark und Tirol, Neustadt 1979.

Siebmacher J., Die Wappen des Adels in Niederösterreich, Bd. 1-2, Neustadt 1983.

Siebmacher J., Die Wappen des Adels in Oberösterreich, Neustadt 1984.

Svatek P., Wolfgang Lazius. Leben und Werke eines Wiener Gelehrten des 16. Jahrhunderts, in: Wiener Geschichtsblätter 61/1 (2006), 1-22.

Svatek P., Wolfgang Lazius und seine kartographischen Werke. Kartenanfertigung – künstlerische Kartenelemente – wissenschaftlicher Stellenwert der Karten im internationalen Vergleich, in: Mensch – Wissenschaft – Magie. Mitteilungen Österreichische Gesellschaft für Wissenschaftsgeschichte 25 (2008), 1-39.

Svatek P., Die Geschichtskarten des Wolfgang Lazius – Die Anfänge der thematischen Kartographie in Österreich, in: Cartographica Helvetica 37 (2008), 35-43.

Vancsa M., Quellen und Geschichtsschreibung. in: Geschichte der Stadt Wien IV, Wien 1911.

Wawrik F., Historische und kulturhistorische Informationen in den Werken österreichischer Kartographen des 16. Jahrhunderts, mit besonderer Berücksichtigung des Wolfgang Lazius, in: Unverhau D. (Hrsg.), Geschichtsdeutung auf alten Karten (Wolfenbütteler Forschungen 101), Wiesbaden 2003, 193-212.

Wawrik F., Von den Anfängen der österreichischen Kartographie bis zur Zweiten Türkenbelagerung Wiens (1683), in: Kretschmer I. – Dörflinger J. – Wawrik F., Österreichische Kartographie. Wiener Schriften zu Geographie und Kartographie 15, Wien 2004, 11-73.

Wegele F., Mathias Bernegger, in: Allgemeine Deutsche Biographie 2, Leipzig 1875, 412 f.

Zur Autorin

Petra Svatek, geb. 1976, Studium der Geschichte und Geographie an der Universität Wien mit Schwerpunkt Wissenschaftsgeschichte, seit 2006 Projektmitarbeiterin am Institut für Geschichte der Universität Wien, Hauptforschungsbereiche: Geschichte der Kartographie (19. und 20. Jahrhundert, Wolfgang Lazius), Geschichte der Geographie (19. und 20. Jahrhundert), Geschichte der Raumforschung und Landesplanung im mitteleuropäischen Raum (20. Jahrhundert).

Kontakt: petra.svatek@univie.ac.at

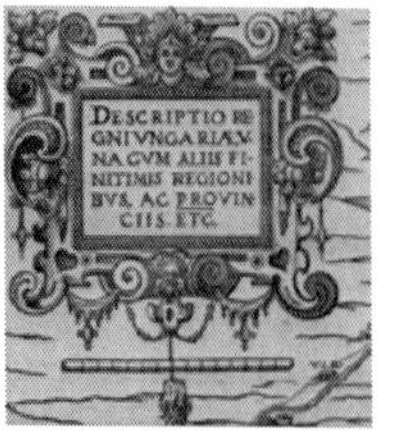

Anonymus,
Descriptio Regni Ungariae,
s.l. ca. 1600

Zsolt Győző Török, Budapest

Die „ungarische" Woldan-Karte und ihre kartographischen Vorgänger im 16. Jahrhundert

Eine unbekannte Kartentradition in Zentraleuropa zur Zeit der Türkenkriege

Die Sammlung Woldan der Österreichischen Akademie der Wissenschaften besitzt zahlreiche Landkartenkuriositäten. Das Werk *Descriptio Regni Ungariae, una cum aliis finitimis Regionibus, ac Provinciis etc.* ist eines davon.[1]

In unserer Studie versuchen wir zu der in der ungarischen Fachliteratur als „Woldan-Karte" bekannten Mappa einige neue Aspekte beizutragen. Der Umfang dieses Artikels gestattet uns jedoch keine eingehende Analyse. Ich bin aber der Meinung, dass der kartenhistorische Kontext zum besseren Verständnis des hervorragenden mitteleuropäischen Kartenwerkes des 16. Jahrhunderts und seiner Vorgänger helfen kann. Meines Erachtens verrät eine solche Karte vieles über der Kartographie des 16. Jahrhunderts, was bisher aber von der Kartographiegeschichte dieser Region nicht beachtet worden ist. Im Folgenden soll deshalb der Versuch unternommen werden, die Verbindung der „Woldan-Karte" mit der Kriegskartierungs-

---

1 Siehe dazu: G. Holzer – J. Dörflinger, Cartographic Rarities of the 'Woldan Collection' late 15th to early 19th centuries. 16th International Conference on the History of Cartography. Exhibition catalogue. Austrian Academy of Sciences, Vienna, 1995, 10, No 22; T. Szathmáry, Desriptio Hungariae. Magyarország és Erdély nyomtatott térképei 1477-1600, Fusignano 1987, 173, No 76; L. Szántai, Atlas Hungaricus 1528-1850, Budapest 1996, 699.

tradition der türkischen Kriege und mit einigen weiteren Karten, die wir während zahlreicher Forschungen in den letzten Jahre näher untersucht haben, aufzuzeigen.[2]

Das einzige bekannte Exemplar der *Descriptio Regni Ungariae* wurde 1936 bei einer Auktion des Wiener Buch- und Kunstantiquariats Gilhofer und Ranschburg für 1200 Schweizer Franken vom österreichischen Sammler Dr. Erich Woldan gekauft.[3] Über den erfolglosen ungarischen Versuch des Ankaufes dieser Karte, und über ihren Verkauf an den „österreichischen Mäzen“ hat in Ungarn Andor Borbély im Dezember desselben Jahres berichtet.[4] 1992 hat der in Paris lebende ungarische Kartensammler Lajos Szántai in der Zeitschrift *Cartographica Hungarica* über „eine wieder entdeckte Wandlandkarte von Ungarn" geschrieben.[5]

Damals, also in den 80er Jahren nach ihrem ersten Auftauchen, wie auch heute, ist das zum Verkauf der Karte von Professor Eugen Oberhummer (1859-1944) verfasste Gutachten gültig: Die Landkarte ist ein Unikat. Dies macht das einzig bekannte Exemplar noch wertvoller. Gleichzeitig soll der interessante Umstand hervorgehoben werden, dass die betreffende Mappa keine Manuskriptkarte, sondern eine gedruckte Arbeit ist. Die Wandkarte von 169 x 89,5 cm Größe wurde von zehn Holzstöcken gedruckt (s. Abb. 4; die Abmessungen der einzelnen Platten betragen 48 cm x 35 cm). Es ist allerdings ungewöhnlich, dass die aus mehreren Teilen bestehende, großformatige Karte für uns nur in einem einzigen Exemplar existiert und es keine anderen Exemplare oder Fragmente von ihr gibt. Auf gleiche Weise ist es merkwürdig, dass die Karte dem Anschein nach mit anderen Kartenwerken keine Verbindung hat und nicht als direkte Quelle für andere Karten herangezogen wurde. Dieses Stück der

---

2 Z. G. Török, Renaissance Cartography in East-Central Europe c.1450-1650, in: D. Woodward (Ed.), Cartography in the European Renaissance. The History of Cartography 3/2, Chicago 2007, 1806-1851.

3 Auktionskatalog von Gilhofer und Ranschburg, Wien 1936, Katalog Nr. 258.

4 A. Borbély, Hazánk ismeretlen térképei a XVI. századból, in: Térképészeti Közlöny 4 (1936), 101-102; vgl. F. Fodor, A magyar térképírás 1, in: Térképészeti Közlöny 15 (1952), 41-42.

5 L. Szántai, Egy újrafelfedezett Magyarország falitérkép (Wiederentdeckung einer unbekannten Wandkarte von Ungarn aus dem 16. Jahrhundert), in: Cartographica Hungarica 1 (1992), 2-3.

Woldan-Sammlung ist also wirklich ein kartenhistorisches Rätsel, obwohl die Forschungen mit der Zeit vieles aufgehellt haben.

Der Titel der Karte ist in der Renaissancenkartusche in der linken Ecke zu finden (allerdings ohne Angaben zum Autor und Erscheinungsjahr). Die erste neuere Studie über diese Karte von Szántai stellt fest, dass sie wahrscheinlich aus der zweiten Hälfte des 16. Jahrhunderts stammt. Er erwähnt darin ein Bild bei Kanizsa, das seiner Meinung nach die Tragödie von Burgkapitän György Thury (1519-1571) darstellt, als er von den Türken in die Falle gelockt wurde. Daher muss die Karte nach 1571 angefertigt worden sein. Auch die durch eine punktierte Linie dargestellte Grenze der türkischen Eroberungen hat die Aufmerksamkeit von Szántai geweckt. Die Datierung hielt er aber noch nicht für eindeutig.[6]

Katalin Plihál hat dagegen in umfangreicheren Studien bewiesen, dass der Vergleich der Grenzlinien mit den zeitgenössischen Kriegsereignissen zu weiteren Informationen führt.[7] An der Darstellung der siebenbürgischen Grenzlinie kann man nämlich sehen, dass sie in südöstlicher Richtung von Kronstadt (*KORONA*) über die Karpaten hinausreicht und bis zur Hauptstadt von Rumänien (*Bukheresk*) geht. Dieses Gebiet wurde 1595 von Zsigmond Báthory und einen Verbündeten, den Vojwoden Mihai Viteazul (1558-1601) und Ştefan Răzvan (?-1595), von den Türken zurückerobert. Die Karte stellt also die politische Lage des Gebietes zwischen 1595-1597 dar.

Szántai sah anhand des Inhalts der Karte eine Ähnlichkeit mit der Karte von Johannes Sambucus aus dem Jahre 1571 als gegeben an, Plihál, die die Ortsnamen gründlich untersuchte, kam zu demselben Ergebnis. Auf Grund der 55, ausschließlich in der Wandkarte vorkommenden Namen und deren Differenzen in der Schriftweise versuchte sie, auf die Hersteller der Karte zu schließen. Sie konnte aber keinen eindeutigen Beweis finden, dass das Werk von einem Ungarn konzipiert worden wäre. Trotzdem vertrat sie die Meinung, dass der Autor dieser Karte zur ungarischen, protestantischen Intelli-

---

6 L. Szántai, Egy újrafelfedezett Magyarország falitérkép, in: Cartographica Hungarica 1 (1992), 3.

7 Plihál, Hazánk ismeretlen térképe a XVI. Század végéről (Eine unbekannte Landkarte Ungarns vom Ende des 16. Jahrhunderts), in: Cartographica Hungarica 3 (1993), 32-41.

genz gehört hat, denn für sie ist es unvorstellbar, dass ein nicht-ungarischer Verfasser dem Werk einen solchen Titel gegeben hätte.[8]

Im Titel der Karte finden wir aber nicht nur das Königreich Ungarn, sondern auch die Nachbarländer ([…] *una cum aliis finitimis Regionibus, ac Provinciis etc.*, vgl. Abb. 1). Das auf der Karte dargestellte Gebiet ist tatsächlich viel größer als das damalige Ungarn und zeigt Regionen – z.B. Bayern oder venezianische Gebiete -, welche nie zum Königreich Ungarn gehörten. Diese Länder waren bis zur türkischen Eroberung des Landes Nachbarländer von Ungarn. Gegen Ende des 16. Jahrhunderts hatte sich die politische Aufteilung dieser Region aber radikal verändert. Die reale, damals aktuelle Lage wird ausschließlich durch punktierte Linien gezeigt. Ansonsten ist die osmanische Eroberung nicht zur Kenntnis genommen.

Abb. 1: Titelkartusche der Karte, ÖAW-Sammlung Woldan

8 Die Übersetzung ihrer Feststellung lautet: *Es gibt nämlich kein Beispiel dafür, dass eine auf österreichischem, bayerischem oder gleich welchem anderen deutschsprachigen Gebiet lebende Person einer Mappa, die eigentlich das Reich des römisch-deutschen Kaisers Rudolf II. darstellt, den Titel gegeben hatte, der sich auf unserer Karte findet*, vgl. Plihál, Hazánk, 40-41.

Die Karte scheint schon auf den ersten Blick mehr zu sein als eine aus dieser Zeit stammende Landkarte. Sie stellt ein weitläufiges Gebiet dar, denn sie erstreckt sich im Westen bis Bayern, im Osten bis Moldawien (*Moldaviae Pars*), im Norden bis Böhmen und im Süden bis Genua. Daher besteht zwischen dem Titel der Karte und der Darstellung ein Widerspruch.

Man muss sich die Frage stellen: Warum hat ein Zeichner einer Ungarnkarte ein Gebiet, das sich weit über das Territorium des damaligen ungarischen Königreichs erstreckte, dargestellt? Offensichtlich wollte er ein viel größeres Gebiet eintragen. Eine Antwort können wir wohl erst dann erwarten, wenn wir die historischen Umstände in Betracht ziehen. Der Anlass für die Anfertigung der Karte ist wahrscheinlich in der europäischen Expansion des Türkischen Reiches und seinen Konflikten mit dem Habsburgerreich zu suchen. Daraus folgt, dass wir bei der Interpretation der Woldan-Karte die im Zusammenhang mit den türkischen Kriegen entstandenen Karten nicht außer Acht lassen dürfen.

Im Folgenden wollen wir einen wichtigen Fund für die ungarische Kartographiegeschichte – die von Hans Brichzin 1992 in einem in Dresden liegenden Manuskript-Befestigungsatlas entdeckte Ungarnkarte – näher untersuchen. Auf diesem Blatt, *Ungariae loca precipua* [...], steht nämlich der Name ihres Verfassers, des im habsburgischen Dienst stehenden italienischen Ingenieurs Nicolo Angielini: *per Nicolau(m) Angielu(m) Italu(m).*[9] Beim Studium dieser bisher unbekannten Karte hat Tibor Szathmáry die auffallende Ähnlichkeit bemerkt, die zwischen der Manuskriptkarte von Angielini und der in Wien 1571 von Johannes Sambucus herausgegebenen, gedruckten Karte *Ungariae loca praecipua* [...] besteht. Ein Vergleich dieser beiden Karten mit der Woldan-Karte zeigt das überraschende Resultat, dass der ungarische Teil der Woldan-Karte den Inhalt nach Sambucus, die Orientierung hingegen nach Angielini wiedergibt.[10]

Mit der Angielini-Karte und mit ihrer Datierung habe ich mich bereits vor einigen Jahren beschäftigt, weil ich das in der Fachliteratur vorgeschlagene Jahr 1566 aufgrund der inhaltlichen Elemente für

---

9 H. Brichzin, Eine Ungarnkarte von Nicolaus Angielus, sowie Grund- und Aufrisse ungarischer Festungen aus dem Jahr 1566 im Sächsischen Hauptstaatsarchiv zu Dresden, in: Cartographica Hungarica 2 (1992), 39-43.

10 T. Szathmáry, Nicolaus Angieli Magyarország-térképe, in: Cartographica Hungarica 3 (1993), 2-13.

nicht korrekt hielt. Die Dresdner Karte zeigt nämlich Grundrisse von Befestigungen, die für eine etwa ein Jahrzehnt spätere Datierung sprechen.[11] Das steht im Gegensatz zu der von Tibor Szathmáry vorgeschlagenen Reihenfolge,[12] denn die 1571 von Sambucus gedruckte Karte geht somit der Handschrift von Angielini voran. Es gilt aber auch weiterhin, dass die Quelle beider Karten dieselbe war.

Das bedeutet zugleich, dass in den 1560er Jahren eine solche Darstellung des ungarischen und kroatischen Kriegsschauplatzes in Wien existiert haben musste. Die Struktur dieser zurzeit verschollenen Karte war aber von der 1552-1556 gedruckten Wandkarte von Wolfgang Lazius offenkundig verschieden.[13] Der Titel der lateinischen Variante der letzteren ist *Regni Hungariae Descriptio Vera*. In der teilweise erhaltenen *Descriptio*, die zu ihrer deutschsprachigen Variante geschrieben wurde, wird das Land mit seinen Teilen vorgestellt (*Des Khünigreichs Hungern sampt seinen eingeleibten Landen grundtliche und warhafftige Chorographica Beschreibung*). Lazius hat 1557 aber auch eine gedruckte Kriegskarte über das Gebiet der ungarischen Militärgrenze, die sich von der Drau bis zur Donau erstreckte, herausgegeben.

Abb. 2: Darstellung der Schlacht bei Kanisia (1601), ÖAW-Sammlung Woldan

---

11 Z. Török, Angielini Magyarország térképe: az 1570-es évekből (Die Ungarnkarte von Angielini: aus den 1570er Jahren), in: Cartographica Hungarica 8 (2004), 2-9.

12 Szathmáry, Nicolaus Angieli Magyarország-térképe, 11.

13 Török, Renaissance Cartography, in: Woodward (Ed.), Cartography in the European Renaissance, 1837.

Die Ereignisse des Türkenkrieges wurden nicht nur in Wien, sondern in ganz Europa aufmerksam verfolgt. Deshalb wurden von den wichtigsten Kriegsereignissen auf dem ungarischen Kriegsschauplatz für das große Publikum zahlreiche Karten, Schlachtengemälde und Veduten gezeichnet (vgl. Abb. 2). Das Interesse in Italien an diesen Kartenprodukten zeigt sich an den in Venedig herausgegebenen Drucken: So hatte Matteo Pagano die von Gastaldi konstruierte Karte schon im Jahr 1546 publiziert. Der lange Titel des großen Holzschnittes – *La vera descrittione di tutta la Ungheria, Transilvania, Valachia* [...] – ist eigentlich die Beschreibung der dargestellten Gebiete.

Die Variante der Angielini-Karte aus dem Jahre 1566, welche das Gebiet von Venedig bis Konstantinopel zeigt, stellt den südöstlichen europäischen Kriegsschauplatz aus venezianischer Sicht dar. Von hohem Interesse ist aber, dass der Ausschnitt der Gastaldi-Karte in nördlicher und südlicher Richtung mit dem von der Woldan-Karte dargestellten Gebiet übereinstimmt. Zugleich bemerkt man auch, dass auf der Woldan-Karte das Delta der Donau und folglich auch Konstantinopel nicht mehr dargestellt werden.

Warum stellt nun die Karte nicht nur den ungarischen und kroatischen Kriegsschauplatz und seine unmittelbare Hinterländer (beispielsweise die österreichischen Erblande) dar, sondern auch das zu den inneren Gebieten des Reiches gehörende Bayern? Die eigentliche Quelle für die Darstellung von Bayern war wahrscheinlich die berühmte Karte von Philipp Apian.[14]

Das alles ist zwar keine Neuigkeit, obwohl man solche Spuren auf gedruckten Karten eher selten nachvollziehen kann. Die Angielini-Manuskriptkarte von Ungarn aber, die am westlichen Rand die Inschrift *Bavaria* trägt, kann man auch in dieser Hinsicht als Vorläufer der Woldan-Karte ansehen. Die Ähnlichkeit zeigt sich zum Beispiel in der Darstellung der Burg Szatmár (*Zakmar;* Satu Mare, Romania). Im Falle der Woldan-Karte ist die Darstellung des fünfeckigen Fortifikationsgrundrisses deswegen seltsam, weil Ortschaften ansonsten bloß von einigen bildhaften Zeichen gekennzeichnet werden (vgl. Abb. 3). Die Grundriss-Darstellung der pentagonalen Szatmárer

---

14 Siehe dazu: K. Brunner, Philipp Apians Landesaufnahme Bayerns, in: Zur Geschichte des Vermessungswesens, VDV-Schriftenreihe 8, Wiesbaden 1995, 38-45.

Fortifikation zeigt ebenso wie jene von Kanizsa, Ivanič und Sziget, eine unmittelbare Verbindung mit der architektonischen und kartographischen Tätigkeit der italienische Baumeister und der Familie Angielini.[15]

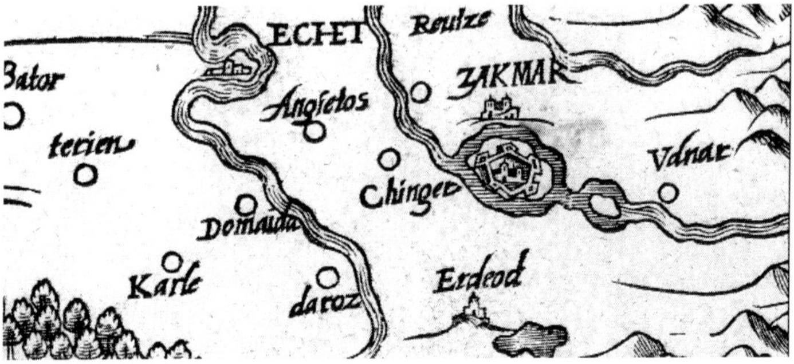

Abb. 3: Grundriss-Darstellung der Festung Szatmár, ÖAW-Sammlung Woldan

Meiner Meinung nach kann man die nach 1595 gezeichnete Woldan-Karte auch in diesen kartographischen Rahmen einfügen. Es scheint, dass der Kartenmacher, der gleichfalls die von Angielini gebrauchte Karte als Grundlage verwendet hatte, dieselbe in westlicher und östlicher Richtung mit jeweils einer Sektion erweitert hat. Diese Erweiterung in die östliche Richtung ermöglichte für den Verfasser, dass er die Ereignisse des siebenbürgischen und rumänischen Feldszuges vom Jahre 1595 darstellen konnte. Auf seiner mit dem Titel *Transylvaniae, Moldaviae, Valachiae, vicinarumque Regionum nova descriptio, Anno MDXCVI ineunte* erschienenen, dekorativen Karte hat Alexander Mair den Siegen von Báthory ein Denkmal gesetzt. Darauf erweitete er das Kartenbild in östlicher Richtung sogar bis zur Krim-Halbinsel. Die Karte von Matthias Quad *Celebrimi Tractus Danubiani pars praecipua ab Austriae Vienna Constantinopoli usque* […] (Köln, 1596) ist eine Zwischenlösung. Die östliche Grenzlinie entspricht bei ihm jener Linie, die ein halbes Jahrhundert früher auf der Gastaldi-Karte zu sehen war. Unten auf dem Blatt, zwischen Kaiser Rudolf II. und dem fälschlicherweise Mohamed II. genannten Sultan, ist das Bildnis von Sigmund Báthory zu sehen. Die Besonderheit dieses Blattes ist, dass der ungarische Teil der Laziuskarte ähnelt, während Quads *Hungaria* das Sambucus-Modell nachahmt.

15 Török, Angielini, in: Cartographica Hungarica 8 (2004), 5.

Die Erweiterung der Karte in westlicher Richtung kann nicht durch historische Ereignisse begründet werden. Aber auch andere derartige Werke wurden in jener Zeit hergestellt, so zum Beispiel die zwei Blätter des *Tractus Danubii* in dem 1578 erschienenen Atlas von Gerard de Jode. Diese Karte hat (wie es ihr Titel auch verrät) die Darstellung der Donau von der Quelle bis zur Mündung zum Thema. In der Kartusche wird der Verfasser der Karte, Christian Sgrooten, der Kartograph Philipps II., erwähnt, dessen Person auf die Ansichten der Habsburger zur Kartenkonstruktion verweisen kann. Unten bei der Kartusche der Woldan-Karte kann man auch die Quelle der Donau (*Danubius Fl.*) sehen.

Zusammenfassend können wir also sagen, dass die Karte *Descriptio Regni Ungariae, una cum aliis finitimis Regionibus, ac Provinciis etc.* ohne Zweifel unter dem Eindruck der Türkenkriege, der Ereignisse der 1595er Jahre und der christlichen Kriegserfolgen hergestellt wurde. Das Unikat aus der Sammlung Woldan ist aber keine Karte eines einzelnen Landes, sondern eine Darstellung des habsburgisch-ottomanischen Kriegsschauplatzes und seiner weiteren Umgebung aus habsburgischer Sicht. Sie trägt somit nicht nur Spuren der gedruckten Werke der Verlagskartographie, sondern auch der früheren Manuskriptkarten der habsburgischen Kriegskartographie aus dem 16. Jahrhundert in sich.

Zum Autor

Zsolt Győző Török, Associate Professor für Kartographie und Geoinformation an der Eötvös Loránd Universität Budapest, Direktor von Imago Muni Ltd. und der *International Commission on the History of Hydrography and Nautical Science.* Forschungsgebiete: Kosmographie der Renaissance, Kartenanfertigung und Fortifikationen im Kontext der Grenzkonflikte zwischen Habsburgern und Türken, frühneuzeitliche Kartographie und Kartographie der Aufklärung in Zentraleuropa, Erforschung und Kartenanfertigung der Libyschen Wüste im 19. und 20. Jahrhundert

Kontakt: zoltorok@ludens.elte.hu

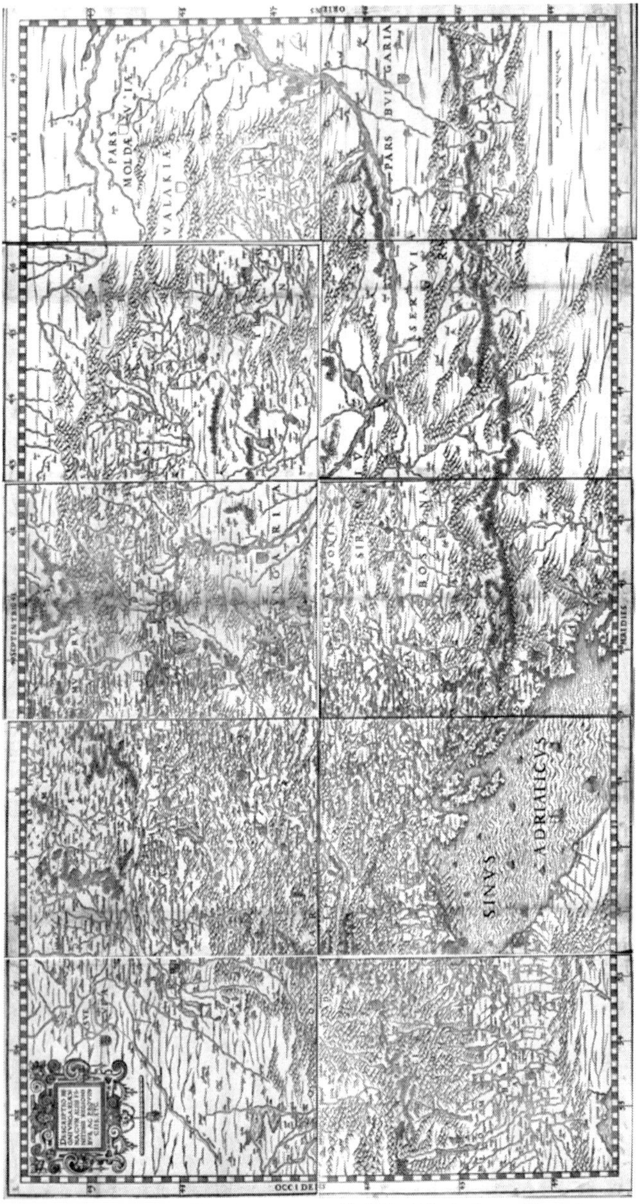

Abb. 4: *Descriptio Regni Ungariae, una cum aliis finitimis Regionibus, ac Provinciis etc.*, ÖAW-Sammlung Woldan

Hartmann Schedel,
Buch der Croniken und geschichten,
Nürnberg 1493

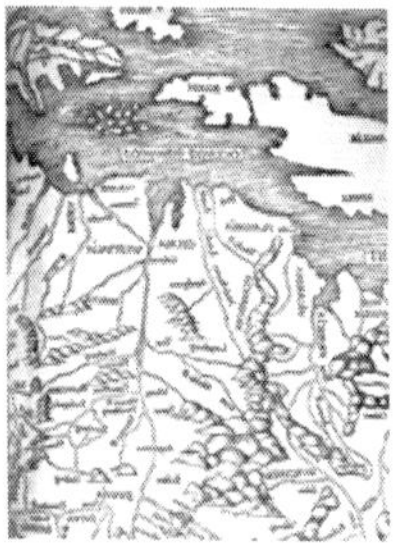

Franz Wawrik, Wien

Hieronymus Münzer

Arzt, Reisender, Geograph, Diplomat

Der aus Vorarlberg stammende, vielseitig begabte Dr. med. Hieronymus Münzer gehörte dem Kreis der Nürnberger Humanisten an, unterstützte Martin Behaim, als dieser in Nürnberg seinen berühmten *Erdapfel* konzipierte, war an der Anfertigung der *Schedelschen Weltchronik* maßgeblich beteiligt und unternahm ausgedehnte Reisen nach Italien, in die Niederlande, vor allem aber auf der Iberischen Halbinsel.[1] Über diese letzte, zum Teil diplomatische Exkursion, die kurz nach der Einverleibung des muslimischen Fürstentums Granada durch das christliche Kastilien erfolgte, verfasste Münzer eine lesenswerte Schilderung. Münzer besaß eine umfangreiche wissenschaftliche Bibliothek, die bedauerlicherweise nicht als Ganzes erhalten blieb: Nur ein Teil von ihr wird in der Stadtbibliothek Feldkirch aufbewahrt.

---

1 J. Fischer, Der Nürnberger Arzt Dr. Hieronymus Münzer († 1508) aus Feldkirch als Mensch und Gelehrter, in: Stimmen der Zeit 96 (1919), 148-168. – ARGE Geschichte und Sozialkunde (Hrsg.), Hieronymus Münzer, ein Kolumbus aus Vorarlberg?, in: Jahresbericht des Bundesgymnasiums Bregenz-Gallusstraße über das Schuljahr 1991/92 (1992), 3-20. – U. Knefelkamp, Artikel ‚Münzer', in: Neue Deutsche Biographie 18 (1997), 557-558.

1. Biographie

Hieronymus Münzer entstammte einer angesehenen und wohlhabenden Feldkircher Familie. Sein Geburtsjahr ist nicht gesichert. In verschiedenen Quellen werden vor allem 1437 oder 1447 angeführt. Für die zweite Angabe spricht, dass Münster demnach sein Universitätsstudium in Leipzig 1464 – also im Alter von siebzehn Jahren begonnen hätte – was in jener Zeit durchaus üblich war – und nicht erst mit siebenundzwanzig. Nach Erlangung des Titels eines *Magisters artium* unterrichtete Münzer an der Leipziger Universität, wobei er eingehend den römischen Dichter Terenz thematisierte. Zusätzlich begann er das Studium der Medizin. 1476 kehrte er nach Feldkirch zurück und war hier kurz als Lehrer an der Lateinschule tätig.[2] Danach begab er sich aber noch im selben Jahr nach Pavia, wo er im ‚Studium generale' sein Studium wieder aufnahm und 1477 mit dem Doktorat der Medizin erfolgreich abschloss.[3]

Bald darauf trat Hieronymus Münzer in Nürnberg den Posten eines *Stadtphysicus* an, eines vom Stadtrat eingesetzten Arztes und Gesundheitsbeauftragten. Die Anstellung erhielt er vermutlich durch Vermittlung seines Bruders Ludwig Münzer († 1518), der in der fränkischen Stadt seit längerem ein florierendes Handelsgeschäft betrieb.[4] Hieronymus war daran finanziell beteiligt. Das Haus der Münzers stand in der Königstraße, schräg gegenüber der Lorenzkirche. Außerdem besaß die Familie das nördlich von Bregenz gele-

---

2 Die Schule, die Vorgängerin des Jesuitengymnasiums, befand sich nahe dem Bludenzer Tor, das 1870 abgerissen wurde. http://www.bgfeldkirch.at/main/geschichte1.htm (eingesehen am 9.9.2009).

3 Bereits seit 825 war Pavia Sitz einer berühmten Rhetorik-Schule, die von Kaiser Lothar I. eröffnet worden war. Während des gesamten Mittelalters blieb sie ein wichtiges Bildungszentrum. Im 11. Jahrhundert ist auch ein rechtswissenschaftlicher Zweig belegt. 1361 schließlich wurde durch Kaiser Karl IV. die Gründung eines *Studium generale* veranlasst, dem Papst Bonifaz IX. dieselben Rechte wie den Universitäten von Bologna und Paris einräumte. Mit kaiserlichem Dekret wurde das Studium generale 1485 förmlich zur Universität erhoben.

4 Vgl. hierzu K. H. Burmeister, Die Brüder Hieronymus und Ludwig Münzer, in: Montfort 53 (2001), 11-28.

gene Schloss Gwiggen (seit der Mitte des 19. Jahrhunderts ein Kloster der Zisterzienserinnen).

Hieronymus verwendete einen beträchtlichen Teil seines Gewinnanteils aus dem Handelsgeschäft für den Ankauf von Büchern, wodurch seine Privatbibliothek auf etwa 250 Bände anwuchs – für die damalige Zeit eine enorme Zahl, wenn sie auch mit dem Umfang der Sammlung des „Bücherfressers" Hartmann Schedel (1440-1514), mit ihren mehr als 720 Druckwerken und Manuskripten, nicht konkurrieren konnte. Münzer besaß allein 24 Bände über Astronomie und Geographie sowie 23 über Medizin.[5] Erste medizinische Werke erwarb er schon 1470 in Leipzig. Seine eigentliche Sammlerleidenschaft ergab sich aber wohl erst während seines Aufenthalts in Pavia, wo 1473 die erste Druckerei eröffnet worden war, als der Buchdruck immer preisgünstigere Bücher ermöglichte. In manche seiner Erwerbungen schrieb Münzer einen Eigentümervermerk, wie z. B.: *Hic liber est mei Hieronymi Monetarii de Feltkirchen, artium et Medicine doctoris, quem mihi comparavi Nuremberge anno Domini 1482.*[6]

1480 erhielt Münzer das Nürnberger Bürgerrecht und heiratete die einer angesehenen „ratsfähigen" Patrizierfamilie angehörende Dorothea Kieffhaber (1457-1505). Aus der Ehe gingen zwei Söhne und eine Tochter hervor, die ebenfalls Dorothea hieß (geb. 1470). Sie ehelichte 1498 Hieronymus Holtzschuher (1469-1529),[7] ein Mitglied einer der angesehensten Nürnberger Familien und späteren Bürgermeister der Stadt. Bis heute berühmte Zeitgenossen Münsters in Nürnberg waren u. a. Veit Stoß (1450-1533), Adam Kraft (1455-1460) und Albrecht Dürer (1471-1528).

Die Zeit vom September 1483 bis zum Jänner 1484 verbrachte Münzer auf der Flucht vor einer in Deutschland extrem wütenden Pestseuche in Italien. Später berichtete er darüber: *Im Jahr 1483 – als eine ansteckende Pestseuche ausbrach, dachte ich, dass daran – wie im Krieg – am ehesten jene nicht sterben, die nicht anwesend sind – und beschloss zu*

---

5 Außerdem 56 Schriften neulateinischer Autoren, 27 theologische und 43 philosophische Werke, 8 Abhandlungen der Jurisprudenz.

6 E. P. Goldschmidt, Hieronymus Muenzer and other fifteenth century bibliophiles, in: Bulletin of the New York Academy of Medicine, 2nd series, 14/8 (1938), 491-508, hier: 491.

7 Artikel ‚Holzschuher' (Familie), in: Neue Deutsche Biographie 9, Berlin 1972, 579.

flüchten.[8] Er verband das Nützliche mit dem Angenehmen, besuchte mehrere Städte, darunter Mailand, Pavia, Bologna, Florenz, Rom und Neapel, und erwarb bei der Gelegenheit auch zahlreiche Bücher. Es ist durchaus möglich, dass Münzer in Italien sein Interesse an Geographie entdeckte; etwa wenn er in Florenz den aus dem Nürnberger Raum gebürtigen Henricus Martellus Germanus († um 1496) und andere wichtige Humanisten, Kartenmaler und Drucker der Stadt, wie Francesco Berlinghieri (1440-1501) und Francesco Rosselli kennenlernte.[9] Zurück in Nürnberg fand Münzer seine zurückgelassene Familie und seine Bediensteten gesund vor und nahm seinen Beruf als Arzt wieder auf. Aber noch 1484 verließ er die Stadt neuerlich. Diesmal reiste er in achtzehn Tagen den Rhein abwärts und dann über Aachen weiter in die Niederlande, die 1477 an Österreich bzw. an das Römisch-Deutsche Reich gefallen waren.

1493 wurde Münzer in den *Größeren Rat* der Stadt gewählt. Diesen ehrenvollen Sitz behielt er bis zu seinem Tod, 1506.

2. Der Nürnberger Humanistenkreis und die ‚Schedelsche Weltchronik'

In den folgenden Jahren wird Münzers Bibliothek mit wichtigen geographischen Druckwerken erweitert, etwa mit der *Geographia* des Ptolemäus, Werken des Pomponius Mela, *Cosmographia sive De situ orbis* (Venedig 1482) und *De chorographia*, sowie mit der *Europa* des Enea Silvio Piccolomini (1405-1464; ab 1458: Papst Pius II.).[10]

Spätestens zu Ende der 1480er Jahre galt Münzer nicht nur bei den Nürnberger Kaufleuten, die Handelsbeziehungen mit ganz Europa unterhielten, und den höchsten Repräsentanten der Stadt, sondern auch beim König (und späteren Kaiser) Maximilian I. als bedeutende Autorität und begehrter Berater in geographischen Fragen. Es kann daher nicht verwundern, dass er um 1490 zur Mitarbeit an zwei Projekten eingeladen wurde: Erstens, zu jenem groß-

8 Goldschmidt, Muenzer, 495-496.

9 L. Böninger, Die deutsche Einwanderung nach Florenz im Spätmittelalter, in: The medieval Mediterranean 60 (2006), 313-348; über Henricus Martellus' Bedeutung als Kartenhersteller siehe 334-348.

10 Goldschmidt, Muenzer, 497. – Pomponius Mela, *De chorographia libri tres* (später *De situ orbis*) – Enea Silvio Piccolomini, *De Europa* (1458).

angelegten christlich-universellen Geschichtswerk, das wir heute unter dem Namen *Schedelsche Weltchronik* kennen, und zweitens, zur Anfertigung des mindestens ebenso berühmten *Behaim-Globus.*[11]

Im Spätmittelalter entwickelte sich Nürnberg zu einer europaweit wichtigen Handelsmetropole und zu einem Zentrum außeruniversitärer naturwissenschaftlicher Forschung. In der Freien Reichsstadt war ein selbstbewusstes Bürgertum entstanden, dessen Oberschicht – beeinflusst von der italienischen Renaissance – in einer lockeren Vereinigung (einer Art Gelehrter Gesellschaft) rege wissenschaftlichen Interessen nachging. Die meisten dieser „Patrizierhumanisten" wohnten in der vornehmen Gegend am Fuß der Burg, *Unter der Veste* (der heutigen Burgstraße). Die Zusammenkünfte fanden häufig im Haus des angesehenen Kaufmanns und Kirchenmeisters der Sebalduskirche, Sebald Schreyer (Clamosus, 1434-1515/1519), statt.

Von ihm scheint um 1489 die Idee zur Anfertigung einer Weltchronik gekommen zu sein. Er und sein Schwager Sebastian Kammermeister waren es auch, die das Unternehmen finanzierten. Sie beauftragten 1491 mit der Herstellung des Werkes den höchst angesehenen „Druckerfürsten" Anton Koberger (um 1450-1513)[12] und für die Anfertigung von mehr als 1800 Illustrationen die beiden hervorragenden Künstler und Holzschneider Michael Wolgemut (1434-1519) und Wilhelm Pleydenwurff (ca. 1460-1494).[13] Möglicherweise arbeitete auch der junge Albrecht Dürer (1471-1528), damals noch Schüler Wolgemuts, an dem Mammutprojekt mit.

Als wissenschaftlicher Leiter konnte Hartmann Schedel (1440-1514) gewonnen werden. Er entstammte ebenfalls einer wohlhabenden Nürnberger Kaufmannsfamilie, hatte als 16jähriger an der Universität Leipzig immatrikuliert und erhielt hier 1463 den Titel eines *Magister artium.* Anschließend studierte er in Padua Medizin und promovierte 1466. Nebenbei besuchte er aber auch astronomische

---

11 H. Schedel, [Chronica] Buch der Cronicen und geschichten mit figure[n] und pildnussen von anbegin[n] der welt bis auf dise vn[n]sere zeit / [Hartmann Schedel]; durch Georgium alten [...] in diss teutsch gebracht, Nürnberg: Anton Koberger [23. Dezember] 1493. – [11], 286, [1] Bl., 1809 Illustrationen. u. 2 Holzschnitt-Karten.

12 Verheiratet in zweiter Ehe mit Margarete Holzschuher, damit verwandt mit Münzer, 1470 Gründung der Druckerei, die er gewinnorientiert betrieb. Filiale auch in Wien.

13 Der Vertrag von 1491 ist noch erhalten.

Vorlesungen und erlernte die hebräische sowie – als einer der ersten Deutschen überhaupt – die altgriechische Sprache (wie dies später auch Willibald Pirckheimer tat). Nach längeren Reisen wirkte Schedel ab 1470 als Stadtarzt in Nördlingen, wo er der Bruderschaft der Kartäuser beitrat, und danach in Amberg. Um 1482 kehrte er endgültig nach Nürnberg zurück und ließ sich hier (wie schon Münzer zuvor) als *Stadtphysicus* nieder. Mit 723 Büchern besaß Schedel die bei weitem größte Privatbibliothek Nürnbergs (Münzer die zweitgrößte), zu der er sogar eigenhändig einen Katalog anlegte. Das Konzept der Arbeitsgruppe für die Weltchronik sah eine breit angelegte Universalgeschichte in „annalistischer" Form vor.[14] Sie sollte, in sechs Zeitalter (*aetates*) gegliedert, die Geschehnisse von der Erschaffung der Welt bis zum Ende des 15. Jahrhunderts (genauer bis 1486) darbieten. Ein siebentes Zeitalter sollte ursprünglich nur eine Vorausschau auf die restliche Zeit bis zum Weltuntergang und Jüngsten Gericht zum Inhalt haben, wurde aber mit Beschreibungen europäischer Städte und Länder erweitert, die erst knapp vor der Drucklegung bei der Redaktion eingelangt waren und allerletzte Geschehnisse schildern. Als Quellen des Gesamtwerkes diente die Bibel, aber auch Werke von Schriftstellern der Antike und des Mittelalters. Aus Schedels Bibliothek könnte auch die ptolemäische Weltkarte stammen, die aus der Inkunabel *Novellae etati ad geographiae* (Venedig 1482) des römischen Kosmographen Pomponius Mela übernommen,[15] aber mit christlichen Elementen, zu einer *Noachiden-*

14 Geschichte der Welt, so wie man sie zu seiner Zeit verstand: Biblische Erzählungen stehen neben geographischen Beschreibungen (Städte, Länder, Inseln, …), die Erwähnung besonderer Naturereignisse (Naturkatastrophen wie etwa der Erdrutsch, der zur Zeit Ludwigs des Heiligen in Burgund 5000 Menschen das Leben kostete) neben manch wundersamer Begebenheit. Der *Liber chronicarum* von Schedel wurde 1493 in zwei verschiedenen Ausgaben veröffentlicht: einer lateinischen und einer deutschen (in der Übersetzung von Georg Alt). Jene wurde etwa in 1500 Exemplaren gedruckt, diese muss etwa 1000 umfasst haben. P. Hoch, Hartmann Schedel, Liber Chronicarum, Nürnberg, Anton Koberger 1493, Metz 2008. – URL: http://www.libri-europae.org/LIBRI/deutsch/buecher/komm_schedel.htm (eingesehen am 23. 5. 2008).

15 Einfache Kegelprojektion wie bei Ptolemäus 1478, 18,5 x 13,5 cm, mit Hinzufügung des „modernen" Skandinaviens. Pomponius Mela wirkte um 40 n. Chr. Seine Arbeit stellt eine Kompilation griechischer Gelehrter dar.

Karte (so benannt nach den Söhnen des biblischen Noah, die die drei Kontinente vertreten) umgestaltet wurde.

Der eigentlichen Weltchronik wurde eine geographisch-historische Beschreibung von Ländern oder Regionen Europas hinzugefügt. Dafür hatte Hartmann Schedel ursprünglich die 1458 von Enea Silvio Piccolomini verfasste Abhandlung *De Europa* vorgesehen, von der er vermutlich ein Exemplar der 1490 in Memmingen in erschienenen Erstausgabe, angefertigt vom angesehenen Drucker Albert Kunne (ca. 1430-1520), erworben hatte. Der italienische Humanist hatte darin kurz vor seiner Wahl zum Papst versucht, geographische und chorographische Vorstellungen der Antike (Ptolemäus, Strabo und Plinius) mit zeitgeschichtlichen Ereignissen (etwa die Bedrohung durch die Osmanen) zu verknüpfen. Schedel dürften jedoch bald Zweifel über die Aktualität der Schrift gekommen sein, so dass er Münzer bat, sie zu überarbeiten. Als Ergänzung zum Text und zu seinem besseren Verständnis fügte Münzer eine Übersichtskarte weiter Teile Europas hinzu.

3. Die Mitteleuropa-Karte des Hieronymus Münzer (1493)

Der doppelseitige Holzschnitt (fol. ccxcvi verso-ccxcvii recto)[16] ist für die Kartographiegeschichte überaus bemerkenswert, obwohl er sehr einfach, ja geradezu spartanisch gestaltet erscheint und nicht einmal einen eigenen Titel aufweist. Dieser findet sich vielmehr auf der Rückseite der Tafel (fol. ccxcvi recto) und gilt offenbar sowohl für die Karte als auch für eine davor platzierte Beschreibung Deutschlands: *Gelegenheit vnd pildnus Germanie oder Teutscher nation.* Ebenso wenig wie diese Bezeichnung entsprechen die heute üblichen Zitier-Titel *Mitteleuropa-Karte*, *Germania-Karte* oder *Deutschland-Karte* dem tatsächlichen Karteninhalt, da die Darstellung eine weitaus größere Region wiedergibt. Viele Fragen im Zusammenhang mit der

16 Diese und alle folgenden Seitenangaben der Schedelschen Chronik beziehen sich auf deren erste deutsche Ausgabe aus dem Jahr 1493. – Zu den Karten der *Schedelschen Weltchronik* der Sammlung Woldan sei auch auf den gleichnamigen Aufsatz von Kurt Brunner in diesem Sammelband S. 63-80 verwiesen.

Münzer-Karte, ihren Vorlagen und ihrem Inhalt sind noch immer ungeklärt.

Die im 15. Jahrhundert verfassten lateinischen Manuskripte der *Geographia* des Claudius Ptolemäus (oder *Cosmographia*, wie die Humanisten die bis 1508 angefertigten Werke nannten) zeigen Mitteleuropa auf den klassischen Tafeln Europa IV (Deutschland von der Ostsee bis zur Donau bzw. vom Rhein bis zur Weichsel) und Europa V (von der Donau bis zur Adria bzw. vom Oberrhein bis zur Theiss). Ein Konzept für eine neue kartographische Darstellung Mitteleuropas, oder besser des Römisch-Deutschen Reiches samt seiner weit in Nachbarregionen reichenden politischen und wirtschaftlichen Interessensphären, wurde zu einer absoluten Notwendigkeit: Es brauchte eine *Tabula moderna Germaniae*!

Einen ersten Versuch zu einer zumindest teilweisen Realisierung unternahm der große deutsche Philosoph und Kurienkardinal Nicolaus Cusanus (Nikolaus von Kues, 1401-1464). Bei den Konzilen, an denen er teilnahm, auf seinen zahlreichen ausgedehnten Reisen (etwa der Visitationsreise durch Deutschland, 1451/1452) und bei seinen Kontakten zu Gelehrten, Politikern und Kirchenfürsten in vielen Teilen Europas dürfte er versucht haben, Unterlagen zu erhalten, die für die Anfertigung einer zeitgemäßen Karte erforderlich waren. Zu seinen Gewährsleuten könnte der polnische Historiograph Jan Długosz gehört haben,[17] dessen zwölfbändiges Hauptwerk *Annales seu Cronicae incliti Regni Poloniae* sowie die *Chorographia Regni* Poloniae (ab 1455) auf kritischen Quellenstudien basierten. Für den südosteuropäischen Raum gab es unter anderem Kontakte zu dem ungarischen Kanzler Janos Vitez.[18] Cusanus erhielt offenbar auch

---

17 Jan Długosz (Johannes Longinus, Joannes Dlugossius, 1415-1480), polnischer Historiker, Diplomat und Chronist, Erzbischof von Lemberg, unter König Kasimir IV. mit diplomatischen Missionen beauftragt. Während des Dreizehnjährigen Krieges verhandelte er mit dem Deutschen Orden (1454-1466), wobei unter seiner Mitwirkung der Zweite Thorner Frieden (1466) zustande kam. Über die Bedeutung Długoszs für die Landeskunde Polens siehe: F. Bujak, Długosz jako geograf [Długosz als Geograph], Studja 34 (Jahresberichte für deutsche Geschichte aus der Zwischenkriegszeit Bde. 1-14, Berichtsjahre 1925-1938), 91-105.

18 Johann Vitez (János Vitéz de Zredna, 1408-1472), als Kanzler des Regenten und späteren Königs Matthias Corvinus mit zahlreichen diplomatischen Aufgaben betraut; 1465 Erzbischof von Esztergom (Gran). Im Auftrag von König Matthias Corvinus, gründete Vitez in Preßburg die Universität

Informationen von Wissenschaftlern, die Mitglieder der Ersten Wiener mathematisch-astronomischen Schule gewesen, oder zumindest in Kontakt zu ihr gestanden waren. So erwarb er 1444 in Nürnberg von einem Nicolas Heybeck († 1444), einem Absolventen der Universität Erfurt, mehrere einschlägige Manuskripte und Vermessungsinstrumente.[19] Gelegentlich lud Cusanus interessante Gäste auch in seine Villa in Rieti ein: 1450 wohnte sogar Georg von Peuerbach (1423-1461) einige Zeit bei ihm.

Höchst wahrscheinlich kannte er auch die Ergebnisse der Koordinatenbestimmungen des ebenfalls der Wiener Schule zuzurechnenden Reinhard Gensfelder (um 1385-1450/57), und auch die prächtige Mappamundi Andreas Walspergers von 1448.[20]

Ob – wie gelegentlich vermutet wurde – der viel beschäftigte Cusanus selbst den Prototyp einer Deutschland-Karte anfertigte, erscheint eher unwahrscheinlich. Manches deutet eher darauf hin, dass er seine Unterlagen später an den bewährten und von Auftraggebern äußerst geschätzten Kosmographen und Kartographen Nicolaus Germanus (ca. 1420-1480), einen aus Deutschland stammenden Benediktinermönch, abgab, der seit etwa 1460 in Florenz tätig war. Dabei hatte er sich innerhalb kürzester Zeit als innovativster Bearbeiter der ptolemäischen *Geographia* (oder *Cosmographia*) beweisen können. Die meisten der zwischen 1460 und 1477 in Florenz angefertigten, meist prächtig ausgeführten ptolemäischen Manuskriptatlanten stammen aus seiner Werkstätte. Nicolaus Germanus, seit 1477 in Rom tätig, schuf tatsächlich die moderne Deutschland-Karte, die vielleicht sogar als *Tabula moderna Germaniae*

Istropolitana. 1471 schloss er sich einer Verschwörung gegen den König an. Dieser ließ ihn in der Burg von Visegrád gefangen setzen und hielt ihn letztlich in Esztergom unter Hausarrest.

19 Zu Nicolas Heybeck siehe: D. B. Durand, Vienna-Klosterneuburger Map Corpus, Leiden 1952, 34-36.

20 Zu Mitgliedern der Ersten Wiener mathematisch-astronomischen Schule siehe: F. Wawrik, Die Beeinflussung der frühen Kartographie durch Johannes von Gmunden und seinen Kreis, in: R. Simek – K Chlench (Hrsg.), Johannes von Gmunden (ca. 1384-1442). Astronom und Mathematiker (Studia Medievalia Septentrionalia 12), Wien 2006, 45-62, 2 Taf. – Andreas Walsperger: http://wwwg.uni-klu.ac.at/kultdoku/kataloge/26/html/2258.htm; http://www.henry-davis.com/MAPS/LMwebpages/245mono.html (beide eingesehen am 15. 12. 2009).

für die für 1478 vorbereitete gedruckte Erstausgabe der ptolemäischen *Cosmographia* gedacht war. Warum sie nie verwendet wurde, ist nicht bekannt.[21] 1491, erwarb die Kupferplatte – noch ungebraucht – der junge Augsburger Patrizier Conrad Peutinger (1465-1547) in Italien und brachte sie nach Deutschland, wo man in einer Offizin in Eichstätt erste Probeabzüge herstellte (*Eichstätt-Karte*). Ein über der Karte befindliches Gedicht, vielleicht von Nicolaus Germanus verfasst, beginnt mit ihrem Namen: *Quod picta est parva Germania tota tabella* [...].[22]

Nachdem Nicolaus Germanus 1477 nach Rom gewechselt und dort 1480 gestorben war, übernahm sein deutscher Landsmann und Vertrauter Henricus Martellus Germanus (selbständig tätig ab ca. 1480) seine Vertretung in Florenz und war unter eigenem Namen noch etwa fünfzehn Jahre als Kartenmacher tätig.[23] Martellus war zwar kein ausgebildeter Astronom, Kosmograph oder Kartograph, jedoch ein talentierter Kartenmaler, der dem Renaissancegeschmack entsprechende, attraktive Werke schuf. Dabei wirkte sich seine jahrelange enge und fruchtbare Zusammenarbeit mit Francesco Rosselli (*Franciscus rosellus florentinus Cosmographus* genannt, 1445-zwischen 1513 und 1527), dem zu seiner Zeit vermutlich bedeutendsten und erfahrensten Kupferstecher, Kartenverleger und Kartenhändler Italiens, äußerst positiv aus. Rosselli war von 1476 bis 1480 in Budapest als Buchmaler für König Matthias Corvinus tätig gewesen, unterhielt aber auch noch nachher intensive Kontakte zu seinem ungarischen Bekanntenkreis.[24] Martellus schuf ab 1489 für seine Manuskriptatlanten jene berühmte herzförmige Weltkarte, die erstmals das von Bartolomeu Dias ein Jahr zuvor entdeckte Kap der Guten Hoffnung und damit die Möglichkeit eines Seeweges nach

---

21 Zu Nicolaus Germanus siehe: Böninger, Einwanderung, 334-342.

22 Vgl. hierzu: Durand, Vienna-Klosterneuburger Map Corpus, 252-256. – Weitere Drucke folgten erst 1514 durch Hans Burgkmair in Nürnberg, 1529 durch den Baseler Buchdrucker Andreas Cratander in Basel sowie 1531 mit Beschreibung durch Sebastian Münster

23 Zu Henricus Martellus siehe Böninger, Einwanderung, 313-348, hier: 338-348.

24 Zu Rosselli siehe F. Hellwig, Roselli, Artikel in: I. Kretschmer – J. Dörflinger – F. Wawrik, Lexikon zur Geschichte der Kartographie (Die Kartographie und ihre Randgebiete C/2), Wien 1986, 679-680.

Indien zeigte.[25] Rosselli veröffentlichte von der Karte etwa zur selben Zeit einen Kupferstich.[26]

Außerdem fertigte Martellus ebenfalls um 1490 für mehrere seiner Manuskriptatlanten, großformatige, erlesen kolorierte Exemplare einer Mitteleuropa-Karte (*Martellus-Karte*)[27] an – in der konservativen Gestalt einer Plattkarte und ohne irgendeine Graduierung. Auch von dieser Karte stellte Rosselli einen Kupferstich her, der die wichtigste Vorlage für die Karte Hieronymus Münzers in der *Schedelschen Weltchronik* werden sollte.

Die *Schedelsche Weltchronik* erschien 1493: im Juli die lateinische und im Dezember die deutscher Version.[28] Für beide wurde dieselbe Mitteleuropa-Karte verwendet. Münzer hatte die Darstellung gegenüber ihren Vorgängerinnen dahingehend abgeändert, dass er das auf ihr wiedergegebene Gebiet insbesondere nach Norden und Westen hin, beträchtlich erweiterte. Seine doppelseitige Tafel musste sich natürlich dem Format der geöffneten Weltchronik anpassen. Dafür musste Münzer beträchtliche, nach Norden hin zunehmend stärker werdende Verzerrungen in Kauf nehmen, da er nicht die zu jener Zeit schon bekannte trapezförmige „Donis-Projektion", sondern die ungünstige rechteckige Plattkarten-Abbildung benutzte. Um das Kartenbild möglichst übersichtlich zu gestalten, reduzierte er den Karteninhalt drastisch und verzichtete auf jegliche dekorative Ausgestaltung.

Im Süden der Karte verläuft die Grenzlinie des dargestellten Gebietes von der Rhône-Mündung durch den Golf von Genua (*Janua*), die nördliche Poebene (mit Verona, Vicenza [*Vincencia*], Padua [*Badua*]), an der dalmatinischen Küste entlang (Triest, Senj [*Zeng*], bis etwa Dubrovnik (*Ragusa*). Sie quert die Balkanhalbinsel

---

25 Vgl. hierzu G. Hamann, Der Eintritt der südlichen Hemisphäre in die europäische Geschichte (Österreichische Akademie der Wissenschaften, phil.-hist. Kl., Sitzungsberichte 260), Wien 1968, 190-192, 339-342.

26 R. Almagiá, On the cartographic work of Francesco Roselli, in: Imago Mundi 8 (1951), 27-34, hier: 30

27 Henricus Martellus Germanus Mitteleuropa-Karte, Florenz um 1490 (Biblioteca Nazionale, Cod. Magliabechi, XII, 26); Kolorierte Federzeichnung, 83,9 x 57,2 cm.

28 *Liber Chronicarum cum figuris et ymaginibus ab inicio mundi* 12. Juli 1493, ca. 47 x 31 cm, 324 Blätter mit 42 großformatige Stadtansichten und an die 2000 kleinere Illustrationen von 645 Holzblöcken; Münzer-Karte fol. 309v-310r 12. Juli 1493.

südlich der Donau, und endet im Schwarzen Meer. Die Kartengrenze gegen Osten beginnt bei Konstantinopel und Edirne (*Andranopel*), der ersten Hauptstadt des Osmanischen Reiches auf europäischem Boden;[29] sie setzt sich fort durch die Walachia, Tartaria, Russia, Riga, Pleskau (*Plesgo*; heute Pskow),[30] und erreicht den Raum um Nowgorod (*Nogradum*). Für den Nordosten war Jan Długosz wegen der Reichhaltigkeit und Genauigkeit seines Werkes (ab 1455), das auch Pommern und Schlesien mit einbezog, ein verlässlicher Informant.

Nach Norden hin erweiterte Münzer seine Tafel – der *Nordland-Karte* des *Ulmer Ptolemäus* folgend, die ihrerseits auf der Karte des Dänen Claudius Clavus aus dem Jahr 1427 beruhte – um beinahe ganz Skandinavien.[31] Um dem Buchformat der *Weltchronik* entsprechen zu können, musste er allerdings die nördlichen Regionen gewaltig nach Süden pressen. Dennoch wirkt die Ostsee (*Mare Germanicum*) einigermaßen wirklichkeitsnahe. Ihre wichtigsten Inseln sind eingetragen, allerdings falsch positioniert: Schonen (*Schonland*, *Scan*[d]*ia*) befindet sich westlich der dänischen Insel Seeland (*Silandia*), Gotland zu nahe bei der Bucht von Riga. Nördlich der Ostsee zieht sich eine Landmasse (*Norwega, Sweden*) hin, die dort, wo eigentlich Finnlands liegen sollte, über eine langgestreckte, schmale Landbrücke (*Wildlappen*[32] und *Grunland*) mit dem europäischen Festland (*Russia, Tartaria*) verbunden ist.[33]

---

29 Edirne wurde 1361 von den Osmanen erobert; Hauptstadt war es von 1368 bis zum Fall Konstantinopels (1453).

30 Pleskau, eine alte bedeutende Hansestadt auf dem Weg von Riga nach Nowgorod, war im 15. Jahrhundert ein Zankapfel zwischen Polen, Litauen und dem Deutschen Orden.

31 Eine Ähnlichkeit mit der *Eichstätt-Karte* ist in dieser Region kaum gegeben. Sie hatte schon im frühen 14. Jahrhundert Pietro Vesconte recht fantasievoll in Form eines Eichenblattes eingezeichnet gehabt; s. exploring Carta Marina (http://cipher.uiah.fi/forum/materials/carta_marina/maps/; eingesehen am 13. 8. 2009). Eine vielleicht ebenfalls auf die *Clavus-Karte* zurückgehende Darstellung Skandinaviens findet sich auf der *Fra Mauro-Karte* (1459).

32 Vgl. hierzu: K. Brunner, Nordeuropa-Darstellungen in Ptolemäus-Ausgaben am Ende des 15. Jahrhunderts, in: W. Scharfe (Hrsg.), 6. Kartographiehistorisches Colloquium Berlin 1992, Vorträge und Berichte, Berlin 1994, 19-29. – Zu den Wildlappen vgl. Durand, Map Corpus (wie Anm. 19), 181, 381.

33 Vgl. u. a. G. Hamann – R. Kinauer, 100 Jahre Franz Josefs-Land. Zur Erinnerung an die Entdeckungsreise der Österreichisch-Ungarischen Nordpol-Expedition 1872-1874 unter Julius Payer und Carl Weyprecht.

Weit weniger Beachtung als die Verlängerung der Münster-Karte nach Norden fand bisher ihre beträchtliche Ausweitung in westlicher Richtung. Hier reicht sie – und diese Neuerung war wahrscheinlich die ureigenste Idee ihres Autors – über die gesamte östliche Hälfte Frankreichs. Dabei zeigt sie die Flusssysteme der Rhône (*Rodan*[us]) mit Lyon (*Lion*), der oberen Loire mit den Orten Orleans und Tours (*Orliens, Tors*), sowie der Seine (*Secana*) mit Paris. Nördlich davon bietet sie außerdem noch die Umrisse weiter Teile der Britischen Inseln: England (*Anglia*) mit London (*Lvndea*), dessen in Großbuchstaben angeführter Name allerdings als Region ausgewiesen wird; Schottland (*Scocia*) und Irland (*Ibernia*). Sogar Island (*Vslant*) ist zu erkennen, allerdings an einem völlig falschen Ort, nämlich nördlich einer Meerenge (!) zwischen dem nördlichsten Schottland und südlichsten Norwegen.

Das hydrographische Netz der *Münzer-Karte* ist, was seine Vollständigkeit anbelangt, nicht schlecht. Die Genauigkeit der Flussverläufe ist jedoch höchst unterschiedlich. Manche von ihnen zeigen noch die von Ptolemäus herrührenden Fehler: etwa die nach Norden gerichteten Unterläufe von Flüssen, die in die Nordsee münden, wie Seine, Schelde (*Tibulla*), Maas (*Masa*) und Rhein (*Renus*), diesen mit etlichen seiner Nebenflüsse: Limath (*Limag*), Neckar (*Necarus*), Main (*Menus*), Mosel (*Mosella*). Der Wirklichkeit entsprechender wirken die Flussläufe von Ems (*Amasus*), Weser (*Visurgis*) und Elbe (*Albis*). Besonders störend wirken die durchgehend in östlicher Richtung verlaufenden Flüsse Donau (*Danubi*[us]) – von Ulm bis Wien – und aller ihrer aus den Alpen kommenden Zubringer, etwa von Mur, Drau (*Draua*) und Save (*Sau*). Offenbar auf Flüchtigkeitsfehler dürften drei Bifurkationen zurückzuführen sein (Etsch – Inn [*Athesis* – *Enos*], Main – Saale [*Menos* – *Sala*] und zwischen Rhône und Loire). Dagegen sind die im östlichen Teil der Karte gezeigten Ströme, etwa Oder (*Ader*) und Weichsel (*Vistula*) sowie deren unbenannten Nebenflüsse Neiße und Bug relativ gut gelungen. Die Geländedarstellung erfolgte in Form schematischer Maulwurfs- oder Fischschuppen, wobei kurioserweise die deutschen Mittelgebirge mächtiger als die eher schmächtig gestalteten Alpen wirken. Bezeichnungen von Bergen bzw. Gebirgen fehlen ausnahmslos.

Ausstellung im Prunksaal der Österreichischen Nationalbibliothek. Katalog, Wien 1973, 3.

Hingegen werden Länder und Siedlungen einzig allein mit ihren Namen ausgewiesen, Signaturen für Siedlungen oder Grenzen fehlen. Länder und Landschaften führen (mit Großbuchstaben geschrieben) lateinische oder latinisierte Bezeichnungen, z. B. *Anglia, Flandria, Germania Magna, Polonia, Astria*(!), *Boemia, Vngaria* und sogar das *Wvrtzlant.*

Bei den spärlich eingetragenen Orten traf Münzer immerhin eine zukunftsweisende Entscheidung, indem er sie überwiegend mit ihren deutschen Namen anführte. Etwas dichter ist die Zahl der Orte in Franken (*Frankonia*) und nahe Meissen (*Mixia*), aber auch entlang der deutschen Ostseeküste. Hier sind es die für den Fernhandel wichtigen Hansestädte, Lübeck (*Lubick*), Rostock, Greifswald (*Grispalt*, seit 1456 auch Universitätsstadt), Danzig (*Dantzg*), Elbing (*Melbing*, polnisch Elblag), Stettin (polnisch: Szczecin, seit 1309 Sitz der pommerschen Herzöge), Riga (wichtig für den Russlandhandel). Im östlichen Hinterland, in Livland und Russland (*Livonia*, *Russia*), sind noch Pleskau (*Plesgo*, russisch Pskov) sowie das einst mächtige, aber 1478 von Moskau unterworfene Nowgorod (*Nogradum*) eingetragen. Dazu kommen noch ausländische Orte mit Kontoren der Hanse: London, Bergen, Brügge (*Prug*). Im südlichen und westlichen Deutschland, der Schweiz, in Italien sowie in Frankreich finden sich unter den Städten einige mit Beziehungen zu der um 1380 gegründeten Ravensburger Handelsgesellschaft, beispielsweise Ulm (*Vlma*), Konstanz (*Constancia*), Zürich (*Zuich*), Luzern (*Lucern*), Genua (*Ianua*), Avignon (*Avinio*) und Lyon.

Auch in anderen Regionen Deutschlands werden etliche wirtschaftlich relevante Orte gezeigt: Nördlingen (*Norling*, mit einer im Spätmittelalter sehr wichtigen Fernhandelsmesse zu Pfingsten), Görlitz (*Gorlitz*, mit Webereien und Bierbrauereien) und Besançon (*Visontz*) wegen der Metall- und Salzvorkommen der Franche Comté.

Auf dem Balkan erscheinen die Hauptstädte des Osmanischen Reiches, die alte: Edirne (*Adrianopel*, 1368-1453), und die neue: Istanbul (*Constantinopel*, ab 1453).

Im Bereich des heutigen Österreichs sind Wien (*Wienna*), außerdem das sehr wichtige Krems (*crems*), Steyr (*Steyer*), Salzburg (*Saltzburg*), Judenburg und Villach angegeben – und Wiener Neustadt als einziger Ort mit seinem lateinischem Namen, *Nova Civitas*. Das Fehlen der habsburgischen Residenzstädte Graz und Innsbruck überrascht hingegen.

Für die Anfertigung seiner Karte dürfte Hieronymus Münster mit folgenden Vorlagen das Auslangen gefunden haben (Tafel S. 560 a):

1. Mitteleuropa-Karte des Francesco Rosselli (Florenz, um 1480; 33 x 47 cm); sie ähnelt, was Hydrographie und Geländedarstellung anbelangt, der Münzer Karte dermaßen, dass eine andere Variante nicht in Frage kommt (Tafeln S. 560 b und 561 a),[34]
2. *Tabula moderna Francie* und
3. *Nordland-Karte*, nach Claudius Clavus.

Die Vorlagen von 2 und 3 sind im *Ulmer Ptolemäus* enthalten, die auf Tafeln in einem von Nicolas Germanus 1477 geschaffenen ptolemäischen Manuskriptatlas basieren.[35] Die erste moderne Frankreich-Karte schufen 1469 Pietro del Massaio (1424-1496) und Hugo Comminelli (tätig 1450-1470).[36] Die erste Version der *Nordland-Karte* erschien 1527 in einem für den französischen Kurienkardinal Guillaume Fillastre angefertigten Ptolemäus-Atlas.[37]

Die *Eichstätt-Karte* dürfte im Zusammenhang mit der Herstellung der Münzer-Karte keine Rolle gespielt haben. Die Ähnlichkeit der beiden Kartenbilder ist eher gering. Dies gilt insbesondere für die Hydrographie und den Verlauf der Meeresküsten (Tafeln S. 561 b-563 a). Dazu zeigt die Münzer-Karte – ebenso wie jene von Martellus und Rosselli – die Pfalzinsel (*Palatina*) im Rhein, nahe bei Kaub, während sie auf der *Eichstätt-Karte* fehlt. Nach dem Tod des Nicolaus Germanus, zu Beginn des Jahres 1480, scheint die Druckplatte in Italien in Vergessenheit geraten zu sein, so dass sie erst 1490 oder 1491 nach Deutschland gelangte.

Münzer dürfte es nicht schwer gefallen sein, die für seine Kartengestaltung erforderlichen Vorlagen zu erhalten: Exemplare des Ulmer

---

34 F. Rosselli, Karte von Mitteleuropa, Florenz um 1490-1492, 33 x 47 cm; (BNCF Landau Finaly – Carte Rosselli). Für die Analyse stand nur eine qualitativ minderwertige Internet-Abbildung des Exemplars zur Verfügung (http://www.bncf.firenze.sbn.it/notizie/Cartografia%20Web/Rinascimento/Rosselli/RosselliEU.htm, eingesehen am 5. 1. 2010).

35 Schloss Wolfegg, Codex Wolfeggianus latinus.

36 Codex Vaticanus latinus 5699.

37 Stadtbibliothek Nancy; vgl. J. Fischer, Die kartographische Darstellung der Entdeckungen der Normannen in Amerika, in: Proceedings. International Congress of Americanists, Stuttgart 1904.

Ptolemäus besaßen sowohl er selbst als auch Hartmann Schedel.[38] Als geographisch und kartographisch interessierter und kunstsinniger Mensch mit weitreichenden privaten und beruflichen Beziehungen und außerdem an einem Fernhandelsunternehmen Beteiligter erwarb er wohl auch öfters einschlägige Publikationen aus Italien.

4. Die Begegnung mit Martin Behaim

Parallel zu seinen Arbeiten an der *Schedelschen Weltchronik* unterstützte Münzer ab 1490 auch Martin Behaim, der wegen der Abwicklung einer Erbschaftsangelegenheit für längere Zeit aus Portugal nach Nürnberg gekommen war, bei der Anfertigung von dessen *Erdapfel*. Dabei erfuhr Münzer Einzelheiten über die seit Jahrzehnten durchgeführten portugiesischen Expeditionen entlang der Westküste Afrikas, auf der Suche nach einem Seeweg in den Indischen Ozean; und er glaubte offenbar den vollmundigen Behauptungen Behaims, an einer der letzen der Reisen, nämlich an jener des Diego Cão (1483-1484 oder 1485-1486), in verantwortungsvoller Position teilgenommen zu haben.

Münzer scheint von dessen Vorstellung beeindruckt worden zu sein, man könnte China (*Katha*) mit einer Seefahrt über den Atlantik erreichen, wie dies auch der große Kosmograph Paolo dal Pozzo Toscanelli dem portugiesischen Hof schon 1474 dargelegt und zu Beginn der 1480er Jahre gegenüber Christoph Columbus wiederholt hatte. Auch der römisch-deutsche König Maximilian I. (der spätere Kaiser), durch seine Mutter Eleonore, der Gattin Friedrichs III., mit dem portugiesischen Königshaus verwandt, und damit ein Cousin König Johanns II. (1481-1495), konnte dieser Idee einiges abgewinnen. Mit Billigung Maximilians verfasste Münzer am 14. Juli 1493 ein Schreiben an den portugiesischen Herrscher, in dem er die Vorteile einer Suche Ostasiens jenseits des Atlantiks zusammenfasste und als

---

38 Goldschmidt, Muenzer, 497. – K. A. Vogel, Neue Horizonte der Kosmographie. Die kosmographischen Bücherlisten Hartmann Schedels (um 1498) und Konrad Peutingers (1523), in: Anzeiger des Germanischen Nationalmuseums [etc.], Nürnberg 1991, 77-85; hier: 79. – Aus Schedels Bücherliste geht hervor, dass es sich um ein Exemplar einer der beiden Ulmer Ptolemäus-Atlanten handelte: *Cosmographia Ptolemei Alexandrini cum figuris novis XXXII.*

geeigneten Kommandanten des Unternehmens Martin Behaim vorschlug:[39]

Auch sind die Beweise zahllos, ja, ich darf sagen, sicher, aus denen sich klar ergibt, dass man jenes Meer in der Richtung auf das östliche Cathay in wenig Tagen(!) durchfahren kann. Und es mag dich nicht Alfraganus stören noch andere Leute ohne Erfahrung, die behauptet haben, nur ein Viertel der Erde sei vom Meere frei und die andren drei Teile der Erde seien unter dem Meer ausgebreitet, denn in Dingen, die sich auf die Besiedlung der Erde beziehen, muss man der Erfahrung und vertrauenswürdigen Berichten mehr glauben als phantastischen Einbildungen.

Gewiss weißt auch du, dass viele angesehene Astronomen das Vorhandensein bewohnbaren Landes in den Tropen und unter den Äquinoktien leugneten. Dass diese Behauptungen eitel und falsch waren, hast du aus eigner Erkundung herausgefunden.

Du sollst für diese Fahrt, falls du es wünschst, auch einen von unsrem König Maximilian abgesandten Landsmann erhalten, Herrn Martin Behaim, der beauftragt worden ist, ganz besonders jenes auszuführen, und viele andere erfahrene Seeleute, welche die Breite des Meeres durchsegeln werden, indem sie ihren Weg mit kühnem Unternehmungsgeist von den Azoren aus nehmen werden unter Mitnahme ihrer Zylinder [gemeint sind kleine, tragbare Sonnenuhren], *ihrer Quadranten, Astrolabien und andrer Instrumente.*

5. Die Reise nach der Pyrenäenhalbinsel (1494/95)

1494 war Nürnberg wieder von einer Pestseuche bedroht und neuerlich flüchtete Münster, diesmal sogar mit einer guten Ausrede. Nachdem auf seinen Brief aus Portugal keinerlei Reaktion erfolgt, dabei aber Münzers Neugier an den portugiesischen Expeditionen geweckt war und er schließlich Botschaften Maximilians an die Herrscher Spaniens und Portugals senden wollte, beschloss Münzer, eine Reise auf die Pyrenäenhalbinsel zu unternehmen (Tafel S. 563 b).[40]

---

[39] Hieronymus Münzer, Brief an König Johann II. von Portugal, Nürnberg 14. Juli 1493, in: R. Hennig (Ed.), Terrae Incognitae 4, Leiden ^{2}1956, 238–239, 345. – Vgl. K. A. Vogel, Sphaera terrae – das mittelalterliche Bild der Erde und die kosmographische Revolution, Göttingen 1995, phil. Diss., 387-419; webdoc.sub.gwdg.de/diss/2000/vogel/kap5.pdf (eingesehen am 15.12.2009).

[40] A. Classen, Die iberische Halbinsel aus der Sicht eines humanistischen Nürnberger Gelehrten. Hieronymus Münzer, Itinerarium Hispanicum (1494-

Er verließ Nürnberg mit zwei Begleitern, Kaspar Fischer und Nikolaus Wolkenstein, am 2. August 1494; in Augsburg kam noch Anton Herwart hinzu. Alle drei waren Söhne reicher Kaufleute der Ravensburger Handelsgesellschaft, die enge Beziehungen zu Spanien unterhielt. Die etwa 7000 km lange Reiseroute verlief über die Schweiz, Lyon, das Rhônetal abwärts, Marseille, Arles, Perpignan, Montserrat, Barcelona, Valencia, Sevilla, Madrid, das erst 1492 von den Arabern eroberte Granada, Malaga, Sevilla, Evora nach Lissabon. Stationen auf dem Rückweg waren Santiago de Compostella, Guadeloupe, Madrid, Saragossa, Pamplona, Roncenvalles), Toulouse, Poitiers, Tours, Paris, Abbeville, Brugge, Gent, Köln, Mainz, Worms, Frankfurt und Würzburg.[41] Im April 1495 waren die vier Reisenden wieder in Franken.

Münzer war ein scharfer Beobachter, der bei Stadtbesichtigungen systematisch vorging. Meist bestieg er zuerst Kirchtürme, um so einen guten Gesamteindruck der Städte zu gewinnen.[42] Sein detailliertes Reisejournal wirkt überraschend modern. Es ist ausführlich, lebendig, gelegentlich kritisch, aber auch humorvoll. Als Vorlage scheint das „Jacobsbuch“ (*Liber Sancti Jacobi)* aus dem 12. Jahrhundert gedient zu haben.[43]

1495), in: Mitteilungen des Instituts für Österreichische Geschichtsforschung 111 (2003), 317-428, hier: 340. – Vgl. auch: K. Herbers – N. Jaspert (Hrsg.), Das kommt mir spanisch vor. Eigenes und Fremdes in den deutsch-spanischen Beziehungen des späten Mittelalters, Münster 2004 (Geschichte und Kultur der iberischen Welt 1). – K. Herbers, Die „ganze“ Hispania: der Nürnberger Hieronymus Münzer unterwegs – seine Ziele und Wahrnehmung auf der Iberischen Halbinsel (1494-1495), in: R. Babel – W. Paravicini (Hrsg.), Grand Tour. Adeliges Reisen und Europäische Kultur vom 14. bis zum 18. Jahrhundert, Ostfildern 2005, 203-308.

41 H. Münzer, Itinerarium Hispanicum Hieronymi Monetarii, 1494-1495. Hrsg. v. Ludwig Pfandl, in: Revue historique 48, New York, Paris 1920; Nachdruck 1964. – Eine Kartenskizze *Reiseweg des Hieronymus Münzer auf der Iberischen Halbinsel 1494/1495* befindet sich in: K. Herbers, Vom Bodensee nach Spanien. Eigenes und Fremdes im Blick eines Reisenden um 1500, in: Oberschwaben und Spanien an der Schwelle zur Neuzeit, Ostfildern 2006, 9-12, hier: 10.

42 Herbers – Jaspert (Hrsg.), Das kommt mir spanisch vor, 10.

43 Das „Jakobsbuch“ (lat. *Liber Sancti Jacobi*, häufiger genannt *Codex Calixtinus* oder *Codex Calixtus*) ist eine Sammlung von Handschriften aus dem 12. Jahrhundert. Zugeschrieben wurde sie fälschlicherweise Papst Calixt II., jedoch vermutlich von dem französischen Gelehrten Aimeric Picaud

Münzers Reisebeschreibung ist in der „Ich"-Form verfasst, hervorragend geschrieben, ausführlich,[44] und detailreich, stellenweise sogar überraschend humorvoll und dabei informativ. Darin enthalten ist z. B. der einzige Bericht über die große Synagoge von Lissabon vor der Vertreibung der Juden, bevor sie, zu einer Kirche umgewidmet, durch das Erbeben von 1755 zerstört wurde.

Überaus kritisch fielen Münzers Aussagen über den Wallfahrtsort Santiago de Compostella aus: Er meinte etwa: *Der Boden ist gut, und die Gärten der Stadt sind voll von Orangen und Zitronen, Pfirsichen, Pflaumen und anderen Früchten. Aber die Leute dort sind schweinisch* [...] *und faul – je weniger sie sich im Landbau anstrengen, umso mehr Mühe geben sie sich, den Pilgern das Geld aus der Tasche zu ziehen.*

Und dauernd ist ein solches Volksgeschrei in der Kirche, dass man es nicht für möglich halten möchte. Mäßig ist das die Ehrfurcht. Der heiligste der Apostel wäre doch würdig genug, inniger verehrt zu werden Man nimmt an, dass er gemeinsam mit zwei Jüngern, dem einen zur Rechten, den anderen zur Linken, unter dem Hochaltar begraben liegt. Sein Leib ist allerdings noch von keinem da gesehen worden. Selbst im Jahr 1487, als der König von Kastilien dort war, hat ihn keiner gesehen. Vertrauensvoll glauben wir, was uns Menschen rettet.[45]

Kritisch äußerte sich Münzer auch über das lockere Benehmen und die aufreizenden Kleider der Frauen von Valencia.[46]

Was die diplomatische Mission Münzers hinsichtlich einer Überquerung des Atlantiks in westlicher Richtung nach Cathay anbelangte, war sie obsolet geworden. Nach der Seeexpedition des Bartolomeu Diaz (1486/1487), über die Behaim in Nürnberg offenbar nicht informiert hatte,[47] und der etwa gleichzeitigen Landreise der Pêro de Covilhão (ca. 1455-ca. 1530) nach Südasien schien der Seeweg in den Indischen Ozean in greifbare Nähe gerückt. Schließlich waren es die Spanier, die als erste in den karibischen Raum

---

zusammengestellt. Neben Predigten, Wunderberichten und liturgischen Texten enthält der Codex einen Pilgerführer, mit Wegbeschreibungen sowie Empfehlungen für die Pilgerfahrt nach Santiago de Compostella.

44 D. F. Lach, Asia in the Making of Europe 2: A Century of Wonder 2 – The Literary Arts, Chicago 1994, 56-57.

45 G. Fleischer (Hrsg.), Jacobusweg I_1. Nürnberg – Schwabach – Abenberg – Kalbensteinberg – Gunzenhausen, Ostfildern ^{3}2007, 94.

46 U. Ganz-Blättler, Andacht und Abenteuer. Berichte europäischer Jerusalem- und Santiago-Pilger (1320-1520), Tübingen ^{3}2000, 192.

47 Behaim dürfte vorgegeben haben, er wäre ein wichtiges Mitglied der zweiten Expedition des Diego Cão (1484-85) gewesen.

gelangten. Christoph Columbus kehrte im März 1493 nach Europa zurück – knappe fünf Monate bevor Münzer sein Schreiben schickte.

In Evora wurde Münzer im November 1494 viermal von König João II. in Audienz empfangen und dabei eingehend über die Situation in den neu entdeckten afrikanischen Ländern informiert. Im Reisebericht des Nürnbergers findet sich eine längere Abhandlung: *De inventione Africae maritimae et occidentalis videlicet genee* [Guineae?] *per infantem Heinricum Portugalliae*. Die Schilderung gibt die Entwicklung bis ins Jahr 1469 wieder.[48]

Münzer fand in Lissabon Unterkunft bei Behaims Schwiegervater, dem aus Brügge stammenden Jobst von Hurter (Josse Van Huerter, Capitao-donatorio der Azoreninseln Faial and Pico). Behaim selbst war nicht in Lissabon. Merkwürdigerweise wird er im gesamten Reisebericht mit keinem Wort erwähnt. Es könnte durchaus sein, dass Münzer erkannte, die Bedeutung und das Ansehen Behaims in Portugal bei weitem überschätzt zu haben.

Zwei Monate später kam es in Madrid zu einem Treffen Münzers mit dem spanischen Königspaar Ferdinand und Isabella. Dabei konnte der Nürnberger Einzelheiten der Reisen von Columbus und anderen Seeleuten in der Neuen Welt in Erfahrung bringen.

Münzer war einer der ersten Mitteleuropäer, die die Pyrenäenhalbinsel nach der Zerschlagung des muslimischen Königreiches Granada besuchten, und vermutlich der erste, der seine Eindrücke schriftlich und weitgehend ohne Ressentiments festhielt. Besonders ausführlich und informativ berichtete er über den Alltag der Bevölkerung, die in den ersten Jahren nach der Eroberung noch ein einigermaßen freies Leben führen konnte.[49]

---

48 F. Kunstmann (Hrsg.), Hieronymus Münzer's Bericht über die Entdeckung der Guinea, in: Abhandlungen der historischen Classe der königlich bayerischen Akademie der Wissenschaften 7, 2 (1854), 291-362.

49 Vgl. hierzu: S. Maislinger, Die Spanische Inquisition und die Moriskos, phil. Diplomarbeit, Wien 2008.

6. Die letzten Jahre des Hieronymus Münzer

Eigene Werke geographischen Inhalts sind das *Itinerarium* (Reise durch Frankreich, Spanien und Portugal, 1494/95),[50] ein Brief *A carta que enviou H. Monetario* (Lissabon, ohne Erscheinungsjahr), Beiträge zum *Behaim-Globus* und zur *Schedelschen Weltchronik* (Mitteleuropa-Karte, bis 1493). Ein Bericht *De inventione insularum Indicarum* über die Entdeckung Amerikas ist nicht erhalten (davon dürfte er 1494 in Spanien erfahren haben), ebenso eine seiner astronomischen Schriften.

Hieronymus Münzer starb in Nürnberg am 27. August 1508 und wurde in der Sebalduskirche bestattet. Schon zuvor hatten er und sein Bruder Ludwig († 1518) der Pfarrkirche St. Nikolaus in Feldkirch eine prächtige silberne Monstranz gestiftet und persönlich übergeben.

Hieronymus Münzer hinterließ ein beträchtliches Vermögen und dazu eine umfangreiche Gelehrtenbibliothek, von der er einen Teil der Lateinschule seiner Geburtsstadt vermachte. Das bei dieser Gelegenheit angefertigte Verzeichnis weist 61 gebundene Bücher mit insgesamt 76 Titeln aus. Sie befinden sich heute in der Stadtbibliothek Feldkirch.[51] Ein anderer Teil der Bibliothek, in schönen Originaleinbänden mit Blinddruck und Beschlägen, gelangte in den Besitz der Fürsten Dietrichstein und befand sich offensichtlich mehrere Jahrhunderte in deren Schloss in Nikolsburg, bis die Bände Ende 1933 bei einer Versteigerung (Gilhofer & Ranschburg) veräußert wurden. Etliche Werke aus dem Besitz Münzers gelangten dabei in die Mährische Landes- und Universitätsbibliothek in Brünn.[52]

---

50 In Deutschland nur in einer von Schedel angefertigten Abschrift vorhanden: München, Bayerische Staatsbibliothek, Clm 431.

51 B. Fabian (Hrsg.), Handbuch der historischen Buchbestände in Deutschland. Digitalisiert von Günter Kükenshöner, Hildesheim 2003; http://134.76.163.162/fabian (eingesehen am 13.5.2008): Stadtbibliothek Feldkirch.

52 B. Fabian (Hrsg.), Handbuch der historischen Buchbestände in Deutschland; http://www.b2i.de/fabian (eingesehen am 13.5.2008): Mährische Landesbibliothek-Universitätsbibliothek (Brünn).

Zum Autor

Franz Wawrik, geb. 1940 in Wien, studierte Geschichte und Geographie an der Universität Wien und befasste sich insbesondere mit der Geschichte der Entdeckungs- und Forschungsreisen, der Kolonial- und Kartographiegeschichte sowie der Geschichte der Naturwissenschaften. 1967 Promotion zum Dr. phil.; ab Dezember 1973 Mitarbeiter der Kartensammlung und des Globenmuseums der ÖNB, ab Dezember 1986 Direktor. Pension ab August 2000.

Autor bzw. Mitherausgeber von etwa 70 Publikationen, darunter *Descriptio Austriae* und *Lexikon zur Geschichte der Kartographie* (1986). Gestaltung mehrerer Ausstellungen, z. B. *Austria Picta. Österreich auf alten Karten und Ansichten* (1989) und *Die Neue Welt. Österreichs Anteil an der Erforschung Amerikas* (1992) sowie Organisation internationaler Kongresse, darunter 16th International Conference on the History of Cartography (1995). Seit 2007 Herausgeber von *Der Globusfreund* und *Globe Studies,* den Zeitschriften der Internationalen Coronelli-Gesellschaft.

Kontakt: franz.wawrik@tele2.at

Franz Schulze,
Balthasar Springers
Indienfahrt 1505/06,
Straßburg 1902

Karl R. Wernhart, Wien

Die Bildquellenwerke der Sammlung Woldan

Grundlage für die Entstehung der *Monumenta Ethnographica* – Erinnerungen eines Zeitzeugen

Wie aus dem Untertitel ersichtlich, ist der Autor vorliegender Darstellung bereits in einem Lebensalter angelangt, dass seine Erinnerungen an die frühen 1960er Jahre ihn zum Zeitzeugen werden lassen. Als Studenten dieses Dezenniums haben wir – meine Studienkollegen und ich – die Entstehung des ersten Bandes eines Monumentalwerkes miterlebt, nämlich des auf 12 Bände und mehr geplanten, aber leider nie fortgesetzten Werkes der *Monumenta Ethnographica.* Darin sollten Bildquellen völkerkundlichen bzw. kulturanthropologischen Inhaltes über außereuropäische Ethnien mit den Originalkommentaren aus den Druckwerken und zusätzlichen Erläuterungen kompetenter Wissenschafter veröffentlicht werden. Leider ist nur der erste Band über „Schwarzafrika“ im Jahre 1962 erschienen, den Walter Hirschberg – damals neu berufener Ordinarius für Völkerkunde II – bearbeitet, kommentiert und ediert hat[1]. Um dieses Vorhaben verwirklichen zu können, mussten Originalveröffentlichungen, d.h. Erstdrucke bzw. Erstausgaben vorhanden sein, denen Illustrationen (Holzschnitte, Metalldrucke bzw. –stiche,

---

1 W. Hirschberg, Monumenta Ethnographica: Schwarzafrika. (Frühe völkerkundliche Bilddokumente 1.) Akademische Druck- und Verlagsanstalt (ADEVA), [269 Tafeln, 84*Seiten Anhang, Sachregister, geographisches Register, Literatur], Graz 1962.

etc.) mit ethnographischem Inhalt zur Erläuterung des Originaltextes beigegeben waren.

Eine Fundgrube für die Realisierung des Projektes stellte die beeindruckend umfangreiche und vielgestaltige Privatbibliothek von Prof. Dr. Erich Woldan (1901-1989) dar, die sich heute auf Grund testamentarischer Verfügung als geschlossene Sammlung in der Bibliothek der Österreichischen Akademie der Wissenschaften befindet[2] und von Mag. Gerhard Holzer (akademischer Mitarbeiter der ÖAW) betreut sowie bibliothekarisch bearbeitet und erschlossen wird.

In der Entstehungszeit dieses ersten Bandes der ethnographischen Bilddokumente über Schwarzafrika vom 15. bis zum 18. Jahrhundert – also Ende der 1950er und zu Beginn der 1960er Jahre – studierte an der Wiener Universität im Fachbereich „Neuere Geschichte“ zunächst bei Univ.-Doz. und später Univ.-Prof. Dr. Günther Hamann (1924-1994) eine Gruppe von Hörern wie Johannes Dörflinger, Alfred Kohler, Herbert Matis, Franz Wawrik und der Autor dieser Zeilen, Karl Wernhart. Diese Gruppe junger Historiker stieg in den Folgejahren zu anerkannten Wissenschaftern der historischen Disziplinen und deren Nachbarfächern auf, wobei letzterer als ausgebildeter Historiker für Themen des Entdeckungs- und Kolonialzeitalters als Mitarbeiter von Prof. Dr. Walter Hirschberg (1904-1996) zur Völkerkunde wechselte und mit diesem für die theoretischen und methodologischen Grundlagen einer Historisierung der Kulturhistorie in der Ethnologie sorgte, die, als „Ethnohistorie“ von Walter Hirschberg eingeführt, heute über international anerkanntes Ansehen verfügt. Das Lehrbuch der Ethnohistorie, das bis dato in drei Auflagen erschienen ist, gibt ein beredtes Zeugnis vom Erfolg dieser Forschungsrichtung[3].

Johannes Dörflinger hatte sich schon als Studierender und dann als Mitarbeiter von Prof. Hamann auf den Bereich der Entdeckungs- bzw. Erschließungs- und Kolonialgeschichte spezialisiert und hier besonders kartographische wie historisch-politische Fragestellungen

---

2 J. Dörflinger, Erich Woldan's Collection of Maps, Atlases and Globes, in: Imago Mundi, The Journal of the International Society for the History of Cartography. London 1990, 117-118.

3 K. R. Wernhart – W. Zips (Hrsg.), Ethnohistorie. Rekonstruktion und Kulturkritik. Eine Einführung, Wien ³2008.

und Themen aufgenommen, die er dann in seiner eigenständigen Forschungs- und Lehrtätigkeit weiter ausbaute und vertiefte. Auf Grund dieses Interesses wurde er durch Günther Hamann mit Erich Woldan und seiner wertvollen Sammlung von Primärwerken kartographischen, topographischen und erdbeschreibenden Inhaltes vertraut und durfte als einer der wenigen Auserwählten in dessen Wohnung – die eigentlich einen Speicher von Erstdrucken und wertvollsten Ausgaben, Atlanten und Globen darstellte – forschen. Zwischen diesen beiden Gelehrten, dem bibliophilen Juristen und Archivar wie Bibliothekar der *Österreichischen Geographischen Gesellschaft* Erich Woldan und dem jungen aufstrebenden und mit Akribie arbeitenden Johannes Dörflinger begann sich eine von gegenseitigem Respekt gekennzeichnete Gedankenfreundschaft zu entwickeln. Dies führte schließlich dazu, dass Dörflinger heute als der beste und versierteste Kenner der Woldan-Sammlung gilt, die wertvollste Quellenwerke des 15. bis 19. Jahrhunderts umfasst. Diese qualifizierten Erstdrucke kartographischen, topographischen, erdkundlichen und auch ethnographischen Inhalts sind für Historiker der Neuzeit, für historisch ausgerichtete Geographen und Kartographen wie auch für historisch orientierte Völkerkundler von Bedeutung[4].

Die wissenschaftliche Bearbeitung und Auswertung der Quellenwerke und Dokumente, u.a. auch der Woldan-Sammlung, wurde durch die vorher angeführte junge Historikergruppe vom Forschungsansatz Günther Hamanns beeinflusst, der als Professor der „Neueren Geschichte“ als Initiator der „Schule der Entdeckungs- und Wissenschaftsgeschichte“ anzusehen ist. Er hatte als humanistisch Gebildeter in seiner Forschungs- und Lehrtätigkeit an der Universität Wien anhand seines umfassenden Wissens über wissenschaftsgeschichtliche Fakten und Erkenntnisse und durch seine Einfühlungsfähigkeit in historische Ereignisse und deren gesellschaftspolitische Strukturen die Begabung besessen, die Ereignisse herauszulösen, zu analysieren und umsichtig zu hinterfragen, um sie dann letztendlich wieder in den Gesamtzusammenhang menschheitsgeschichtlicher Dimensionen einbetten zu können. Günther Hamann war als Historiker befähigt, „Visionen“ zu entwickeln, ohne dabei jemals die „historische Realität“ zu verlieren. „Vision“ und „historische Realität“ bilden ein Begriffspaar, das sich gegenseitig bedingt

---

4 Dörflinger, Erich Woldan’s Collection, 117f.

und den Historiker in einem gewissen Spannungsverhältnis hält. Jeder Historiker muss also Visionen für Interpretation, Auslotung und integrative Darstellung des nach den Kriterien der Quellenkritik überprüften Faktenmaterials besitzen und dabei die unterschiedliche Bedeutung und Aussagekraft der verschiedenen Quellentypen bzw. -gattungen aus der eigenen Disziplin und den Nachbarfächern vor Augen haben. Die historische Realität holt ihn wieder auf den Boden der in Zeit und Raum stattgefundenen Ereignisse. Ohne beflügelnde visionäre Dimension kann die beste reale Darstellung bestimmter Fakten nicht erfolgen. Der Historiker muss in diesen Diskurs eintreten, die Analyse im interaktionalen Dialog reflektiverweise erarbeiten. Die angestrebten und entwickelten Deutungsversuche können durch das Fortschreiten bzw. Vertiefen der Erkenntnisse und mittels politischer bzw. gesellschaftspolitischer Kurskorrekturen von der Vision zur neuen geschichtlich verstandenen Realität werden. Ein ausformuliertes Gedankengebäude betreffend die „Entdeckungs- und Wissenschaftsgeschichte" hat Günther Hamann nie erstellt, aber sowohl in seinem Œuvre wie auch in den Seminaren und Vorlesungen hat er Ansätze und Gedanken über verschiedene Quellentypen und deren Materialien und vor allem über Quellenkritik und Interpretation vorgelegt, die Hörer und Leser gleichermaßen beeindruckten[5], so auch Johannes Dörflinger und den Autor dieses Beitrages.

Die Fachkenntnisse Günther Hamanns und die Tatsache des Vorhandenseins zahlreicher wertvoller öffentlicher und besonders privater Sammlungen in Österreich, vor allem auf Wiener Boden, wie auch die kollegiale Verbundenheit Günther Hamanns mit Erich Woldan und Walter Hirschberg, der stets um eine Historisierung der Völkerkunde als historische Wissenschaft bemüht war, hat ein Netzwerk von Forschungsrichtungen entstehen lassen. Der inter- bzw. transdisziplinäre Dialog lag Günther Hamann stets am Herzen, sowohl im Bereich der Geschichte der Naturwissenschaften, wie zum Beispiel mit Vertretern der Disziplinen der Mathematik, Astronomie,

---

5 K. R. Wernhart, Nachruf, Günther Hamann, in: Mitteilungen des Instituts für Österreichische Geschichtsforschung (MIÖG) 104/1-2, Wien-München 1996, 233-236, bes. 233, 234. – K. R. Wernhart, Günther Hamann. Des Historikers Vision und die historische Realität, in: G. Hamann, Die Welt begreifen und erfahren. Aufsätze zur Wissenschafts- und Entdeckungsgeschichte (H. Grössing – K. Kadletz – M. Klemun [Hrsg.], Perspektiven der Wissenschaftsgeschichte 1), Wien 1993, 13-20.

Physik, als auch im Rahmen der Reflexion über Entdeckungs- und Kolonialgeschichte. Hier wurde intensiver Gedankenaustausch und kollegiale Zusammenarbeit mit Geographen (außereuropäische Länderkunde), den historisch orientierten Kartographen und den Ethnologen, besonders mit den Ethnohistorikern, gepflogen. Das gemeinsam geleitete Privatissimum von Günther Hamann, Ernest Troger und Walter Hirschberg, welches über viele Semester hinweg veranstaltet wurde, belegt die offene und die Fachgrenzen überschreitende Diskussion, bei der Professoren, Assistenten und Dissertanten gemeinsam inhaltlich befruchtende Diskurse zu regional und historisch begrenzten Fragestellungen führten, die sich u.a. auch in „grenzüberschreitenden" Dissertationen manifestierten. Die Entdeckungs- und auch die Kolonialgeschichte in Verbindung mit der Wissenschaftsgeschichte, die den Erkenntnisniederschlag widerspiegelt, ist eigentlich „kein Fach am Rande", wie Günther Hamann bereits 1957 feststellte, sondern nimmt einen zentralen Platz in der Erklärung des Welt- und Wissenschaftsbildes für das Verständnis der Gegenwart ein[6]. Dieser transdisziplinäre Forschungs- und Arbeitsansatz und die spezielle günstige personelle Situation von exzellenten Wissenschaftlern sowohl im Bereich der Hohen Schulen wie auch in dem der Fachbibliotheken und Sammlungen seit den 1960er Jahren in Wien ermöglichten, den wissenschaftstheoretischen wie methodologischen Ansatz in der Praxis zu verwirklichen. Ein Forum dafür waren die oben erwähnten Kolloquien der Folgejahre, an denen auch Johannes Dörflinger, Erich Woldan und die weiteren Mitarbeiter der professoralen Veranstalter engagiert teilnahmen.

Als Anstoß für diese spezielle Entwicklung des transdisziplinären Ansatzes auf Wiener Boden ist auch die Arbeit am ersten Band der *Monumenta Ethnographica* über das südliche Afrika, also über „Schwarzafrika" anzusehen.

Im Vorwort des Verlages, verfasst von Dr. Hans Biedermann, kann man lesen: *Diese erste Zeit wissenschaftlicher Erforschung außereuropäischer Erdteile und seiner Bewohner fällt mit einem Ereignis zusammen, das geeignet ist, die Ergebnisse der Entdeckungsfahrten nicht nur leichter festzuhalten, sondern sie auch weiten Kreisen zugänglich zu machen: die Erfindung des Buch-*

6 Wernhart, Nachruf, Günther Hamann, 235f. – K. R. Wernhart, o. Univ. Prof. Mag. Dr. Ernest Troger zum 60. Geburtstag, in: Mitteilungen der Österreichischen Geographischen Gesellschaft 128, Wien 1986, 199-202.

druckes. Sehr bald entsteht eine Literatur wertvollster Reiseberichte in Text und Bild, welche den ersten Kontakt Europas mit den außereuropäischen Völkern und Kulturen dokumentiert. Die ADEVA stellt sich zur Aufgabe, diese frühen Reisewerke, die zum großen Teil zu den Rarissima nicht nur des antiquarischen Büchermarktes, sondern auch wohlfundierter Bibliotheken zählen, neu zu edieren, und zwar mit der speziellen Heraushebung der Bilddokumente ethnographischen Inhalts, ergänzt durch den das Bild erläuternden Originaltext [7]. Den ethnographischen Bildtafeln der Originalwerke, die mittels fotografischem Reproduktionsverfahren gemeinsam mit den korrespondierenden Textpassagen der Originalquellen als Faksimile dargestellt wurden (269 Tafeln) fügte Walter Hirschberg als Bearbeiter und Analysator eine ethnographisch-anthropologische Interpretation im Anhang (84* Seiten) bei. So entstand ein Quellenwerk, das der Erforschung ursprünglichen Kulturgutes außereuropäischer Völker oder Ethnien und der „Wiedererweckung" schwer zugänglicher Quellen in repräsentativer Form dient. Und weiter heißt es im Vorwort: *Das Unternehmen verdankt seine Durchführung an hervorragender Stelle Dr. Erich Woldan, Bibliothekar der Österreichischen Geographischen Gesellschaft, Wien, durch Vorlage der Quellen und ihrer chronologischen Anordnung als Grundlage des Tafelwerkes. Seine auf profundem Fachwissen begründete Sammlertätigkeit auf dem Gebiet früher Reiseberichte, speziell von Erstausgaben, ließ eine mehrere tausend Bände umfassende Privatbibliothek entstehen, aus welcher der vorliegende Band der Monumenta Ethnographica wertvolles Material zur Verfügung gestellt erhielt. Dr. Erich Woldan nahm sich auch der Erstellung der im wissenschaftlichen Anhang abgedruckten Biographien der Forschungsreisenden an. HR Dr. Rudolf Kinauer, Direktor der Kartensammlung der Österreichischen Nationalbibliothek, nahm sich der kartographischen Bearbeitung des Sachregisters an und war jederzeit bereit sein Wissen auf dem Gebiet der historischen Geographie zur Verfügung zu stellen*[8]. Das Vorwort schließt mit dem Wunsch: *Wir hoffen, dass es uns* [dem Verlag] *gelingen möge, mit den Monumenta Ethnographica ein unerlässliches Quellenwerk für die Völkerkunde zu schaffen, das jedoch auch für Anthropologen, Geographen, Historiker, Kunsthistoriker, Soziologen, Kulturpolitiker und Wirtschaftswissenschaftler ein wertvolles Inventarium darstellt.*[9]

---

7 Hirschberg, Monumenta Ethnographica, Vorwort, III u. IV.
8 Ebd., Vorwort, IV.
9 Hirschberg, Monumenta Ethnographica, V.

Dieser Wunsch ging leider nicht in Erfüllung, da das Werk wegen der zu aufwändigen verlagstechnischen Produktionsweise der Akademischen Druck- und Verlagsanstalt (ADEVA) in Graz und dem Rückzug der Sponsoren (Staats- und Landessubventionen) nicht fortgesetzt werden konnte.

Die Erarbeitung dieses ersten und leider zugleich auch letzten Bandes der *Monumenta Ethnographica* brachte einige „Begleiterscheinungen" mit sich, wie z.B. die Erstellung thematisch geordneter bibliographischer Nachschlagewerke, die wissenschaftliches Forschen erleichtern konnten. Ebenso wurde eine Klärung des Selbstverständnisses der historischen Völkerkunde als Folge der methodischen Arbeit mit den Bild- und Schriftquellen aus dem Entdeckungszeitalter bei Walter Hirschberg herbeigeführt, was zur Entwicklung der „Ethnohistorie" als „Neue Wiener Schule der historischen Ethnologie" führte.

Das sehr umfangreiche und ins Detail gehende bibliographische Wissen von Erich Woldan bezüglich der Quellenwerke geographischen, kartographischen und ethnographischen Inhalts, das auf unzähligen Katalogzettelchen und Notizblättern vermerkt war, sollte auf Anregung von Walter Hirschberg und dem Autor vorliegenden Beitrages auch der wissenschaftlichen Fachwelt zur Verfügung gestellt werden. Als Publikationsorgan wurden die von Karl R. Wernhart im Jahre 1970 gegründeten *Wiener Ethnohistorischen Blätter* vorgesehen[10]. Erich Woldan erklärte sich bereit, in diesen *Wiener Ethnohistorischen Blättern* (WEB) in unregelmäßiger Folge Zusammenstellungen von ältester Reiseliteratur in Erstdrucken zu veröffentlichen. Diese erschienen in sieben Beiträgen unter Angabe von Autor (Herausgeber), Titel, Ort und Jahreszahl in chronologischer Reihenfolge in dem Zeitraum von 1970 bis 1977. Der erste Beitrag trug den Titel *Die älteste Literatur über die portugiesischen Entdeckungen und Eroberungen in Originalausgaben* (WEB Heft 1, Wien 1970); in Heft 3 (1971) folgte die *Geographisch-ethnographische Quellenliteratur über Afrika des 15. bis 17. Jahrhunderts in Originalausgaben.* Ab Heft 5 der WEB (1972) bis zu Heft 14 (1977), also für weitere fünf von Erich Woldan

10 Diese „Blätter" wurden 1998 (ab Heft 43) dem Wiener Ethnologie-Institut (Institut für Kultur- und Sozialanthropologie) übergeben und von jungen Wissenschaftern weitergeführt, das letzte Heft, hrsg. v. Marie-France Chevron, erschien als Nummer 47/48, Wien 2008 im LIT-Verlag (Wien, Berlin, Münster).

erarbeitete Zusammenstellungen wurde ein Einheitstitel gewählt: *Die erd- und völkerkundlichen Quellenwerke über* [...] Die regionale Ausrichtung betraf in Heft 5 (1972) das nördliche und östliche Südamerika des 16. und 17. Jahrhunderts. Einen Schwerpunkt bildete der geographische Raum Ozeanien und Australien, dem insgesamt vier Beiträge in den Heften 6 (1973), 9 (1974), 10 (1975) und 14 (1977) gewidmet waren. Diese Zusammenstellungen bibliographisch erschlossener Titel, die auch die Bestände der Woldan-Bibliothek berücksichtigten, chronologisch geordnet und gegliedert nach Erdgebieten vom 15. bis zum 19. Jahrhundert, stellen für Fachwissenschafter eine Fundgrube für die transdisziplinäre Forschung dar. Die Druckvorlagen wurden von Erich Woldan handschriftlich der Redaktion der *Ethnohistorischen Blätter* übergeben und von meiner Frau Elisabeth in Druckfahnen umgesetzt. Für seine Arbeit und Mühen wurde Erich Woldan beim Korrekturlesen der Fahnen mit einem reichhaltigen Essen bedankt. Die Sonderdrucke seiner Beiträge in den WEB wurden dem Autor Erich Woldan übergeben und befinden sich im Nachlassfundus. Ein Überblick über die Leistungen von Erich Woldan auf dem Gebiet der Bibliographierung von Originalausgaben findet sich im Generalindex der Wiener *Ethnohistorischen Blätter* für die Jahre 1970 bis 1986 aus dem Jahre 1986[11].

Der andere Impuls, den die Arbeit und Auseinandersetzung mit den *Monumenta Ethnographica* auslöste, betraf die Etablierung der Forschungsrichtung der „Ethnohistorie“ als neuen Schwerpunkt am Wiener Institut für Völkerkunde, heute Institut für Kultur- und Sozialanthropologie. Walter Hirschberg musste sich auf Grund des Arbeitskonzeptes und des thematischen Schwerpunktes der in Rede stehenden Buchpublikation wieder mit dem Grundwissen über Quellenkritik und Quellensequenz für die ethnographischen Analysen der zu interpretierenden Holzschnitte und Kupferstiche auseinandersetzen. Die physisch-anthropologischen Merkmale der dargestellten Menschen waren bedeutungslos, da die Abbildungen fast ausschließlich Europäer mit fremdländischem Inventar zeigten. Eine Ausnahme bildet die Kohlezeichnung von Albrecht Dürer (1471-

11 Wiener Ethnohistorische Blätter, Index der Hefte 1 (1970) – 30 (1986), Titelverzeichnis und Autorenregister, in: Wiener Ethnohistorische Blätter 30, Wien 1986, 83-91.

1528) aus dem Jahre 1508, die das Porträt eines Afrikaners zeigt[12]. Damit lag der Schwerpunkt der Analysen auf den ethnographischen Details, seien es Objekte oder spezifische Szenen eines Arbeitsvorganges oder einer Festivität eines speziellen Regionalgebietes (meist Küstennähe) Afrikas in Verbindung mit den korrespondierenden Textpassagen. Durch die Beschäftigung mit den *Monumenta Ethnographica* und der Bildquellenforschung wie der Auswertung dieser nach den Kriterien der Quellenkritik wurde Hirschberg vor Augen geführt, dass die Kulturen stets dynamisch sind und durch historische Einflüsse ebenso wie durch Eigenentwicklung sich verändern. Die wissenschaftliche und methodologische Problematik der Quelleninterpretation und deren Kritik wurde von ihm richtig erkannt und durch seine Auseinandersetzungen mit dem Quellenmaterial, seine Ausführungen und Diskussionen auch in der Fachwelt der historisch arbeitenden Ethnologen verankert[13].

Seit seiner Studienzeit in den 1920er Jahren bis zu seinem Lebensende 1996 stand Walter Hirschberg in Opposition zur sogenannten „Alten Wiener Schule der Ethnologie", deren pseudohistorische Arbeitsweise der freien und vor allem der gebundenen Parallelisierung von gegenwärtigen ethnographischen Daten und Kenntnissen über Jäger-, Sammler- und Fischerethnien mit den Frühformen der Menschheit in Vergleichsverfahren er zurecht ablehnte. Weil diese Forschungsrichtung die Dynamik der Veränderung der Kulturen und damit die Quellensequenz nicht berücksichtigte, sondern die Statik der Kulturen bzw. die Kulturkonstanz in den Vordergrund rückte[14], betrieb Hirschberg die Umsetzung der Arbeitsweise des Fachhistorikers in der Völkerkunde. Durch diese Einführung der adäquaten historischen Arbeitsweise in den Bereich der historischen Völkerkunde sollte eine theoretische und methodologische Neuorientierung der historisch ausgerichteten Ethnologie erfolgen, die Walter Hirschberg schließlich als „Ethnohistorie" bezeichnete und ab

---

12 Hirschberg, Monumenta Ethnographica, Tafel 8 u. Anhang 5*.

13 W. Hirschberg, Der Quellenwert früher ethnographischer Bilddokumente, in: Anthropos 63, St. Augustin bei Bonn 1968, 139-155. – W. Hirschberg, Early historical Illustrations of West and Central African Music, in: African Music. Journal of the African Music Society 4/3, London 1969, 6-18.

14 R. K. Wernhart, Walter Hirschberg (1904-1996). Leben und Werk eines Anthropologen. In: Mitteilungen der Anthropologischen Gesellschaft in Wien 127, Wien 1997, 1-19, bes. 4-6.

1962 als Forschungsprogramm am Institut einführte. Heute gilt sie als die „Neue Schule der historischen Ethnologie" und stellt eine bedeutende Lehr- und Forschungsrichtung am Wiener Institut dar[15].

Ein Blick in die Wissenschaftsgeschichte zeigt uns, dass schon Prof. Fritz Röck (1879-1953), damals Direktor des Museums für Völkerkunde in Wien und Spezialist für die Ethnologie Amerikas (Altamerikanistik) schon im Jahre 1932 auf Wiener Boden den Begriff „ethnohistory" von Clark Wissler aus dem Jahre 1909 übernahm und ihn auf „Ethnohistorie" eindeutschte[16]. Auch Röck und die Jugendgruppe der *Arbeitsgemeinschaft für afrikanische Kulturgeschichte (WAFAK)*, deren Kopf und Sprachrohr Walter Hirschberg war, stellten sich gegen die katholisch apologetisch ausgerichtete Wiener Kulturhistorie der Völkerkunde, die von den Padres des Ordens SVD aus St. Gabriel bei Wien (P. Wilhelm Schmidt (1868-1954) – der geistig-politische Führer, P. Wilhelm Koppers, P. Paul Schebesta, P. Martin Gusinde) getragen wurde. Die Idee der Ethnohistorie, also eine Etablierung einer korrekten historischen Arbeitsweise, konnte sich gegen die Macht der Padres und die Bedeutung des politisch engagierten Staatskatholizismus der Ersten Republik Österreichs nicht durchsetzen, da das evolutive Konzept der Alten Wiener Schule der Völkerkunde, den Beweis des Urmonotheismus am Beginn der Menschwerdung im Sinne der Bibel und des Lehramtes der Katholischen Kirche zu erbringen (vgl. P. Wilhelm Schmidt, Der Ursprung der Gottesidee, 12 Bände, 1912-1955), zu dominant war[17]. Auf

---

15 W. Hirschberg, Ethnohistorische Arbeitsweise in Wien. Ein Beitrag zu ihrer Geschichte, in: K. R. Wernhart (Hrsg.), Ethnohistorie und Kulturgeschichte (Böhlau, Studienbücher – Aspekte der Ethnologie). Wien, Köln 1986, 19-29. – W. Hirschberg, Das „dynamische Kulturbild" in der Ethnohistorie, in: Wiener Ethnohistorische Blätter 14, Wien, 1977, 3-21.

16 K. R. Wernhart – W. Zips, Einführung in die theoretischen und methodologischen Grundlagen der Ethnohistorie, in: Ethnohistorie (2008), 13-40, bes. 13.

17 W. Hirschberg, Die Wiener Arbeitsgemeinschaft für Afrikanische Kulturgeschichte in den Jahren 1930 bis 1932. Ein Beitrag zu den Anfängen der Ethnohistorie in Wien, in: Wiener Ethnohistorische Blätter 13, Wien 1977, 3-41. – W. Hirschberg, Der Gottesname Nyambi im Lichte alter westafrikanischer Reiseberichte, in: Zeitschrift für Ethnologie 88, Berlin 1963, 163-179. – W. Hirschberg, Zur Frage „Kontinuität und Quellensequenz" in der historischen Völkerkunde, in: Ethnologische Zeitschrift Zürich 1975/1, Zürich 1975, 43-48. – R. K. Wernhart, Aspekte

Grund innerer (eigener Mitarbeiter) und vor allem äußerer theoriebezogener und methodologischer Kritik musste Prof. Josef Haekel (1907-1973) im Jahre 1956 die „Alte“ Schule der Wiener Völkerkunde aufgeben. Wie bereits erwähnt begann Walter Hirschberg 1962 als neu berufener Professor die Ethnohistorie als neue Lehr- und Forschungsrichtung am Wiener Institut zu etablieren. Ihre Wurzeln lagen in den Ausführungen Fritz Röcks und dessen Studenten, die in der Arbeitsgemeinschaft damals den Aufstand gegen Pater Schmidt probten[18], sowie in seiner Auseinandersetzung mit den Bildquellen der *Monumenta Ethnographica*. Bis heute gelten vorrangig die Bildquellen als besondere Quellenkategorie, zu der nicht nur Holzschnitte und Kupferstiche aus dem 15. bis zum 18. Jahrhundert zählen, sondern ebenso die historischen Fotografien des 19. und 20. Jahrhunderts wie auch Filmsequenzen der jüngsten Vergangenheit[19].

So gesehen waren die Bildquellenwerke der Erich Woldan Sammlung – heute in der Österreichischen Akademie der Wissenschaften verwahrt und benutzerfreundlich aufgestellt – Ausgangspunkt bzw. wissenschaftlicher Impetus für die Erweiterung der Wissenschaftslandschaft auf Wiener Boden. Als Zeitzeuge dieser Entwicklung wurde es mir möglich, durch die vorgelegten Seiten einen Beitrag zur Wissenschaftsgeschichte zu erstellen, der auch zahlreiche Bezugspunkte zu Johannes Dörflinger und seinem wissenschaftlichen Werdegang aufweist.

Zum Autor

Karl R. Wernhart, geb. 1941, o. Univ. Prof. i. R., Studium der Ethnologie, Neueren Geschichte, Ur- und Frühgeschichte, Anthropologie und Volkskunde an der Univ. Wien (Dr. phil. 1967); Habilitation 1974; 1968-1980 Assistent, 1980-2002 o. Univ. Prof. am Inst. f. Völkerkunde der Univ. Wien, Gastvorlesungen: Univ. Innsbruck 1975-1985; Gastprofessur:

der Gründung des Institutes für Ethnologie an der Universität Wien, in: Mitteilungen der Österreichischen Gesellschaft für Wissenschaftsgeschichte 20, Wien 2000. 243-254, bes. 247-250.

18 K. R. Wernhart, Walter Hirschberg, 4-6.

19 Für die Bedeutung der Bildquellen siehe K. R. Wernhart, Die Quellengattungen und Nachbarwissenschaften der Ethnohistorie, in: Ethnohistorie (2008), 57-72, bes. 60-62.

Humboldt Universität in Berlin 1986, Kunstuniv. Linz 1994, Univ. Bayreuth u. Univ. Mainz 1996. Dekan der GRUWI-Fakultät 1985-1989; Rektor d. Universität Wien 1989-1991; Prorektor d. Universität Wien 1991-1994; Institutsvorstand 1996-2002.

Forschungsbereich: Ethnohistorie und Kulturgeschichte; Religionsethnologie; Wissenschaftsgeschichte, Dialog mit den Nachbarfächern; regionale Schwerpunkte: Ozeanien (Polynesien), Afrika, Karibik, Inselwelt des indischen Ozeans. Publikationen: 16 Bücher, über 150 wissenschaftliche Abhandlungen in Fachzeitschriften.

Kontakt: karl.r.wernhart@univie.ac.at

Tafeln / Plates

zu Band 2

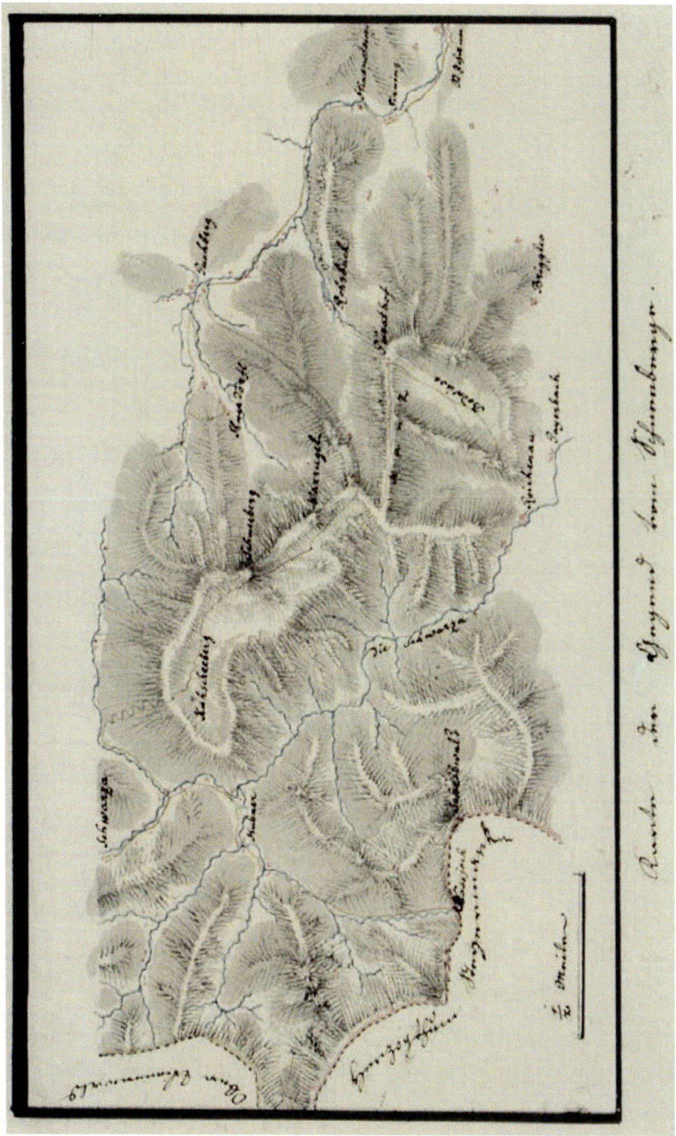

Karte der Gegend vom Schneeberge,
ÖNB/HAD Cod. Ser. n. 2023 Anhang 1

Gebiet um Neuberg zwischen Schneealpe, Mürzzuschlag und der Veitschalpe
ÖNB/HAD Cod. Ser. n. 2023 Anhang 2

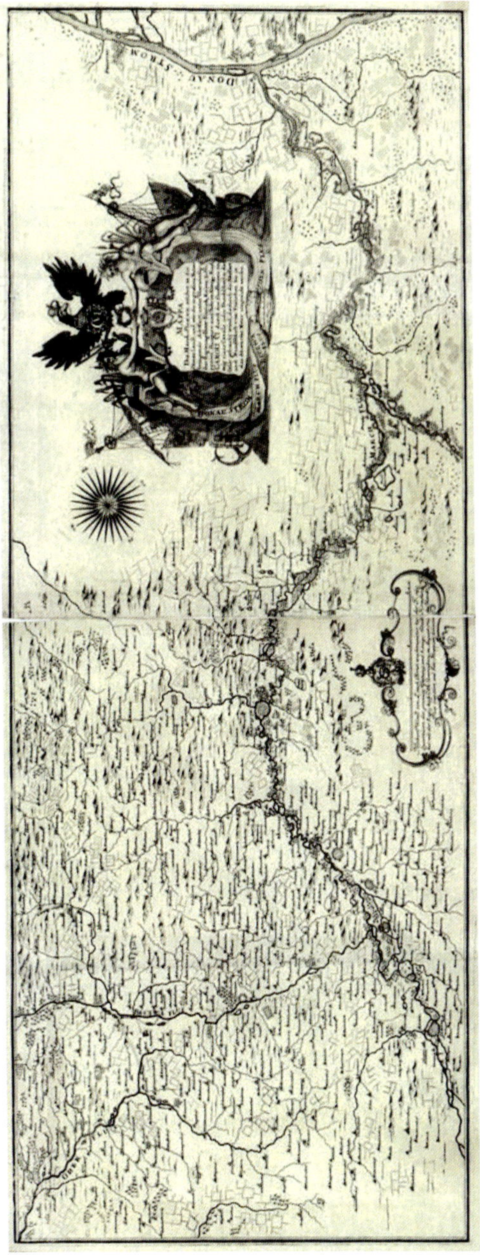

Norbert Wend von Linck, *Mappa des March-Flus*
ÖAW-Sammlung Woldan, K-V(Bl): OE/Boh 94 (1)

a) Handgezeichnete Vedute der Stadt *Nepagedel* (Napajedla), Moravský zemský archiv, Palachovo námestí 1, 626 00 Brno. Fond D 22, Signatur 96

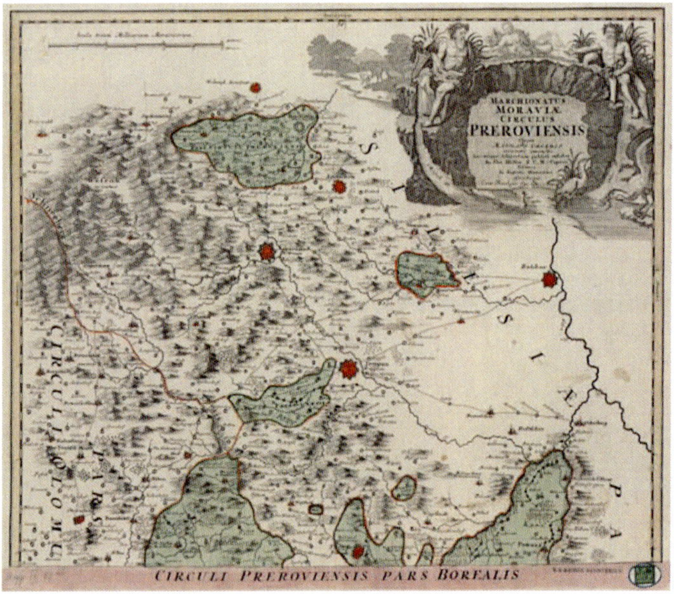

b) Homannsche Karte des Prerauer Kreises, Bayerische Staatsbibliothek München, Abt. für Musik, Karten und Bilder, Mapp. IX, 82ab

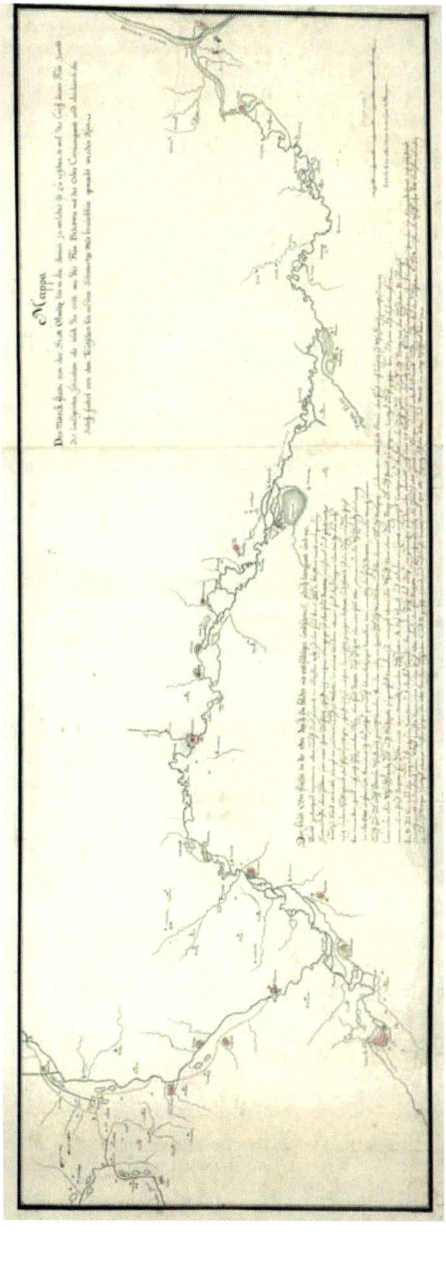

Mappa Des March fluss von der Statt Olmitz bis in die donau
Österreichisches Staatsarchiv, Abteilung Kriegsarchiv,, B IXb 169

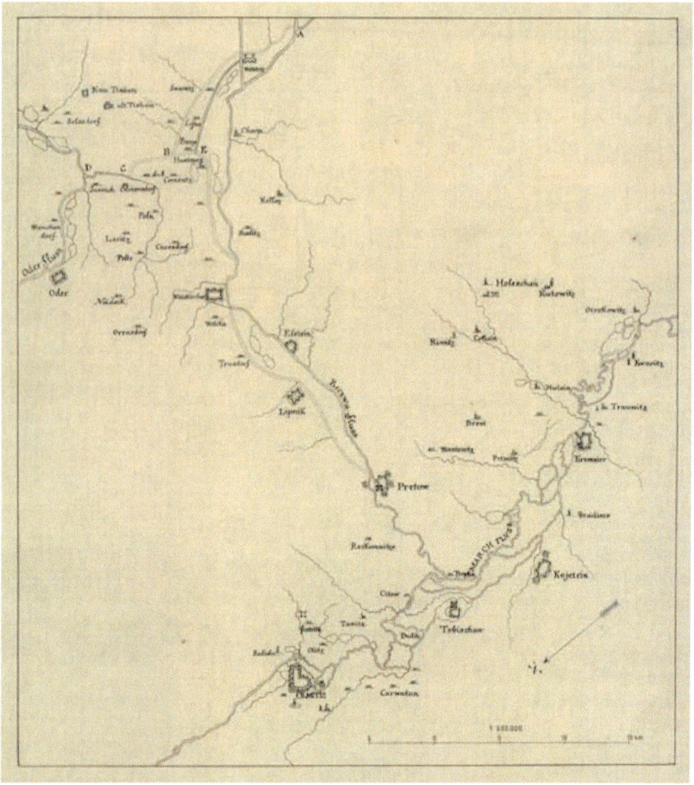

Mappa Des March fluss von der Statt Olmitz bis in die donau
Österreichisches Staatsarchiv, Abteilung Kriegsarchiv, B IXb 169

Friedrich Strass: Der Strom der Zeiten;
Archiv für Bildforschung, Salzburg

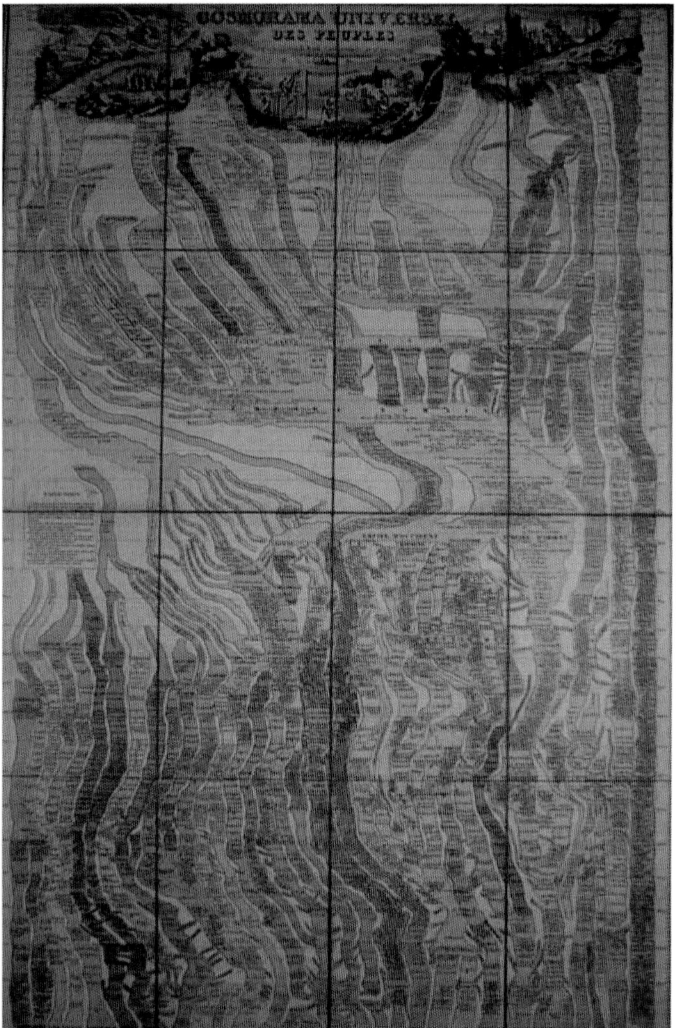

[Victor Clérot]: Cosmorama universel des peuples;
Archiv für Bildforschung, Salzburg

a) Friedrich Strass: Der Strom der Zeiten (Ausschnitt);
Archiv für Bildforschung, Salzburg

b) Le Cours des tems ou tableau de l'histoire (Ausschnitt);
Archiv für Bildforschung, Salzburg

Dresdnern: Historischer-Zeitstrom des heutigen Kaiserthums Oesterreich; Archiv für Bildforschung, Salzburg

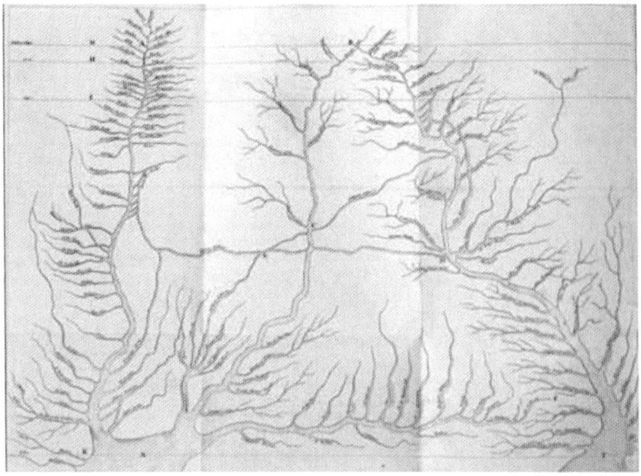

a) Clarkson: History of the Rise, Progress, and Accomplishment of the Abolition of the African Slave Trade;
Archiv für Bildforschung, Salzburg

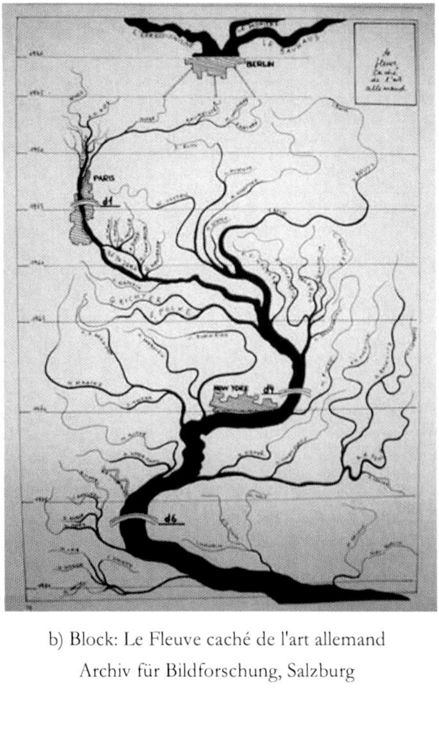

b) Block: Le Fleuve caché de l'art allemand
Archiv für Bildforschung, Salzburg

a) Hieronymus Münzer: Mitteleuropa-Karte (1493);
mit Blattschnitten ihrer Vorlagen

b) Francesco Roselli, Mitteleuropa-Karte (um 1450)

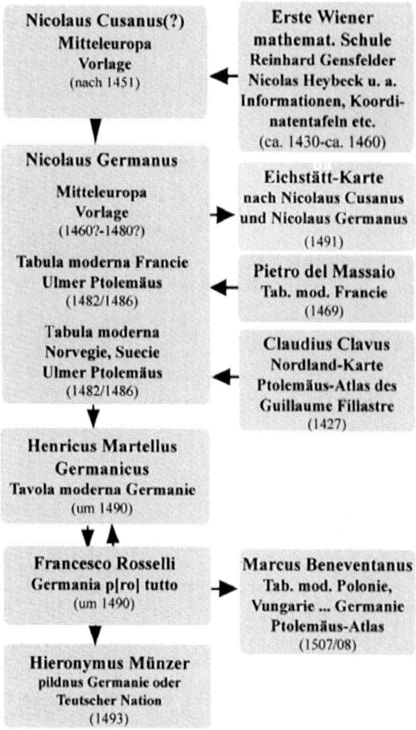

a) Stammbaum von H.Münzers Mitteleurpa-Karte

b) Nicolaus Cusanus (?), Nicolaus Germanus (?): Norddeutschland, Ausschnitt aus der "Eichstätt-Karte" (1491)

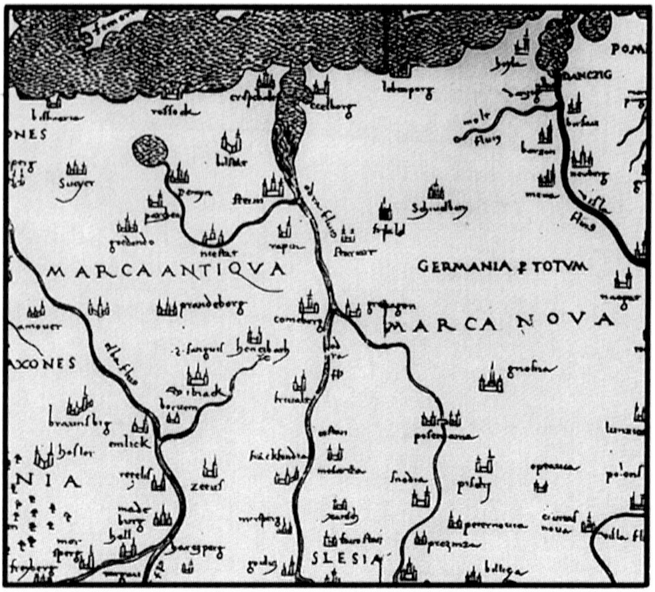

a) Francesco Rosselli: Norddeutschland,
Ausschnitt aus der Mitteleuropa-Karte (um 1490)

b) Henricus Martellus: Norddeutschland,
Ausschnitt aus der Mitteleuropa-Karte (um 1490)

a) Hieronymus Münzer: Norddeutschland,
Ausschnitt aus der Mitteleuropa-Karte (um 1490)

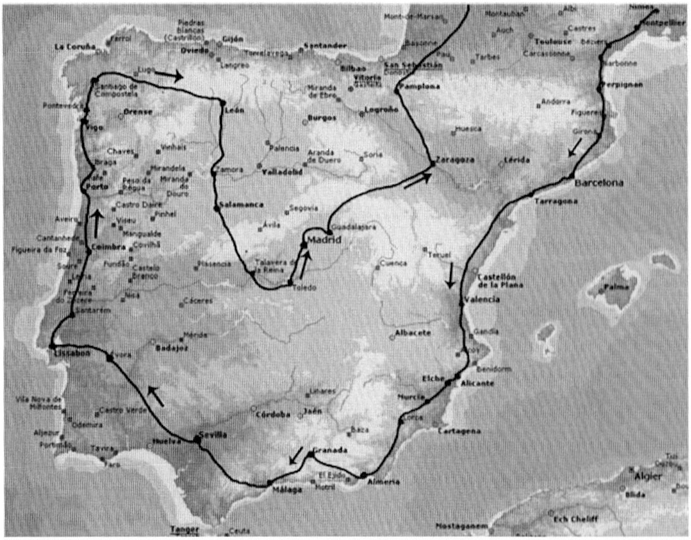

b) Route von Hieronymus Münzers Reise in Spanien und Portugal (1494/95)